Cell Death in Biology an

Series Editors

Xiao-Ming Yin
Zheng Dong

For further volumes:
http://www.springer.com/series/8908

Daniel E. Johnson
Editor

Cell Death Signaling in Cancer Biology and Treatment

Editor
Daniel E. Johnson
Division of Hematology/Oncology
Department of Medicine
University of Pittsburgh
Pittsburgh
PA, USA

ISBN 978-1-4939-0107-4 ISBN 978-1-4614-5847-0 (eBook)
DOI 10.1007/978-1-4614-5847-0
Springer New York Heidelberg Dordrecht London

Softcover reprint of the hardcover 1st edition 2013

Printed on acid-free paper

Humana Press is a brand of Springer
Springer is part of Springer Science+Business Media (www.springer.com)

This book is dedicated to my loving wife Sylvia Chan Johnson, and to the one who makes all things possible

Series Preface

Cell death, or conversely cell survival, is a major biological phenomenon. Just as with cell proliferation and cell differentiation, cell death is a choice that a cell has to make, sometimes voluntarily, other times accidentally. As such, cell death serves a purpose in the biology of a multicellular organism. The machinery of cell death and that of cell protection are evolutionarily conserved and their elements can even be found in single-celled organism. The disruption of cell death mechanisms can often cause developmental abnormalities. Factors that can trigger cell death are diverse and the cell death process is intricately connected with other biological processes. Cell death directly contributes to the pathogenesis of many diseases, including cancer, neurodegenerative diseases, and tissue injury in organ failure.

The study of cell death and cell survival has become a multidisciplinary subject, which requires expertise from all fields of the modern biology. Exploring the role of cell death in disease development and the modulation of cell death for the prevention and treatment of devastating disease demands constant updating of our knowledge through the broadest interactions among all investigators, basic and clinical. The rapid expansion of our knowledge in this field has gone beyond what could be summarized in a single book. Thus, this timely series *Cell Death in Biology and Diseases* summarizes new developments in different areas of cell death research in an elaborate and systemic way. Each volume of this series addresses a particular topic of cell death that either has a broad impact on the field or that has an in-depth development in a unique direction. As a whole, this series provides a current and encyclopedic view of cell death.

We would like to sincerely thank the editors of each volume in the series and the authors of each chapter in these volumes for their strong commitment and great effort towards making this mission possible. We are also grateful to our team of professional Springer editors. They have worked with us diligently and creatively from the initiation and continue this on the development and production of each volume of the series. Finally, we hope the readers will enjoy the reading, find the content helpful to their work, and consider this series an invaluable resource.

Xiao-Ming Yin MD, Ph.D.
Department of Pathology and Laboratory Medicine
Indiana University School of Medicine
Indianapolis, IN 46202, USA

Zheng Dong Ph.D.
Department of Cellular Biology and Anatomy
Georgia Health Science University and Charlie Norwood VA Medical Center
Augusta, GA 30912, USA

Book Preface

Over the past few decades, it has become widely appreciated that genetically programmed, active cell death processes are critically important for immune-mediated removal of infected or transformed cells, as well as self-elimination of damaged cells. This includes removal via cellular suicide of cells that have been damaged by conventional anti-cancer therapies such as chemotherapy and radiation. Concurrently, it has been determined that defects in cell death pathways can promote tumor development and progression, including the development of resistance to chemotherapy, radiation, immunotherapies, and biologic agents. These resistance mechanisms have devastating consequences for the successful treatment of cancer patients. A key goal in the development of new therapeutic strategies and agents for the treatment of cancer is to achieve selective and efficient killing of tumor cells. Assisting this goal are numerous ongoing studies and discoveries that have elucidated the molecular mechanisms of cell death pathways, including the extrinsic and intrinsic apoptosis pathways, and the proteins that act to regulate these pathways. Understanding of these cell death pathways is now making it possible to identify the specific cell death defects that occur in cancer cells, and is leading to the development of novel strategies and agents to overcome or correct these defects in cancer patients.

This volume summarizes current understanding of the molecular mechanisms that govern cell death in normal and malignant cells, and introduces cutting-edge opportunities for achieving selective killing of cancer cells by targeting these mechanisms in patient's tumors. A particular emphasis is placed on data emerging from the translation of basic research findings to clinical trials. The book begins by describing many of the common cell death defects that have been identified in primary patient specimens, including aberrant overexpression of anti-apoptotic proteins and oncoproteins, and mutation or loss of expression of pro-apoptotic proteins and tumor suppressors. The unique bioenergetics of cancer cells and unique characteristics of the tumor microenvironment are then discussed, along with opportunities for novel therapeutic interventions that are afforded by these distinctive features. The role of autophagy in regulating cancer cell death and current progress in targeting autophagy to improve responsiveness to conventional anti-cancer therapies is also presented. Additionally, a rapidly expanding field implicates microRNAs in the regulation of cell death proteins and pathways.

Recent descriptions of aberrant microRNA expression in cancer cells and the potential for targeting microRNAs are summarized. Further chapters describe preclinical and clinical approaches currently being used to target DNA repair pathways and protein chaperones as a means to provoke tumor cell death. New discoveries regarding the complex interplay between dying cancer cells and the immune system are also discussed, with an eye towards optimizing these interactions for therapeutic advantage. Finally, recent progress in targeting specific components of cell death signaling pathways is reviewed. These chapters highlight exciting advancements in the targeting of sphingolipid signaling, Bcl-2 family members, IAPs, death receptor signaling, the proteasome, and survival signaling mediated by the PI3K/AKT and RAS/RAF/MEK/ERK pathways.

Cell Death Signaling in Cancer Biology and Treatment will be particularly beneficial to biochemists, molecular biologists, and systems biologists interested in basic mechanisms of cell death signaling, the nature of cell death defects in cancer, and the impact of basic biological processes on cell death regulation. Scientists interested in the translation of findings from basic cell death research to clinical trials will appreciate the emphasis each chapter places on these important advances. Moreover, industry and academic scientists interested in anti-cancer drug development will learn of unique opportunities and approaches being taken to develop and evaluate highly selective agents with potent killing activities against tumors cells resistant to radiation or conventional chemotherapy drugs.

The work presented in this book is built on the foundation of many excellent studies and discoveries from a vast number of scientists. I wish to acknowledge those whose work is not presented, or whose area is not a specific focus of this book. In addition, I would like to thank the inspirational guidance of the late Stanley Korsmeyer, who introduced me to the fascinating field of cell death. I am particularly thankful for the authors who are respected leaders in their areas of research. Special thanks also go to my close colleagues in this field, with whom I have had many helpful discussions, including Pam Hershberger, John Lazo, Changyou Li, Hannah Rabinowich, Shivendra Singh, Xiao-Ming Yin, Jian Yu, Yan Zang, and Lin Zhang. This book represents the first in a new book series entitled *Cell Death in Biology and Diseases* (Springer) with Series co-editors Xiao-Ming Yin and Zheng Dong. I am indebted to the co-editors for helping to develop the content of this book, as well as to Aleta Kalkstein and Renata Hutter, editors from Springer. Lastly, I wish to thank my children, Rachel, Josiah, and Matthew, for their patience, inspiration, and laughter.

Daniel E. Johnson Ph.D.
Departments of Medicine and Pharmacology
and Chemical Biology
University of Pittsburgh and the University
of Pittsburgh Cancer Institute
Pittsburgh
PA 15213 USA

Contents

Contributors

Cori Abikoff Division of Clinical Research, Fred Hutchinson Cancer Research Center, Seattle, WA, USA

Stephen L. Abrams Department of Microbiology and Immunology, Brody School of Medicine at East Carolina University, Greenville, NC, USA

Ravi Amaravadi Department of Medicine, Perelman School of Medicine, Abramson Cancer Center, University of Pennsylvania, Philadelphia, PA, USA

Jörg Bäsecke Department of Medicine, University of Göttingen, Göttingen, Germany

Matthew F. Brown Department of Pathology and Molecular and Cellular Pathology Graduate Training Program, University of Pittsburgh School of Medicine, University of Pittsburgh Cancer Institute, Pittsburgh, PA, USA

Melchiorre Cervello Istituto di Biomedicina e Immunologia Molecolare, "Alberto Monroy", Consiglio Nazionale delle Ricerche, Palermo, Italy

William H. Chappell Department of Microbiology and Immunology, Brody School of Medicine at East Carolina University, Greenville, NC, USA

Francesca Chiarini Institute of Molecular Genetics, National Research Council-Rizzoli Orthopedic Institute, Bologna, Italy

Gabriela Chiosis Breast Cancer Service, Department of Medicine, Memorial Sloan-Kettering Cancer Center, New York, NY, USA; Department of Molecular Pharmacology and Chemistry, Sloan-Kettering Institute, New York, NY, USA

Lucio Cocco Cell Signalling Laboratory, Department of Human Anatomy, University of Bologna, Bologna, Italy

Q. Ping Dou Department of Pharmacology, Karmanos Cancer Institute, Wayne State University School of Medicine, Detroit, MI, USA

Camilla Evangelisti Institute of Molecular Genetics, National Research Council-Rizzoli Orthopedic Institute, Bologna, Italy

Richard A. Franklin Department of Microbiology and Immunology, Brody School of Medicine at East Carolina University, Greenville, NC, USA

Simone Fulda Institute for Experimental Cancer Research in Pediatrics, Goethe-University, Frankfurt, Germany

Anthony W. Gebhard Molecular Pharmacology and Physiology Program and Molecular Oncology Program, H Lee Moffitt Cancer Center, Tampa, FL, USA

Erica DaGama Gomes Breast Cancer Service, Department of Medicine, Memorial Sloan-Kettering Cancer Center, New York, NY, USA

Monica L. Guzman Division of Hematology and Oncology, Weill Cornell Medical College, New York, NY, USA

Yusuf. A. Hannun Stony Brook University, Stony Brook, NY, USA

Lori A. Hazlehurst Molecular Pharmacology and Physiology Program and Molecular Oncology Program, H Lee Moffitt Cancer Center, Tampa, FL, USA; Cancer Biology Program, University of South Florida, Tampa, FL, USA

Kan He Department of Pathology, University of Pittsburgh School of Medicine, University of Pittsburgh Cancer Institute, Pittsburgh, PA, USA

David M. Hockenbery Division of Clinical Research, Fred Hutchinson Cancer Research Center, Seattle, WA, USA

Daniel E. Johnson Departments of Medicine and Pharmacology and Chemical Biology, University of Pittsburgh, University of Pittsburgh Cancer Institute, Pittsburgh, PA, USA

John Koren III Breast Cancer Service, Department of Medicine, Memorial Sloan-Kettering Cancer Center, New York, NY, USA

Richard A. Lake School of Medicine and Pharmacology, National Centre for Asbestos Related Diseases, The University of Western Australia, Crawley, WA, Australia; St John of God Health Care, Subiaco, WA, Australia

Massimo Libra Department of Biomedical Sciences, University of Catania, Catania, Italy

Maeve A. Lowery Gastrointestinal Service, Department of Medicine, Memorial Sloan-Kettering Cancer Center, New York, NY, USA

Graziella Malaponte Department of Biomedical Sciences, University of Catania, Catania, Italy

Daciana Margineantu Division of Clinical Research, Fred Hutchinson Cancer Research Center, Seattle, WA, USA

Alberto M. Martelli Institute of Molecular Genetics, National Research Council-Rizzoli Orthopedic Institute, Bologna, Italy; Cell Signalling Laboratory, Department of Human Anatomy, University of Bologna, Bologna, Italy

Clorinda Massarino Department of Biomedical Sciences, University of Catania, Catania, Italy

Michele Milella Regina Elena National Cancer Institute, Rome, Italy

Melanie J. McCoy School of Medicine and Pharmacology, National Centre for Asbestos Related Diseases, The University of Western Australia, Crawley, WA, Australia ; St John of God Health Care, Subiaco, WA, Australia

James A. McCubrey Department of Microbiology and Immunology, Brody School of Medicine at East Carolina University, Greenville, NC, USA

Giuseppe Montalto Department of Internal Medicine and Specialties, University of Palermo, Palermo, Italy

Rajesh R. Nair Molecular Oncology Program, H Lee Moffitt Cancer Center, Tampa, FL, USA

Ferdinando Nicoletti Department of Biomedical Sciences, University of Catania, Catania, Italy

Anna K. Nowak School of Medicine and Pharmacology, The University of Western Australia, Crawley, WA, Australia ; St John of God Health Care, Subiaco, WA, Australia; Department of Medical Oncology, National Centre for Asbestos Related Diseases, Sir Charles Gairdner Hospital, Nedlands, WA, Australia

Jong Kook Park Division of Pharmaceutics, College of Pharmacy, Ohio State University, Columbus, OH, USA

Vinodh Rajagopalan Department of Biochemistry and Molecular biology, University of South Carolina, Charleston, SC, USA

Reshma Rangwala Department of Medicine, Perelman School of Medicine, Abramson Cancer Center, University of Pennsylvania, Philadelphia, PA, USA

Thomas D. Schmittgen Division of Pharmaceutics, College of Pharmacy, Ohio State University, Columbus, OH, USA

Min Shen Department of Pharmacology, Karmanos Cancer Institute, Wayne State University School of Medicine, Detroit, MI, USA

Robert W. Sobol Department of Pharmacology and Chemical Biology, Hillman Cancer Center, University of Pittsburgh Cancer Institute, University of Pittsburgh School of Medicine, Pittsburgh, PA, USA; Department of Human Genetics, University of Pittsburgh School of Public Health, Pittsburgh, PA, USA

Linda S. Steelman Department of Microbiology and Immunology, Brody School of Medicine at East Carolina University, Greenville, NC, USA

Agostino Tafuri Department of Cellular Biotechnology and Hematology, Sapienza, University of Rome, Rome, Italy

Tony Taldone Breast Cancer Service, Department of Medicine, Memorial Sloan-Kettering Cancer Center, New York, NY, USA

Mark Tom Division of Clinical Research, Fred Hutchinson Cancer Research Center, Seattle, WA, USA

Conchita Vens Division of Biological Stress Response, The Netherlands Cancer Institute, Amsterdam, NL, The Netherlands

Victor Y. Yazbeck Department of Medicine, University of Pittsburgh and the University of Pittsburgh Cancer Institute, Pittsburgh, USA

Jian Yu Department of Pathology and Molecular and Cellular Pathology Graduate Training Program, University of Pittsburgh School of Medicine, University of Pittsburgh Cancer Institute, Pittsburgh, PA, USA

Chapter 1
Defective Apoptosis Signaling in Cancer

Daniel E. Johnson

Abstract Apoptosis is critically important during development, facilitating the sculpting and molding of tissues, and in the adult, acting to maintain homeostasis of cell numbers. Apoptosis also plays a key role in immune-mediated elimination of infected or transformed target cells. In addition, apoptosis drives cellular suicide following damage to DNA or other cell components, including damage resulting from treatment with chemotherapy or radiation. In view of the fundamental importance of apoptotic cell death, it is, perhaps, not surprising that defects in apoptosis signaling are involved in a number of human diseases, including the development and progression of human malignancies. Efforts to promote therapeutic elimination of cancer cells via induction of apoptosis will benefit from more complete understanding of normal apoptosis signaling and the defects in apoptosis which frequently occur in human tumors. This chapter will describe the elucidation of the intrinsic and extrinsic apoptosis signaling pathways and focus on the defects in these pathways that have commonly been observed in cancers.

1.1 Characterization of a Human Disease and Biochemical Studies Unravel Apoptosis Signaling Pathways

The elimination of normal or neoplastic cells via induction of a cell death program has been recognized since the 1960s, with the term "apoptosis" first being used to describe this process in 1972 [1]. Early studies defined a number of ordered morphologic changes in cells undergoing apoptosis, including cell shrinkage, chromatin condensation, membrane blebbing, and eventual breakup of the cell into membrane encapsulated apoptotic bodies [1, 2]. In vivo, these apoptotic bodies were observed to be engulfed and degraded by macrophages or neighboring cells [1, 3, 4]. The early characterization of apoptotic cells also identified a few

D. E. Johnson (✉)
Departments of Medicine and Pharmacology and Chemical Biology, University of Pittsburgh and the University of Pittsburgh Cancer Institute, Pittsburgh, PA, USA
e-mail: johnsond@pitt.edu

D. E. Johnson (ed.), *Cell Death Signaling in Cancer Biology and Treatment*,
Cell Death in Biology and Diseases, DOI: 10.1007/978-1-4614-5847-0_1,

X.-M. Yin and Z. Dong (Series eds.), *Cell Death in Biology and Diseases*

biochemical changes, including externalization of plasma membrane phospholipids, activation of cellular DNAses, and degradation of genomic DNA to oligonucleosomal-length fragments visualized as "apoptotic DNA ladders" on agarose gels [5]. More detailed elucidation of the molecular mechanisms and pathways of apoptosis has involved a remarkable convergence of investigations, including examination of an apoptosis-deficient human cancer, genetic studies in *C. elegans*, and biochemical studies in multiple organisms.

Follicular lymphomas are a form of B-cell malignancy and represent one of the most common cancers of the human immune system. Initially, follicular lymphomas exhibit a relatively indolent phenotype. However, later during the course of disease they become highly aggressive, which is likely due to the acquisition of secondary genetic alterations. Cytogenetic analyses discovered that greater than 85 % of follicular lymphomas, as well as approximately 20 % of diffuse B-cell lymphomas, harbor a specific chromosomal translocation, the t(14;18) translocation [6, 7]. Molecular cloning of the t(14;18) breakpoint revealed juxtaposition of the *IgH* chain locus with sequences encoding a gene that was designated *bcl-2*, resulting in overexpression of wild-type Bcl-2 protein [8–11]. It was initially assumed that Bcl-2 would act as a typical oncoprotein, promoting rapid proliferation of cells. However, enforced overexpression of Bcl-2 in transfected cell lines did not markedly impact cell cycle status or proliferation. Instead Bcl-2 overexpression acted to markedly inhibit cellular apoptosis, including apoptosis induced by withdrawal of essential cytokines or neurotrophic factors, or treatment with apoptotic stimuli such as chemotherapy or radiation [12–22]. Overexpression of Bcl-2 in the B cells of transgenic mice resulted in extended B-cell survival and an expanded B-cell compartment [23]. Initially, these mice exhibited an indolent phenotype, followed by development of more aggressive malignant lymphomas [24]. Collectively, these studies identified Bcl-2 as the first oncoprotein that acts by inhibiting cellular apoptosis.

In unrelated studies, Horovitz and colleagues were performing mutational screens to identify genes that regulate apoptosis in a population of neuronal cells that consistently undergo apoptosis during *C. elegans* development. These studies identified several genes that were important for negatively regulating apoptosis and several that were required for apoptosis to occur. Among the genes that inhibited apoptosis was *ced-9* [25]. Sequence analyses determined that the protein encoded by *ced-9* bore close homology to human Bcl-2. Moreover, the human *bcl-2* gene effectively replaced *ced-9* function in mutant *C. elegans* worms [26]. Among the genes that were required for efficient apoptosis were *ced-3* and *ced-4* [27]. Sequencing revealed that the protein encoded by *ced-3* was closely related to a previously identified human protease responsible for processing of IL-1β [28–30]. Those who had previously been working with this protease, termed ICE (for interleukin-1β converting enzyme), had determined that it was a cysteine protease with a specificity for cleaving after aspartate residues and had developed a number of peptide inhibitors of ICE [29, 30]. It was quickly demonstrated that ICE promoted apoptotic cell death when overexpressed in cell culture [31]. In addition, peptide inhibition of cellular ICE was found to potently inhibit cell death resulting

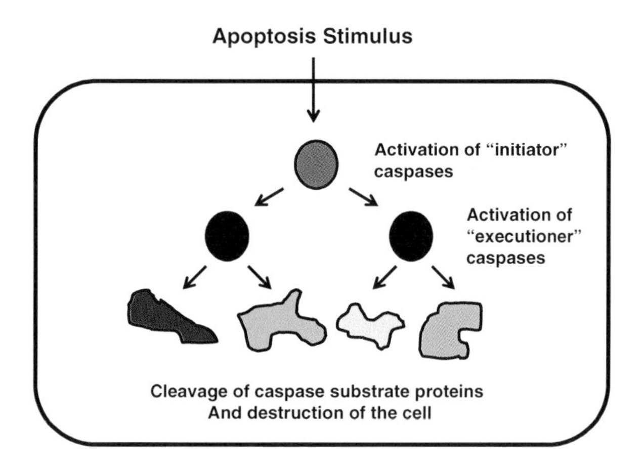

Fig. 1.1 Activation of the caspase protease cascade during apoptosis

from a variety of apoptotic stimuli, underscoring the importance of ICE and ICE-like proteases in apoptosis execution [32–37]. Subsequently, more than a dozen proteases related to ICE have been cloned and have come to be known as caspases (reviewed in [38]). The term caspases refers to the fact that these proteases are "c-asp-ases". The caspases have conveniently been assigned names of caspase-1 through caspase-14 [38]. Normally, caspases exist as inactive zymogens in cells. However, in response to an apoptotic stimulus, a subset of caspases, called the initiator caspases (e.g., caspase-8 and caspase-9), undergo processing to active enzyme forms (Fig. 1.1). The activated initiator caspases then cleave and activate executioner caspases, of which the most common are caspase-3 and caspase-7 [39, 40]. Activated executioner caspases cleave specific cellular substrate proteins promoting the eventual destruction of the cell.

Following the identification of caspases as executioners of apoptosis, it was soon determined that overexpression of Bcl-2 could act to prevent caspase protease activation in the cell. But how was Bcl-2 preventing caspase activation, and what are the mechanisms for caspase activation in the absence of Bcl-2? To address these questions, Wang and colleagues performed cellular fractionation studies to identify proteins that could promote activation of a procaspase in cell-free extracts [41]. Remarkably, cytochrome c, a mitochondrial protein, was found to be capable of promoting caspase activation in conjunction with a second protein that was given the name Apaf-1 (apoptotic protease activating factor 1) [41, 42]. Interestingly, Apaf-1 was found to bear close homology with Ced-4 protein from *C. elegans* [43]. Further studies revealed that during the course of apoptosis caused by chemotherapy, radiation, or cytokine withdrawal, cytochrome c is released into the cytosol, and this release is prevented in cells overexpressing Bcl-2 [41, 42, 44, 45].

A further foundational development in the understanding of apoptosis signaling has come with the identification of the tumor necrosis factor (TNF) family of death ligands. This family includes proteins such as TNF, Fas ligand, and TRAIL (TNF-related apoptosis-inducing ligand), which bind to cognate plasma membrane

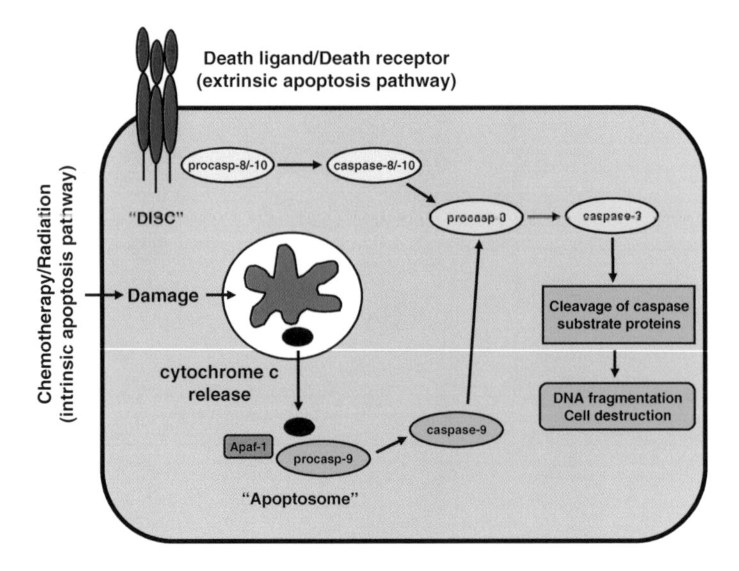

Fig. 1.2 The extrinsic and intrinsic apoptosis pathways

receptors, promoting caspase activation and cell death (see Chap. 11). The identification of this receptor-mediated pathway of apoptosis, in conjuction with the early discoveries regarding Bcl-2, caspases, Apaf-1, and cytochrome c, has provided the foundation for detailed delineation of the extrinsic and intrinsic pathways of apoptosis described in the section below.

1.2 The Intrinsic and Extrinsic Apoptosis Pathways

Based on the seminal discoveries described above, two pathways of apoptosis execution have subsequently been delineated, the extrinsic (or death receptor-mediated) pathway and the intrinsic (or mitochondrial-mediated) pathway (Fig. 1.2). The extrinsic apoptosis pathway is activated following binding of a death ligand to its cognate cell surface receptor. The simplest example of this pathway is illustrated by Fas-mediated signaling (Figs. 1.2 and 1.3). Expression of Fas ligand is primarily restricted to activated immune cells, as well as immune-privileged sites such as testis and eye. Fas ligand is predominantly expressed as a membrane-spanning protein. The binding of Fas ligand to its receptor, Fas, results in trimerization of the receptor protein, bringing together the three cytoplasmic regions (Fig. 1.3; reviewed in Chap. 11). Within the cytoplasmic region of Fas, and other death receptors, is a domain called the death domain (DD). Trimerization of the Fas DDs results in recruitment of the DD in the adaptor protein FADD (Fas-associated death domain protein), beginning a process of forming a larger complex referred to as the death-inducing signaling complex (DISC; Fig. 1.2).

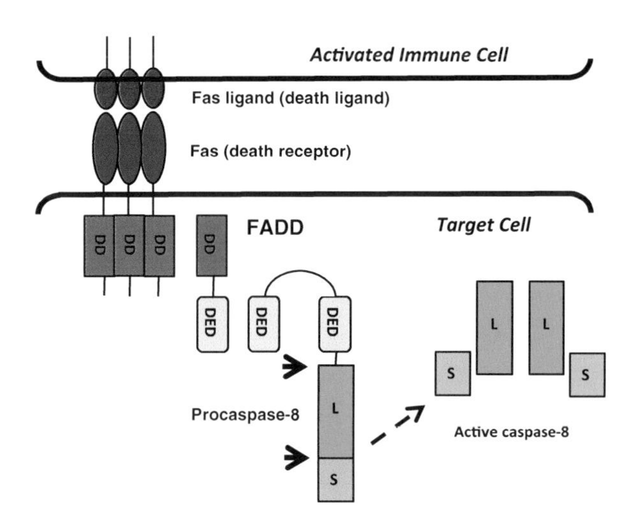

Fig. 1.3 DISC formation and caspase-8 activation

FADD also contains a domain called a death-effector domain (DED) which then recruits the DED found in the prodomain of the zymogen form of the initiator caspase, caspase-8 (or the initiator caspase-10). Recruitment of procaspase-8 to the DISC results in a slight conformational change in the zymogen protein, resulting in modest activation of the enzyme activity and proximity-induced proteolytic processing of procaspase-8 proteins present in the DISC (Fig. 1.3) [46–48]. This processing removes the inhibitory prodomain and produces large and small caspase-8 subunits. A heterotetrameric complex consisting of two large subunits and two small subunits comprises the fully activated caspase enzyme (Fig. 1.3) [38]. Once the initiator caspases, caspase-8 or caspase-10, are activated, they then cleave and activate executioner caspases, including the primary executioner, caspase-3 (Fig. 1.2) [38]. The activated executioners cleave specific substrate proteins, resulting in activation of cellular DNAses and proteolytic destruction of the cell.

The intrinsic apoptosis pathway is activated by a variety of cellular insults, including withdrawal of essential cytokines or neurotrophic factors, or treatment with chemotherapy or radiation. In the case of agents causing cellular damage (e.g., chemotherapy drugs), the damage is detected by the cell, and a signal (described in Sect. 1.4.3) is relayed to the mitochondria, causing release of cytochrome c into the cytosol (Fig. 1.2) [41]. Once released, the cytosolic cytochrome c associates with the cytoplasmic adapter protein Apaf-1 and the zymogen form of the initiator caspase, caspase-9, forming a complex referred to as the apoptosome [42]. Formation of the apoptosome results in a slight conformational change in procaspase-9, causing sufficient activation to promote autoprocessing to active caspase-9 [48]. Active caspase-9 cleaves and activates downstream executioner caspases, and at this point, the intrinsic and extrinsic apoptosis pathways converge. In some instances, earlier cross-talk between the intrinsic and extrinsic pathways can occur. Active caspase-8 from the extrinsic pathway is known to cleave Bid protein, and the Bid cleavage product (tBid) can

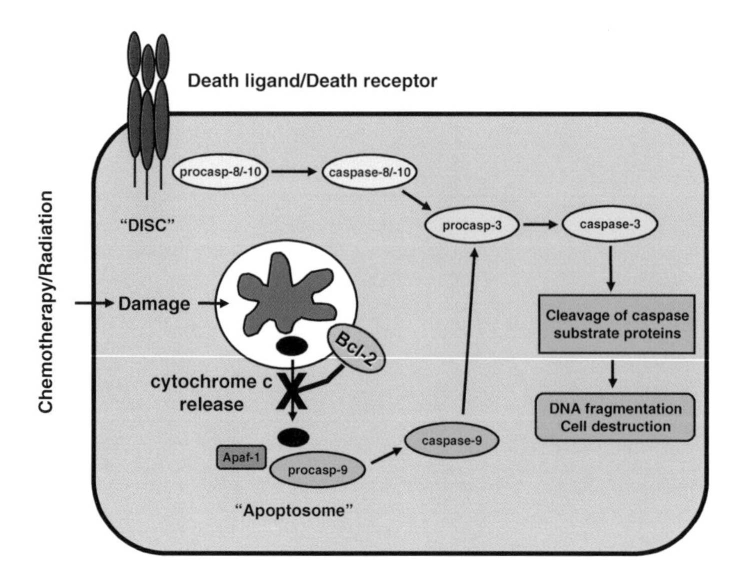

Fig. 1.4 Bcl-2 inhibition of the intrinsic apoptosis pathway

migrate to the mitochondria and stimulate cytochrome c release and activation of the intrinsic pathway [49, 50].

The Bcl-2 oncoprotein acts to negatively regulate the intrinsic apoptosis pathway, in part due to its prominent localization as an integral membrane protein in the outer mitochondrial membrane (Fig. 1.4). In preventing the release of cytochrome c from mitochondria (discussed in further detail in Sect. 1.4.1), Bcl-2 prevents formation of the apoptosome and activation of caspase-9 [44, 45].

In subsequent sections of this chapter, we will focus on genetic and epigenetic alterations that directly impact the expression or activity of specific components of the extrinsic and intrinsic apoptosis pathways. We will focus, in particular, on alterations that have been observed or confirmed in primary human tumor specimens.

1.3 Defects in Caspase Signaling

In view of the critical importance of caspases to both the extrinsic and intrinsic apoptosis pathways, one might predict that the genes encoding these proteases would be targets for genetic alterations impacting the function or expression of caspases in human tumors. Indeed, a considerable body of evidence indicates that this is likely the case, particularly for the genes encoding the initiator caspases, caspase-8 and caspase-10. A smaller number of publications also point to mutations or reduced expression of the initiator, caspase-9, and the executioners, caspase-3 and caspase-7.

1.3.1 *Mutation or Dysregulated Expression of Caspase Proteases*

Mandruzzato et al. [51] reported in 1997 a mutant form of caspase-8 expressed in a patient with head and neck carcinoma. In this mutant, the stop codon was altered, and an additional 88 amino acids were added to the C-terminus of the protein. Functional testing demonstrated reduced ability of the mutant protein to trigger apoptosis. Subsequently, Soung et al. [52] identified a specific frameshift mutation in the caspase-8 gene (1224_1226delTG) which resulted in premature termination within small subunit domain of the procaspase-8 protein. In hepatocellular carcinoma (HCC), 9/69 (13 %) patients were found to harbor this inactivating mutation. Mutation of caspase-8 has also been observed in 5/98 (5.1 %) colorectal carcinoma specimens and 13/162 (8.0 %) gastric carcinomas [53, 54].

A large number of publications have reported reduced expression or hypermethylation of the caspase-8 gene in human tumors. In neuroblastoma, several groups have observed loss of caspase-8 expression or gene hypermethylation in a majority of patients [55–60]. However, the relationship in neuroblastoma between reduced caspase-8 expression or gene methylation and either MYCN amplification or poor prognosis remains controversial, as some reports support a strong correlation while others indicate a lack of correlation [55, 58, 59]. A majority of medulloblastoma tumors also exhibit hypermethylation of the caspase-8 gene [61, 62], and loss of caspase-8 expression has been shown to correlate with poor prognosis in childhood medulloblastoma [63]. In pituitary adenomas, 19/35 (54 %) patients exhibited hypermethylation of the caspase-8 gene [64]. Hyerpmethylation of the caspase-8 gene, albeit in a minority of patients, has also been reported in bladder cancer [65], HCC [66, 67], and small-cell lung cancer (SCLC) [68]. In pediatric malignancies, hypermethylation of the caspase-8 gene has been observed in a majority of neuroblastomas, medulloblastomas, retinoblastomas, and rhabdomyosarcomas [61].

Mutations in the caspase-10 gene have been detected in autoimmune diseases as well as tumors. In 1999, Wang et al. identified inactivating caspase-10 point mutations in autoimmune lymphoproliferative syndrome (ALPS) type II [69]. A frameshift mutation resulting in premature caspase-10 termination has been identified by Tadaki et al. [70] in a patient with systemic juvenile idiopathic arthritis. In solid tumors, caspase-10 mutations have been observed in 3/99 (3.0 %) gastric cancers [71] and 2/47 (4.3 %) colon cancers [72]. Mutations in caspase-10 have also been detected in hematopoietic malignancies, including non-Hodgkin's lymphomas (17/117; 14.5 %) [73], T-acute lymphoblastic leukemia (1/13; 7.7 %) [74], and multiple myeloma (1/22; 4.5 %) [74].

Considerably less information has been reported on the mutational status of caspase-9, caspase-3, and caspase-7 in human tumors, perhaps reflecting a reduced prevalence of mutations of these caspases in cancer. An extensive analysis of the human caspase-9 gene by Soung et al. [75] in 180 gastric cancers, 104 colorectal cancers, and 69 lung adenocarcinomas detected a total of only three mutations, but none of these mutations resulted in an amino acid change. A similar analysis

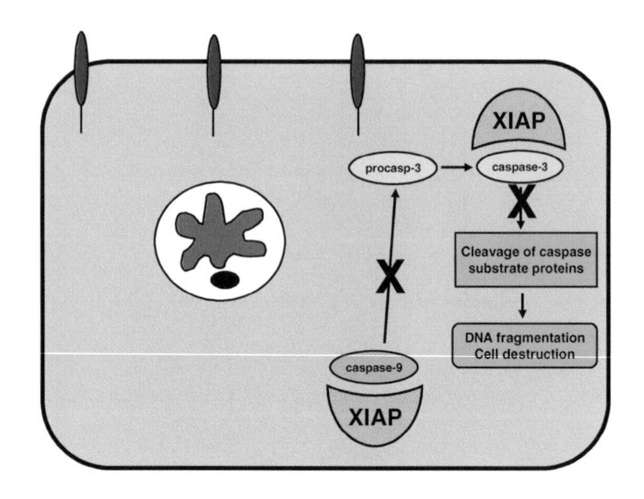

Fig. 1.5 XIAP inhibition of caspases in healthy cells

detected somatic mutations of the caspase-3 gene in 4/98 (4.1 %) colon carcinomas, 4/181 (2.2 %) non-small-cell lung cancers (NSCLCs), 2/129 (1.6 %) non-Hodgkin's lymphomas, 1/28 (3.6 %) multiple myelomas, 1/80 (1.3 %) HCCs, and 2/165 (1.2 %) gastric carcinomas [76]. Inactivating mutations of caspase-7 have been reported in 1/33 (3.0 %) head and neck carcinomas, 1/50 (2.0 %) esophageal carcinomas, and 2/98 (2.0 %) colon carcinomas [77].

1.3.2 Aberrant Expression of IAPs

The intracellular activities of the initiator, caspase-9, and the executioners, caspase-3 and caspase-7, are negatively regulated by certain members of the inhibitor of apoptosis (IAP) protein family (reviewed in [78, 79] and Chap. 10). The IAP family is comprised of eight members (XIAP, cIAP1, cIAP2, NAIP, survivin, MLIAP, BRUCE, and ILP2), with each characterized by the presence of 1 or 3 baculovirus IAP repeat (BIR) domains (see Chap. 10). Four members of this family (XIAP, cIAP1, cIAP2, MLIAP2, and ILP2) also contain carboxy-terminal RING domains, which possess E3 ubiquitin ligase activity. Among the IAPs, XIAP has potent and direct inhibitory activity against caspase-9, caspase-3, and caspase-7, while cIAP1 and cIAP2 can act to indirectly inhibit the activities of these caspases [80].

XIAP directly binds active caspase-3 and caspase-7 with high affinity via two interaction sites. The BIR2 domain of XIAP binds to an IBM (IAP-binding motif) that becomes exposed in the processed/active proteases [81, 82]. In addition, a linker region between the BIR1 and BIR2 domains binds to the active site of caspase-3 and caspase-7 and prevents the access of caspase substrate proteins [81–84]. In the case of caspase-9, the BIR3 domain of XIAP binds to the monomeric form of caspase-9, preventing dimerization and full activation of the enzyme [85, 86].

As shown in Fig. 1.5, XIAP-mediated inhibition of any caspase-9, caspase-3, or caspase-7 activities present in healthy growing cells prevents undesirable

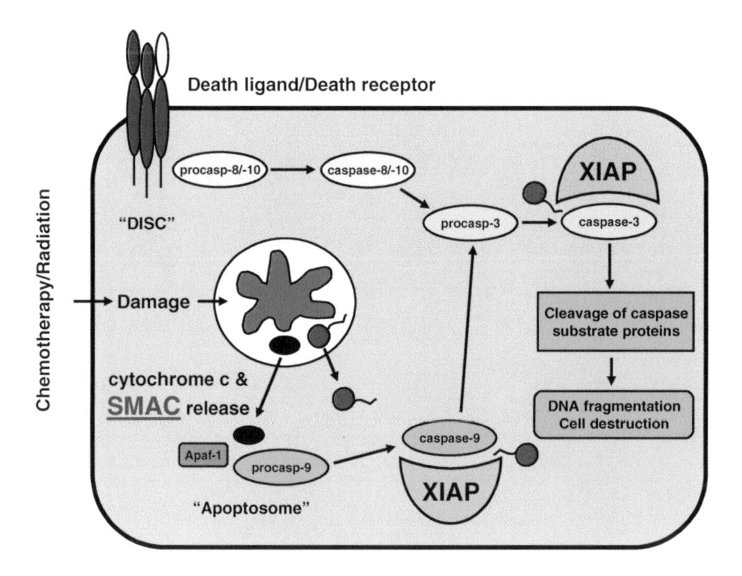

Fig. 1.6 SMAC overcomes XIAP to promote the intrinsic apoptosis pathway

activation of the caspase protease cascade and death of the cells. However, during activation of the intrinsic apoptosis pathway, the mitochondrial protein SMAC (second mitochondria-derived activator of caspases), is released from the mitochondria in conjunction with cytochrome c. The amino-terminal domain of the processed SMAC protein contains a tetrapeptide sequence (AVPI) that binds tightly to XIAP and displaces XIAP from bound caspases, leading to the activation of caspase-9, as well as caspase-3 and caspase-7 (see Fig. 1.6; reviewed in Chap. 10). While this is the scenario in cells that efficiently undergo apoptosis via the intrinsic apoptosis pathway, it is now evident that many tumors overexpress XIAP (discussed below), thwarting the ability of limiting levels of SMAC to overcome XIAP inhibition of the caspase cascade (see Fig. 1.7).

cIAP1 and cIAP2 act to indirectly inhibit caspase-9, caspase-3, and caspase-7 via two distinct mechanisms. First, the cIAP1 and cIAP2 proteins are capable of binding to SMAC and can thereby sequester SMAC protein and prevent displacement of XIAP from bound caspases [80]. Second, acting via their RING domains, cIAP1, cIAP2, as well as XIAP, can promote ubiquitination of caspase-3 and caspase-7 [87–90]. The ubiquitination of these executioner caspases results in inhibition of their activities and their eventual degradation via the proteasome.

Aberrant overexpression of XIAP, as well as other IAPs, has been observed in multiple solid tumor and hematopoietic malignancies. In several cases, overexpression of XIAP has been shown to correlate with treatment resistance or poor prognosis. XIAP is overexpressed in HCC, where its expression has been linked to metastasis, recurrence, and poor prognosis [91–93]. In advanced head and neck cancer, XIAP correlates with resistance to cisplatin-based regimens and poor prognosis [94]. Similarly, XIAP overexpression correlates with chemoresistance

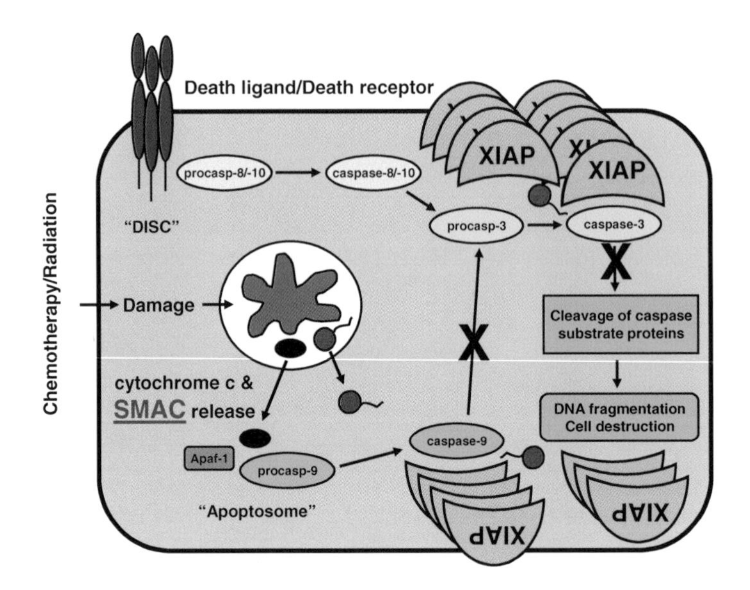

Fig. 1.7 XIAP overexpression in cancer overwhelms SMAC to prevent caspase activation

in pancreatic cancer [95]. XIAP also serves as a negative prognostic indicator of survival in both colorectal cancer [96] and renal cell carcinoma [97–99]. Overexpression of XIAP has also been detected in NSCLC [100–102], but the significance of this expression remains unclear. In one study, XIAP expression in radically resected NSCLC was found to be inversely correlated with proliferation and directly correlated with favorable prognosis [100], while another study reported a lack of correlation between XIAP and response to chemotherapy in advanced NSCLC patients [101]. Additional studies have identified XIAP overexpression in testicular germ cell tumors [103] and papillary thyroid carcinomas [104].

Among hematopoietic malignancies, overexpression of XIAP is associated with poor clinical outcome in diffuse large B-cell lymphomas (DLBCL) [105]. XIAP overexpression has also been observed in myelodysplastic syndromes transforming to acute leukemia [106]. High levels of XIAP in adult acute myeloid leukemia (AML) correlate with poor response to chemotherapy, unfavorable cytogenetics, and poor prognosis [107–109]. In childhood de novo AML, XIAP overexpression also correlates with chemoresistance and poor prognosis [110, 111]. Poor response to glucocorticoid therapy is observed in childhood T-cell acute lymphoblastic leukemias exhibiting XIAP overexpression [112].

Dysregulated expression of cIAP1, cIAP2, and survivin has been demonstrated in a large variety of tumors and frequently correlates with poor prognosis (reviewed in [78, 79, 113]). Interestingly, cIAP2 is overexpressed in MALT lymphomas harboring the t(11;18) translocation, which fuses the 3 BIR domains of cIAP2 with sequences encoding the carboxyl terminus of MALT1 protein [78]. Additionally, amplification of 11q21-q23, which encodes both cIAP1 and cIAP2, is frequently detected in multiple forms of cancer [78, 79].

1.4 Defects Affecting the Intrinsic Apoptosis Pathway

Activation of the intrinsic apoptosis pathway is fundamentally important for cell death induced by chemotherapy or radiation. Hence, defects in components of this pathway can have a profound impact on the sensitivities of tumor cells to these therapeutic agents. By extension, resistance to chemo- and radiotherapy has a huge impact on prognosis and survival for cancer patients.

The intrinsic apoptosis pathway is tightly regulated by members of the Bcl-2 protein family [114–117]. Members of this family share homology via the presence of conserved domains called Bcl-2 homology, or BH domains. While some Bcl-2 family members contain up to 4 BH domains (BH1-4), others contain only a single BH domain, typically a single BH3 domain. On a functional basis, members of this family can be divided into 2 major categories, anti-apoptotic proteins and proapoptotic proteins. The anti-apoptotic Bcl-2 family members include Bcl-2 [10, 118], Bcl-X_L [119], Mcl-1 [120], A1/Bfl-1 [121, 122], and Bcl-w [123]. These proteins contain multiple BH domains, as well as carboxyl-terminal membrane-anchoring domains, and act to inhibit cytochrome c release from the mitochondria. The proapoptotic Bcl-2 family members can be subdivided into two groups, those containing multiple BH domains (e.g., Bax and Bak), referred to as "multi-domain" proteins, and those containing only a single BH3 domain (e.g., PUMA, Bik, Noxa, Bid, Bad, etc.), referred to as "BH3 domain-only" proteins (reviewed in [116, 117]). Genetic studies have revealed that cells lacking Bax and Bak are highly resistant to stimuli that activate the intrinsic apoptosis pathway [124, 125], underscoring the importance of Bax and Bak for chemotherapy- and radiation-induced apoptosis.

1.4.1 Overexpression of Anti-Apoptotic Bcl-2 Family Members

Overexpression of anti-apoptotic Bcl-2 has been reported in a wide variety of hematopoietic and solid tumor malignancies, a subject that has been reviewed extensively [116, 117, 126–128]. Among hematopoietic tumors, Bcl-2 overexpression has been observed in follicular B-cell lymphoma [7, 129], DLBCL [130], AML [131–135], acute lymphoblastic leukemia (ALL) [127, 136], chronic lymphocytic leukemia (CLL) [137, 138], anaplastic large cell lymphoma (ALCL) [139], and multiple myeloma [140]. Solid tumor malignancies exhibiting overexpression of Bcl-2 include melanoma [141–143], glioblastoma [144], SCLC [145], and cancers of the bladder [146–149], colon [150], and prostate [151–153]. Overexpression of anti-apoptotic Bcl-X_L has also been detected in a wide range of cancers, including AML [154, 155], multiple myeloma [156], melanoma [157, 158], squamous cell carcinoma of the head and neck (SCCHN) [159, 160], and cancers of the liver [161], breast [162], bladder [163], colon [164], and pancreas [165, 166]. Importantly, in several of these malignancies, overexpression of Bcl-2 and/or Bcl-X_L has been closely correlated with resistance to chemotherapy or radiation, and poor overall survival [116, 126].

Less information is available regarding potential overexpression of anti-apoptotic Mcl-1, A1/Bfl-1, and Bcl-w in human cancers. However, overexpression of Mcl-1 has been reported AML [167], ALL [167], multiple myeloma [168, 169], CLL [170], and ovarian cancer [171]. Moreover, Mcl-1 is rapidly degraded via the proteasome. Treatment of tumors with proteasome inhibitors results in elevated Mcl-1 expression, which can act to attenuate the efficacies of these therapeutic agents [172]. Overexpression of A1/Bfl-1 has been reported in B-cell lymphomas and gastric carcinomas [122, 173, 174], while Bcl-w overexpression occurs in colorectal carcinomas [175].

Several different mechanisms have been described which account for Bcl-2 overexpression in cancer. In hematopoietic malignancies harboring the t(14;18) chromosomal translocation, overexpression results from juxtaposition of the intact Bcl-2 coding sequence next to cis transcriptional elements from the *IgH* chain locus [8–11]. Amplification of chromosomal segments encompassing the *bcl-2* gene has been reported to play a role in Bcl-2 overexpression in DLBCL and SCLC [176, 177]. Gene amplification has also been observed for the genes encoding Bcl-X_L and Mcl-1 [117, 178]. In B-CLL, hypomethylation of the *bcl-2* gene promoter represents an additional mechanism that may contribute to Bcl-2 overexpression [137].

Recent evidence indicates that expression of Bcl-2 family members can be controlled via expression of microRNAs (see Chap. 5). The microRNAs miR-15a and miR-16-1 negatively regulate expression of Bcl-2 [179]. In CLL, expression of miR-15a and miR-16-1 is frequently lost due to deletion of the *miR-15a* and *miR-16-1* genetic locus encoding these microRNAs [180, 181]. Loss of miR-15a and miR-16-1 is thought to play a key role in the overexpression of Bcl-2 in CLL, and potentially other malignancies where this genetic locus is deleted [182, 183].

1.4.2 Mutation or Reduced Expression of Proapoptotic Bcl-2 Family Members

Early efforts to uncover the molecular mechanism whereby Bcl-2 acts to inhibit apoptosis were stymied by the failure to identify any enzymatic activity associated with the Bcl-2 protein. The first clue regarding the mechanism of Bcl-2 action came with the discovery that Bcl-2 binds tightly to the closely related proapoptotic protein Bax [184]. Subsequent studies have revealed a complex network of physical interactions between anti-apoptotic Bcl-2 family members and proapoptotic members of the Bcl-2 protein family [114, 115, 126, 185]. Additional studies determined that heterodimerization between pro- and anti-apoptotic Bcl-2 family members served to neutralize the apoptosis-inducing activity of the proapoptotic protein, and led to the proposal of the "rheostat model" by Korsmeyer and colleagues [186]. The rheostat model proposed that the relative levels of pro- versus anti-apoptotic Bcl-2 family members expressed in a cell would determine whether that cell is prone to undergo apoptosis (i.e., chemotherapy sensitive), or prone to be apoptosis resistant. The second major clue regarding the mechanism of Bcl-2 action came with the discovery discussed previously that Bcl-2 (and Bcl-X_L/Mcl-1) acts to prevent cytochrome c release from the mitochondria [44, 45]. The third major clue came from biophysical and biochemical studies of proapoptotic Bax and Bak. These studies revealed that Bax and Bak can homooligomerize and form small pores in artificial lipid membranes, as well as mitochondrial membranes in intact cells [187–191]. The pores formed by Bax and Bak were found to be sufficient in size to allow the release of cytochrome c from mitochondria. The binding of anti-apoptotic Bcl-2 proteins to Bax or Bak prevents their homooligomerization, explaining how these anti-apoptotic proteins can prevent cytochrome c release and activation of the intrinsic apoptosis pathway. A large number of additional studies have shown that certain "BH3 domain-only" proapoptotic proteins, such as PUMA, tBid, and Bim, can bind directly to Bax or Bak and induce pore formation. These "BH3 domain-only" proteins are commonly referred to as "activators" [117, 192–195]. The other "BH3 domain-only" proapoptotic proteins are called "derepressors" (or "sensitizers") and exert their effects by binding to the anti-apoptotic Bcl-2 family members (e.g., Bcl-2, Bcl-X_L, and Mcl-1) [117, 192–195]. In binding to anti-apoptotic proteins, the "derepressors," as well as "activators," cause release of sequestered Bax and Bak, freeing these proteins up to form pores. In addition, "derepressor" binding can cause the release of bound "activators," freeing them up to directly bind and activate Bax and Bak [196–198].

Just as anti-apoptotic Bcl-2 family members are found to be frequently overexpressed in human cancers, it is reasonable to predict that the proapoptotic Bcl-2 proteins might be underexpressed or functionally inactivated in tumors. Indeed, this appears to be the case [117, 199]. Frameshift and inactivating mutations in the *bax* gene have been detected in colorectal and hematopoietic malignancies [200–202]. Mutation of *noxa* gene has been reported in DLBCL [203], while *bak* gene mutations are present in gastric and colon tumors [204]. In mantle cell lymphomas (MCL), homozygous deletion of *bim* gene has been observed [205]. Loss of heterozygosity in *bik* gene occurs in renal cell carcinoma [206].

Promoter methylation also appears to play an important role in reducing the expression of proapoptotic proteins in cancer. Hypermethylation of *bim* gene promoter has been reported in Burkitt's lymphoma (BL) and DLBCL [203, 207]; hypermethylation of *puma* gene occurs in BL [208]; and hypermethylation of *noxa* gene is found in DLBCL [203]. Hypermethylation of *bik* gene is present in renal cell carcinomas [206], while colorectal tumors exhibit hypermethylation of *hrk* gene [209].

Emerging evidence indicates that expression of proapoptotic proteins is likely regulated by the expression of specific microRNAs (see also Chap. 5). Expression of Bim has been found to be downregulated in several types of cancer via expression of microRNAs encoded by the *miR-17-92* and *miR-106b-25* clusters [210–213]. It appears highly likely that regulation of the expression of pro- and anti-apoptotic proteins by microRNAs will remain an active area of research in the coming years.

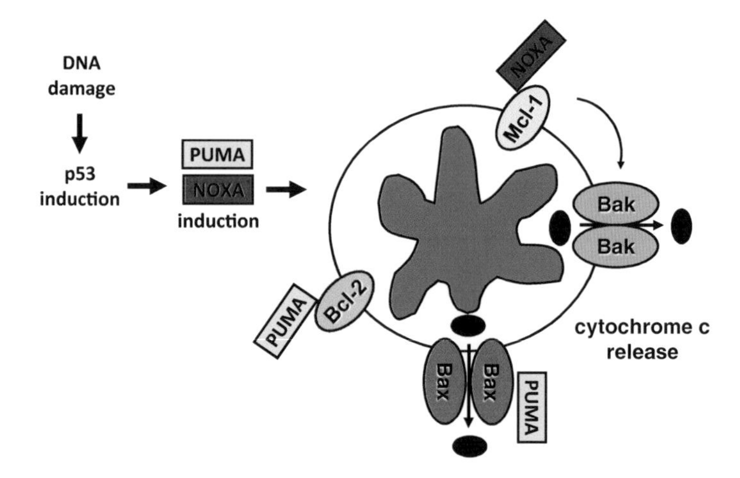

Fig. 1.8 p53 induces PUMA and NOXA to activate the intrinsic apoptosis pathway

1.4.3 Mutation/Deletion of p53

The tumor suppressor protein p53 is often cited as the most commonly mutated protein in human cancers [214]. Mutations in the *p53* gene lead to loss of p53 function as a transcription factor, with frequent conversion to a dominant-negative inhibitor, or loss of expression of the p53 protein [214]. The prevalence of p53 mutation or loss of expression in cancer is likely due to the critical roles that p53 plays in the regulation of cell cycle arrest, DNA repair, apoptosis, and senescence.

p53 actively promotes apoptosis induction via the intrinsic apoptosis pathway, particularly in response to DNA damage. In damaged cells, activation of ATM and ATR kinases leads to phosphorylation of p53 and MDM2, disrupting p53/MDM2 interactions and abrogating MDM2-mediated degradation of p53 via the proteasome [215–221]. This leads to upregulation of the p53 protein, with the majority of the protein being expressed in the nucleus, but some also appearing in the cytosol and at the mitochondria.

Upregulated p53 protein in damaged cells activates the intrinsic apoptosis pathway via both transcription-dependent and transcription-independent mechanisms. The transcription-dependent mechanism requires functional DNA binding and transcription factor function of the p53 protein. The genes encoding Bax and Apaf-1 have been shown to be p53 target genes, and their transcription is moderately induced in p53-expressing cells [222, 223]. However, although Bax and Apaf-1 are central factors in the intrinsic pathway, p53 induction of these proteins may play only a modest role in the proapoptotic activity of p53. By contrast, p53 induction of select proapoptotic "BH3 domain-only" proteins appears to play a major role in the transcription-dependent mechanism of apoptosis induction by p53. p53 markedly induces the expression of the "BH3 domain-only" proteins PUMA [224, 225], NOXA [226], and BID [227]. As illustrated in Fig. 1.8, p53-mediated induction of PUMA and NOXA results in migration of these proteins to the mitochondria, where PUMA can directly activate Bax or Bak, or indirectly activate these proteins via inhibitory interaction with anti-apoptotic Bcl-2 family members, including Bcl-2 and Bcl-X_L [197, 198, 228–230]. Induced NOXA demonstrates specificity for binding to anti-apoptotic Mcl-1 and functionally inactivates Mcl-1 by promoting the release of bound proapoptotic proteins [193, 196–198, 229–231].

Transcription-independent activation of the intrinsic apoptosis pathway by p53 occurs via several different mechanisms. Cytosolic p53 directly activates proapoptotic Bax, leading to mitochondrial membrane permeabilization and cytochrome c release [232, 233]. p53 present at the mitochondria physically associates with proapoptotic Bak, disrupting Bak sequestration by Mcl-1 and resulting in Bak activation [234]. In addition, mitochondrial-localized p53 can bind to Bcl-2 or Bcl-X_L, promoting the release of sequestered proapoptotic proteins [233, 235].

Collectively, the combination of transcription-dependent mechanisms mediated by nuclear p53 and transcription-independent mechanisms mediated by cytosolic/mitochondrial p53 confers a potent impact of p53 upregulation on cellular apoptosis. Moreover, it is readily apparent why loss of p53 can provide an important survival advantage for tumors and may enhance resistance to chemotherapy and radiation. However, it is important to note that activation of the intrinsic apoptosis pathway is not entirely dependent on p53, as activation of this pathway can also be observed in some cell types lacking p53. The mechanisms of p53-independent activation of the intrinsic apoptosis pathway are less well understood, and this remains an area of active investigation.

1.5 Defects Affecting the Extrinsic Apoptosis Pathway

Similar to what is seen with the intrinsic apoptosis pathway, multiple defects in the extrinsic (or death receptor-mediated) pathway have been observed in human tumors. In healthy individuals, the extrinsic apoptosis pathway plays a central role in immune-mediated elimination of infected or transformed cells. Therefore, defects in the extrinsic pathway most likely contribute to tumorigenesis primarily by negatively impacting the efficiency of immune surveillance. However, it is also important to note that in certain cell types, cross-talk occurs between the extrinsic and intrinsic pathways, meaning defects in one pathway may affect killing initiated by the other pathway. For example, in cells referred to as type I cells, stimulation with a death ligand leads to efficient killing via a linear pathway involving caspase-8 activation, followed by caspase-3 activation. By contrast, in cells referred to as type II cells, efficient killing by a death ligand also requires caspase-8-mediated cleavage of Bid, followed by tBid induction of cytochrome c release and activation of the intrinsic pathway [49, 50, 236]. Thus, in type II cells, defects in the intrinsic pathway can impair killing initiated by the extrinsic pathway. In addition, other reports have indicated that defects in the extrinsic pathway may reduce the sensitivity of leukemic cells to chemotherapy [237–240].

As described previously (Sect. 1.3.1), the initiator caspases for the extrinsic pathway, caspase-8 and caspase-10, are frequently mutated or underexpressed in human cancers. In the sections below, we will focus on dysregulated expression or mutation of the death receptors for Fas ligand and TRAIL, the DISC adaptor protein FADD, and the negative regulator c-FLIP.

1.5.1 Loss or Mutation of Death Receptors

Similar to what is seen for caspase-8 and caspase-10, the expression and function of Fas death receptor is commonly altered in autoimmune disease and malignant tumors (reviewed in [241]; see also Chap. 11). In the *lpr* strain of mice, sequences from the 3′ LTR of a transposable element are inserted into the second exon of the *fas* gene, resulting in expression of defective Fas protein [242–245]. The *lpr* mice are characterized by ineffective deletion of self-reactive T cells, lymphoproliferation, and development of autoimmune disease [242]. In humans, somatic Fas mutations are also commonly detected in patients with autoimmune lymphoproliferative syndrome (ALPS) [241, 246, 247]. Loss of heterozygosity in *fas* gene has also been reported in human ALPS [248, 249]. Mutations in Fas, typically leading to an inactive protein, have been detected in both hematologic and solid tumor malignancies. In hematologic malignancies, Fas mutations have been reported in cutaneous T cell [250, 251], MALT-type [252, 253], nasal NK/T cell [254], non-Hodgkin's [255], DLBC [256] and thyroid lymphomas [257], as well as multiple myeloma [258]. Solid tumors harboring Fas mutations include bladder carcinoma [259], malignant melanoma [260], NSCLC [261, 262], squamous cell carcinoma [263], and cancers of the stomach [264] and testis [265]. Methylation of the *fas* gene may be an important mechanism for downregulation of Fas expression in human tumors, with hypermethylation having been reported in Sezary syndrome [266] and cancers of the bladder [267], colon [268], and prostate [269].

Apoptosis induced by the death ligand TRAIL is mediated by the death receptors TRAIL-R1 (also called DR4) and TRAIL-R2 (also called DR5) (reviewed in [270–272]; see also Chap. 11). As might be expected, reduced expression and mutation of TRAIL-R1 and TRAIL-R2 has been reported in a variety of cancers. Mutations in TRAIL-R1 have been detected in non-Hodgkin's lymphoma [273], metastatic breast cancer [274], gastric carcinoma [275], head and neck cancer [275], NSCLC [275], and osteosarcoma [276]. Mutations in TRAIL-R2 occur in non-Hodgkin's lymphoma [273], metastatic breast cancer [274], gastric carcinoma [277], and NSCLC [278]. Loss of heterozygosity in the TRAIL-R2 has been reported in colorectal carcinoma [279], gastric carcinoma [277], and NSCLC [278]. Hypermethylation of the *TRAIL-R1* gene has been observed in gastric carcinoma [280], glioma [281], SCLC [282], and ovarian cancer [283].

1.5.2 *Regulation of Death Receptor Signaling by Decoy Receptors, Aberrant Subcellular Trafficking, and p53*

Although mutation, loss of heterozygosity, and gene hypermethylation appear to play important roles in the inactivation of death receptor signaling in tumors, other mechanisms that contribute to inhibition of the extrinsic apoptosis pathway also exist. In particular, decoy receptors for both Fas ligand and TRAIL have been extensively described (reviewed in [270–272]; see also Chap. 11). It remains somewhat unclear, however, what role the decoy receptors for TRAIL play in tumor progression. Additionally, work in cell line models indicates that aberrant subcellular trafficking of TRAIL-R1 and TRAIL-R2 may contribute to resistance to TRAIL. In breast cancer cells, reduced surface expression of TRAIL-R1 and TRAIL-R2 was found to correlate with constitutive endocytosis of these receptors [284]. In colon cancer cells selected for resistance to TRAIL, deficient transport of TRAIL-R1 and TRAIL-R2 to the cell surface was observed [285].

Further evidence indicates that the p53 status of tumor cells may play a role in sensitivity to TRAIL. The genes encoding TRAIL-R1 and TRAIL-R2 have been shown to be p53 target genes [286–291]. In both cases, p53 induces gene expression via binding to intronic sequences [289, 291]. The ability of p53 to induce TRAIL-R1 and TRAIL-R2 underscores the significance of p53 loss in human tumors. However, the situation may not be as simple as a direct correlation between wild-type p53 expression and sensitivity to TRAIL. Upregulation of TRAIL receptors can also occur via p53-independent pathways [287]. Moreover, the genes encoding decoy receptors for TRAIL also appear to be p53 target genes [292, 293].

1.5.3 *FADD Mutation*

The FADD adaptor protein plays an important role in caspase-8 activation following stimulation of the receptors for Fas ligand, TNF, and TRAIL. During Fas- and TRAIL receptor-mediated activation of caspase-8, FADD is the first protein recruited to the DISC. During TNF receptor-mediated activation of caspase-8, FADD recruitment to the DISC follows the recruitment of RIP (receptor-interacting protein), TRADD (TNF receptor-associated death domain protein), and TRAF-1/-2 (TNF receptor-associated factor). Mutation of FADD protein has been occasionally observed. Bolze et al. identified a missense mutation in FADD in an ALPS patient that leads to reduced FADD protein levels and impaired apoptosis induction following Fas stimulation [294]. In addition, FADD mutations have been reported in 4/80 NSCLC patients [262] and 1/98 colon cancer patients [295].

1.5.4 Dysregulated c-FLIP Expression

Just as anti-apoptotic Bcl-2 family members and IAPs act as endogenous negative regulators of cytochrome c release and caspase activities, respectively, the c-FLIP (cellular FLICE-inhibitory protein) proteins are endogenous inhibitors of death receptor-mediated caspase-8/caspase-10 activation. Three major c-FLIP protein isoforms have been reported, c-FLIP$_L$, c-FLIP$_S$, and c-FLIP$_r$ (reviewed in [116, 296, 297]). Similar to procaspase-8 and procaspase-10, all 3 c-FLIP isoforms contain two amino-terminal DED domains. The c-FLIP$_s$ and c-FLIP$_r$ contain only a short carboxyl-terminal sequence following the DED domains. By contrast, the DED domains of c-FLIP$_L$ are followed by sequences closely homologous to the caspase-8 enzyme, although c-FLIP$_L$ is catalytically inactive. A large number of publications have shown that high levels of c-FLIP$_L$, c-FLIP$_S$, or c-FLIP$_r$ effectively compete with procaspase-8 and procaspase-10 for binding to FADD or TRADD proteins present in the forming DISC of activated death receptors [296, 297]. In so doing, high levels of the c-FLIP proteins effectively prevent caspase-8/caspase-10 activation and execution of the extrinsic apoptosis pathway. Conversely, when expressed at only low levels, c-FLIP$_L$ has been shown to heterodimerize with procaspase-8, promoting caspase-8 activation [298, 299].

Overexpression of c-FLIP proteins, particularly c-FLIP$_L$, has been detected in a variety of human tumors, rendering the cancer cells more resistant to apoptosis induction by death ligands. Specifically, c-FLIP is overexpressed in breast, colon, endometrial, liver, ovarian, and prostate cancers, as well as melanomas, glioblastomas, and NSCLCs (reviewed in [116, 296, 297]). Importantly, high levels of c-FLIP$_L$ expression have been determined to correlate with tumor progression or poor prognosis in colon, endometrial, liver, ovarian, and prostate cancers, as well as BL [116, 296, 297, 300–302]. The prevalence of c-FLIP overexpression in human tumors, and the impact of these proteins on the extrinsic apoptosis pathway, has made the development of c-FLIP inhibitors an area of active investigation.

1.6 Conclusions

A remarkable convergence of genetic and biochemical studies in humans, mice, and *C. elegans* has led to our current understanding of the mechanisms whereby cells kill themselves in response to apoptotic stimuli, as well as the mechanisms responsible for regulation of cell death pathways. Elucidation of the components of the extrinsic and intrinsic apoptosis pathways has provided the foundation for investigating cell death defects occurring in human tumors. Multiple components have been identified which are important for promoting or facilitating apoptosis via these pathways (e.g., caspases, p53, cytochrome c, SMAC, proapoptotic Bcl-2 family members, Apaf-1, death ligands and receptors, FADD, etc.), and multiple components have been identified that act to negatively regulate apoptosis (e.g., anti-apoptotic Bcl-2 family members, IAPs, c-FLIP, etc.). Detailed analyses have determined that several of the proapoptotic proteins involved in the extrinsic and intrinsic pathways commonly exhibit reduced expression in a wide variety of human tumors via loss of gene heterozygosity or promoter methylation. Additionally, the proapoptotic proteins also exhibit frequent functional inactivation via mutation. By contrast, many malignancies overexpress the anti-apoptotic proteins that negatively regulate the extrinsic and intrinsic pathways. These findings are stimulating a flurry of drug discovery efforts (described in subsequent Chapters) aimed at restoring the expression or activities of proapoptotic proteins, or reducing the expression or inhibiting the activities of anti-apoptotic proteins in patient tumors. Moreover, by identifying the location of defects within the apoptosis pathways, treatment options can focus on either correcting the defect, or activating the pathway downstream from the molecular block. Ultimately, the rapid determination of the apoptosis defects present in a patient's tumor should allow the development of effective personalized treatments to restore sensitivity to cell death–inducing therapies such as chemotherapy, radiation, and therapeutic death ligands.

Acknowledgments This work was supported by National Institutes of Health grants R01 CA137260 and P50 CA097190. A vast number of researchers have made important contributions to the work described in this review. We apologize to those authors whose work we have not cited.

References

1. Kerr JF, Wyllie AH, Currie AR (1972) Apoptosis: a basic biological phenomenon with wide-ranging implications in tissue kinetics. Br J Cancer 26(4):239–257
2. Wyllie AH, Kerr JF, Currie AR (1980) Cell death: the significance of apoptosis. Int Rev Cytol 68:251–306
3. Savill J, Fadok V, Henson P, Haslett C (1993) Phagocyte recognition of cells undergoing apoptosis. Immunol Today 14(3):131–136
4. Savill J, Fadok V (2000) Corpse clearance defines the meaning of cell death. Nature 407(6805):784–788
5. Wyllie AH, Morris RG, Smith AL, Dunlop D (1984) Chromatin cleavage in apoptosis: association with condensed chromatin morphology and dependence on macromolecular synthesis. J Pathol 142(1):67–77
6. Fukuhara S, Rowley JD, Variakojis D, Golomb HM (1979) Chromosome abnormalities in poorly differentiated lymphocytic lymphoma. Cancer Res 39(8):3119–3128
7. Yunis JJ, Frizzera G, Oken MM, McKenna J, Theologides A, Arnesen M (1987) Multiple recurrent genomic defects in follicular lymphoma. A possible model for cancer. N Engl J Med 316(2):79–84
8. Cleary ML, Sklar J (1985) Nucleotide sequence of a t(14;18) chromosomal breakpoint in follicular lymphoma and demonstration of a breakpoint-cluster region near a transcriptionally active locus on chromosome 18. Proc Natl Acad Sci U S A 82(21):7439–7443
9. Tsujimoto Y, Cossman J, Jaffe E, Croce CM (1985) Involvement of the bcl-2 gene in human follicular lymphoma. Science 228(4706):1440–1443
10. Tsujimoto Y, Finger LR, Yunis J, Nowell PC, Croce CM (1984) Cloning of the chromosome breakpoint of neoplastic B cells with the t(14;18) chromosome translocation. Science 226(4678):1097–1099

11. Cleary ML, Smith SD, Sklar J (1986) Cloning and structural analysis of cDNAs for bcl-2 and a hybrid bcl-2/immunoglobulin transcript resulting from the t(14;18) translocation. Cell 47(1):19–28
12. Vaux DL, Cory S, Adams JM (1988) Bcl-2 gene promotes haemopoietic cell survival and cooperates with c-myc to immortalize pre-B cells. Nature 335(6189):440–442
13. Nunez G, London L, Hockenbery D, Alexander M, McKearn JP, Korsmeyer SJ (1990) Deregulated Bcl-2 gene expression selectively prolongs survival of growth factor-deprived hemopoietic cell lines. J Immunol 144(9):3602–3610
14. Hockenbery D, Nunez G, Milliman C, Schreiber RD, Korsmeyer SJ (1990) Bcl-2 is an inner mitochondrial membrane protein that blocks programmed cell death. Nature 348(6299):334–336
15. Garcia I, Martinou I, Tsujimoto Y, Martinou JC (1992) Prevention of programmed cell death of sympathetic neurons by the bcl-2 proto-oncogene. Science 258(5080):302–304
16. Allsopp TE, Wyatt S, Paterson HF, Davies AM (1993) The proto-oncogene bcl-2 can selectively rescue neurotrophic factor-dependent neurons from apoptosis. Cell 73(2): 295–307
17. Mah SP, Zhong LT, Liu Y, Roghani A, Edwards RH, Bredesen DE (1993) The protooncogene bcl-2 inhibits apoptosis in PC12 cells. J Neurochem 60(3):1183–1186
18. Sentman CL, Shutter JR, Hockenbery D, Kanagawa O, Korsmeyer SJ (1991) bcl-2 inhibits multiple forms of apoptosis but not negative selection in thymocytes. Cell 67(5):879–888
19. Strasser A, Harris AW, Cory S (1991) bcl-2 transgene inhibits T cell death and perturbs thymic self-censorship. Cell 67(5):889–899
20. Alnemri ES, Fernandes TF, Haldar S, Croce CM, Litwack G (1992) Involvement of BCL-2 in glucocorticoid-induced apoptosis of human pre-B-leukemias. Cancer Res 52(2):491–495
21. Walton MI, Whysong D, O'Connor PM, Hockenbery D, Korsmeyer SJ, Kohn KW (1993) Constitutive expression of human Bcl-2 modulates nitrogen mustard and camptothecin induced apoptosis. Cancer Res 53(8):1853–1861
22. Miyashita T, Reed JC (1993) Bcl-2 oncoprotein blocks chemotherapy-induced apoptosis in a human leukemia cell line. Blood 81(1):151–157
23. McDonnell TJ, Deane N, Platt FM, Nunez G, Jaeger U, McKearn JP et al (1989) bcl-2-immunoglobulin transgenic mice demonstrate extended B cell survival and follicular lymphoproliferation. Cell 57(1):79–88
24. McDonnell TJ, Korsmeyer SJ (1991) Progression from lymphoid hyperplasia to high-grade malignant lymphoma in mice transgenic for the t(14; 18). Nature 349(6306):254–256
25. Hengartner MO, Ellis RE, Horvitz HR (1992) Caenorhabditis elegans gene ced-9 protects cells from programmed cell death. Nature 356(6369):494–499
26. Vaux DL, Weissman IL, Kim SK (1992) Prevention of programmed cell death in Caenorhabditis elegans by human bcl-2. Science 258(5090):1955–1957
27. Ellis HM, Horvitz HR (1986) Genetic control of programmed cell death in the nematode *C. elegans*. Cell 44(6):817–829
28. Yuan J, Shaham S, Ledoux S, Ellis HM, Horvitz HR (1993) The *C. elegans* cell death gene ced-3 encodes a protein similar to mammalian interleukin-1 beta-converting enzyme. Cell 75(4):641–652
29. Thornberry NA, Bull HG, Calaycay JR, Chapman KT, Howard AD, Kostura MJ et al (1992) A novel heterodimeric cysteine protease is required for interleukin-1 beta processing in monocytes. Nature 356(6372):768–774
30. Cerretti DP, Kozlosky CJ, Mosley B, Nelson N, Van Ness K, Greenstreet TA et al (1992) Molecular cloning of the interleukin-1 beta converting enzyme. Science 256(5053):97–100
31. Miura M, Zhu H, Rotello R, Hartwieg EA, Yuan J (1993) Induction of apoptosis in fibroblasts by IL-1 beta-converting enzyme, a mammalian homolog of the *C. elegans* cell death gene ced-3. Cell 75(4):653–660
32. Nicholson DW, Ali A, Thornberry NA, Vaillancourt JP, Ding CK, Gallant M et al (1995) Identification and inhibition of the ICE/CED-3 protease necessary for mammalian apoptosis. Nature 376(6535):37–43

33. Datta R, Banach D, Kojima H, Talanian RV, Alnemri ES, Wong WW et al (1996) Activation of the CPP32 protease in apoptosis induced by 1-beta-D-arabinofuranosylcytosine and other DNA-damaging agents. Blood 88(6):1936–1943
34. Kondo S, Barna BP, Morimura T, Takeuchi J, Yuan J, Akbasak A et al (1995) Interleukin-1 beta-converting enzyme mediates cisplatin-induced apoptosis in malignant glioma cells. Cancer Res 55(24):6166–6171
35. Zhu H, Fearnhead HO, Cohen GM (1995) An ICE-like protease is a common mediator of apoptosis induced by diverse stimuli in human monocytic THP.1 cells. FEBS Lett 374(2):303–308
36. An B, Dou QP (1996) Cleavage of retinoblastoma protein during apoptosis: an interleukin 1 beta-converting enzyme-like protease as candidate. Cancer Res 56(3):438–442
37. Dou QP, An B, Antoku K, Johnson DE (1997) Fas stimulation induces RB dephosphorylation and proteolysis that is blocked by inhibitors of the ICE protease family. J Cell Biochem 64(4):586–594
38. Riedl SJ, Shi Y (2004) Molecular mechanisms of caspase regulation during apoptosis. Nat Rev Mol Cell Biol 5(11):897–907
39. Stennicke HR, Jurgensmeier JM, Shin H, Deveraux Q, Wolf BB, Yang X et al (1998) Pro-caspase-3 is a major physiologic target of caspase-8. J Biol Chem 273(42):27084–27090
40. Slee EA, Harte MT, Kluck RM, Wolf BB, Casiano CA, Newmeyer DD et al (1999) Ordering the cytochrome c-initiated caspase cascade: hierarchical activation of caspases-2, -3, -6, -7, -8, and -10 in a caspase-9-dependent manner. J Cell Biol 144(2):281–292
41. Liu X, Kim CN, Yang J, Jemmerson R, Wang X (1996) Induction of apoptotic program in cell-free extracts: requirement for dATP and cytochrome c. Cell 86(1):147–157
42. Li P, Nijhawan D, Budihardjo I, Srinivasula SM, Ahmad M, Alnemri ES et al (1997) Cytochrome c and dATP-dependent formation of Apaf-1/caspase-9 complex initiates an apoptotic protease cascade. Cell 91(4):479–489
43. Zou H, Henzel WJ, Liu X, Lutschg A, Wang X (1997) Apaf-1, a human protein homologous to *C. elegans* CED-4, participates in cytochrome c-dependent activation of caspase-3. Cell 90(3):405–413
44. Yang J, Liu X, Bhalla K, Kim CN, Ibrado AM, Cai J et al (1997) Prevention of apoptosis by Bcl-2: release of cytochrome c from mitochondria blocked. Science 275(5303):1129–1132
45. Kluck RM, Bossy-Wetzel E, Green DR, Newmeyer DD (1997) The release of cytochrome c from mitochondria: a primary site for Bcl-2 regulation of apoptosis. Science 275(5303):1132–1136
46. Muzio M, Stockwell BR, Stennicke HR, Salvesen GS, Dixit VM (1998) An induced proximity model for caspase-8 activation. J Biol Chem 273(5):2926–2930
47. Martin DA, Siegel RM, Zheng L, Lenardo MJ (1998) Membrane oligomerization and cleavage activates the caspase-8 (FLICE/MACHalpha1) death signal. J Biol Chem 273(8):4345–4349
48. Yang X, Chang HY, Baltimore D (1998) Autoproteolytic activation of pro-caspases by oligomerization. Mol Cell 1(2):319–325
49. Li H, Zhu H, Xu CJ, Yuan J (1998) Cleavage of BID by caspase 8 mediates the mitochondrial damage in the Fas pathway of apoptosis. Cell 94(4):491–501
50. Luo X, Budihardjo I, Zou H, Slaughter C, Wang X (1998) Bid, a Bcl2 interacting protein, mediates cytochrome c release from mitochondria in response to activation of cell surface death receptors. Cell 94(4):481–490
51. Mandruzzato S, Brasseur F, Andry G, Boon T, van der Bruggen P (1997) A CASP-8 mutation recognized by cytolytic T lymphocytes on a human head and neck carcinoma. J Exp Med 186(5):785–793
52. Soung YH, Lee JW, Kim SY, Sung YJ, Park WS, Nam SW et al (2005) Caspase-8 gene is frequently inactivated by the frameshift somatic mutation 1225_1226delTG in hepatocellular carcinomas. Oncogene 24(1):141–147

53. Kim HS, Lee JW, Soung YH, Park WS, Kim SY, Lee JH et al (2003) Inactivating mutations of caspase-8 gene in colorectal carcinomas. Gastroenterology 125(3):708–715
54. Soung YH, Lee JW, Kim SY, Jang J, Park YG, Park WS et al (2005) CASPASE-8 gene is inactivated by somatic mutations in gastric carcinomas. Cancer Res 65(3):815–821
55. Teitz T, Wei T, Valentine MB, Vanin EF, Grenet J, Valentine VA et al (2000) Caspase 8 is deleted or silenced preferentially in childhood neuroblastomas with amplification of MYCN. Nat Med 6(5):529–535
56. Banelli B, Casciano I, Croce M, Di Vinci A, Gelvi I, Pagnan G et al (2002) Expression and methylation of CASP8 in neuroblastoma: identification of a promoter region. Nat Med 8(12):1333–1335, author reply 5
57. Lazcoz P, Munoz J, Nistal M, Pestana A, Encio I, Castresana JS (2006) Frequent promoter hypermethylation of RASSF1A and CASP8 in neuroblastoma. BMC Cancer 6:254
58. Fulda S, Poremba C, Berwanger B, Hacker S, Eilers M, Christiansen H et al (2006) Loss of caspase-8 expression does not correlate with MYCN amplification, aggressive disease, or prognosis in neuroblastoma. Cancer Res 66(20):10016–10023
59. Yang Q, Kiernan CM, Tian Y, Salwen HR, Chlenski A, Brumback BA et al (2007) Methylation of CASP8, DCR2, and HIN-1 in neuroblastoma is associated with poor outcome. Clin Cancer Res 13(11):3191–3197
60. Kamimatsuse A, Matsuura K, Moriya S, Fukuba I, Yamaoka H, Fukuda E et al (2009) Detection of CpG island hypermethylation of caspase-8 in neuroblastoma using an oligonucleotide array. Pediatr Blood Cancer 52(7):777–783
61. Harada K, Toyooka S, Shivapurkar N, Maitra A, Reddy JL, Matta H et al (2002) Deregulation of caspase 8 and 10 expression in pediatric tumors and cell lines. Cancer Res 62(20):5897–5901
62. Gonzalez-Gomez P, Bello MJ, Inda MM, Alonso ME, Arjona D, Aminoso C et al (2004) Deletion and aberrant CpG island methylation of Caspase 8 gene in medulloblastoma. Oncol Rep 12(3):663–666
63. Pingoud-Meier C, Lang D, Janss AJ, Rorke LB, Phillips PC, Shalaby T et al (2003) Loss of caspase-8 protein expression correlates with unfavorable survival outcome in childhood medulloblastoma. Clin Cancer Res 9(17):6401–6409
64. Bello MJ, De Campos JM, Isla A, Casartelli C, Rey JA (2006) Promoter CpG methylation of multiple genes in pituitary adenomas: frequent involvement of caspase-8. Oncol Rep 15(2):443–448
65. Malekzadeh K, Sobti RC, Nikbakht M, Shekari M, Hosseini SA, Tamandani DK et al (2009) Methylation patterns of Rb1 and Casp-8 promoters and their impact on their expression in bladder cancer. Cancer Invest 27(1):70–80
66. Liedtke C, Zschemisch NH, Cohrs A, Roskams T, Borlak J, Manns MP et al (2005) Silencing of caspase-8 in murine hepatocellular carcinomas is mediated via methylation of an essential promoter element. Gastroenterology 129(5):1602–1615
67. Cho S, Lee JH, Cho SB, Yoon KW, Park SY, Lee WS et al (2010) Epigenetic methylation and expression of caspase 8 and survivin in hepatocellular carcinoma. Pathol Int 60(3):203–211
68. Shivapurkar N, Toyooka S, Eby MT, Huang CX, Sathyanarayana UG, Cunningham HT et al (2002) Differential inactivation of caspase-8 in lung cancers. Cancer Biol Ther 1(1):65–69
69. Wang J, Zheng L, Lobito A, Chan FK, Dale J, Sneller M et al (1999) Inherited human Caspase 10 mutations underlie defective lymphocyte and dendritic cell apoptosis in autoimmune lymphoproliferative syndrome type II. Cell 98(1):47–58
70. Tadaki H, Saitsu H, Kanegane H, Miyake N, Imagawa T, Kikuchi M et al (2011) Exonic deletion of CASP10 in a patient presenting with systemic juvenile idiopathic arthritis, but not with autoimmune lymphoproliferative syndrome type IIa. Int J Immunogenet 38(4):287–293
71. Park WS, Lee JH, Shin MS, Park JY, Kim HS, Kim YS et al (2002) Inactivating mutations of the caspase-10 gene in gastric cancer. Oncogene 21(18):2919–2925
72. Oh JE, Kim MS, Ahn CH, Kim SS, Han JY, Lee SH et al (2010) Mutational analysis of CASP10 gene in colon, breast, lung and hepatocellular carcinomas. Pathology 42(1):73–76

73. Shin MS, Kim HS, Kang CS, Park WS, Kim SY, Lee SN et al (2002) Inactivating mutations of CASP10 gene in non-Hodgkin lymphomas. Blood 99(11):4094–4099
74. Kim MS, Oh JE, Min CK, Lee S, Chung NG, Yoo NJ et al (2009) Mutational analysis of CASP10 gene in acute leukaemias and multiple myelomas. Pathology 41(5):484–487
75. Soung YH, Lee JW, Kim SY, Park WS, Nam SW, Lee JY et al (2006) Mutational analysis of proapoptotic caspase-9 gene in common human carcinomas. APMIS 114(4):292–297
76. Soung YH, Lee JW, Kim SY, Park WS, Nam SW, Lee JY et al (2004) Somatic mutations of CASP3 gene in human cancers. Hum Genet 115(2):112–115
77. Soung YH, Lee JW, Kim HS, Park WS, Kim SY, Lee JH et al (2003) Inactivating mutations of CASPASE-7 gene in human cancers. Oncogene 22(39):8048–8052
78. Gyrd-Hansen M, Meier P (2010) IAPs: from caspase inhibitors to modulators of NF-kappaB, inflammation and cancer. Nat Rev Cancer 10(8):561–574
79. Fulda S, Vucic D (2012) Targeting IAP proteins for therapeutic intervention in cancer. Nat Rev Drug Discov 11(2):109–124
80. Eckelman BP, Salvesen GS, Scott FL (2006) Human inhibitor of apoptosis proteins: why XIAP is the black sheep of the family. EMBO Rep 7(10):988–994
81. Scott FL, Denault JB, Riedl SJ, Shin H, Renatus M, Salvesen GS (2005) XIAP inhibits caspase-3 and -7 using two binding sites: evolutionarily conserved mechanism of IAPs. EMBO J 24(3):645–655
82. Riedl SJ, Renatus M, Schwarzenbacher R, Zhou Q, Sun C, Fesik SW et al (2001) Structural basis for the inhibition of caspase-3 by XIAP. Cell 104(5):791–800
83. Huang Y, Park YC, Rich RL, Segal D, Myszka DG, Wu H (2001) Structural basis of caspase inhibition by XIAP: differential roles of the linker versus the BIR domain. Cell 104(5):781–790
84. Chai J, Shiozaki E, Srinivasula SM, Wu Q, Datta P, Alnemri ES et al (2001) Structural basis of caspase-7 inhibition by XIAP. Cell 104(5):769–780
85. Srinivasula SM, Hegde R, Saleh A, Datta P, Shiozaki E, Chai J et al (2001) A conserved XIAP-interaction motif in caspase-9 and Smac/DIABLO regulates caspase activity and apoptosis. Nature 410(6824):112–116
86. Shiozaki EN, Chai J, Rigotti DJ, Riedl SJ, Li P, Srinivasula SM et al (2003) Mechanism of XIAP-mediated inhibition of caspase-9. Mol Cell 11(2):519–527
87. Huang H, Joazeiro CA, Bonfoco E, Kamada S, Leverson JD, Hunter T (2000) The inhibitor of apoptosis, cIAP2, functions as a ubiquitin-protein ligase and promotes in vitro monoubiquitination of caspases 3 and 7. J Biol Chem 275(35):26661–26664
88. Suzuki Y, Nakabayashi Y, Takahashi R (2001) Ubiquitin-protein ligase activity of X-linked inhibitor of apoptosis protein promotes proteasomal degradation of caspase-3 and enhances its anti-apoptotic effect in Fas-induced cell death. Proc Natl Acad Sci U S A 98(15):8662–8667
89. Schile AJ, Garcia-Fernandez M, Steller H (2008) Regulation of apoptosis by XIAP ubiquitin-ligase activity. Genes Dev 22(16):2256–2266
90. Choi YE, Butterworth M, Malladi S, Duckett CS, Cohen GM, Bratton SB (2009) The E3 ubiquitin ligase cIAP1 binds and ubiquitinates caspase-3 and -7 via unique mechanisms at distinct steps in their processing. J Biol Chem 284(19):12772–12782
91. Shiraki K, Sugimoto K, Yamanaka Y, Yamaguchi Y, Saitou Y, Ito K et al (2003) Overexpression of X-linked inhibitor of apoptosis in human hepatocellular carcinoma. Int J Mol Med 12(5):705–708
92. Shi YH, Ding WX, Zhou J, He JY, Xu Y, Gambotto AA et al (2008) Expression of X-linked inhibitor-of-apoptosis protein in hepatocellular carcinoma promotes metastasis and tumor recurrence. Hepatology 48(2):497–507
93. Augello C, Caruso L, Maggioni M, Donadon M, Montorsi M, Santambrogio R et al (2009) Inhibitors of apoptosis proteins (IAPs) expression and their prognostic significance in hepatocellular carcinoma. BMC Cancer 9:125
94. Yang XH, Feng ZE, Yan M, Hanada S, Zuo H, Yang CZ et al (2012) XIAP is a predictor of cisplatin-based chemotherapy response and prognosis for patients with advanced head and neck cancer. PLoS One 7(3):e31601

95. Lopes RB, Gangeswaran R, McNeish IA, Wang Y, Lemoine NR (2007) Expression of the IAP protein family is dysregulated in pancreatic cancer cells and is important for resistance to chemotherapy. Int J Cancer 120(11):2344–2352
96. Xiang G, Wen X, Wang H, Chen K, Liu H (2009) Expression of X-linked inhibitor of apoptosis protein in human colorectal cancer and its correlation with prognosis. J Surg Oncol 100(8):708–712
97. Ramp U, Krieg T, Caliskan E, Mahotka C, Ebert T, Willers R et al (2004) XIAP expression is an independent prognostic marker in clear-cell renal carcinomas. Hum Pathol 35(8):1022–1028
98. Yan Y, Mahotka C, Heikaus S, Shibata T, Wethkamp N, Liebmann J et al (2004) Disturbed balance of expression between XIAP and Smac/DIABLO during tumour progression in renal cell carcinomas. Br J Cancer 91(7):1349–1357
99. Mizutani Y, Nakanishi H, Li YN, Matsubara H, Yamamoto K, Sato N et al (2007) Overexpression of XIAP expression in renal cell carcinoma predicts a worse prognosis. Int J Oncol 30(4):919–925
100. Ferreira CG, van der Valk P, Span SW, Ludwig I, Smit EF, Kruyt FA et al (2001) Expression of X-linked inhibitor of apoptosis as a novel prognostic marker in radically resected non-small cell lung cancer patients. Clin Cancer Res 7(8):2468–2474
101. Ferreira CG, van der Valk P, Span SW, Jonker JM, Postmus PE, Kruyt FA et al (2001) Assessment of IAP (inhibitor of apoptosis) proteins as predictors of response to chemotherapy in advanced non-small-cell lung cancer patients. Ann Oncol 12(6):799–805
102. Krepela E, Dankova P, Moravcikova E, Krepelova A, Prochazka J, Cermak J et al (2009) Increased expression of inhibitor of apoptosis proteins, survivin and XIAP, in non-small cell lung carcinoma. Int J Oncol 35(6):1449–1462
103. Kempkensteffen C, Jager T, Bub J, Weikert S, Hinz S, Christoph F et al (2007) The equilibrium of XIAP and Smac/DIABLO expression is gradually deranged during the development and progression of testicular germ cell tumours. Int J Androl 30(5):476–483
104. Gu LQ, Li FY, Zhao L, Liu Y, Zang XX, Wang TX et al (2009) BRAFV600E mutation and X-linked inhibitor of apoptosis expression in papillary thyroid carcinoma. Thyroid 19(4):347–354
105. Hussain AR, Uddin S, Ahmed M, Bu R, Ahmed SO, Abubaker J et al (2010) Prognostic significance of XIAP expression in DLBCL and effect of its inhibition on AKT signalling. J Pathol 222(2):180–190
106. Yamamoto K, Abe S, Nakagawa Y, Suzuki K, Hasegawa M, Inoue M et al (2004) Expression of IAP family proteins in myelodysplastic syndromes transforming to overt leukemia. Leuk Res 28(11):1203–1211
107. Tamm I, Richter S, Scholz F, Schmelz K, Oltersdorf D, Karawajew L et al (2004) XIAP expression correlates with monocytic differentiation in adult de novo AML: impact on prognosis. Hematol J 5(6):489–495
108. Chen GH, Lin FR, Ren JH, Chen J, Zhang JN, Wang Y et al (2006) Expression and significance of X-linked inhibitor of apoptosis protein and its antagonized proteins in acute leukemia. Zhongguo Shi Yan Xue Ye Xue Za Zhi 14(4):639–643
109. Ibrahim AM, Mansour IM, Wilson MM, Mokhtar DA, Helal AM, Al Wakeel HM (2012) Study of survivin and X-linked inhibitor of apoptosis protein (XIAP) genes in acute myeloid leukemia (AML). Lab Hematol 18(1):1–10
110. Tamm I, Richter S, Oltersdorf D, Creutzig U, Harbott J, Scholz F et al (2004) High expression levels of x-linked inhibitor of apoptosis protein and survivin correlate with poor overall survival in childhood de novo acute myeloid leukemia. Clin Cancer Res 10(11):3737–3744
111. Sung KW, Choi J, Hwang YK, Lee SJ, Kim HJ, Kim JY et al (2009) Overexpression of X-linked inhibitor of apoptosis protein (XIAP) is an independent unfavorable prognostic factor in childhood de novo acute myeloid leukemia. J Korean Med Sci 24(4):605–613

112. Hundsdoerfer P, Dietrich I, Schmelz K, Eckert C, Henze G () XIAP expression is post-transcriptionally upregulated in childhood ALL and is associated with glucocorticoid response in T-cell ALL. Pediatr Blood Cancer 55(2):260–266
113. LaCasse EC, Mahoney DJ, Cheung HH, Plenchette S, Baird S, Korneluk RG (2008) IAP-targeted therapies for cancer. Oncogene 27(48):6252–6275
114. Danial NN, Korsmeyer SJ (2004) Cell death: critical control points. Cell 116(2):205–219
115. Cory S, Huang DC, Adams JM (2003) The Bcl-2 family: roles in cell survival and oncogenesis. Oncogene 22(53):8590–8607
116. Plati J, Bucur O, Khosravi-Far R (2011) Apoptotic cell signaling in cancer progression and therapy. Integr Biol (Camb) 3(4):279–296
117. Kelly GL, Strasser A (2011) The essential role of evasion from cell death in cancer. Adv Cancer Res 111:39–96
118. Tsujimoto Y, Croce CM (1986) Analysis of the structure, transcripts, and protein products of bcl-2, the gene involved in human follicular lymphoma. Proc Natl Acad Sci U S A 83(14):5214–5218
119. Boise LH, Gonzalez-Garcia M, Postema CE, Ding L, Lindsten T, Turka LA et al (1993) bcl-x, a bcl-2-related gene that functions as a dominant regulator of apoptotic cell death. Cell 74(4):597–608
120. Kozopas KM, Yang T, Buchan HL, Zhou P, Craig RW (1993) MCL1, a gene expressed in programmed myeloid cell differentiation, has sequence similarity to BCL2. Proc Natl Acad Sci U S A 90(8):3516–3520
121. Lin EY, Orlofsky A, Berger MS, Prystowsky MB (1993) Characterization of A1, a novel hemopoietic-specific early-response gene with sequence similarity to bcl-2. J Immunol 151(4):1979–1988
122. Choi SS, Park IC, Yun JW, Sung YC, Hong SI, Shin HS (1995) A novel Bcl-2 related gene, Bfl-1, is overexpressed in stomach cancer and preferentially expressed in bone marrow. Oncogene 11(9):1693–1698
123. Gibson L, Holmgreen SP, Huang DC, Bernard O, Copeland NG, Jenkins NA et al (1996) bcl-w, a novel member of the bcl-2 family, promotes cell survival. Oncogene 13(4):665–675
124. Wei MC, Zong WX, Cheng EH, Lindsten T, Panoutsakopoulou V, Ross AJ et al (2001) Proapoptotic BAX and BAK: a requisite gateway to mitochondrial dysfunction and death. Science 292(5517):727–730
125. Zong WX, Lindsten T, Ross AJ, MacGregor GR, Thompson CB (2001) BH3-only proteins that bind pro-survival Bcl-2 family members fail to induce apoptosis in the absence of Bax and Bak. Genes Dev 15(12):1481–1486
126. Shangary S, Johnson DE (2003) Recent advances in the development of anticancer agents targeting cell death inhibitors in the Bcl-2 protein family. Leukemia 17(8):1470–1481
127. Reed JC (2008) Bcl-2-family proteins and hematologic malignancies: history and future prospects. Blood 111(7):3322–3330
128. Kelly PN, Strasser A () The role of Bcl-2 and its pro-survival relatives in tumourigenesis and cancer therapy. Cell Death Differ 18(9):1414–1424
129. Crisan D (1996) BCL-2 gene rearrangements in lymphoid malignancies. Clin Lab Med 16(1):23–47
130. Hill ME, MacLennan KA, Cunningham DC, Vaughan Hudson B, Burke M, Clarke P et al (1996) Prognostic significance of BCL-2 expression and bcl-2 major breakpoint region rearrangement in diffuse large cell non-Hodgkin's lymphoma: a British national lymphoma investigation study. Blood 88(3):1046–1051
131. Campos L, Rouault JP, Sabido O, Oriol P, Roubi N, Vasselon C et al (1993) High expression of bcl-2 protein in acute myeloid leukemia cells is associated with poor response to chemotherapy. Blood 81(11):3091–3096
132. Maung ZT, MacLean FR, Reid MM, Pearson AD, Proctor SJ, Hamilton PJ et al (1994) The relationship between bcl-2 expression and response to chemotherapy in acute leukaemia. Br J Haematol 88(1):105–109

133. Porwit-MacDonald A, Ivory K, Wilkinson S, Wheatley K, Wong L, Janossy G (1995) Bcl-2 protein expression in normal human bone marrow precursors and in acute myelogenous leukemia. Leukemia 9(7):1191–1198
134. Karakas T, Maurer U, Weidmann E, Miething CC, Hoelzer D, Bergmann L (1998) High expression of bcl-2 mRNA as a determinant of poor prognosis in acute myeloid leukemia. Ann Oncol 9(2):159–165
135. Bincoletto C, Saad ST, da Silva ES, Queiroz ML (1999) Haematopoietic response and bcl-2 expression in patients with acute myeloid leukaemia. Eur J Haematol 62(1):38–42
136. Campana D, Coustan-Smith E, Manabe A, Buschle M, Raimondi SC, Behm FG et al (1993) Prolonged survival of B-lineage acute lymphoblastic leukemia cells is accompanied by overexpression of bcl-2 protein. Blood 81(4):1025–1031
137. Hanada M, Delia D, Aiello A, Stadtmauer E, Reed JC (1993) bcl-2 gene hypomethylation and high-level expression in B-cell chronic lymphocytic leukemia. Blood 82(6):1820–1828
138. Robertson LE, Plunkett W, McConnell K, Keating MJ, McDonnell TJ (1996) Bcl-2 expression in chronic lymphocytic leukemia and its correlation with the induction of apoptosis and clinical outcome. Leukemia 10(3):456–459
139. Ten Berge RL, Meijer CJ, Dukers DF, Kummer JA, Bladergroen BA, Vos W et al (2002) Expression levels of apoptosis-related proteins predict clinical outcome in anaplastic large cell lymphoma. Blood 99(12):4540–4546
140. Harada N, Hata H, Yoshida M, Soniki T, Nagasaki A, Kuribayashi N et al (1998) Expression of Bcl-2 family of proteins in fresh myeloma cells. Leukemia 12(11):1817–1820
141. Grover R, Wilson GD (1996) Bcl-2 expression in malignant melanoma and its prognostic significance. Eur J Surg Oncol 22(4):347–349
142. Selzer E, Schlagbauer-Wadl H, Okamoto I, Pehamberger H, Potter R, Jansen B (1998) Expression of Bcl-2 family members in human melanocytes, in melanoma metastases and in melanoma cell lines. Melanoma Res 8(3):197–203
143. Vlaykova T, Talve L, Hahka-Kemppinen M, Hernberg M, Muhonen T, Collan Y et al (2002) Immunohistochemically detectable bcl-2 expression in metastatic melanoma: association with survival and treatment response. Oncology 62(3):259–268
144. Deininger MH, Weller M, Streffer J, Meyermann R (1999) Antiapoptotic Bcl-2 family protein expression increases with progression of oligodendroglioma. Cancer 86(9):1832–1839
145. Jiang SX, Sato Y, Kuwao S, Kameya T (1995) Expression of bcl-2 oncogene protein is prevalent in small cell lung carcinomas. J Pathol 177(2):135–138
146. Gazzaniga P, Gradilone A, Vercillo R, Gandini O, Silvestri I, Napolitano M et al (1996) Bcl-2/bax mRNA expression ratio as prognostic factor in low-grade urinary bladder cancer. Int J Cancer 69(2):100–104
147. Pollack A, Wu CS, Czerniak B, Zagars GK, Benedict WF, McDonnell TJ (1997) Abnormal bcl-2 and pRb expression are independent correlates of radiation response in muscle-invasive bladder cancer. Clin Cancer Res 3(10):1823–1829
148. Kong G, Shin KY, Oh YH, Lee JJ, Park HY, Woo YN et al (1998) Bcl-2 and p53 expressions in invasive bladder cancers. Acta Oncol 37(7–8):715–720
149. Ye D, Li H, Qian S, Sun Y, Zheng J, Ma Y (1998) bcl-2/bax expression and p53 gene status in human bladder cancer: relationship to early recurrence with intravesical chemotherapy after resection. J Urol 160(6 Pt 1):2025–2028, discussion 9
150. Sinicrope FA, Hart J, Michelassi F, Lee JJ (1995) Prognostic value of bcl-2 oncoprotein expression in stage II colon carcinoma. Clin Cancer Res 1(10):1103–1110
151. McDonnell TJ, Troncoso P, Brisbay SM, Logothetis C, Chung LW, Hsieh JT et al (1992) Expression of the protooncogene bcl-2 in the prostate and its association with emergence of androgen-independent prostate cancer. Cancer Res 52(24):6940–6944
152. Colombel M, Symmans F, Gil S, O'Toole KM, Chopin D, Benson M et al (1993) Detection of the apoptosis-suppressing oncoprotein bc1-2 in hormone-refractory human prostate cancers. Am J Pathol 143(2):390–400

153. Bauer JJ, Sesterhenn IA, Mostofi FK, McLeod DG, Srivastava S, Moul JW (1996) Elevated levels of apoptosis regulator proteins p53 and bcl-2 are independent prognostic biomarkers in surgically treated clinically localized prostate cancer. J Urol 156(4):1511–1516
154. Pallis M, Zhu YM, Russell NH (1997) Bcl-x(L) is heterogenously expressed by acute myeloblastic leukaemia cells and is associated with autonomous growth in vitro and with P-glycoprotein expression. Leukemia 11(7):945–949
155. Deng G, Lane C, Kornblau S, Goodacre A, Snell V, Andreeff M et al (1998) Ratio of bcl-xshort to bcl-xlong is different in good- and poor-prognosis subsets of acute myeloid leukemia. Mol Med 4(3):158–164
156. Tu Y, Renner S, Xu F, Fleishman A, Taylor J, Weisz J et al (1998) BCL-X expression in multiple myeloma: possible indicator of chemoresistance. Cancer Res 58(2):256–262
157. Tang L, Tron VA, Reed JC, Mah KJ, Krajewska M, Li G et al (1998) Expression of apoptosis regulators in cutaneous malignant melanoma. Clin Cancer Res 4(8):1865–1871
158. Leiter U, Schmid RM, Kaskel P, Peter RU, Krahn G (2000) Antiapoptotic bcl-2 and bcl-xL in advanced malignant melanoma. Arch Dermatol Res 292(5):225–232
159. Aebersold DM, Kollar A, Beer KT, Laissue J, Greiner RH, Djonov V (2001) Involvement of the hepatocyte growth factor/scatter factor receptor c-met and of Bcl-xL in the resistance of oropharyngeal cancer to ionizing radiation. Int J Cancer 96(1):41–54
160. Trask DK, Wolf GT, Bradford CR, Fisher SG, Devaney K, Johnson M et al (2002) Expression of Bcl-2 family proteins in advanced laryngeal squamous cell carcinoma: correlation with response to chemotherapy and organ preservation. Laryngoscope 112(4):638–644
161. Watanabe J, Kushihata F, Honda K, Mominoki K, Matsuda S, Kobayashi N (2002) Bcl-xL overexpression in human hepatocellular carcinoma. Int J Oncol 21(3):515–519
162. Olopade OI, Adeyanju MO, Safa AR, Hagos F, Mick R, Thompson CB et al (1997) Overexpression of BCL-x protein in primary breast cancer is associated with high tumor grade and nodal metastases. Cancer J Sci Am 3(4):230–237
163. Kirsh EJ, Baunoch DA, Stadler WM (1998) Expression of bcl-2 and bcl-X in bladder cancer. J Urol 159(4):1348–1353
164. Krajewska M, Moss SF, Krajewski S, Song K, Holt PR, Reed JC (1996) Elevated expression of Bcl-X and reduced Bak in primary colorectal adenocarcinomas. Cancer Res 56(10):2422–2427
165. Friess H, Lu Z, Andren-Sandberg A, Berberat P, Zimmermann A, Adler G et al (1998) Moderate activation of the apoptosis inhibitor bcl-xL worsens the prognosis in pancreatic cancer. Ann Surg 228(6):780–787
166. Miyamoto Y, Hosotani R, Wada M, Lee JU, Koshiba T, Fujimoto K et al (1999) Immunohistochemical analysis of Bcl-2, Bax, Bcl-X, and Mcl-1 expression in pancreatic cancers. Oncology 56(1):73–82
167. Kaufmann SH, Karp JE, Svingen PA, Krajewski S, Burke PJ, Gore SD et al (1998) Elevated expression of the apoptotic regulator Mcl-1 at the time of leukemic relapse. Blood 91(3):991–1000
168. Zhang B, Gojo I, Fenton RG (2002) Myeloid cell factor-1 is a critical survival factor for multiple myeloma. Blood 99(6):1885–1893
169. Le Gouill S, Podar K, Amiot M, Hideshima T, Chauhan D, Ishitsuka K et al (2004) VEGF induces Mcl-1 up-regulation and protects multiple myeloma cells against apoptosis. Blood 104(9):2886–2892
170. Kitada S, Andersen J, Akar S, Zapata JM, Takayama S, Krajewski S et al (1998) Expression of apoptosis-regulating proteins in chronic lymphocytic leukemia: correlations with In vitro and In vivo chemoresponses. Blood 91(9):3379–3389
171. Shigemasa K, Katoh O, Shiroyama Y, Mihara S, Mukai K, Nagai N et al (2002) Increased MCL-1 expression is associated with poor prognosis in ovarian carcinomas. Jpn J Cancer Res 93(5):542–550
172. Li C, Li R, Grandis JR, Johnson DE (2008) Bortezomib induces apoptosis via Bim and Bik up-regulation and synergizes with cisplatin in the killing of head and neck squamous cell carcinoma cells. Mol Cancer Ther 7(6):1647–1655

173. Feuerhake F, Kutok JL, Monti S, Chen W, LaCasce AS, Cattoretti G et al (2005) NFkappaB activity, function, and target-gene signatures in primary mediastinal large B-cell lymphoma and diffuse large B-cell lymphoma subtypes. Blood 106(4):1392–1399
174. Brien G, Trescol-Biemont MC, Bonnefoy-Berard N (2007) Downregulation of Bfl-1 protein expression sensitizes malignant B cells to apoptosis. Oncogene 26(39):5828–5832
175. Wilson JW, Nostro MC, Balzi M, Faraoni P, Cianchi F, Becciolini A et al (2000) Bcl-w expression in colorectal adenocarcinoma. Br J Cancer 82(1):178–185
176. Monni O, Joensuu H, Franssila K, Klefstrom J, Alitalo K, Knuutila S (1997) BCL2 overexpression associated with chromosomal amplification in diffuse large B-cell lymphoma. Blood 90(3):1168–1174
177. Olejniczak ET, Van Sant C, Anderson MG, Wang G, Tahir SK, Sauter G et al (2007) Integrative genomic analysis of small-cell lung carcinoma reveals correlates of sensitivity to bcl-2 antagonists and uncovers novel chromosomal gains. Mol Cancer Res 5(4):331–339
178. Beroukhim R, Mermel CH, Porter D, Wei G, Raychaudhuri S, Donovan J et al (2010) The landscape of somatic copy-number alteration across human cancers. Nature 463(7283):899–905
179. Cimmino A, Calin GA, Fabbri M, Iorio MV, Ferracin M, Shimizu M et al (2005) miR-15 and miR-16 induce apoptosis by targeting BCL2. Proc Natl Acad Sci U S A 102(39):13944–13949
180. Calin GA, Dumitru CD, Shimizu M, Bichi R, Zupo S, Noch E et al (2002) Frequent deletions and down-regulation of micro- RNA genes miR15 and miR16 at 13q14 in chronic lymphocytic leukemia. Proc Natl Acad Sci U S A 99(24):15524–15529
181. Calin GA, Ferracin M, Cimmino A, Di Leva G, Shimizu M, Wojcik SE et al (2005) A MicroRNA signature associated with prognosis and progression in chronic lymphocytic leukemia. N Engl J Med 353(17):1793–1801
182. Calin GA, Cimmino A, Fabbri M, Ferracin M, Wojcik SE, Shimizu M et al (2008) MiR-15a and miR-16-1 cluster functions in human leukemia. Proc Natl Acad Sci U S A 105(13):5166–5171
183. Aqeilan RI, Calin GA, Croce CM (2010) miR-15a and miR-16-1 in cancer: discovery, function and future perspectives. Cell Death Differ 17(2):215–220
184. Oltvai ZN, Milliman CL, Korsmeyer SJ (1993) Bcl-2 heterodimerizes in vivo with a conserved homolog, Bax, that accelerates programmed cell death. Cell 74(4):609–619
185. Cory S, Adams JM (2002) The Bcl2 family: regulators of the cellular life-or-death switch. Nat Rev Cancer 2(9):647–656
186. Korsmeyer SJ, Shutter JR, Veis DJ, Merry DE, Oltvai ZN (1993) Bcl-2/Bax: a rheostat that regulates an anti-oxidant pathway and cell death. Semin Cancer Biol 4(6):327–332
187. Annis MG, Soucie EL, Dlugosz PJ, Cruz-Aguado JA, Penn LZ, Leber B et al (2005) Bax forms multispanning monomers that oligomerize to permeabilize membranes during apoptosis. EMBO J 24(12):2096–2103
188. Korsmeyer SJ, Wei MC, Saito M, Weiler S, Oh KJ, Schlesinger PH (2000) Pro-apoptotic cascade activates BID, which oligomerizes BAK or BAX into pores that result in the release of cytochrome c. Cell Death Differ 7(12):1166–1173
189. Wei MC, Lindsten T, Mootha VK, Weiler S, Gross A, Ashiya M et al (2000) tBID, a membrane-targeted death ligand, oligomerizes BAK to release cytochrome c. Genes Dev 14(16):2060–2071
190. Sharpe JC, Arnoult D, Youle RJ (2004) Control of mitochondrial permeability by Bcl-2 family members. Biochim Biophys Acta 1644(2–3):107–113
191. Kuwana T, Mackey MR, Perkins G, Ellisman MH, Latterich M, Schneiter R et al (2002) Bid, Bax, and lipids cooperate to form supramolecular openings in the outer mitochondrial membrane. Cell 111(3):331–342
192. Letai A, Bassik MC, Walensky LD, Sorcinelli MD, Weiler S, Korsmeyer SJ (2002) Distinct BH3 domains either sensitize or activate mitochondrial apoptosis, serving as prototype cancer therapeutics. Cancer Cell 2(3):183–192

193. Kuwana T, Bouchier-Hayes L, Chipuk JE, Bonzon C, Sullivan BA, Green DR et al (2005) BH3 domains of BH3-only proteins differentially regulate Bax-mediated mitochondrial membrane permeabilization both directly and indirectly. Mol Cell 17(4):525–535
194. Certo M, Del Gaizo Moore V, Nishino M, Wei G, Korsmeyer S, Armstrong SA et al (2006) Mitochondria primed by death signals determine cellular addiction to antiapoptotic BCL-2 family members. Cancer Cell 9(5):351–365
195. Cartron PF, Gallenne T, Bougras G, Gautier F, Manero F, Vusio P et al (2004) The first alpha helix of Bax plays a necessary role in its ligand-induced activation by the BH3-only proteins Bid and PUMA. Mol Cell 16(5):807–818
196. Chen L, Willis SN, Wei A, Smith BJ, Fletcher JI, Hinds MG et al (2005) Differential targeting of prosurvival Bcl-2 proteins by their BH3-only ligands allows complementary apoptotic function. Mol Cell 17(3):393–403
197. Willis SN, Chen L, Dewson G, Wei A, Naik E, Fletcher JI et al (2005) Proapoptotic Bak is sequestered by Mcl-1 and Bcl-xL, but not Bcl-2, until displaced by BH3-only proteins. Genes Dev 19(11):1294–1305
198. Willis SN, Fletcher JI, Kaufmann T, van Delft MF, Chen L, Czabotar PE et al (2007) Apoptosis initiated when BH3 ligands engage multiple Bcl-2 homologs, not Bax or Bak. Science 315(5813):856–859
199. Fernandez-Luna JL (2008) Regulation of pro-apoptotic BH3-only proteins and its contribution to cancer progression and chemoresistance. Cell Signal 20(11):1921–1926
200. Rampino N, Yamamoto H, Ionov Y, Li Y, Sawai H, Reed JC et al (1997) Somatic frameshift mutations in the BAX gene in colon cancers of the microsatellite mutator phenotype. Science 275(5302):967–969
201. Meijerink JP, Mensink EJ, Wang K, Sedlak TW, Sloetjes AW, de Witte T et al (1998) Hematopoietic malignancies demonstrate loss-of-function mutations of BAX. Blood 91(8):2991–2997
202. Miquel C, Borrini F, Grandjouan S, Auperin A, Viguier J, Velasco V et al (2005) Role of bax mutations in apoptosis in colorectal cancers with microsatellite instability. Am J Clin Pathol 123(4):562–570
203. Mestre-Escorihuela C, Rubio-Moscardo F, Richter JA, Siebert R, Climent J, Fresquet V et al (2007) Homozygous deletions localize novel tumor suppressor genes in B-cell lymphomas. Blood 109(1):271–280
204. Kondo S, Shinomura Y, Miyazaki Y, Kiyohara T, Tsutsui S, Kitamura S et al (2000) Mutations of the bak gene in human gastric and colorectal cancers. Cancer Res 60(16):4328–4330
205. Tagawa H, Karnan S, Suzuki R, Matsuo K, Zhang X, Ota A et al (2005) Genome-wide array-based CGH for mantle cell lymphoma: identification of homozygous deletions of the proapoptotic gene BIM. Oncogene 24(8):1348–1358
206. Sturm I, Stephan C, Gillissen B, Siebert R, Janz M, Radetzki S et al (2006) Loss of the tissue-specific proapoptotic BH3-only protein Nbk/Bik is a unifying feature of renal cell carcinoma. Cell Death Differ 13(4):619–627
207. Richter-Larrea JA, Robles EF, Fresquet V, Beltran E, Rullan AJ, Agirre X et al (2010) Reversion of epigenetically mediated BIM silencing overcomes chemoresistance in Burkitt lymphoma. Blood 116(14):2531–2542
208. Garrison SP, Jeffers JR, Yang C, Nilsson JA, Hall MA, Rehg JE et al (2008) Selection against PUMA gene expression in Myc-driven B-cell lymphomagenesis. Mol Cell Biol 28(17):5391–5402
209. Obata T, Toyota M, Satoh A, Sasaki Y, Ogi K, Akino K et al (2003) Identification of HRK as a target of epigenetic inactivation in colorectal and gastric cancer. Clin Cancer Res 9(17):6410–6418
210. Fontana L, Fiori ME, Albini S, Cifaldi L, Giovinazzi S, Forloni M et al (2008) Antagomir-17-5p abolishes the growth of therapy-resistant neuroblastoma through p21 and BIM. PLoS ONE 3(5):e2236

211. Inomata M, Tagawa H, Guo YM, Kameoka Y, Takahashi N, Sawada K (2009) MicroRNA-17-92 down-regulates expression of distinct targets in different B-cell lymphoma subtypes. Blood 113(2):396–402
212. Kan T, Sato F, Ito T, Matsumura N, David S, Cheng Y et al (2009) The miR-106b-25 polycistron, activated by genomic amplification, functions as an oncogene by suppressing p21 and Bim. Gastroenterology 136(5):1689–1700
213. Li Y, Tan W, Neo TW, Aung MO, Wasser S, Lim SG et al (2009) Role of the miR-106b-25 microRNA cluster in hepatocellular carcinoma. Cancer Sci 100(7):1234–1242
214. Lane DP (1992) Cancer p53, guardian of the genome. Nature 358(6381):15–16
215. Shieh SY, Ikeda M, Taya Y, Prives C (1997) DNA damage-induced phosphorylation of p53 alleviates inhibition by MDM2. Cell 91(3):325–334
216. Mayo LD, Turchi JJ, Berberich SJ (1997) Mdm-2 phosphorylation by DNA-dependent protein kinase prevents interaction with p53. Cancer Res 57(22):5013–5016
217. Lakin ND, Hann BC, Jackson SP (1999) The ataxia-telangiectasia related protein ATR mediates DNA-dependent phosphorylation of p53. Oncogene 18(27):3989–3995
218. Tibbetts RS, Brumbaugh KM, Williams JM, Sarkaria JN, Cliby WA, Shieh SY et al (1999) A role for ATR in the DNA damage-induced phosphorylation of p53. Genes Dev 13(2):152–157
219. Chen L, Gilkes DM, Pan Y, Lane WS, Chen J (2005) ATM and Chk2-dependent phosphorylation of MDMX contribute to p53 activation after DNA damage. EMBO J 24(19):3411–3422
220. Maya R, Balass M, Kim ST, Shkedy D, Leal JF, Shifman O et al (2001) ATM-dependent phosphorylation of Mdm2 on serine 395: role in p53 activation by DNA damage. Genes Dev 15(9):1067–1077
221. Sancar A, Lindsey-Boltz LA, Unsal-Kacmaz K, Linn S (2004) Molecular mechanisms of mammalian DNA repair and the DNA damage checkpoints. Annu Rev Biochem 73:39–85
222. Miyashita T, Reed JC (1995) Tumor suppressor p53 is a direct transcriptional activator of the human bax gene. Cell 80(2):293–299
223. Moroni MC, Hickman ES, Lazzerini Denchi E, Caprara G, Colli E, Cecconi F (2001) Apaf-1 is a transcriptional target for E2F and p53. Nat Cell Biol 3(6):552–558
224. Yu J, Zhang L, Hwang PM, Kinzler KW, Vogelstein B (2001) PUMA induces the rapid apoptosis of colorectal cancer cells. Mol Cell 7(3):673–682
225. Nakano K, Vousden KH (2001) PUMA, a novel proapoptotic gene, is induced by p53. Mol Cell 7(3):683–694
226. Oda E, Ohki R, Murasawa H, Nemoto J, Shibue T, Yamashita T et al (2000) Noxa, a BH3-only member of the Bcl-2 family and candidate mediator of p53-induced apoptosis. Science 288(5468):1053–1058
227. Sax JK, Fei P, Murphy ME, Bernhard E, Korsmeyer SJ, El-Deiry WS (2002) BID regulation by p53 contributes to chemosensitivity. Nat Cell Biol 4(11):842–849
228. Jeffers JR, Parganas E, Lee Y, Yang C, Wang J, Brennan J et al (2003) Puma is an essential mediator of p53-dependent and -independent apoptotic pathways. Cancer Cell 4(4):321–328
229. Villunger A, Michalak EM, Coultas L, Mullauer F, Bock G, Ausserlechner MJ et al (2003) p53- and drug-induced apoptotic responses mediated by BH3-only proteins puma and noxa. Science 302(5647):1036–1038
230. Michalak EM, Villunger A, Adams JM, Strasser A (2008) In several cell types tumour suppressor p53 induces apoptosis largely via Puma but Noxa can contribute. Cell Death Differ 15(6):1019–1029
231. Shibue T, Takeda K, Oda E, Tanaka H, Murasawa H, Takaoka A et al (2003) Integral role of Noxa in p53-mediated apoptotic response. Genes Dev 17(18):2233–2238
232. Chipuk JE, Maurer U, Green DR, Schuler M (2003) Pharmacologic activation of p53 elicits Bax-dependent apoptosis in the absence of transcription. Cancer Cell 4(5):371–381
233. Chipuk JE, Kuwana T, Bouchier-Hayes L, Droin NM, Newmeyer DD, Schuler M et al (2004) Direct activation of Bax by p53 mediates mitochondrial membrane permeabilization and apoptosis. Science 303(5660):1010–1014

234. Leu JI, Dumont P, Hafey M, Murphy ME, George DL (2004) Mitochondrial p53 activates Bak and causes disruption of a Bak-Mcl1 complex. Nat Cell Biol 6(5):443–450
235. Mihara M, Erster S, Zaika A, Petrenko O, Chittenden T, Pancoska P et al (2003) p53 has a direct apoptogenic role at the mitochondria. Mol Cell 11(3):577–590
236. Yin XM, Wang K, Gross A, Zhao Y, Zinkel S, Klocke B et al (1999) Bid-deficient mice are resistant to Fas-induced hepatocellular apoptosis. Nature 400(6747):886–891
237. Friesen C, Fulda S, Debatin KM (1997) Deficient activation of the CD95 (APO-1/Fas) system in drug-resistant cells. Leukemia 11(11):1833–1841
238. Antoku K, Liu Z, Johnson DE (1997) Inhibition of caspase proteases by CrmA enhances the resistance of human leukemic cells to multiple chemotherapeutic agents. Leukemia 11(10):1665–1672
239. Fulda S, Los M, Friesen C, Debatin KM (1998) Chemosensitivity of solid tumor cells in vitro is related to activation of the CD95 system. Int J Cancer 76(1):105–114
240. Fulda S, Meyer E, Friesen C, Susin SA, Kroemer G, Debatin KM (2001) Cell type specific involvement of death receptor and mitochondrial pathways in drug-induced apoptosis. Oncogene 20(9):1063–1075
241. Tauzin S, Debure L, Moreau JF, Legembre P (2010) CD95-mediated cell signaling in cancer: mutations and post-translational modulations. Cell Mol Life Sci 69(8):1261–1277
242. Watanabe-Fukunaga R, Brannan CI, Copeland NG, Jenkins NA, Nagata S (1992) Lymphoproliferation disorder in mice explained by defects in Fas antigen that mediates apoptosis. Nature 356(6367):314–317
243. Chu JL, Drappa J, Parnassa A, Elkon KB (1993) The defect in Fas mRNA expression in MRL/lpr mice is associated with insertion of the retrotransposon, ETn. J Exp Med 178(2):723–730
244. Wu J, Zhou T, He J, Mountz JD (1993) Autoimmune disease in mice due to integration of an endogenous retrovirus in an apoptosis gene. J Exp Med 178(2):461–468
245. Adachi M, Watanabe-Fukunaga R, Nagata S (1993) Aberrant transcription caused by the insertion of an early transposable element in an intron of the Fas antigen gene of lpr mice. Proc Natl Acad Sci U S A 90(5):1756–1760
246. Dowdell KC, Niemela JE, Price S, Davis J, Hornung RL, Oliveira JB et al (2010) Somatic FAS mutations are common in patients with genetically undefined autoimmune lymphoproliferative syndrome. Blood 115(25):5164–5169
247. Hsu AP, Dowdell KC, Davis J, Niemela JE, Anderson SM, Shaw PA et al (2012) Autoimmune lymphoproliferative syndrome due to FAS mutations outside the signal-transducing death domain: molecular mechanisms and clinical penetrance. Genet Med 14(1):81–89
248. Kuehn HS, Caminha I, Niemela JE, Rao VK, Davis J, Fleisher TA et al (2011) FAS haploinsufficiency is a common disease mechanism in the human autoimmune lymphoproliferative syndrome. J Immunol 186(10):6035–6043
249. Magerus-Chatinet A, Neven B, Stolzenberg MC, Daussy C, Arkwright PD, Lanzarotti N et al (2011) Onset of autoimmune lymphoproliferative syndrome (ALPS) in humans as a consequence of genetic defect accumulation. J Clin Invest 121(1):106–112
250. Wu J, Siddiqui J, Nihal M, Vonderheid EC, Wood GS (2011) Structural alterations of the FAS gene in cutaneous T-cell lymphoma (CTCL). Arch Biochem Biophys 508(2):185–191
251. Dereure O, Levi E, Vonderheid EC, Kadin ME (2002) Infrequent Fas mutations but no Bax or p53 mutations in early mycosis fungoides: a possible mechanism for the accumulation of malignant T lymphocytes in the skin. J Invest Dermatol 118(6):949–956
252. Seeberger H, Starostik P, Schwarz S, Knorr C, Kalla J, Ott G et al (2001) Loss of Fas (CD95/APO-1) regulatory function is an important step in early MALT-type lymphoma development. Lab Invest 81(7):977–986
253. Wohlfart S, Sebinger D, Gruber P, Buch J, Polgar D, Krupitza G et al (2004) FAS (CD95) mutations are rare in gastric MALT lymphoma but occur more frequently in primary gastric diffuse large B-cell lymphoma. Am J Pathol 164(3):1081–1089

254. Takakuwa T, Dong Z, Nakatsuka S, Kojya S, Harabuchi Y, Yang WI et al (2002) Frequent mutations of Fas gene in nasal NK/T cell lymphoma. Oncogene 21(30):4702–4705
255. Gronbaek K, Straten PT, Ralfkiaer E, Ahrenkiel V, Andersen MK, Hansen NE et al (1998) Somatic Fas mutations in non-Hodgkin's lymphoma: association with extranodal disease and autoimmunity. Blood 92(9):3018–3024
256. Scholl V, Stefanoff CG, Hassan R, Spector N, Renault IZ (2007) Mutations within the 5′ region of FAS/CD95 gene in nodal diffuse large B-cell lymphoma. Leuk Lymphoma 48(5):957–963
257. Takakuwa T, Dong Z, Takayama H, Matsuzuka F, Nagata S, Aozasa K (2001) Frequent mutations of Fas gene in thyroid lymphoma. Cancer Res 61(4):1382–1385
258. Landowski TH, Qu N, Buyuksal I, Painter JS, Dalton WS (1997) Mutations in the Fas antigen in patients with multiple myeloma. Blood 90(11):4266–4270
259. Lee SH, Shin MS, Park WS, Kim SY, Dong SM, Pi JH et al (1999) Alterations of Fas (APO-1/CD95) gene in transitional cell carcinomas of urinary bladder. Cancer Res 59(13):3068–3072
260. Shin MS, Park WS, Kim SY, Kim HS, Kang SJ, Song KY et al (1999) Alterations of Fas (Apo-1/CD95) gene in cutaneous malignant melanoma. Am J Pathol 154(6):1785–1791
261. Lee SH, Shin MS, Park WS, Kim SY, Kim HS, Han JY et al (1999) Alterations of Fas (Apo-1/CD95) gene in non-small cell lung cancer. Oncogene 18(25):3754–3760
262. Shin MS, Kim HS, Lee SH, Lee JW, Song YH, Kim YS et al (2002) Alterations of Fas-pathway genes associated with nodal metastasis in non-small cell lung cancer. Oncogene 21(26):4129–4136
263. Lee SH, Shin MS, Kim HS, Park WS, Kim SY, Jang JJ et al (2000) Somatic mutations of Fas (Apo-1/CD95) gene in cutaneous squamous cell carcinoma arising from a burn scar. J Invest Dermatol 114(1):122–126
264. Park WS, Oh RR, Kim YS, Park JY, Lee SH, Shin MS et al (2001) Somatic mutations in the death domain of the Fas (Apo-1/CD95) gene in gastric cancer. J Pathol 193(2):162–168
265. Takayama H, Takakuwa T, Tsujimoto Y, Tani Y, Nonomura N, Okuyama A et al (2002) Frequent Fas gene mutations in testicular germ cell tumors. Am J Pathol 161(2):635–641
266. Jones CL, Wain EM, Chu CC, Tosi I, Foster R, McKenzie RC et al (2010) Downregulation of Fas gene expression in Sezary syndrome is associated with promoter hypermethylation. J Invest Dermatol 130(4):1116–1125
267. Watson CJ, O'Kane H, Maxwell P, Sharaf O, Petak I, Hyland PL et al (2012) Identification of a methylation hotspot in the death receptor Fas/CD95 in bladder cancer. Int J Oncol 40(3):645–654
268. Petak I, Danam RP, Tillman DM, Vernes R, Howell SR, Berczi L et al (2003) Hypermethylation of the gene promoter and enhancer region can regulate Fas expression and sensitivity in colon carcinoma. Cell Death Differ 10(2):211–217
269. Carvalho JR, Filipe L, Costa VL, Ribeiro FR, Martins AT, Teixeira MR et al (2010) Detailed analysis of expression and promoter methylation status of apoptosis-related genes in prostate cancer. Apoptosis 15(8):956–965
270. Lavrik I, Golks A, Krammer PH (2005) Death receptor signaling. J Cell Sci 118(Pt 2):265–267
271. Ozoren N, El-Deiry WS (2003) Cell surface death receptor signaling in normal and cancer cells. Semin Cancer Biol 13(2):135–147
272. Abdulghani J, El-Deiry WS (2010) TRAIL receptor signaling and therapeutics. Expert Opin Ther Targets 14(10):1091–1108
273. Lee SH, Shin MS, Kim HS, Lee HK, Park WS, Kim SY et al (2001) Somatic mutations of TRAIL-receptor 1 and TRAIL-receptor 2 genes in non-Hodgkin's lymphoma. Oncogene 20(3):399–403
274. Shin MS, Kim HS, Lee SH, Park WS, Kim SY, Park JY et al (2001) Mutations of tumor necrosis factor-related apoptosis-inducing ligand receptor 1 (TRAIL-R1) and receptor 2 (TRAIL-R2) genes in metastatic breast cancers. Cancer Res 61(13):4942–4946
275. Fisher MJ, Virmani AK, Wu L, Aplenc R, Harper JC, Powell SM et al (2001) Nucleotide substitution in the ectodomain of trail receptor DR4 is associated with lung cancer and head and neck cancer. Clin Cancer Res 7(6):1688–1697

276. Dechant MJ, Fellenberg J, Scheuerpflug CG, Ewerbeck V, Debatin KM (2004) Mutation analysis of the apoptotic "death-receptors" and the adaptors TRADD and FADD/MORT-1 in osteosarcoma tumor samples and osteosarcoma cell lines. Int J Cancer 109(5):661–667
277. Park WS, Lee JH, Shin MS, Park JY, Kim HS, Kim YS et al (2001) Inactivating mutations of KILLER/DR5 gene in gastric cancers. Gastroenterology 121(5):1219–1225
278. Lee SH, Shin MS, Kim HS, Lee HK, Park WS, Kim SY et al (1999) Alterations of the DR5/TRAIL receptor 2 gene in non-small cell lung cancers. Cancer Res 59(22):5683–5686
279. Arai T, Akiyama Y, Okabe S, Saito K, Iwai T, Yuasa Y (1998) Genomic organization and mutation analyses of the DR5/TRAIL receptor 2 gene in colorectal carcinomas. Cancer Lett 133(2):197–204
280. Lee KH, Lim SW, Kim HG, Kim DY, Ryu SY, Joo JK et al (2009) Lack of death receptor 4 (DR4) expression through gene promoter methylation in gastric carcinoma. Langenbecks Arch Surg 394(4):661–670
281. Elias A, Siegelin MD, Steinmuller A, von Deimling A, Lass U, Korn B et al (2009) Epigenetic silencing of death receptor 4 mediates tumor necrosis factor-related apoptosis-inducing ligand resistance in gliomas. Clin Cancer Res 15(17):5457–5465
282. Hopkins-Donaldson S, Ziegler A, Kurtz S, Bigosch C, Kandioler D, Ludwig C et al (2003) Silencing of death receptor and caspase-8 expression in small cell lung carcinoma cell lines and tumors by DNA methylation. Cell Death Differ 10(3):356–364
283. Horak P, Pils D, Haller G, Pribill I, Roessler M, Tomek S et al (2005) Contribution of epigenetic silencing of tumor necrosis factor-related apoptosis inducing ligand receptor 1 (DR4) to TRAIL resistance and ovarian cancer. Mol Cancer Res 3(6):335–343
284. Zhang Y, Zhang B (2008) TRAIL resistance of breast cancer cells is associated with constitutive endocytosis of death receptors 4 and 5. Mol Cancer Res 6(12):1861–1871
285. Jin Z, McDonald ER 3rd, Dicker DT, El-Deiry WS (2004) Deficient tumor necrosis factor-related apoptosis-inducing ligand (TRAIL) death receptor transport to the cell surface in human colon cancer cells selected for resistance to TRAIL-induced apoptosis. J Biol Chem 279(34):35829–35839
286. Wu GS, Burns TF, McDonald ER 3rd, Jiang W, Meng R, Krantz ID et al (1997) KILLER/DR5 is a DNA damage-inducible p53-regulated death receptor gene. Nat Genet 17(2):141–143
287. Sheikh MS, Burns TF, Huang Y, Wu GS, Amundson S, Brooks KS et al (1998) p53-dependent and -independent regulation of the death receptor KILLER/DR5 gene expression in response to genotoxic stress and tumor necrosis factor alpha. Cancer Res 58(8):1593–1598
288. Wu GS, Burns TF, McDonald ER 3rd, Meng RD, Kao G, Muschel R et al (1999) Induction of the TRAIL receptor KILLER/DR5 in p53-dependent apoptosis but not growth arrest. Oncogene 18(47):6411–6418
289. Takimoto R, El-Deiry WS (2000) Wild-type p53 transactivates the KILLER/DR5 gene through an intronic sequence-specific DNA-binding site. Oncogene 19(14):1735–1743
290. Guan B, Yue P, Clayman GL, Sun SY (2001) Evidence that the death receptor DR4 is a DNA damage-inducible, p53-regulated gene. J Cell Physiol 188(1):98–105
291. Liu X, Yue P, Khuri FR, Sun SY (2004) p53 upregulates death receptor 4 expression through an intronic p53 binding site. Cancer Res 64(15):5078–5083
292. Sheikh MS, Huang Y, Fernandez-Salas EA, El-Deiry WS, Friess H, Amundson S et al (1999) The antiapoptotic decoy receptor TRID/TRAIL-R3 is a p53-regulated DNA damage-inducible gene that is overexpressed in primary tumors of the gastrointestinal tract. Oncogene 18(28):4153–4159
293. Liu X, Yue P, Khuri FR, Sun SY (2005) Decoy receptor 2 (DcR2) is a p53 target gene and regulates chemosensitivity. Cancer Res 65(20):9169–9175
294. Bolze A, Byun M, McDonald D, Morgan NV, Abhyankar A, Premkumar L et al (2010) Whole-exome-sequencing-based discovery of human FADD deficiency. Am J Hum Genet 87(6):873–881

295. Soung YH, Lee JW, Kim SY, Nam SW, Park WS, Kim SH et al (2004) Mutation of FADD gene is rare in human colon and stomach cancers. APMIS 112(9):595–597
296. Safa AR, Day TW, Wu CH (2008) Cellular FLICE-like inhibitory protein (C-FLIP): a novel target for cancer therapy. Curr Cancer Drug Targets 8(1):37–46
297. Bagnoli M, Canevari S, Mezzanzanica D (2010) Cellular FLICE-inhibitory protein (c-FLIP) signalling: a key regulator of receptor-mediated apoptosis in physiologic context and in cancer. Int J Biochem Cell Biol 42(2):210–213
298. Chang DW, Xing Z, Pan Y, Algeciras-Schimnich A, Barnhart BC, Yaish-Ohad S et al (2002) c-FLIP(L) is a dual function regulator for caspase-8 activation and CD95-mediated apoptosis. EMBO J 21(14):3704–3714
299. Yu JW, Jeffrey PD, Shi Y (2009) Mechanism of procaspase-8 activation by c-FLIPL. Proc Natl Acad Sci U S A 106(20):8169–8174
300. Korkolopoulou P, Saetta AA, Levidou G, Gigelou F, Lazaris A, Thymara I et al (2007) c-FLIP expression in colorectal carcinomas: association with Fas/FasL expression and prognostic implications. Histopathology 51(2):150–156
301. Du X, Bao G, He X, Zhao H, Yu F, Qiao Q et al (2009) Expression and biological significance of c-FLIP in human hepatocellular carcinomas. J Exp Clin Cancer Res 28:24
302. Bagnoli M, Ambrogi F, Pilotti S, Alberti P, Ditto A, Barbareschi M et al (2009) c-FLIPL expression defines two ovarian cancer patient subsets and is a prognostic factor of adverse outcome. Endocr Relat Cancer 16(2):443–453

Chapter 2
The Warburg Effect and Beyond: Metabolic Dependencies for Cancer Cells

David M. Hockenbery, Mark Tom, Cori Abikoff and Daciana Margineantu

Abstract Current definitions of cancer are best realized as a list of traits or hallmarks, such as tissue invasion, metastasis, cell-autonomous growth, and resistance to apoptosis. A recent update included deregulated cellular energetics as an emerging hallmark. However, debate about tumor cell metabolism occupied center stage in the pre-oncogene era. Over the last 15 years, direct links of oncogenes and tumor suppressor genes to cell metabolism have brought cancer metabolism to the forefront once again. Current tools provide much greater opportunities for probing metabolic differences between normal and cancer cells, in some cases revealing flux through unexpected metabolic pathways. Metabolic networks may also be truncated, presenting opportunities for selective growth inhibition or death by targeting non-redundant pathways in cancer cells. These "metabolic dependencies" are not likely to be associated with classically defined oncogenes or computationally derived drivers and thus may require novel strategies for discovery.

2.1 Pre-Molecular Biology Research

Two giants of biochemistry, Otto Warburg and Herbert Crabtree, developed novel insights into cancer metabolism in the 1920s. Warburg improved on previous manometric techniques to study respiratory quotients in a variety of cell types and tissue slices. Based on his demonstration of a sixfold increase of oxygen uptake in sea urchin eggs following fertilization, he entertained the notion that tumor growth could be explained by increased bioenergetic metabolism. In opposition to his original predictions, the respiratory rate in Flexner rat carcinomas was similar to normal rat tissues. However, the rate of glycolysis was increased up to 30-fold in the carcinoma compared with rat liver.

Comparing glycolysis in air and nitrogen, Warburg observed that both tissues had higher rates of lactate generation in the absence of oxygen. In normoxic

D. M. Hockenbery (✉) · M. Tom · C. Abikoff · D. Margineantu
Division of Clinical Research Seattle, Fred Hutchinson Cancer Research Center, WA, USA
e-mail: dhockenb@fhcrc.org

D. E. Johnson (ed.), *Cell Death Signaling in Cancer Biology and Treatment*,
Cell Death in Biology and Diseases, DOI: 10.1007/978-1-4614-5847-0_2,

X.-M. Yin and Z. Dong (Series eds.), *Cell Death in Biology and Diseases*

samples, glycolytic rates were at or below the limit of detection in most normal tissues. In contrast, glycolysis in normoxic cancer tissues still accounted for the majority of glucose turnover. The ratio of aerobic glycolysis to respiration was substantially elevated in cancers, expressed in various ways as mole percent glucose metabolized to lactate (~90 %), ATP generated by glycolysis (35–50 %), or mass of lactate produced (10 % of tissue weight per hour) [1].

Warburg extended these results to human cancer tissues and also demonstrated that although many adult tissues did not produce lactate even under hypoxic conditions, normal growing tissue (e.g., embryonic samples) had intermediate rates of aerobic glycolysis. According to the Pasteur effect, oxygen should inhibit fermentation (lactate production). Warburg interpreted his findings as demonstrating that respiration was insufficient to suppress glycolysis, either due to limitations in oxygen supply or due to mitochondrial activity. Since his aerobic experiments were conducted in thin tissue slices in which oxygen diffusion was not rate limiting, the problem seemed to Warburg to be intrinsic to mitochondria. Furthermore, if respiration was insufficient for the cellular energy demand, selection of cells with high glycolytic rates for survival was possible. Several normal tissues, including retina, kidney medulla, cartilage, bone marrow, skin, fibroblasts, intestinal mucosa, placenta, and proliferating thymocytes, have been demonstrated to have high rates of aerobic glycolysis. In fact, Warburg argued against using aerobic glycolysis as a specific test for cancer cells [2].

The central observation that cancer cells have high rates of aerobic glycolysis, now known as the Warburg effect , has been repeated numerous times and is the basis for the use of 2-deoxy-2-(^{18}F) fluoro-D-glucose positron emission tomography (PET) scans for tumor staging. However, mitochondrial dysfunction as a basis for the Warburg effect is still debatable. The Warburg effect is inhibited by treatment with mitochondrial uncouplers, indicating a well functioning Krebs cycle and electron transport chain. Anaerobic glycolysis occurs at higher rates in cancer cells than aerobic glycolysis, while the ratio of the decrease in lactic acid production to oxygen consumption in normoxia is similar in cancer and normal tissues. Thus, if energy demand remains constant, the efficiency of oxidative phosphorylation at suppressing glycolysis is similar in tumors and normal tissues.

However, it has been suggested that cancer cells may have reduced mitochondrial reserve in response to glycolytic inhibition [3]. Isolated mitochondria from cancer cell lines with pronounced Warburg effects have been shown to exhibit specific defects in substrate utilization, respiratory control ratios, and mitochondrial content [4, 5]. Reduced expression of the β subunit of the mitochondrial F_1F_O ATPase has been linked to breast, lung, and colon adenocarcinomas with poor prognosis [6–8]. Upregulated expression of the ATPase inhibitor IF1 has also been reported in cancers [9, 10]. Aside from intrinsic differences in mitochondrial function, several regulatory mechanisms restricting mitochondrial oxidative phosphorylation in cancer cells have been identified.

The Crabtree effect refers to the ability of glucose to inhibit respiration in cancer cells, as opposed to the stimulation of respiration noted in normal tissues [11]. This dynamic regulation provides another perspective on the glycolytic shift in cancer cells identified by Warburg. Typically observed in cells in which glucose metabolism by glycolysis equals or exceeds oxidation in the TCA cycle, the Crabtree effect has been proposed to represent competition between glycolysis and oxidative phosphorylation for limiting substrates such as ADP and phosphate, and the regulation of respiration by the energy charge [ATP]/[ADP][Pi] [12, 13]. Although ATP is also a negative feedback regulator of glycolysis at phosphofructokinase-1 (PFK-1), the allosteric activator fructose-2,6-bisphosphate (F-2,6-BP) is overproduced in many cancer cells and lessens the effect of ATP [14]. Thus, high rates of ATP synthesis can suppress oxidative phosphorylation to a greater extent than glycolysis under these circumstances. Like aerobic glycolysis, the Crabtree effect is also observed in normal cell types, such as proliferating thymocytes [15].

2.2 Glycolytic Regulation in Cancer

Current hypotheses rationalize the association of aerobic glycolysis and cancer by the high demand for glycolytic intermediates in rapidly growing cells. Glucose is utilized for the production of nucleosides (the pentose phosphate pathway produces ribose; 3-phosphoglycerate is precursor for glycine; serine and glycine are utilized in one carbon folate metabolism), amino acids (pyruvate is transaminated to produce alanine, 3-phosphoglycerate yields serine, glycine, and cysteine), and reducing power (NADPH from pentose phosphate pathway). Paradoxically, several steps in the glycolytic pathway are regulated in cancer to increase the levels of glycolytic intermediates by slowing downstream enzyme turnover. However, some cancer cells can have similar growth rates on slowly metabolized carbohydrates, such as fructose and galactose, in which case most metabolism is diverted to the pentose phosphate pathway, with minimal lactate production [16]. This suggests that glycolysis in the presence of glucose operates in large excess to the requirements for anabolic reactions.

The final enzyme in glycolysis, lactate dehydrogenase, produces lactic acid, which is transported out of the cell rather than serving as a synthetic intermediate. Two ideas to explain the high rates of lactic acid production have been proposed. The first involves an "overflow" phenomenon, in which the rate of pyruvate generation exceeds the capacity of the mitochondrial TCA cycle metabolism [17]. Alternatively, LDH may fulfill the requirement for regeneration of NAD+ for glycolysis to proceed, in the setting of low activity of malate–aspartate and glycerol-3-phosphate mitochondrial shuttles for reducing equivalents [18].

In vivo, poor perfusion due to the disordered tumor microvasculature may evoke a hypoxic response associated with expression of the hypoxia-inducible transcription factors (HIFs), upregulation of glycolysis, and diversion of substrate fuels away from mitochondrial metabolism. However, unless there is some mechanism for "fixing" this response, it seems unlikely to account for aerobic glycolysis measured under normoxic conditions.

The alternative explanation that increased glycolysis is matched to an increased ATP demand is less favored in the face of evidence that tumor cells exhibit adaptations that diminish ATP generation. Pyruvate kinase catalyzes the transfer of phosphate from phosphoenolpyruvate (PEP) to ADP, generating pyruvate and ATP. Originally described by Sato in 1978 [19], the pyruvate kinase splicing isoform M2 (PK-M2) is highly expressed in cancers, embryonic tissues, and several adult tissues, including adipocytes, retina, and lung [20]. PK-M2 is active as a tetramer, but mostly occurs as inactive dimer in cancer cells. Tetramer formation is induced allosterically by the proximal glycolytic intermediate, fructose-1,6-bisphosphate (F-1,6-BP), but opposed by tyrosine phosphorylation of PK-M2 or interactions with phosphotyrosine-bearing proteins at the F-1,6-BP-binding site [21–24]. Reduced activity of PK-M2 increases the diversion of upstream glycolytic intermediates into anabolic pathways, including the pentose phosphate pathway [25], as well as slowing ATP generation. Interestingly, substitution of the PK-M1 splicing isoform increases glucose oxidation and reduces lactate production by an unexplained mechanism [23].

Reduction in the PK-M2 activity in cancer cells raises questions as to the source of pyruvate utilized for lactate generation in aerobic glycolysis. Analogous to bacterial pathways, the substrate for PK-M2, PEP can be used as a phosphate donor in protein kinase reactions [26, 27]. One of the protein substrates for PEP-dependent phosphorylation is the glycolytic enzyme phosphoglycerate mutase (PGAM1), which is phosphorylated on the catalytic histidine (H11) involved in transferring phosphate to the C-2 position of 3-phosphoglycerate. This reaction produces pyruvate as a product in the absence of ATP generation. Thus, cancer cells can switch to an alternative glycolytic pathway that does not produce net ATP.

Finally, the concept that the predominant function of glycolysis is ATP production can be questioned by a study of glucose deprivation in IL-3-dependent cells [28]. Glucose is necessary for cell growth and proliferation in many cell types. In this example, glucose was shown to be required for N-linked glycosylation of the IL-3 receptor, required for its cell surface localization and in turn, uptake of an alternative substrate fuel, glutamine. Glucose could be replaced by the specific product of the hexosamine biosynthetic pathway, N-acetylglucosamine, enabling glutamine uptake and cell growth.

Three major pathways are involved in deregulating glycolysis in cancer, the transcription factors c-Myc and HIF-1, and the serine/threonine kinase, Akt . The c-Myc oncoprotein is overexpressed in more than 50 % of human cancers and binds to 15 % of all gene promoters in the genome. Among the pathways controlled by c-Myc is glycolysis, with 9/10 glycolytic enzymes and glucose transporters type 1, 2, and 4 identified as Myc targets. Myc-dependent cell cycle entry and transformation are associated with increased lactic acid production [29, 30]. Stable isotope labeling studies indicate that Myc directs glucose carbons to multiple anabolic pathways, including nucleotide, amino acid, and lipid biosynthesis [31, 32]. N-Myc appears to share similar targets in glycolysis [33].

The hypoxia-inducible transcription factor, HIF-1α, is degraded by the ubiquitin–proteasome system following oxygen-dependent proline hydroxylation. HIF-1α is overexpressed in 13/19 common tumor types [34]. HIF-1 targets include glycolytic enzymes, glucose transporters and the pyruvate dehydrogenase kinase

(PDK1) that inactivates pyruvate dehydrogenase, required for pyruvate oxidation in the TCA cycle. HIF-1 can also be stabilized in normoxia by mutations in succinate dehydrogenase and fumarate hydratase that lead to product inhibition of proline hydroxylases [35], increased concentrations of pyruvate and lactate [36], and mTOR -dependent translation [37].

Although HIF-1 and c-Myc may have antagonistic functions [38], studies in a Burkitt's lymphoma model demonstrated a cooperative effect on glycolysis [39].

The Akt serine/threonine kinases are activated downstream of phosphoinositide 3-kinase (PI3K) and direct both glucose uptake and metabolism. Glucose transporters Glut 1 and 4 recycle between insulin-responsive storage compartments and the plasma membrane via Akt phosphorylation of the Rab GTPase activity TBC1D proteins [40, 41]. Akt also activates 6-phosphofructo-2-kinase (PFK-2), the enzyme responsible for synthesis of fructose-2,6-bisphosphate, by phosphorylation [42]. Finally, Akt promotes association of hexokinases I and II with mitochondria [43], increasing glucose phosphorylation, by preventing GSK3β phosphorylation of the voltage-dependent anion channel (VDAC) [44]. Overexpression of c-Myc , HIF-1α, and constitutively active Akt stimulates aerobic glycolysis [29, 45–47].

2.3 Therapeutic Strategies Based on Glycolysis Inhibition

Inhibition of glycolysis is often lethal in tumor cell lines in vitro, prompting both early and renewed interest in this strategy for cancer treatment. The glucose analog, 2-deoxy-D-glucose (2-DG), is phosphorylated by hexokinase, but in the absence of further glycolytic transformations, 2-DG-6-P accumulates in cells and inhibits hexokinase. 2-DG was first administered to human cancer patients [48] with a variety of cancers and has since been studied in early stage clinical trials in patients with glioblastomas [49] and advanced solid malignancies [50]. Novel LDH-A inhibitors have been reported [51, 52]. A peptide inhibitor of PK-M2 is in clinical trials for melanoma and renal cell carcinoma (www.thallion.com). Several classes of inhibitors of the PI3K–Akt signaling pathway are in clinical trials, including PI3K inhibitors, Akt inhibitors, and dual PI3K–mTOR inhibitors [53]. A common theme in cell death following glucose deprivation or glycolytic inhibition is an increase in mitochondrial generation of reactive oxygen species, with decreased generation of NADPH reducing equivalents in the pentose phosphate pathway [54, 55].

2.4 Alternative Mitochondrial Fuels

Mitochondrial substrate fuels are not restricted to pyruvate-derived acetyl-CoA. The role of glutamine as a major fuel for cancer cell mitochondria was recognized more than 50 years ago [16]. Glutamine is initially metabolized to glutamate by

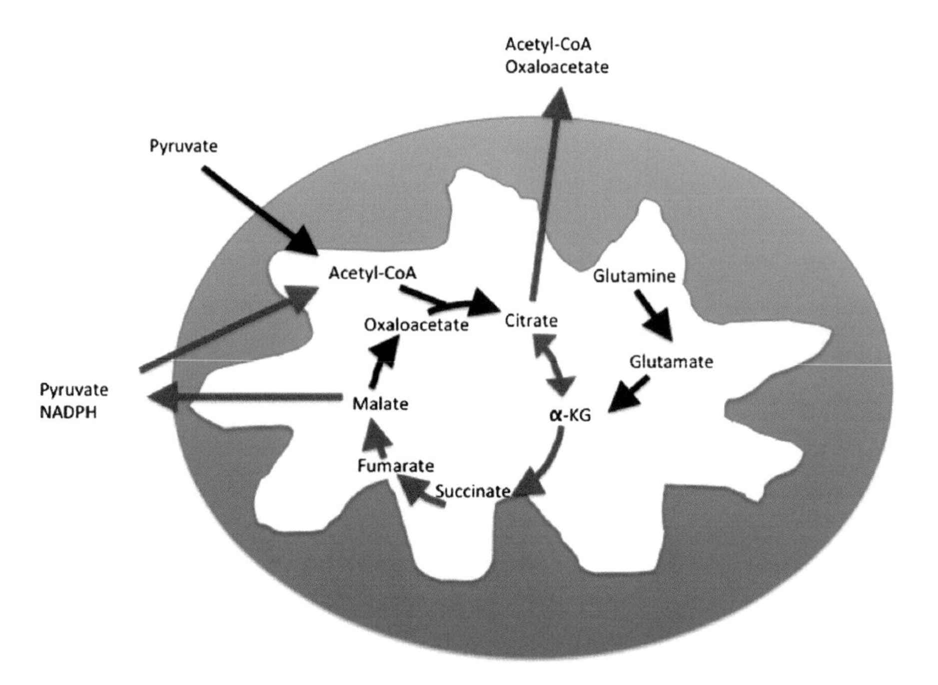

Fig. 2.1 Truncated and reversed TCA cycles (*red*) with glutaminolysis, involving malic enzyme producing pyruvate and NADPH, and reductive carboxylation of α-ketoglutarate to citrate

phosphate-dependent glutaminase, which can be converted to α-ketoglutarate using either glutamate dehydrogenase or transaminase enzymes. Glutamine entry into the TCA cycle can replenish intermediates (anaplerosis) or serve as a source of acetyl-CoA from pyruvate generated by malic enzyme from malate (Fig. 2.1). The latter reaction is one of the major sources of NADPH generation in cancer cells, used for fatty acid and cholesterol biosynthesis, ribonucleotide reduction, and reduction of GSSG to GSH, in addition to the pentose phosphate pathway. Glutamine can also be utilized for lipid biosynthesis, with conversion of mitochondrial citrate to cytosolic acetyl-CoA by ATP citrate lyase. Two paths for citrate production from glutamate are available, a truncated TCA cycle (α-ketoglutarate to citrate) or reductive carboxylation to isocitrate. ATP citrate lyase activity is activated by Akt -dependent phosphorylation. The energy yield from glutamine metabolism is 6-12 ATP/glutamine, depending on the inclusion of NADP+ -dependent glutamate and malate dehydrogenases in the reaction pathway. Glutamine metabolism accounts for 65 % or higher of ATP production in many cancer cells. Chemical inhibition of glutaminase slows growth or kills outright cells dependent on glutamine, especially hypoxic cells or cells that lack other pathways to generate α-ketoglutarate [56, 57].

Several nonessential amino acids can be derived from glutamate (proline, arginine, aspartate, and asparagine). Glutamine contributes its γ-nitrogen to nucleotide and hexosamine synthesis and its α-nitrogen to alanine and aspartate in transamination reactions. Glutamate is utilized in glutathione synthesis, and glutamine exchange with leucine and other essential amino acids is essential for mTOR activity [58]. However, a significant portion of glutamine nitrogen and carbon is transported extracellularly, as glutamate, alanine, and lactate, indicating that, like glycolysis, glutaminolysis is an inefficient process in cancer cells [59].

c-Myc has emerged as a major driver of glutamine metabolism. Glutamine deprivation is sufficient to kill Myc-overexpressing cells [60, 61]. Surprisingly, glutamine-deprivation-induced cell death involves depletion of TCA cycle intermediates, but not ATP deficiency, glutathione depletion, or DNA damage, and can be partially rescued with antioxidants [60, 62]. The anaplerotic requirement for glutamine is highlighted by the glutamine-independent growth of tumor cells expressing pyruvate carboxylase, an anaplerotic enzyme associated with glucose metabolism [63]. Glutaminase translation is suppressed by two microRNAs, miR-23a and miR-23b, that are in turn targets for Myc transcriptional repression [62]. Myc also increases the expression of glutamine transporters [61].

Although there are differences in the regulation of glucose and glutamine metabolism, one mechanism for dual control is the basic helix-loop-helix leucine zipper transcription factor, MondoA (also known as MLXIP), related to Myc [64]. In glucose-containing growth media, MondoA functions as a transcriptional activator for TXNIP, which suppresses glucose uptake and aerobic glycolysis. However, the addition of glutamine converts MondoA to a transcriptional repressor, increasing glucose uptake. The ability of a cell-permeable analog of α-ketoglutarate to similarly affect MondoA transcriptional activity suggests that the levels of TCA cycle intermediates may be monitored to adjust glucose uptake and metabolism. Recent data also indicate that Nrf2, a transcription factor responding to oxidative stress, directs both glucose and glutamine utilization in anabolic pathways, such as the pentose phosphate pathway and nucleotide synthesis [65].

Tumor cells engage in de novo fatty acid synthesis, in contrast to importation of circulating lipids by most normal cells in the body [66, 67]. Several recent reports have indicated that cancer cells may also exhibit high rates of peroxisomal and mitochondrial β-oxidation [68, 69]. Prostate cancers with low 2-deoxy-2-fluoro-D-glucose (FDG) avidity on PET scans may engage in fatty acid β-oxidation as a principal bioenergetic pathway [69]. Respiration in glioblastoma cells is inhibited 30–40 % by treatment with the carnitine palmitoyltransferase-1 (CPT1) inhibitor, etomoxir, with a 50 % decrease in ATP levels [70]. Fatty acid β-oxidation has been associated with uncoupling protein-2 expression and chemoresistance [71, 72]. The brain CPT1C isoform is upregulated in non-small-cell lung cancers and confers resistance to hypoxia, glucose deprivation, and the mTOR inhibitor, rapamycin [73]. Adipocytes promote the growth of ovarian cancer cells by mobilizing free fatty acids for β-oxidation by tumor cells [74]. Inhibition of fatty acid β-oxidation (etomoxir, ranolazine) or fatty acid biosynthesis (cerulenin, C75, orlistat) is selectively cytotoxic to some cancer cells [70, 75].

2.5 Screening for Additional Metabolic Dependencies

The Recon 1 genome-scale metabolic reconstruction includes 1,496 open reading frames and 3,311 metabolic and transport reactions [76]. Enzymes are optimal targets for drug development, with active sites and allosteric binding pockets well suited for drug interactions. The extent of metabolic differences between cancer and normal cells is currently not known, although it is evident that the Warburg effect is only one aspect of "cancer metabolism". In addition to glutaminolysis and lipid metabolism, cancer cells exhibit auxotrophy for specific amino acids and increased NAD+ turnover [77–80].

Several approaches have been employed recently to identify metabolic pathways that are required for growth or survival in cancer cells, but not normal cell types. Analogous to other strategies for identifying "driver" genes in cancer, analysis of gene expression may reveal candidate targets for functional genomic studies. Possemato et al. generated a high-priority set of 133 metabolic and transporter genes based on high expression in cancers versus normal tissues, high expression in aggressive breast cancers, and expression in stem cells [81]. Tumorigenic breast cancer cells were transduced with pools of shRNAs targeting the 133 genes and grown as orthotopic tumors in mice. 16 genes were identified with significant depletion of targeting shRNAs during 28 days of in vivo growth. One of the genes, phosphoglycerate dehydrogenase (PHGDH), is amplified in breast cancers and melanomas, and PHGDH expression is associated with poor prognosis in ER-negative breast cancers. PHGDH catalyzes the oxidation of the glycolytic intermediate, 3-phosphoglycerate, to phospho-hydroxypyruvate, which can subsequently be converted to serine. Serine is used for the synthesis of phosphatidylserine, sphingosine, cysteine, and glycine, as well as protein synthesis. Additional stable isotope analysis demonstrated increased serine biosynthetic flux in cancer cells overexpressing PHGDH, accounting for 8–9 % of total glycolytic flux. RNAi-mediated suppression of PHGDH in these cells induced non-apoptotic cell death and reduced tumor growth in vivo. Finally, metabolomic studies demonstrated large decreases in α-ketoglutarate concentrations in these cells, indicating that phosphoserine aminotransferase provides a major route for conversion of glutamate to α-ketoglutarate.

Unlike transcription factors or signal transduction pathways, metabolic enzyme activity is not manifested by accumulation of downstream gene products or post-translationally modified substrates, but rather by increased substrate to product flux. Furthermore, enzyme regulation often includes allosteric effects. Locasale et al. used [U-^{13}C] glucose stable isotope metabolic flux measurements to identify high levels of glycine enrichment in cancer, but not immortalized, cell lines [82]. As in the previous study, PHGDH amplification was observed to correlate with high flux in the serine biosynthetic pathway.

Finally, *in silico* network models of cancer metabolism based on flux balance analysis have been used to predict metabolic pathways required for growth in specific genetic contexts [83]. Germline mutations of the TCA cycle enzyme, fumarate hydratase (Fh1), cause hereditary leiomyomatosis and renal cell cancer (HLRCC). The accumulation of fumarate inhibits prolyl hydroxylases, leading to an increase in HIF expression under normoxic conditions. Incorporation of cellular growth as consumption of biosynthetic precursors in the *in silico* model allowed testing of specific gene knockouts for effects on growth. Frezza et al. predicted 24 reactions to be synthetically lethal with Fh1 deletion, 18 of which involved heme metabolism. Heme biosynthesis utilizes TCA-derived succinate, providing an alternative route to fumarate synthesis. Subsequent analysis of $Fh1^{-/-}$ cells demonstrated increased excretion of bilirubin, the degradation product of heme. Only three enzymes involved in heme degradation were overexpressed in $Fh1^{-/-}$ cells, highlighting the power of the computational approach to identify significant pathways in the absence of expression criteria. Several strategies to inhibit heme pathway flux reduced growth of $Fh^{-/-}$ cells.

Toward an unbiased screening strategy, we have generated a chemical library with inhibitors for enzymes of intermediary metabolism, as validated in the literature by direct enzymatic assay. Inhibitors were chosen from the BRENDA enzyme database (www.brenda-enzymes.org) and review of published literature, purchased from Sigma-Aldrich or other chemical supply houses, or donated by academic labs. Inhibitors of 585 enzymes have been identified. Each inhibitor is plated at 100X EC50 concentration in DMSO. Transcriptome analysis in mouse and human databases indicates ~750 enzymes of intermediary metabolism encoded in the genome with unique EC numbers [84]. Such a library has several applications relevant to cancer research. Screening cancer cell lines versus normal cellular counterparts for loss of viability or cell growth can reveal critical metabolic dependencies associated with the transformed phenotype. Altered expressions of metabolic enzymes, in particular, as revealed by Gene Set Enrichment analyses, are frequently observed in RNA microarray profiles. As yet, there is no rapid method to determine the essentiality of a metabolic pathway for cancer cell viability. Metabolomic studies of cancer cells are increasingly being reported. A chemical inhibitor library can rapidly ascertain which pathways connecting to a given metabolite are casually linked to cell growth or viability.

We have tested 217 causally compounds out of a total of 585 in our enzyme inhibitor compound library in the MCF10A cell line expressing Myc-ER^{TAM}. As shown in Fig. 2.2, inhibitors are available for a high proportion of enzymes in major metabolic pathways. MCF10A-MycER^{TAM} cells were treated with tamoxifen to activate Myc for 24 h and compared with comparably treated vehicle controls. Cell metabolism was evaluated using a Seahorse analyzer, which demonstrated a shift toward glycolytic metabolism in tamoxifen-treated, but not control, cells.

Cells were tested with each compound in triplicate at a single dose, based on published IC_{80} values for enzymatic inhibition. Cell viability was assessed after 24 h by Alamar Blue assay. Log_2-transformed results were analyzed according to SSMD scores (strictly standardized mean difference; [85]) with respect to negative vehicle controls. The results are graphed in Fig. 2.3 (high SSMD values correspond to selective killing of tamoxifen-treated cells). A secondary assay for cell number (Hoechst 33342 fluorescence of adherent cells) confirmed these results.

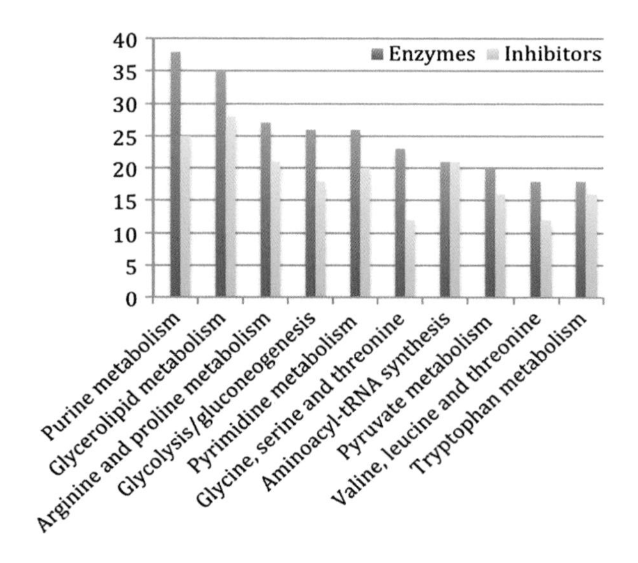

Fig. 2.2 Comparison of enzymes in major primary metabolic pathways and availability of specific inhibitors

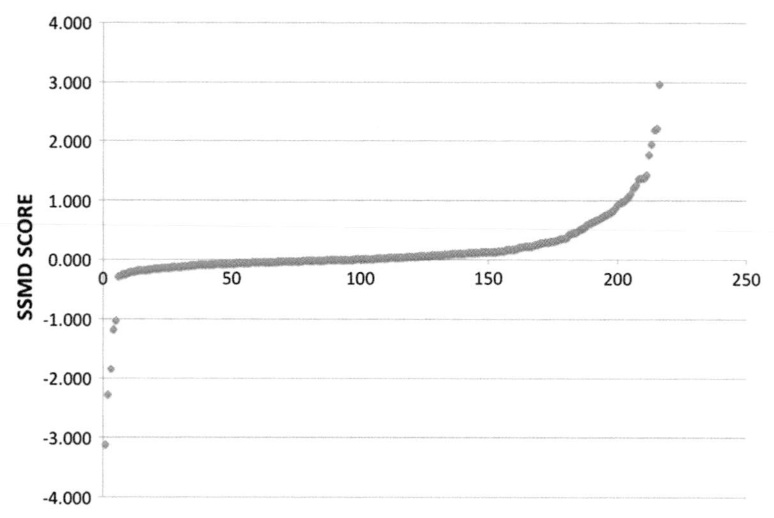

Fig. 2.3 Plot of SSMD scores for 217 compounds tested against MCF10A-MycER cells +/− tamoxifen, in relation to negative (vehicle) controls

The top hits include aldehyde dehydrogenase, malic enzyme, aromatic L-amino acid decarboxylase, and threonine tRNA ligase. Aldehyde dehydrogenases are markers for cancer stem cells. Disulfiram, the aldehyde dehydrogenase inhibitor in the screen, is reported to have anticancer activity [86], although the relevant target has not been validated. Aromatic L-amino acid decarboxylase (also known as L-dopa decarboxylase) is expressed in several epithelial cancers in addition to neuroendocrine tumors and has co-activator activity for the androgen receptor [87, 88].

Borrelidin, the threonine tRNA ligase inhibitor, induces apoptosis in acute lymphoblastic leukemia cells associated with the activation of the GCN2 stress kinase and proapoptotic CHOP transcription factor [89].

2.6 Summary

It is not surprising, based on the intimate role of catabolic and anabolic metabolism in furnishing biochemical energy and building blocks for macromolecular synthesis and repair, that metabolic pathways are vital to cancer cell growth and survival in different tumor microenvironments. What we have learned more recently is that metabolic pathways can also direct chromatin structure and gene expression, differentiation, and stemness [90–95], typified by the discovery of the oncometabolite, D-2-hydroxyglutarate (2HG). Specific gain-of-function mutations in the isocitrate dehydrogenase-1 and dehydrogenase-2 enzymes occur in glioblastomas and acute myeloid leukemia and alter enzymatic function to produce 2HG from α-ketoglutarate (α-KG) [96]. 2HG is an inhibitor of α-KG-dependent Jumonji-C domain histone demethylases [97, 98] and is associated with increased histone methylation and altered gene expression [99, 100]. Comprehensive mapping of cancer cell metabolism is at an early stage, but current knowledge suggests that the Warburg effect may be the "tip of the iceberg". The ultimate value of cancer therapies directed against metabolic dependencies is still to be determined, as the links between metabolism and cell death, and extent of metabolic flexibility are largely unknown. Further exploration of metabolic networks in cancer will require novel strategies, but seems likely to yield novel targets for cancer therapy.

References

1. Warburg O (1956) On the origin of cancer cells. Science 123(3191):309–314
2. Warburg O (1956) On respiratory impairment in cancer cells. Science 124(3215):269–270
3. Wu M, Neilson A, Swift AL, Moran R, Tamagnine J, Parslow D, Armistead S, Lemire K, Orrell J, Teich J, Chomicz S, Ferrick DA (2007) Multiparameter metabolic analysis reveals a close link between attenuated mitochondrial function and enhanced glycolytic dependency in human tumor cells. Am J Physiol Cell Physiol 292(1):C125–C136
4. Pedersen PL, Greenawalt JW, Chan TL, Morris HP (1970) A comparison of some ultrastructural and biochemical properties of mitochondria from Morris hepatomas 9618A, 7800, and 3924A. Cancer Res 30(11):2620–2626
5. Eboli ML, Paradies G, Galeotti T, Papa S (1977) Pyruvate transport in tumour-cell mitochondria. Biochim Biophys Acta 460(1):183–187
6. Isidoro A, Casado E, Redondo A, Acebo P, Espinosa E, Alonso AM, Cejas P, Hardisson D, Fresno Vara JA, Belda-Iniesta C, Gonzalez-Baron M, Cuezva JM (2005) Breast carcinomas fulfill the warburg hypothesis and provide metabolic markers of cancer prognosis. Carcinog 26(12):2095–2104
7. Cuezva JM, Krajewska M, de Heredia ML, Krajewski S, Santamaria G, Kim H, Zapata JM, Marusawa H, Chamorro M, Reed JC (2002) The bioenergetics signature of cancer: a marker of tumor progression. Cancer Res 62(22):6674–6681

8. Cuezva JM, Chen G, Alonso AM, Isidoro A, Misek DE, Hanash SM, Beer DG (2004) The bioenergetics signature of lung carcinomas is a molecular marker of cancer diagnosis and prognosis. Carcinog 25(7):1157–1163
9. Sanchez-Cenizo L, Formentini L, Aldea M, Ortega AD, Garcia-Huerta P, Sanchez-Arago M, Cuezva JM (2010) Up-regulation of the ATPase inhibitory factor (IF1) of the mitochondrial H+ -ATP synthase in human tumors mediates the metabolic shift of cancer cells to a Warburg phenotype. J Biol Chem 285(33):25308–25313
10. Crabtree JG (1929) Observations on the carbohydrate metabolism of tumours. Biochem J 23(3):536–545
11. Formentini L, Sanchez-Arago M, Sanchez-Cenizo L, Cuezva JM (2012) The mitochondrial ATPase inhibitory factor 1 triggers a ROS-mediated retrograde prosurvival and proliferative response. Mol Cell 45(6):731–742
12. Sussman I, Erecinska M, Wilson DF (1980) Regulation of cellular energy metabolism: the crabtree effect. Biochim Biophys Acta 591(2):209–223
13. Wu R, Racker E (1959) Regulatory mechanisms in carbohydrate metabolism. IV. Pasteur effect and crabtree effect in ascites tumor cells. J Biol Chem 234(5):1036–1041
14. Chesney J, Mitchell R, Benigni F, Bacher M, Spiegel L, Al-Abed Y, Han JH, Metz C, Bucala R (1999) An inducible gene product for 6-phosphofructo-2-kinase with an AU-rich instability element: role in tumor cell glycolysis and the warburg effect. Proc Natl Acad Sci U S A 96(6):3047–3052
15. Guppy M, Greiner E, Brand K (1993) The role of the crabtree effect and an endogenous fuel in the energy metabolism of resting and proliferating thymocytes. Eur J Biochem 212(1):95–99
16. Reitzer LJ, Wice BM, Kennell D (1979) Evidence that glutamine, not sugar, is the major energy source for cultured HeLa cells. J Biol Chem 254(8):2669–2676
17. Vazquez A, Liu J, Zhou Y, Oltvai ZN (2010) Catabolic efficiency of aerobic glycolysis: the warburg effect revisited. BMC Syst Biol 6(5):58
18. Boxer GE, Devlin TM (1961) Pathways of intracellular hydrogen transport. Science 134(3489):1495–1501
19. Sato K, Takaya S, Imai F, Hatayama I, Ito N (1978) Different deviation patterns of carbohydrate-metabolizing enzymes in primary rat hepatomas induced by different chemical carcinogens. Cancer Res 38(9):3086–3093
20. Mazurek S (2011) Pyruvate kinase type M2: a key regulator of the metabolic budget system in tumor cells. Int J Biochem Cell Biol 43(7):969–980
21. Hitosugi T, Kang S, Vander Heiden MG, Chung TW, Elf S, Lythgoe K, Dong S, Lonial S, Wang X, Chen GZ, Xie J, Gu TL, Polakiewicz RD, Roesel JL, Boggon TJ, Khuri FR, Gilliland DG, Cantley LC, Kaufman J, Chen J (2009) Tyrosine phosphorylation inhibits PKM2 to promote the warburg effect and tumor growth. Sci Signal 2(97):ra73
22. Christofk HR, Vander Heiden MG, Wu N, Asara JM, Cantley LC (2008) Pyruvate kinase M2 is a phosphotyrosine-binding protein. Nature 452(7184):181–186
23. Christofk HR, Vander Heiden MG, Harris MH, Ramanathan A, Gerszten RE, Wei R, Fleming MD, Schreiber SL, Cantley LC (2008) The M2 splice isoform of pyruvate kinase is important for cancer metabolism and tumour growth. Nature 452(7184):230–233
24. Presek P, Reinacher M, Eigenbrodt E (1988) Pyruvate kinase type M2 is phosphorylated at tyrosine residues in cells transformed by rous sarcoma virus. FEBS Lett 242(1):194–198
25. Anastasiou D, Poulogiannis G, Asara JM, Boxer MB, Jiang JK, Shen M, Bellinger G, Sasaki AT, Locasale JW, Auld DS, Thomas CJ, Vander Heiden MG, Cantley LC (2011) Inhibition of pyruvate kinase M2 by reactive oxygen species contributes to cellular antioxidant responses. Science 334(6060):1278–1283
26. Vander Heiden MG, Locasale JW, Swanson KD, Sharfi H, Heffron GJ, Amador-Noguez D, Christofk HR, Wagner G, Rabinowitz JD, Asara JM, Cantley LC (2010) Evidence for an alternative glycolytic pathway in rapidly proliferating cells. Science 329(5998):1492–1499
27. Gao X, Wang H, Yang JJ, Liu X, Liu ZR (2012) Pyruvate kinase M2 regulates gene transcription by acting as a protein kinase. Mol Cell 45(5):598–609

28. Wellen KE, Lu C, Mancuso A, Lemons JM, Ryczko M, Dennis JW, Rabinowitz JD, Coller HA, Thompson CB (2010) The hexosamine biosynthetic pathway couples growth factor-induced glutamine uptake to glucose metabolism. Genes Dev 24(24):2784–2799
29. Shim H, Dolde C, Lewis BC, Wu CS, Dang G, Jungmann RA, Dalla-Favera R, Dang CV (1997) c-Myc transactivation of LDH-A: implications for tumor metabolism and growth. Proc Natl Acad Sci U S A 94(13):6658–6663
30. Morrish F, Neretti N, Sedivy JM, Hockenbery DM (2008) The oncogene c-Myc coordinates regulation of metabolic networks to enable rapid cell cycle entry. Cell Cycle 7(8):1054–1066
31. Morrish F, Isern N, Sadilek M, Jeffrey M, Hockenbery DM (2009) c-Myc activates multiple metabolic networks to generate substrates for cell-cycle entry. Oncogene 28(27):2485–2491
32. Morrish F, Noonan J, Perez-Olsen C, Gafken PR, Fitzgibbon M, Kelleher J, VanGilst M, Hockenbery D (2010) Myc-dependent mitochondrial generation of acetyl-CoA contributes to fatty acid biosynthesis and histone acetylation during cell cycle entry. J Biol Chem 285(47):36267–36274
33. Qing G, Skuli N, Mayes PA, Pawel B, Martinez D, Maris JM, Simon MC (2010) Combinatorial regulation of neuroblastoma tumor progression by N-Myc and hypoxia inducible factor HIF-1 alpha. Cancer Res 70(24):10351–10361
34. Zhong H, De Marzo AM, Laughner E, Lim M, Hilton DA, Zagzag D, Buechler P, Isaacs WB, Semenza GL, Simons JW (1999) Overexpression of hypoxia-inducible factor 1 alpha in common human cancers and their metastases. Cancer Res 59(22):5830–5835
35. Selak MA, Armour SM, MacKenzie ED, Boulahbel H, Watson DG, Mansfield KD, Pan Y, Simon MC, Thompson CB, Gottlieb E (2005) Succinate links TCA cycle dysfunction to oncogenesis by inhibiting HIF-alpha prolyl hydroxylase. Cancer Cell 7(1):77–85
36. Lu H, Forbes RA, Verma A (2002) Hypoxia-inducible factor 1 activation by aerobic glycolysis implicates the Warburg effect in carcinogenesis. J Biol Chem 277(26):23111–23115
37. Treins C, Giorgetti-Peraldi S, Murdaca J, Semenza GL, Van Obberghen E (2002) Insulin stimulates hypoxia-inducible factor 1 through a phosphatidylinositol 3-kinase/target of rapamycin-dependent signaling pathway. J Biol Chem 277(31):27975–27981
38. Koshiji M, Kageyama Y, Pete EA, Horikawa I, Barrett JC, Huang LE (2004) HIF-1 alpha induces cell cycle arrest by functionally counteracting Myc. EMBO J 23(9):1949–1956
39. Kim JW, Gao P, Liu YC, Semenza GL, Dang CV (2007) Hypoxia-inducible factor 1 and dysregulated c-Myc cooperatively induce vascular endothelial growth factor and metabolic switches hexokinase 2 and pyruvate dehydrogenase kinase 1. Mol Cell Biol 27(21):7381–7393
40. Zhou QL, Jiang ZY, Holik J, Chawla A, Hagan GN, Leszyk J, Czech MP (2008) Akt substrate TBC1D1 regulates GLUT1 expression through the mTOR pathway in 3T3-L1 adipocytes. Biochem J 411(3):647–655
41. Sakamoto K, Holman GD (2008) Emerging role for AS160/TBC1D4 and TBC1D1 in the regulation of GLUT4 traffic. Am J Physiol Endocrinol Metab 295(1):E29–E37
42. Deprez J, Vertommen D, Alessi DR, Hue L, Rider MH (1997) Phosphorylation and activation of heart 6-phosphofructo-2-kinase by protein kinase B and other protein kinases of the insulin signaling cascades. J Biol Chem 41(3):533–535
43. Gottlob K, Majewski N, Kennedy S, Kandel E, Robey RB, Hay N (2001) Inhibition of early apoptotic events by Akt/PKB is dependent on the first committed step of glycolysis and mitochondrial hexokinase. Genes Dev 15(11):1406–1418
44. Pastorino JG, Hoek JB, Shulga N (2005) Activation of glycogen synthase kinase 3 beta disrupts the binding of hexokinase II to mitochondria by phosphorylating voltage-dependent anion channel and potentiates chemotherapy-induced cytotoxicity. Cancer Res 65(22):10545–10554
45. Robey IF, Lien AD, Welsh SJ, Baggett BK, Gillies RJ (2005) Hypoxia-inducible factor-1 alpha and the glycolytic phenotype in tumors. Neoplasia 7(4):324–330
46. Lum JJ, Bui T, Gruber M, Gordan JD, DeBerardinis RJ, Covello KL, Simon MC, Thompson CB (2007) The transcription factor HIF-1 alpha plays a critical role in the growth factor-dependent regulation of both aerobic and anaerobic glycolysis. Genes Dev 21(9):1037–1049

47. Elstrom RL, Bauer DE, Buzzai M, Karnauskas R, Harris MH, Plas DR, Zhuang H, Cinalli RM, Alavi A, Rudin CM, Thompson CB (2004) Akt stimulates aerobic glycolysis in cancer cells. Cancer Res 64(11):3892–3899
48. Landau BR, Laszlo J, Stengle J, Burk D (1958) Certain metabolic and pharmacologic effects in cancer patients given infusions of 2-deoxy-D-glucose. J Natl Cancer Inst 21(3):475–483
49. Dwarakanath BS, Singh D, Banerji AK, Sarin R, Venkataramana NK, Jalali R, Vishwanath PN, Mohanti BK, Tripathi RP, Kalia VK, Jain V (2009) Clinical studies for improving radiotherapy with 2-deoxy-D-glucose: present status and future prospects. J Cancer Res Ther Suppl 1:S21–S26
50. Garber K (2010) Oncology's energetic pipeline. Nat Biotechnol 28(9):888–891
51. Le A, Cooper CR, Gouw AM, Dinavahi R, Maitra A, Deck LM, Royer RE, Vander Jagt DL, Semenza GL, Dang CV (2010) Inhibition of lactate dehydrogenase A induces oxidative stress and inhibits tumor progression. Proc Natl Acad Sci U S A 107(5):2037–2042
52. Granchi C, Bertini S, Macchia M, Minutolo F (2010) Inhibitors of lactate dehydrogenase isoforms and their therapeutic potentials. Curr Med Chem 17(7):672–697
53. Courtney KD, Corcoran RB, Engelman JA (2010) The PI3K pathway as drug target in human cancer. J Clin Oncol 28(6):1075–1083
54. Shutt DC, O'Dorisio MS, Aykin-Burns N, Spitz DR (2010) 2-deoxy-D-glucose induces oxidative stress and cell killing in human neuroblastoma cells. Cancer Biol Ther 9(11):8953–8961
55. Aykin-Burns N, Ahmad IM, Zhu Y, Oberley LW, Spitz DR (2009) Increased levels of superoxide and H_2O_2 mediate the differential susceptibility of cancer cells versus normal cells to glucose deprivation. Biochem J 418(1):29–37
56. Le A, Lane AN, Hamaker M, Bose S, Gouw A, Barbi J, Tsukamoto T, Rojas CJ, Slusher BS, Zhang H, Zimmerman LJ, Liebler DC, Slebos RJ, Lorkiewicz PK, Higashi RM, Fan TW, Dang CV (2012) Glucose-independent glutamine metabolism via TCA cycling for proliferation and survival in B cells. Cell Metab 15(1):110–121
57. Seltzer MJ, Bennett BD, Joshi AD, Gao P, Thomas AG, Ferraris DV, Tsukamoto T, Rojas CJ, Slusher BS, Rabinowitz JD, Dang CV, Riggins GJ (2010) Inhibition of glutaminase preferentially slows growth of glioma cells with mutant IDH1. Cancer Res 70(22):8981–8987
58. Nicklin P, Bergman P, Zhang B, Triantafellow E, Wang H, Nyfeler B, Yang H, Hild M, Kung C, Wilson C, Myer VE, MacKeigan JP, Porter JA, Wang YK, Cantley LC, Finan PM, Murphy LO (2009) Bidirectional transport of amino acids regulates mTOR and autophagy. Cell 136(3):521–534
59. Deberardinis RJ, Mancuso A, Daikhim E, Nissim I, Yudkoff M, Wehrli S, Thompson CB (2007) Beyond aerobic glycolysis: transformed cells can engage in glutamine metabolism that exceeds the requirement for protein and nucleotide synthesis. Proc Natl Acad Sci U S A 104(49):19345–19350
60. Yuneva M, Zamboni N, Oefner P, Sachidanandam R, Lazebnik Y (2007) Deficiency in glutamine but not glucose induces MYC-dependent apoptosis in human cells. J Cell Biol 78(1):93–105
61. Wise DR, De Berardinis RJ, Mancuso A, Sayed N, Zhang XY, Pfeiffer HK, Nissim I, Daikhin E, Yudkoff M, McMahon SB, Thompson CB (2008) Myc regulates a transcriptional program that stimulates mitochondrial glutaminolysis and leads to glutamine addiction. Proc Natl Acad Sci U S A 105(48):18782–18787
62. Gao P, Tchernyshyov I, Chang TC, Lee YS, Kita K, Ochi T, Zeller KI, De Marzo AM, Van Eyk JE, Mendell JT, Dang CV (2009) c-Myc suppression of miR-23a/b enhances mitochondrial glutaminase expression and glutamine metabolism. Nature 458(7239):762–765
63. Cheng T, Sudderth J, Yang C, Mullen AR, Jin ES, Mates JM, DeBerardinis RJ (2011) Pyruvate carboxylase is required for glutamine-independent growth of tumor cells. Proc Natl Acad Sci U S A 108(21):8674–8679

64. Kaadige MR, Looper RE, Kamalanaadhan S, Ayer DE (2009) Glutamine-dependent anapleurosis dictates glucose uptake and cell growth by regulating MondoA transcriptional activity. Proc Natl Acad Sci U S A 106(35):14878–14883
65. Mitsuishi Y, Taguchi K, Kawatani Y, Shibata T, Nukiwa T, Aburatani H, Yamamoto M, Motohashi H (2012) Nrf2 redirects glucose and glutamine into anabolic pathways in metabolic reprogramming. Cancer Cell 22(1):66–79
66. Kuhajda FP, Jenner K, Wood FD, Hennigar RA, Jacobs LB, Dick JD, Pasternack GR (1994) Fatty acid synthesis: a potential selective target for antineoplastic therapy. Proc Natl Acad Sci U S A 91(14):6379–6383
67. Menendez JA, Lupu R (2007) Fatty acid synthase and the lipogenic phenotype in cancer pathogenesis. Nat Rev Cancer 7(10):763–777
68. Zha S, Ferdinandusse S, Hicks JL, Denis S, Dunn TA, Wanders RJ, Luo J, De Marzo AM, Isaacs WB (2005) Peroxisomal branched chain fatty acid beta-oxidation pathway is upregulated in prostate cancer. Prostate 63(4):316–323
69. Liu Y (2006) Fatty acid oxidation is a dominant bioenergetic pathway in prostate cancer. Prostate Cancer Prostatic Dis 9(3):230–234
70. Pike LS, Smift AL, Croteau NJ, Ferrick DA, Wu M (2011) Inhibition of fatty acid oxidation by etomoxir impairs NADPH production and increases reactive oxygen species resulting in ATP depletion and cell death in human glioblastoma cells. Biochim Biophys Acta 1807(6):726–734
71. Harper ME, Antoniou A, Villalobos-Menuey E, Russo A, Trauger R, Vendemelio M, George A, Bartholomew R, Carlo D, Shaikh A, Kupperman J, Newell EW, Bespalov IA, Wallace SS, Liu Y, Rogers JR, Gibbs GL, Leahy JL, Camley RE, Melamede R, Newell MK (2002) Characterization of a novel metabolic strategy used by drug-resistant tumor cells. FASEB J 16(12):1550–1557
72. Samudio I, Harmancey R, Fiegl M, Kantarjian H, Konopleva M, Korchin B, Kaluarachchi K, Bornmann W, Duvvuri S, Taegtmeyer H, Andreeff M (2010) Pharmacologic inhibition of fatty acid oxidation sensitizes human leukemia cells to apoptosis induction. J Clin Invest 120(1):142–156
73. Zaugg K, Yao Y, Reilly PT, Kannan K, Kiarash R, Mason J, Huang P, Sawyer SK, Fuerth B, Faubert B, Kalliomaki T, Elia A, Luo X, Nadeem V, Bungard D, Yalavarthi S, Growney JD, Wakeham A, Moolani Y, Silvester J, Ten AY, Bakker W, Tsuchihara K, Berger SL, Hill RP, Jones RG, Tsao M, Robinson MO, Thompson CB, Pan G, Mak TW (2011) Carnitine palmitoyltransferase 1 C promotes cell survival and tumor growth under conditions of metabolic stress. Genes Dev 25(10):1041–1051
74. Niemen KM, Kenny HA, Penicka CV, Ladanyi A, Buell-Gutbrod R, Zilihardt MR, Romero IL, Carey MS, Mills GB, Hotamisligil GS, Yamada SD, Peter ME, Gwin K, Lengyel E (2011) Adipocytes promote ovarian cancer metastasis and provide energy for rapid tumor growth. Nat Med 17(11):1498–1503
75. Pizer ES, Wood FD, Pasternack GR, Kuhajda FP (1996) Fatty acid synthase (FAS): a target for cytotoxic antimetabolites in HL60 promyelocytic leukemia cells. Cancer Res 56(4):745–751
76. Duarte NC, Becker SA, Jamshidi N, Thiele I, Mo ML, Vo TD, Srivas R, Palsson BO (2007) Global reconstruction of the human metabolic network based on genomic and bibliomic data. Proc Natl Acad Sci U S A 104(6):1777–1782
77. Jain M, Nilsson R, Sharma S, Madhusudhan N, Kitami T, Souza AL, Kafri R, Kirschner MW, Clish CB, Mootha VK (2012) Metabolite profiling identifies a key role for glycine in rapid cancer cell proliferation. Science 336(6064):1040–1044
78. Kriegler MP, Pawlowski AM, Livingston DM (1981) Cysteine auxotrophy of human leukemic lymphoblasts is associated with decreased amounts of intracellular cystathionase messenger ribonucleic acid. Biochemistry 20(5):1312–1318
79. Ensor CM, Holtsberg FW, Bomalaski JS, Clark MA (2002) Pegylated arginine deiminase (ADi-SS PEG 20,000 mw) inhibits human melanomas and hepatocellular carcinomas in vitro and in vivo. Cancer Res 62(19):5443–5450

80. Watson M, Roulston A, Belec L, Billot X, Marcellus R, Bedard D, Bernier C, Branchaud S, Chan H, Dairi K, Gilbert K, Goulet D, gratton MO, Isakau H, Jang A, Khadir A, Koch E, Lavoie M, Lawless M, Nguyen M, Paquette D, Turcotte E, Berger A, Mitchell M, Shore GC, Beauparlant P (2009) The small molecule GMX1778 is a potent inhibitor of NAD+ biosynthesis: strategy for enhanced therapy in nicotinic acid phosphoribosyltransferase 1-deficient tumors. Mol Cell Biol 29(21):5872–5888
81. Possemato R, Marks KM, Shaul YD, Pacold ME, Kim D, Birsoy K, Sethumadhavan S, Woo HK, Jang HG, Jha AK, Chen WW, Barrett FG, Stransky N, Tsun ZY, Cowley GS, Barretina J, Kalaany NY, Hsu PP, Ottina K, Chan AM, Yuan B, Garraway LA, Root DE, Mino-Kenudson M, Brachtel EF, Driggers EM, Sabatini DM (2011) Functional genomics reveal that the serine synthesis pathway is essential in breast cancer. Nature 476(7360):346–360
82. Locasale JW, Grassian AR, Melman T, Lyssiotis CA, Mattaini KR, Bass AJ, Heffron G, Metallo CM, Muranen T, Sharfi H, Sasaki AT, Anastasiou D, Mullarky E, Vokes NI, Sasaki M, Beroukhim R, Stephanopoulos G, Ligon AH, Meyerson M, Richardson AL, Chin L, Wagner G, Asara JM, Brugge JS, Cantley LC, Vander Heiden MG (2011) Phosphoglycerate dehydrogenase diverts glycolytic flux and contributes to oncogenesis. Nat Genet 43(9):869–874
83. Frezza C, Zheng L, Folger O, Rajagopalan KN, MacKenzie ED, Jerby L, Micaroni M, Chaneton B, Adam J, Hedley A, Kalna G, Tomlinson IP, Pollard PJ, Watson DG, DeBerardinis RJ, Shlomi T, Ruppin E, Gottlieb E (2011) Haem oxygenase is synthetically lethal with the tumour suppressor fumarate hydratase. Natur 477(7363):225–228
84. http:fantom2.gsc.riken.go.jp/metabolome/
85. Zhang XD (2011) Illustration of SSMD, Z score, SSMD*, z* score and t statistic for hit selection in RNAi high-throughput screens. J Biomol Screen 16(7):775–785
86. Lin LZ, Lin J (2011) Antabuse (disulfiram) as an affordable and promising anticancer drug. Int J Cancer 129(5):1285–1286
87. Kontos CK, Papadopoulos IN, Fragoulis EG, Scorilas A (2010) Quantitative expression analysis and prognostic significance of L-DOPA decarboxylase in colorectal adenocarcinoma. Br J Cancer 102(9):1384–1390
88. Wafa LA, Cheng H, Rao MA, Nelson CC, Cox M, Hirst M, Sadowski I, Rennie PS (2003) Isolation and identification of L-dopa decarboxylase as a protein that binds to and enhances transcriptional activity of the androgen receptor using the repressed transactivator yeast two-hybrid system. Biochem J 375(Pt 2):373–383
89. Habibi D, Ogloff N, Jalili RB, Yost A, Weng AP, Ghahary A, Ong CJ (2012) Borrelidin, a small molecule nitrile-containing macrolide inhibitor of threonyl-tRNA synthetase, is a potent inducer of apoptosis in acute lymphoblastic leukemia. Invest New Drugs 30(4):1361–1370
90. Teperino R, Schoonjans K, Auwerx J (2010) Histone methyl transferases and demethylases: can they link metabolism and transcription? Cell Metab 12(4):321–327
91. Albaugh BN, Arnold KM, Denu JM (2011) KAT(ching) metabolism by the tail: insight into the links between lysine acetyltransferases and metabolism. Chem Biochem 12(2):290–299
92. Rajendran P, Williams DE, Ho E, Dashwood RH (2011) Metabolism as a key to histone deacetylase inhibition. Crit Rev Biochem Mol Biol 46(3):181–199
93. Hanover JA, Krause MW, Love DC (2012) Bittersweet memories: linking metabolism to epigenetics through O-GlcNAcylation. Nat Rev Mol Cell Biol 13(5):312–321
94. Pearce EL (2010) Metabolism in T cell activation and differentiation. Curr Opin Immunol 22(3):314–320
95. Zhou W, Choi M, Margineantu D, Margaretha L, Hesson J, Cavanaugh C, Blau CA, Horwitz MS, Hockenbery D, Ware C, Ruohola-Baker H (2012) HIF alpha induced switch from bivalent to exclusively glycolytic metabolism during ESC-to-EpiSC/hESC transition. EMBO J 31(9):2103–2116
96. Ward PS, Patel J, Wise DR, Abdel-Wahab O, Bennett BD, Coller HA, Cross JR, Fantin VR, Hedvat CV, Perl AE, Rabinowitz JD, Carroll M, Su SM, Sharp KA, Levine RL, Thompson

CB (2010) The common feature of leukemia-associated IDH1 and IDH2 mutations is a neomorphic enzyme activity converting alpha-ketoglutarate to 2-hydroxyglutarate. Cancer Cell 17(3):225–234

97. Xu W, Yang H, Liu Y, Yang Y, Wang P, Kim SH, Ito S, Yang C, Wang P, Xiao MT, Liu LX, Jiang WQ, Liu J, Zhang JY, Wang B, Frye S, Zhang Y, Xu YH, Lei QY, Guan KL, Zhao SM, Xiong Y (2011) Oncometabolite 2-hydroxyglutarate is a competitive inhibitor of alpha-ketoglutarate-dependent dioxygenases. Cancer Cell 19(1):17–30
98. Chowdhury R, Yeoh KK, Tian YM, Hillringhaus L, Bagg EA, Rose NR, Leung IK, Li XS, Woon EC, Yang M, McDonough MA, King ON, Clifton IJ, Klose RJ, Claridge TD, Ratcliffe PJ, Schofield CJ, Kawamura A (2011) The oncometabolite 2-hydroxyglutarate inhibits histone lysine demethylases. EMBO Rep 12(5):463–469
99. Lu C, Ward PS, Kapoor GS, Rohle D, Turcan S, Abdel-Wahab O, Edwards CR, Khanin R, Figueroa ME, Melnick A, Wellen KE, O'Rourke DM, Berger SL, Chan TA, Levine RL, Mellinghoff IK, Thompson CB (2012) IDH mutation impairs histone demethylation and results in a block to cell differentiation. Nature 483(7390):474–478
100. Turcan S, Rohle D, Goenka A, Walsh LA, Fang F, Yilmaz E, Campos C, Fabius AW, Lu C, Ward PS, Thompson CB, Kaufman A, Guryanova O, Levine R, Heguy A, Viale A, Morris LG, Huse JT, Mellinghoff IK, Chan TA (2012) IDH1 mutation is sufficient to establish the glioma hypermethylator phenotype. Nature 483(7390):479–483

Chapter 3
Emerging Opportunities for Targeting the Tumor–Stroma Interactions for Increasing the Efficacy of Chemotherapy

Rajesh R. Nair, Anthony W. Gebhard and Lori A. Hazlehurst

Abstract It has become evident that tumor cells utilize survival signals that emanate from the tumor microenvironment to aid in survival and disease progression. Experimental evidence indicates that these same pathways contribute to de novo drug resistance. Identification of the mechanisms underlying the recruitment of accessory cells and survival signals provided by normal cells has provided a novel area for drug discovery for increasing the efficacy of cancer therapy.

3.1 Introduction

Cancer is characterized as a disease that is initially driven by the expression of oncogenes and attenuation of tumor suppression genes expressed in the malignant tumor cell. From these observations came the term oncogene addiction which implies that cancer cells require the activity of specific oncogenes for survival and growth. Furthermore, this premise infers that if we can specifically target oncogenes, then we could more effectively treat cancer. Early chemotherapeutic agents typically target rapidly dividing cells, a phenotype typically driven by oncogene expression. However, DNA-damaging agents are associated with toxicity and the emergence of resistance can limit the overall success of these agents. Resistance to DNA-damaging agents includes overexpression of drug transporters, increased DNA repair, increased metabolism, and failure to eliminate dormant cancer cells, as well as the cancer stem cell population [1]. It was envisioned that targeting specific oncogenes may reduce toxicity and increase efficacy over standard therapy

R. R. Nair · A. W. Gebhard · L. A. Hazlehurst
Molecular Oncology Program, H Lee Moffitt Cancer Center, Tampa, FL, USA

A. W. Gebhard · L. A. Hazlehurst
Molecular Pharmacology and Physiology Program, Tampa, FL, USA

L. A. Hazlehurst (✉)
Cancer Biology Program, University of South Florida, Tampa, FL, USA
e-mail: Lori.Hazlehurst@moffitt.org

D. E. Johnson (ed.), *Cell Death Signaling in Cancer Biology and Treatment*,
Cell Death in Biology and Diseases, DOI: 10.1007/978-1-4614-5847-0_3,

X.-M. Yin and Z. Dong (Series eds.), *Cell Death in Biology and Diseases*

due to increased specificity for the cancer cell as well as reduce the emergence of multi-drug resistant phenotypes associated with standard therapy.

In support of this potential promise of targeted agents are data generated in chronic myeloid leukemia (CML) which is driven by the BCR-ABL oncogene [2, 3]. A myriad of experimental data support the findings that the expression of the fusion oncogene BCR-ABL is sufficient for transformation and disease progression using transplant and transgene in vivo model systems [4, 5]. The identification of BCR-ABL as driving transformation of CML provided an ideal target for drug discovery, and first- and second-generation BCR-ABL inhibitors have provided proof-of-principle that this approach may indeed lead to clinical success [6]. However, it became clear that although BCR-ABL inhibitors are very effective when treating patients in the early chronic stage of the disease, BCR-ABL inhibitors were not curative and similar to DNA-damaging agents did not bypass the emergence of drug resistance. Because CML is initially driven by one driver oncogene, it represents an ideal disease to delineate de novo drug resistance and the emergence of acquired resistance. Acquired resistance to BCR-ABL inhibitors includes point mutations in what is commonly referred to as the gatekeeper region of the molecule [7]. These mutations result in reduced affinity of drug binding and thereby render the cells insensitive to the compound. The identification of the mutations, along with structural information, led to the development of second-generation compounds predicted to be active in the presence of specific gatekeeper mutations, and indeed, these molecules do show promising activity in the subset of patients harboring the T315I gatekeeper mutation [8]. Other mechanisms reported to contribute toward resistance to BCR-ABL inhibitors include increased expression of transporters, quiescence, and progression of the disease to blast phase which typically is the result of expression of additional driver mutations and subsequent resistance to BCR-ABL inhibitors.

Of particular interest for this review is the observation that even in patients exquisitely sensitive to BCR-ABL inhibitors, minimal residual disease can be detected at the molecular level. Subsequent to the identification of BCR-ABL came the identification of activating EGFR mutations in lung cancer and head and neck cancer and expression of BRAF in melanoma. Again, the identification of these driver oncogenes led to the rapid translation of specific kinase inhibitors. However, within the landscape of the complexity associated with lung cancer, colon cancer, and melanoma, these inhibitors, albeit effective, demonstrated shorter clinical response time before the emergence of resistance was a clinical concern [9–12]. Thus, it became clear that targeted therapies which potently inhibit the activity of a driver oncogene do not circumvent the emergence of drug resistance. The emergence of drug resistance can be caused by mutations and overamplification of targets which are mechanisms that contribute to acquired resistance. However, more recently, attention has shifted toward the role of the tumors inherent ability for co-opting and recruiting normal cells into the niche to provide an adaptive environment that favors the survival of tumor cells [13], and hence, as the niche is conditioned by the tumor, the dependency of cancer cells on the activity of a specific oncogene is diminished. This review will focus on mechanisms associated with components of the tumor cell microenvironment which contribute the maintenance of minimal residual disease and resistance to cancer chemotherapy.

The tumor microenvironment is a complex network consisting of multiple cell types including cancer-associated fibroblasts, immune accessory cells, and endothelial cells that support growth and invasion through production of cytokines, growth factors extracellular matrixes, and the formation of angiogenic blood vessels. Some of these same pathways which tumor cells initially co-opt to support growth and survival can also contribute to resistance to standard therapy and targeted therapy [14]. Components of the tumor microenvironment known to contribute to drug resistance will be discussed in the first half of this chapter. These interactions provide novel targets for drug discovery for increasing the efficacy of cancer therapies, and these strategies will be discussed in the second half of this review.

3.2 Cellular Components of the Tumor Microenvironment Aiding Drug Resistance

3.2.1 Mesenchymal Stroma Cells and Cancer-Associated Fibroblasts

Mesenchymal stem cells (MSCs) are characterized by their ability to differentiate into osteoblasts, chondrocytes, adipocytes, muscle cells, and reticular fibroblasts. MSCs are considered a rare population contained within the bone marrow estimated at 0.01–0.001 % of all mononuclear cells in the bone marrow [15]. Co-culturing of hematopoietic cells with either MSCs or bone marrow stroma cells has been shown to induce a multi-drug resistant phenotype in a variety of hematopoietic tumors including myeloma, acute myeloid leukemia, chronic myeloid leukemia, and chronic lymphoblastic leukemia (CLL) [16–20]. The mechanism of resistance varies depending on the drug investigated and the tumor type. For example, in CML, resistance toward BCR-ABL inhibitors is associated with the activation of the JAK/STAT3 pathway [16, 20]. In contrast, resistance associated with topoisomerase II inhibitors correlates with a reduction in the formation of cleavable complex and DNA double-strand breaks [21, 22]. Multiple cytokines and growth factors can initiate the JAK/STAT3 pathway, and correspondingly, MSCs are known to secrete many of these same factors including IL-6 and VEGF (See Table 3.1 for a complete list).

STAT3 can be activated by cytokines and growth factors [23–26]. Furthermore, the activation of STAT3 can be amplified in the context of β1 integrin–mediated adhesion [27]. Thus, the activation of STAT3 can be adaptively activated by multiple components of the tumor microenvironment, which is typically rich in the levels of inflammatory cytokines and extracellular matrixes. The activation of STAT3 induces increased expression of anti-apoptotic Bcl-2 family members such as Bcl-2, Bcl-X_L, and Mcl-1. Additionally, STAT3 increases the expression of inhibitors of apoptotic machinery like survivin and c-IAP2 [28–32]. STAT3 is proximally activated via the tyrosine kinase receptor–associated Janus kinase (JAK) [33–35], resulting in the

Table 3.1 Factors secreted by cultured stromal cells that activate STAT3

Effect	Molecule	References
Survival and angiogenic	VEGF	[197–199]
Adhesion, survival, and angiogenic	IL-6	[27, 198, 200]
Immunomodulation and differentiation	LIF	[201–203]
Adhesion and survival	SDF-1	[204–207]
Chemoattractant and survival	IL-8	[208, 209]
Chemoattractant	MCP-1	[210–212]
Survival	SCF	[200, 205, 213, 214]
Survival	G-CSF	[215–217]
Survival	GM-CSF	[216, 218, 219]
Immunomodulation and survival	HGF	[220–222]
Survival and angiogenic	FGF-1,2	[223–226]

phosphorylation at the Tyr705 residue in the C-terminal domain [33–35]. The activated phosphorylated STAT3 then undergoes homodimerization by reciprocal interaction between the SH2 domain of one monomer and the phosphorylated Tyr705 residue of its dimerizing partner. STAT3 dimer can translocate to the nucleus and bind specific DNA sequences and regulate the transcription of the responsive gene [36, 37].

Targeting of STAT3 could potentially occur at multiple points of regulation including inhibition of the activating receptor kinase or cytokine receptor, via inhibition of JAK kinases or by disrupting the dimerization or subsequent DNA binding. Specifically, disrupting the dimerization and DNA binding has proven to be a difficult task using small molecule design. Experimental evidence indicates that in some tumors, IL-6 plays a dominant role in conferring resistance. For example, in neuroblastoma, Ara et al. found that blocking IL-6 signaling using a blocking IL-6Rα antibody reversed resistance to the topoisomerase II inhibitor, etoposide when neuroblastoma cells were cultured in the presence of MSCs [38]. In multiple myeloma, IL-6 has been shown to be critical for cellular survival via activation of the JAK/STAT3 pathway and subsequent increased expression of the anti-apoptotic family member Bcl-X_L [39]. More recently, several groups have provided evidence that targeting IL-6 or JAKs in the context of co-culturing multiple myeloma cells with bone stroma cells enhances the efficacy of standard therapy including melphalan and dexamethasone [40–44]. Thus, in myeloma, it appears that IL-6 is the predominant cytokine-driving drug resistance in the bone marrow compartment. In contrast, in CML, experimental evidence indicates that multiple soluble factors can activate STAT3 and induce resistance to BCR-ABL inhibitors [20]. Due to the plethora of cytokines and growth factors present in the bone marrow milieu capable of activating the JAK/STAT3 pathway, it is likely that targeting downstream of the cytokine receptor at the level of JAK or STAT3 will be the most effective strategy and may circumvent resistance due to the selection of cells that utilize an alternative receptor for activating the JAK/STAT3 pathway [20] (for a complete list of drugs targeting JAK/STAT3 pathway see Table 3.2). MSCs are specifically relevant to tumors which home to the bone such as multiple myeloma or metastasize to the bone such as lung, breast, and prostate.

Table 3.2 JAK-STAT pathway inhibitors

Drugs	Function	Indication	Reference
Siltuximab	Monoclonal antibody against IL-6	Prostrate cancer	[227]
		Renal cancer	[228]
		Multiple myeloma	[41]
AZD1480	JAK2 inhibition	Solid tumors	[229]
		Myeloproliferative neoplasma	[230]
TG101209	JAK2 inhibition	Multiple myeloma	[231]
SB1518	JAK2 inhibition	Myeloid and lymphoid cancer	[232]
Lestaurtinib	JAK2 inhibition	Hodgkin's lymphoma	[233]
		Myelofibrosis and polycythemia vera	[234]
MS-1020	JAK3 inhibition	JAK3-driven cancer	[235]
Ruxolitinib	JAK1/JAK2 inhibition	Myeloproliferative neoplasma	[236]
INCB1656	JAK1/JAK2 inhibition	JAK2V617F-driven neoplasma	[237]
		Multiple myeloma	[43]
CYT387	JAK1/JAK2 inhibition	Multiple myeloma	[238]
AG490	Pan-JAK inhibition	Laryngeal cancer	[239]
		Acute lymphoblastic leukemia	[240] [241]
		Pancreatic cancer	
Gö6976	JAK2/FLT3 inhibition	Myeloproliferative neoplasma and acute myeloid leukemia	[242]
TG02	CDKs/JAK2/FLT3 inhibition	Acute myeloid leukemia	[243]
Sorafenib	JAK/STAT3 signaling inhibition	Cholangiocarcinoma	[244]
Trichostatin A	JAK/STAT3 signaling inhibition	Colorectal cancer	[245]
AUH-6-96	JAK/STAT3 signaling inhibition	Hodgkin's lymphoma	[246]
WP1193	JAK/STAT3 signaling inhibition	Glioblastoma	[247]
CEP-33779	IL6/STAT3 signaling inhibition	Colitis-induced colorectal cancer	[248]
Quercetin	IL6/STAT3 signaling inhibition	Glioblastoma	[249]

The origin of cancer-associated fibroblasts (CAFs) found in solid tumors is due to the active recruitment of bone marrow-derived MSCs by the tumor as well as recruitment of localized tissue-associated fibroblasts. CAFs are characterized as expressing "activated" fibroblast markers including fibroblast-specific protein

(FSP) and fibroblast-activated protein (FAP). Experimental evidence indicates that CAFs are modified by the tumor via epigenetic mechanisms [45]. The end result is an increased expression of inflammatory cytokines such as IL-6 and IL-8 and growth factor expression such as EGF and SDF-1 which can further promote tumor progression and drug resistance. Co-culturing cancer cells with activated fibroblasts has been shown to induce resistance to targeted therapies such as ER inhibitors for the treatment for breast cancer and EGFR inhibitors for the treatment for lung cancer [46, 47].

3.2.2 Immune Cells

In cancer, immune cells can participate in surveillance and elimination of tumor cells. However, paradoxically, immune cells can also participate in conditioning of the pre-metastatic niche which favors cancer cell initiation and progression of cancer as well as contributing to drug resistance. The premalignant niche resembles chronic inflammatory models and wound healing models, as characterized by significant infiltration of leukocytes into the tumor microenvironment. The leukocytes found in the tumor site include T_{reg}, immature monocytes, and alternative activated macrophages. T_{reg} cells are a subset of the CD4-positive T-cell population that expresses lineage-specific Foxp3, IL-2 receptor (CD25), and CTLA-4. T_{reg} cells are immunosuppressive and critical for curbing autoimmunity but also attenuate anti-tumor immunity. Increased expression of the T_{reg} population is associated with poor clinical outcomes for several tumor types and thus inhibiting the trafficking or depleting T_{reg} cells represents a possible strategy for improving clinical outcomes [48, 49].

Recently, immature myeloid progenitor cells have gained attention for their immunosuppressive function. In murine models, myeloid-derived suppressor cells (MDSCs) are identified as Gr1 and CD11b positive [50]. Functionally, MDSCs are characterized by their ability to suppress T and NK cell proliferation and promote the growth of T_{reg} cells. The functional phenotype is due to arginase I, inducible nitric oxide synthase expression and subsequent generation of NO [51] and peroxynitrite [52]. Similar to T_{reg} cells, the abundance of MDSCs is increased in the tumor and metastatic site. Moreover, MDSCs have been show to confer resistance to anti-VEGF therapies used to target angiogenesis [53]. Certainly, either selective elimination of MDSCs or blocking trafficking into tumor sites provides a novel approach for combination therapies. Recently, Schmid et al. [54] showed that blocking VLA-4 integrin via inhibiting inside-out activation (blocking SDF-1 or IL-1β) of VLA-4 and/or use of VLA-4 blocking antibody reduced the infiltration of myeloid-derived cells in a murine model of lung cancer and simultaneously reduced tumor burden. Interestingly, these studies highlight the need to utilize immune-competent mice when considering strategies for the development of novel chemotherapeutic agents.

3.2.3 Endothelial Cells

Once tumors have evaded the immune system and established growth that exceeds a few millimeters, the tumor must quickly establish a blood supply to sustain survival and growth. This is achieved by the production of stimulatory factors that attract and activate endothelial cells to form angiogenic vesicles. As the tumor grows, localized areas of hypoxia arise and results in induced HIF-1α expression in the tumor, endothelial, and stroma cells. HIF-1α expression is a major player for driving the expression and secretion of multiple angiogenic factors including VEGF, PDGF, FGF, angiopoietins, and SDF-1α [55]. Additionally, integrins are known to augment growth factor receptor signaling and thus may provide an additional target for inhibiting angiogenesis [56]. One soluble factor that is critical for the angiogenic switch is vascular endothelial growth factor (VEGF). VEGF stimulates the sprouting and proliferation of endothelial cells. However, it is becoming clear that targeting VEGF alone will likely not be sufficient for inhibiting angiogenesis, as resistance has been observed preclinically and clinically [57]. Overall, due to the ability of tumor cells to quickly adapt to a reduction in blood flow, it is currently not clear how effective targeting angiogenesis will be in controlling cancer. Although it is feasible, as we learn more about the bidirectional communications between the tumor, myeloid population, stroma, and endothelial cells that such strategies may be fully realized.

3.3 The Tumor Microenvironment Milieu and Drug Resistance and Tumor Progression

3.3.1 Galectins

Cell surface receptors like cytokine and growth factor receptors are N-glycosylated transmembrane glycoproteins that can bind glycan-binding proteins called lectins [58, 59]. Galectins are members of the lectin family that has an affinity for β-galactoside sugars, especially N-acetyllactosamine [60]. Galectins have been demonstrated to crosslink receptor glycoproteins at the cell surface and form lattices and thereby dictate the distribution and the retention time of the receptor on the cell surface [61, 62]. Interestingly, the strength of the interaction between the galectins and the glycoproteins, and thus the lattice structures, can be modulated by either regulating the cell surface protein glycosylation or by altering the expression of galectins [63, 64]. This is important because the distribution pattern of the cell surface receptors and their residency time on the plasma membrane can modulate the cells response to its microenvironment and is a prevailing feature of oncogenesis [65]. For example, *Mgat5* gene encodes a Golgi-localized enzyme that initiates N-glycan synthesis on newly formed cell surface receptors that is recognized by galectins [66]. Granovsky et al. have shown that *Mgat5* knockdown inhibits tumor progression and upregulation of *Mgat5* increases tumorigenesis in mouse tumor models [67]. Furthermore, Partridge et al. demonstrated that in Mgat5$^{-/-}$ tumor cells, cell surface density of EGFR and TGF-β receptor was reduced and activation of these receptors increased their endocytosis, making the cells less sensitive to stimulation by EGF, IGF, PDGF, bFGF, and TGF-β [68]. Also, these receptor glycoproteins were dependent on galectin-mediated lattice formation for surface retention after receptor activation, thereby prolonging the sensitivity of the tumor cells to EGF and TGF-β [68]. This observation was later confirmed by Lajoie et al., who showed that EGFR localization, and thus, the magnitude of EGFR signaling in tumor cells was very much dependent upon the galectin lattice structures opposing and prevailing over the oligomerized caveolin-1 microdomains that mediate endocytosis [69]. Finally, regulated glycosylation of cell surface receptors can create or abolish binding domains for galectins and thereby control important cellular response such as immune responses like T-cell activation, homing, and survival [70].

Presently, there are 15 members of the galectin family identified that all contain a conserved globular domain called carbohydrate recognition domain (CRD) that recognizes and binds to β-galactosidase [71]. The specificity of the galectins for their substrates is defined by the ligand-binding groove contained within the CRD of the protein [72]. The 15 family members are divided into three sub-families on the basis of their molecular structure (see Fig. 3.1): (1) Galectin-1, Galectin-2, Galectin-5, Galectin-7, Galectin-10, Galectin-11, Galectin-13, Galectin-14, and Galectin-15 are categorized under the name "proto-type" galectins as they are comprised of a single polypeptide chain that is able to dimerize; (2) Galectin-3 is categorized under the name "chimera-type" galectin as it consists of one C-terminal CRD linked to an N-terminal peptide that is able to pentamerize; and finally, (3) Galectin-4, Galectin-6, Galectin-8, Galectin-9 and Galectin-12 are categorized under the name "tandem repeat-type" galectins as they are composed of two CRDs connected by a linker peptide [72–74]. Irrespective of their type, all galectins have the ability to recognize and bind galactosyl residues without the requirement for any other catalyst like cations [71]. Additionally, these lectins are widely localized ranging from within the cytoplasmic and nuclear compartment to the extracellular compartment where they modulate a myriad of biologic process (for review see, [75]). Below, we will briefly discuss the role of galectins in tumorigenesis by highlighting the role of each type of the galectin family.

3.3.2 Galectin-1

This protein was first discovered as a low molecular weight, β-galactoside-binding soluble protein purified from the calf heart and lung [76]. It is a prototype galectin with a single CRD that can form dimers and has been shown to bind to oncogenic H-Ras to mediate Ras membrane anchorage and cell transformation, thus

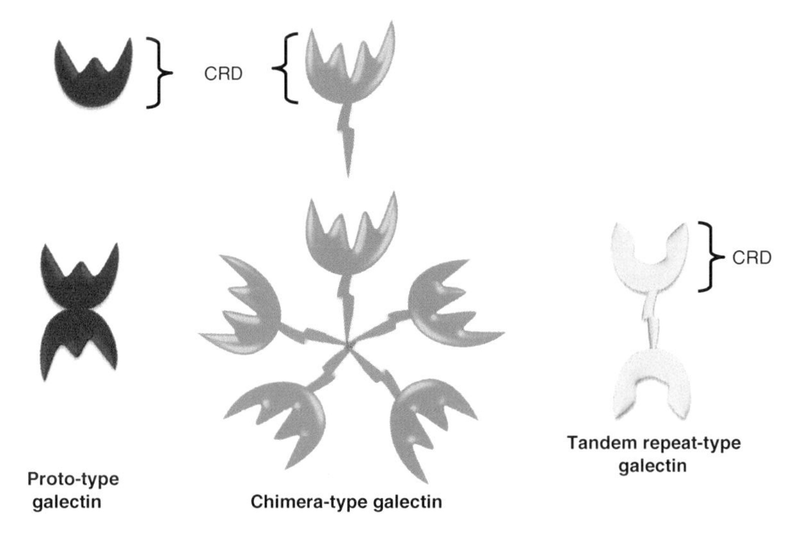

Fig. 3.1 Model structures of the different types of galectins. Galectins are divided into three major types according to their structures. The prototype galectin contains one CRD and can exist as a monomer or a dimer. The chimeric type galectin contains one CRD that is covalently linked to a non-carbohydrate tail. The chimeric type of galectin, like galectin-3, can exist as a pentamer and a monomer. Finally, the last type of galectin is the tandem repeat-type galectin that harbors CRD domains that are covalently linked by a small peptide linker

demonstrating its role in tumor initiation and progression [77]. In addition to tumor progression, in prostrate and ovarian cancer cells, overexpression of galectin-1 and its resulting accumulation in the extracellular space resulted in increased adhesion of these tumor cells to the extracellular matrix like fibronectin and laminin-1 [78, 79]. However, whether this increased cell adhesion led to drug resistance was not tested in these studies. Furthermore, galectin-1 was shown to be upregulated in capillaries associated with prostrate tumor cells where it modulates angiogenesis by mediating the interaction between prostrate carcinoma cells and endothelial cells [80]. In contrast to above-stated studies, in colon cancer Colo201 cells, overexpression of galectin-1 in the cytoplasm of the cells is associated with increased cell apoptosis [81]. Due to these opposing effects, depending on the localization of galectin 1 and the tumor type, it is important to be cautious while advocating the use of galectin-1-specific inhibitor for the treatment for cancer.

3.3.3 Galectin-3

This protein is a unique chimera-type galectin that has been associated with several models of immune disease and cancer [82–84]. Jeon et al. [85] have shown that galectin-3 can exert cytokine-like regulatory actions by the activation of the

JAK2-STAT1/3/5 pathway in an IFN-γ-independent manner. Indeed, the activation of the JAK-STAT pathway required that glial cells secrete galectin-3 in the extracellular compartment under inflammatory conditions, thus demonstrating galectin-3 to be an endogenous danger signaling molecule leading to inflammatory events [85]. In addition to inflammation, galectin-3 also enhances proliferation and angiogenesis of endothelial cells differentiated from MSCs, thereby showing an ability to promote tumor progression [86]. In MM, inhibiting galectin-3 with the use of a modified citrus pectin called GCS-100 resulted in specific induction of cell death in multiple myeloma (MM) cells resistant to conventional and bortezomib therapies [87]. The cell death induced by GCS-100 was impressive since it could overcome growth and survival benefits conferred by the BM microenvironment. Similarly, Streetly et al. have further confirmed the conclusion of the abovementioned study by demonstrating that GCS-100 was able to block the activation of AKT after cytokine stimulation in MM cells and led to caspase-8 and caspase-9 dependent cell death [88]. These preclinical studies provide a rationale for the clinical evaluation of GCS-100 to kill MM cell and to overcome drug resistance. Galectin-3 has also been shown to protect CML cells from apoptotic stimuli through stabilization of anti-apoptotic Bcl-2 family proteins [89]. Interestingly, galectin-3 was shown to be specifically induced in HS-5 BM stromal cells when these cells were co-cultured with various CML cell lines [90]. Secreted galectin-3 was shown to activate ERK and AKT, induce accumulation of Mcl-1, and promote cell proliferation and resistance to BCR-ABL inhibitors and genotoxic agents in CML cells. Also, an in vivo mice transplantation model demonstrated that galectin-3 overexpression promoted long-term BM lodgement of CML, supporting the hypothesis that galectin-3 may be a viable therapeutic target to help overcome the BM microenvironment-mediated drug resistance in CML [90].

3.3.4 Galectin-9

This protein is the tandem repeat-type of galectin which, in contrast to galectin-3, has been shown to have anticancer activity [91, 92]. Kuroda et al. [93] have demonstrated that modified humanized galectin-9 inhibits proliferation of CML cells in the nM range. Proliferation was inhibited due to apoptosis initiated by the activation of ATF3 and its downstream effector molecule Noxa, but not Bim [93]. Importantly, this galectin-9-mediated cell death was not influenced by the absence of p53, overexpression of P-glycoprotein, and the presence of mutant T315I ABL [93]. On the other hand, Kobayashi et al. [94] have shown that protease-resistant humanized galectin-9 can induce apoptosis in MM cells by the activation of a caspase-dependent pathway via JNK- and p38-activated proteins. The ability of galectin-9 to induce cell death could be replicated in primary MM cells with very poor disease prognosis and overt manifestation of drug resistance [94]. The above studies suggest that galectin-9 is a new therapeutic target for hematological malignancies that may overcome resistance to conventional therapies.

3.4 Hypoxia and Drug Resistance

Hypoxia can occur at the site of the primary tumor due to a decreased oxygen content found in the center of the tumor or at specialized niches found in metastatic sites. For example, the bone marrow microenvironment is characteristically hypoxic; direct measurement of oxygen levels has estimated the oxygen content to be 1 to 2 % in the bone marrow [95]. Thus, hypoxia represents an initial selection pressure that contributes to necrotic cell death but also provides pressure for adaptation and tumor progression. Indeed, exposure to hypoxia can induce resistance in multiple tumor types including leukemia and lung cancer cell line models [96, 97]. Moreover, in CML, hypoxic conditions that mimic the O_2 content of the bone marrow confer resistance to BCR-ABL inhibitors, suggesting that hypoxia may contribute to failure to eradicate minimal residual disease [98]. Hypoxia induces a myriad of effects on tumor cells including selection for p53 mutations [99] and increased genomic instability due to increases in reactive oxygen species and attenuation of DNA repair pathways [100]. Experimental findings suggest that hypoxia can contribute to angiogenesis, invasiveness, metastasis as well as immune suppression [101–103]. Many of the phenotypes associated with hypoxia are the result of the induction of the hypoxia-inducible factor (HIF) family of transcription factors. Three members constitute the family (HIF-1, HIF-2, and HIF-3). HIF-1α is targeted for degradation by ubiquitination, which is regulated by the E3 ligase von Hippel–Lindau (VHL) tumor suppressor. Under normal oxygen levels, HIF-1α is hydroxylated on proline 402 and 564 by prolyl-4-hydroxylases (PHDs). The reaction requires oxygen and 2-oxoglutarate as substrates [104, 105]. Thus, under normal oxygen concentration, HIF family members are hydroxylated and subsequently targeted for ubquitination and degradation. In contrast, under hypoxic conditions, HIF-1α is stabilized and can form a heterodimer with HIF-1β and drive transcription of target genes [106].

A hypoxic environment allows for two independent targeting strategies: (1) development of bioreductive prodrugs and (2) use of drugs that inhibits targets that are specifically expressed under hypoxic conditions. Bioreductive prodrugs typically take advantage of redox cycling such that the levels of the prodrug radical are kept low in oxic cells at the expense of generation of superoxide. Common moieties that have the potential to be metabolized under hypoxic conditions include nitro groups, quinones, aromatic N-oxides, aliphatic N-oxides, and transition metals. One example of a hypoxia-activated prodrug is TH-302. TH-302 is nitrogen mustard based prodrug with promising activity in a phase I clinical trial for advanced solid tumors [107]. Because the bone marrow is hypoxic and nitrogen mustards such as melphalan are a mainstay of therapy for myeloma which homes to the bone, it would be of interest to test prodrugs in the context of tumors which home or metastasize to the bone. The identification of pathways that are activated upon hypoxia that contributes to cellular survival represents an additional strategy for specifically targeting the hypoxic tumor microenvironment. Three main pathways activated by hypoxia are the HIF family of transcription factors, the unfolded protein response (UPR), and autophagy. Severe hypoxia disrupts the formation of disulfide bond formation leading to misfolded proteins and activation of the UPR. Thus, targeting the UPR pathway is attractive for selectively eliminating hypoxic tumors. One promising target in the UPR pathway is the inositol-requiring enzyme 1 or IRE1 [108]. Inhibitors of IRE1 are currently in development and show activity in myeloma models which are tipped toward ER stress due to the large production of monoclonal antibodies in these cancerous cells [109, 110]. Another strategy is the use of proteasome inhibitors which also can trigger additional stress on the UPR response, and correspondingly, bortezomib shows activity in hypoxic tumor sites [111]. Thus, inhibition of the proteasome exaggerates the UPR and tips the threshold toward cell death under hypoxic conditions and may represent an attractive strategy for eliminating cells residing in hypoxic microenvironments.

3.5 Molecules Involved in the Interaction of the Tumor Cell with the Microenvironment

3.5.1 Integrins

Damiano et al. [112] reported in 1999 that adhesion of myeloma cells via $\alpha4\beta1$ integrin was sufficient to cause resistance to structurally and mechanistically diverse chemotherapeutic agents. Furthermore, they reported that selection for acquired drug resistance selected for cells with increased expression of $\beta1$ and $\beta7$ integrin. This report suggested that cell adhesion could contribute to an adaptive response for cells to evade cytotoxic insult. More recently, $\alpha4$ integrin expression was shown to be increased in ex vivo samples collected from relapsed myeloma patients compared with specimens obtained from newly diagnosed myeloma patients, albeit only a small sample size was reported [113]. VLA-4 integrin is known to be important for homing, and more recent evidence indicates that VLA-4/VCAM-1 interactions between osteoclasts and breast cancer cells drive the cycle of bone destruction and tumor growth in metastatic models of breast cancer [114]. A similar finding was reported for myeloma where adhesion to stroma induced MIP-1alpha and beta which in turn activated osteoclasts [115] and contributed to further bone destruction and tumor growth. Collectively, these data indicate that VLA-4 integrin is an attractive target for myeloma and perhaps solid tumors which metastasize to the bone. Integrins are heterodimeric proteins comprised of an α and β subunit. Eighteen α and 8 β subunits have been identified and complex to form 24 known $\alpha\beta$ heterodimers. Integrins bind to a diverse set of extracellular ligands which are found in the extracellular matrix. Integrins do not contain intrinsic kinase activity; however, integrin ligation allows for the formation of focal adhesions containing adaptor proteins critical for the activation of signaling complexes (see Fig. 3.2). Molecules contained within the focal

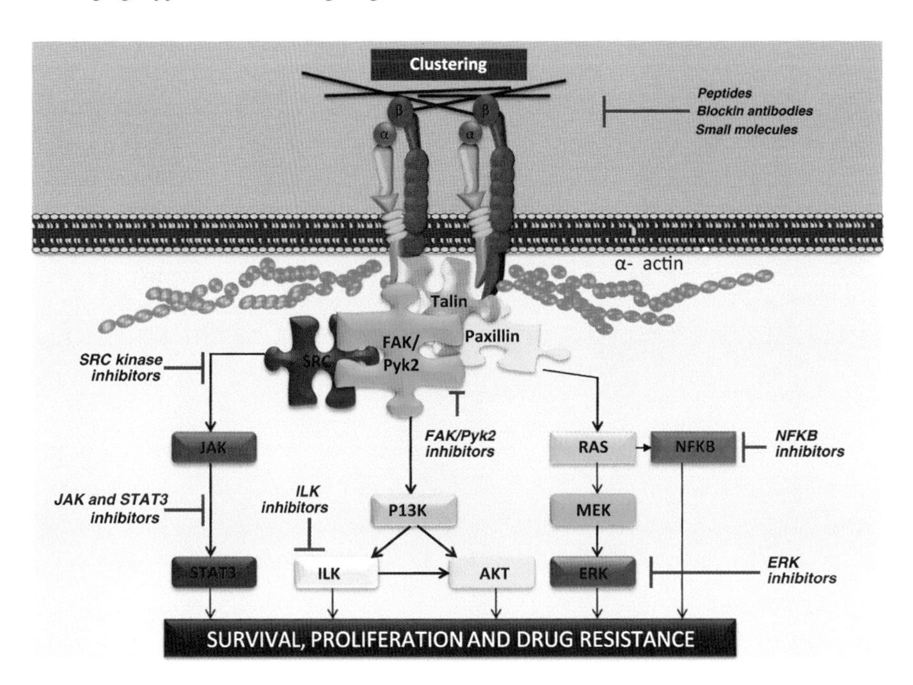

Fig. 3.2 Signal Transduction events triggered by clustering of integrins. Ligand binding induces clustering of integrins and causes the aggregation of adaptor proteins like paxillin and talin, which in turn recruits a large variety of signaling molecules including FAK/Pyk2, Src, vinculin, and polymerized actin filaments. This complex then forms the platform for signal transduction and crosstalk events. Some of the main signaling events include Src-JAK-STAT pathway activation; FAK-RAS-MEK-ERK pathway activation; FAK-RAS-NFKB pathway activation; FAK-PI3 K-AKT pathway activation; and FAK-ILK pathway activation. All this leads to rapid downstream modulation of cell proliferation-inducing proteins, upregulation of anti-apoptotic proteins, cell adhesion, and cell motility, which ultimately helps the cells to survive, divide, and resist the actions of anti-neoplastic drugs

adhesions include integrins, FAK/Pyk2, Src, and several scaffold proteins including talin, actin, and paxillin. Downstream signaling that is amplified following the ligation of integrins includes activation of the MAPK-, Src- and AKT-mediated survival pathways [116, 117]. Finally, integrin signaling is known to augment growth factor– and cytokine-mediated signaling [27, 118]. Due to the importance of integrins in sensing and transmitting external cues and contribution to survival, migration, and growth, integrins remain an attractive target for cancer as well as autoimmune diseases [32]. Integrins can be effectively targeted by utilizing blocking antibodies and peptidomimetics which typically compete with consensus ligand sequences such as RGD peptides. The development of small molecules demonstrating specificity for integrins has been difficult due to inherent difficulties of targeting protein–protein interactions with small molecules.

The structure of integrins has recently been elucidated utilizing crystallography to inform structural models [119–121]. The α subunit is comprised of a short cytoplasmic tail, a single-pass transmembrane domain, and two calf domains immediately proximal to the transmembrane region which make up the lower leg region. The upper leg region is followed by 7 repeating units which comprise the β-propeller region. In nine of the α integrin subunits, an interactive domain (I) is inserted between two and three of the α-helical repeating units found in the β-propeller region. The interactive domain modulates the ligand-binding domain and confers specificity for ligand binding. The β subunit has a longer cytoplasmic domain which is critical for recruitment of adaptor molecules and signaling. The β subunit contains a transmembrane domain, four epidermal growth factor-like domains which constitute the lower leg region and two hybrid domains which lie on either side of the β-I domain. The N-terminus folds over exposing the β-I domain and contributes to the formation of the upper leg region. Integrins can reside on the cell surface in 3 distinct confirmations. A resting state results in a bent confirmation and conceals the ligand-binding domain. An intermediate state has an extended confirmation, while a fully extended molecule is competent to bind ligand. The confirmation of the integrin is regulated by a process referred to as inside-out signaling (see Fig. 3.3). As the name infers, intracellular signaling is required to change the conformation of integrins, and thus, inside-out signaling represents another potential point for pharmacological intervention. Talin represents a central molecule for regulating integrin activation. Talin binds to the β integrin subunit resulting in the tail separation of the α and β subunits leading to conformational changes in the extracellular domain and extension of the

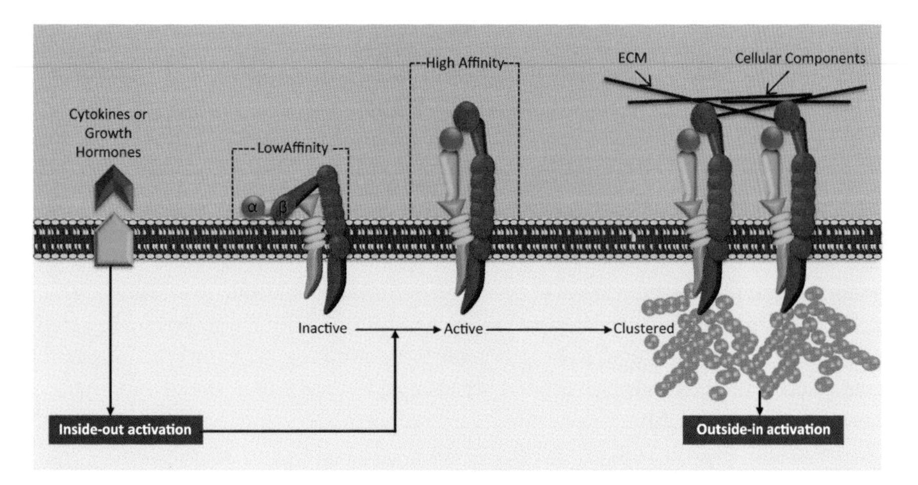

Fig. 3.3 Steps leading to the activation and clustering of integrins. In resting cells, the integrins are in a *bent, low-affinity state*. However, once the cell is exposed to an environment with an abundance of cytokines or growth factors, an inside-out activation pathway is triggered that culminates in an extended, *high-affinity state* of integrins. The integrins are now receptive to ligand binding, either in the form of adhesion to the extracellular matrix or adhesion to other cells within its microenvironment. Ligand binding causes clustering of integrins, with polymerization of actin, and induction of an outside-in activation cascade

external domain, allowing for ligand binding [122–124]. Talin is comprised of a 50-kDa N-terminal domain and a 220-kDa rod domain structure. Talin binds the β1 integrin tail via the head domain (THD), resulting in the activation of integrins [125]. Binding of talin to the β1 integrin tail is regulated by phosphorylation at the NPXY motif. Phosphorylation of this motif results in an increased affinity for Dok1, which acts as a competitive inhibitor of talin for access to the β1 integrin tail [126]. The C-terminal rod domain binds to vinculin and actin participating in the cytoskeletal rearrangement that is a hallmark of integrin ligation [127]. In hematopoietic lineages, Rap1 facilitates the targeting of talin to the β1 integrin tail. Rap1 is a small GTPase and is a member of the RAS superfamily. Rap1 can be activated by extracellular signals including IL-3- and erythropoietin-mediated signaling. Moreover, the activation of Rap1 leads to the activation of VLA-4 and VLA-5 in hematopoietic cells [128]. Both VLA-4- and VLA-5-mediated adhesions are known to confer a cell adhesion–mediated resistance phenotype or CAM-DR in hematopoietic lineages [21, 112, 129]. Thus, targeting intracellular pathways, albeit not specific due to the multitude of downstream signaling that typically are modulated by GTPases such as Rap1, represents a potential point of intervention.

An attractive strategy for targeting specific integrins is the recent development of humanized blocking antibodies and peptides which mimic the ligand and thus represent competitive inhibitors. Natalizumab binds to the α4 subunit and blocks adhesion of VLA-4 (α4β1)-mediated adhesion to VCAM-1 and fibronectin-mediated and α4β7-mediated adhesion to fibronectin. Natalizumab is approved for the treatment for multiple sclerosis [130] and Crohn's disease [131]. However, this agent is associated in rare instances with progressive multifocal leukoencephalopathy, which can be lethal and is thought to be related to the immunosuppressive effects of the antibody. Moreover, in murine models of myeloma and AML, VLA-4 antibodies have demonstrated good anti-tumor activity [132, 133], making it attractive for pursuing in cancer indications for diseases that home or metastasize to the bone. Peptides represent another potential strategy for blocking integrin–matrix interactions. The RGD sequence is a consensus recognition site found in multiple extracellular matrix ligands [134]. Disintegrins are RGD peptides initially found in snake venom [135]. RGD-containing disintegrins have been found to inhibit tumor growth and angiogenesis [136]. However, natural product–derived peptides are limited by their relatively large size and poor bioavailability. Efforts to design more drug-like candidates have included incorporation of D-amino acids to decrease proteolytic cleavage of the peptide. To this end, Cress and colleagues screened all D-amino acid peptide library using cell adhesion of prostate cells overexpressing α6β1 integrin. These investigators identified the sequence KIKMVISWKG and showed that this peptide allows for adhesion, binds α3 and α6 integrin, and blocked migration and invasion [137–139]. More recently, Nair et al. showed that this same peptide induced necrotic cell death in myeloma using both in vitro and in vivo models [140]. Another strategy for increasing the drug-like properties of peptides is via cyclization strategies. Linear peptides can exist in multiple conformations in solution, and cyclization allows for constraining the peptide in one conformation. Cilengitide is an RGD cyclic peptide c(RGDf(nMe)V) which binds αvβ3/αvβ5 integrin and competes with RGD sequences found in components of the matrix [141]. RGD sequences are contained in multiple ligands, yet integrins demonstrate specificity which cannot be entirely explained by flanking sequences. Thus, it was postulated that secondary structure was critical for sequence recognition. Cyclization represents a strategy to constrain the recognition sequence in a well-defined secondary structure. Kessler and colleagues performed extensive structure activity relationship prior to knowledge of crystal structures, and thus, they applied the following strategies for optimizing RGD sequences for inhibition of $\alpha V\beta 3$-mediated adhesion. They reduced the conformation space using cyclization strategies, spatial screening of peptides, and N-methyl scanning [141]. N-methylation of peptide bonds often increases their metabolic stability and bioavailability [142]. Merck-Serona has published on the phase II trial and just completed the accrual of a phase III clinical trial with cilengitide for the treatment for glioblastoma [143]. Preclinically, cilengitide was shown to have activity by inhibiting growth directly, as well as inhibiting invasion and angiogenesis [144]. These studies will provide proof-of-principle whether targeting cell adhesion receptors with cyclized peptides is a viable strategy for drug discovery and development in oncology.

3.5.2 *CD44*

CD44 is a family of cell surface receptors that similar to integrins mediate cell–matrix adhesion. One gene encodes the CD44 family, but alternative splicing leads to multiple variants which are thought to be, in part, responsible for the apparent diverse functions attributed to this single-pass membrane-spanning cell surface protein. CD44 facilitates cell adhesion and metastasis, augments growth factor signaling, and expression of CD44 protects against hypoxia-induced lung injury [145–155]. CD44 can be shed from the cell membrane and soluble CD44 is associated with advanced disease in chronic lymphocytic leukemia [156–160]. Hyaluronic acid (HA) is the most predominant ligand for CD44. HA is an abundant polysaccharide found in extracellular matrixes. However, CD44 can also bind fibronectin and osteopontin [161]. The structure of CD44 consists of a constant region spanning the first five exons which defines the HA-binding domain and is found in all splice variants. Exon 7 through 15 (corresponding to v2–v10) is the variable region of the molecule. This region is localized on the external domain and adds to the stem region of CD44. The highly conserved cytoplasmic region has part of exon 18 combined with exon 19 and 20 [162]. CD44 variant expression associates with markers of cancer stem cells and is a poor prognostic indicator for many cancer types including multiple myeloma [163–165]. Paradoxically, the activation of CD44, depending on the cell context, is associated with both cell death and survival [64, 166–168]. Moreover, CD44v associates with $\alpha 4\beta 1$ integrin in CLL and positively regulates adhesion of myeloma cells to stroma and extracellular matrices [148, 165, 169]. Finally, it is known that CD44 associates with VLA-4 integrin in T cells and CLL [169, 170]. Along with VLA-4, CD44 is considered the major homing receptor for hematopoietic cells for trafficking to the bone marrow compartment [171] and thus, similar to VLA-4, may play a role in the recruitment of accessory cells to the tumor site as well as metastasis of the tumor. CD44 has an N-terminal ligand-binding ectodomain containing the Link module; comprised of three intradisulfide bonds, a stalk region containing one or multiple variant exons, a transmembrane domain, and a short C-terminal tail which is important for cytoskeletal attachment and signal transduction. CD44 has a highly conserved amino acid sequence among mammalian and avian species sharing between 47–93 % sequence homology with humans and an almost identical homology in the Link module, transmembrane, and cytoplasmic tail regions [172]. The CD44 receptor also contains many posttranslational modifications. It has been estimated that 25–40 % of CD44 molecules are phosphorylated at Ser325 and that CD44 is constitutively phosphorylated at this site in cultured cells [172]. While the exact role of Ser325 phosphorylation is unclear, mutations at this site have been shown to inhibit HA binding and cell migration, as well as modulating the interaction of CD44 with the ERM (ezrin, radixin, moesin) family of proteins [173]. CD44 is also palmitoylated at Cys286 and Cys295, and while the exact roles these modifications play are not yet fully understood, it appears that palmitoylation enhances the association between CD44 and ankyrin and may also induce clustering of CD44 into membrane subdomains [172]. The CD44 receptor is also highly glycosylated and linked to chondroitin sulfate; with at least 5 conserved N-glycosylation sites in the ectodomain as well as being highly O-glycosylated in the extracellular region, both of which have been shown to affect ligand-binding affinity [173]. The crystal structure of HA bound to murine CD44 has been solved, and it has been observed that CD44 adopts two conformational binding states and this binding is the result of hydrogen bonding [174]. Further studies by nuclear magnetic resonance (NMR) and surface plasmon resonance (SPR) have further elucidated the two binding configurations of the CD44 receptor, designated as ordered (O) and partial disordered (PO) [175, 176]. It has been shown that the binding of HA oligomers at an allosteric site on the ectodomain leads to conformational change in the Link module and switches the module from the O to PO configuration. Furthermore, while the minimum binding requires an HA hexamer, recent studies have demonstrated that there is a positive correlation with the length of HA and binding affinity. It has long been speculated that this is the result of multiple CD44 molecules binding to one HA polysaccharide. This was recently confirmed by Wolny et al. using an artificial membrane system in which they demonstrated that multiple CD44 molecules can bind a single HA polysaccharide through a multivalent nature [176]. While the crystal structures of CD44 bound to other ligands have yet to be solved, it is postulated that other ligands bind in the ectodomain adjacent to the Link module. The CD44 transcript contains twenty exons that give rise to seventeen known isoforms through alternative splicing and posttranslational modifications. The most ubiquitously expressed isoform is the standard form (CD44s), which is comprised of exons 1–5 and 16–20 [172]. CD44 expression is regulated through receptor-mediated endocytosis as well as the proteolytic cleavage of the extracellular domain by various matrix metalloproteinases (MMPs). CD44 activation through ligand binding has been shown to have diverse cellular effects. CD44s activation has been shown to activate apoptosis in Jurkat cell lines, whereas CD44v induction has been suggested to do the converse via sequestering FasR [147]. Similar to integrins, CD44 has no intrinsic kinase activity; however, it gains access to kinases such as focal adhesion kinase (FAK), protein kinase C (PKC), and phosphatidylinositol 3-kinase (PI3K) through adaptor proteins such as the ERM family [177]. In addition to CD44 gaining access to kinases through adapter proteins, it has been demonstrated that CD44 can also gain access to other signaling complexes through association with other adhesion receptors, most notably α4 integrin in lymphocyte extravasation as well as cell growth and proliferation signaling through the human epidermal growth factor receptor 2 (HER2), c-Met, and VEGF [178–180]. CD44-deficient mice exhibit normal development; however, the mice are resistant to T-cell activation-induced cell death (AICD) [181]. Recently, several in vitro studies targeting CD44 with monoclonal antibodies alone and in combination with low molecular weight HA have shown induction of apoptosis in myeloid cells as well as a novel caspase-independent pathway in erythroleukemia cells. Together, these data suggest that agonistic signals may actually activate cell death in tumor cells. More specifically, one study was able to eradicate human acute myeloid leukemia stem cells in non-diabetic severe combined immune-deficient (SCID) mice by targeting CD44 receptor with a CD44 monoclonal antibody that recognized all isoforms and was characterized as an activating antibody [64]. There has also been an interest in targeting CD44 via peptidomimetics. In one study, a group using a CD44v6-derived peptide has indicated a role of CD44 in VEGF signaling and angiogenesis [182]. A second peptide named A6 (acetyl-KPSSPPEE-amino) has been shown to activate CD44 and inhibit migration and metastasis in CD44-expressing cells [183]. Furthermore, this peptide has shown promise in phase I and phase II clinical trials in patients with ovarian cancer. Another molecule that has shown promise in early stage clinical trials is the immunoconjugate bivatuzumab, a human anti-CD44v6 monoclonal antibody coupled with a benzoansamacrolide moiety, which has been tested in head and neck squamous cell carcinomas [184]. While these molecules have shown promise in the clinic, targeting CD44 will remain a difficult task until the diverse expression and signaling of the many CD44 isoforms is more fully understood in both healthy and diseased cell lineages.

3.6 Signaling Pathways Activated Due to the Interaction of the Tumor Cell with the Microenvironment

3.6.1 FAK/Pyk2

Due to the existence of multiple cell adhesion receptors and potential redundancy, it is attractive to speculate that downstream targets will be more efficacious for targeting cell adhesion mediated drug resistance or CAM-DR. Immediately, downstream of multiple cell adhesion receptors, as well as cytokine and growth factor receptors, is FAK and/or Pyk2 which is a FAK homolog. FAK is a non-receptor tyrosine kinase, is ubiquitously expressed, and localizes to focal adhesions. FAK contains three major structural domains: (1) an N-terminal FERM domain, (2) a tyrosine kinase catalytic domain, and (3) a C-terminal focal adhesion–targeting domain [185]. The activation of FAK requires autophosphorylation at Y397 following the integrin ligation [186]. Phosphorylation of FAK leads to the recruitment and activation of Src family members as well as other components of the focal adhesion including p130Cas and paxillin [187–189]. The activation of FAK leads to the integration and spatial organization of signaling pathways including activation of AKT, Src, and the MAPK pathway. Pyk2 is a closely related homolog to FAK, with a more restricted tissue distribution as expression is limited predominately to hematopoietic cells, vascular smooth muscle, endothelium, spleen, kidney, and the central nervous system [190]. Although Pyk2 and FAK are functionally related, they are not redundant with respect to the phenotype observed in the respective knockout mice. FAK knockout mice are embryonic lethal due to defects in the axial mesodermal tissue and cardiovascular system [191]. In contrast, Pyk2 knockout mice develop normally, but do exhibit defective macrophage migration and have increased bone density [192, 193]. PF-562,271 is a promising dual FAK/Pyk2 inhibitor being developed by Pfizer. This compound competes for the ATP-binding domain and is a reversible inhibitor of FAK and Pyk2 catalytic activity. This compound has shown anti-tumor activity in multiple in vivo xenograft tumor models [194]. Moreover, in models of lytic bone disease involving implantation of breast cancer cells into the tibia, investigators have shown that the FAK/Pyk2 inhibitor reduced the tumor burden and lytic lesions [195]. This is consistent with data showing that *Pyk2*$^{-/-}$ knockout mice show increased bone mass, indicating that the inhibition of Pyk2 may be helpful for diseases such as myeloma, breast, and lung that either home or metastasize to the bone and induce lytic lesions [193]. Recently, a phase 1 clinical trial was reported that showed the compound was well tolerated, and 12 % of patients showed prolonged stable disease for six or more cycles [196].

3.7 Conclusion

Historically, drug development has focused on developing strategies that will target the Achilles heel of tumors. Strategies with a track record of success in the clinic include targeting rapidly dividing cells with DNA-damaging agents or radiation. More recently, focus has centered on targeting oncogene addiction intrinsic to the tumor cells. However, emerging evidence indicates tumors are very efficient at using opportunistic signals provided by the microenvironment, and in fact, current therapies may select for cells more fit to utilize survival signals contained within the microenvironment allowing for survival during exposure to therapy. The challenge with targeting the microenvironment will be to a) determine redundancy and delineate whether multiple targets will need to be inhibited and b) determine patient variability contained within the tumor microenvironment and whether a personalized medicine approach will be required for targeting the tumor microenvironment. Targeting the microenvironment will present challenges due to the complexity and the inherent ability of tumors to provide adaptive signals that continue to inform the surrounding niche to favor tumor progression in a dynamic rather than static fashion. However, it is clear that target validation and drug discovery will need to consider both the intrinsic oncogenic signals and the tumor microenvironment for increasing the efficacy of standard therapy in cancer indications that are currently incurable.

Acknowledgments We are grateful to Deepa G Rathod for her assistance in preparation of the tables & figures.

References

1. Gottesman MM, Ling V (2006) The molecular basis of multidrug resistance in cancer: the early years of P-glycoprotein research. FEBS Lett 580:998–1009
2. Buchdunger E, Zimmermann J, Mett H, Meyer T, Muller M, Druker BJ, Lydon NB (1996) Inhibition of the Abl protein-tyrosine kinase in vitro and in vivo by a 2-phenylaminopyrimidine derivative. Cancer Res 56:100–104
3. Druker BJ, Tamura S, Buchdunger E, Ohno S, Segal GM, Fanning S, Zimmermann J, Lydon NB (1996) Effects of a selective inhibitor of the Abl tyrosine kinase on the growth of Bcr-Abl positive cells. Nat Med 2:561–566
4. Carlesso N, Griffin JD, Druker BJ (1994) Use of a temperature-sensitive mutant to define the biological effects of the p210BCR-ABL tyrosine kinase on proliferation of a factor-dependent murine myeloid cell line. Oncogene 9:149–156
5. Zhang X, Ren R (1998) Bcr-Abl efficiently induces a myeloproliferative disease and production of excess interleukin-3 and granulocyte-macrophage colony-stimulating factor in mice: a novel model for chronic myelogenous leukemia. Blood 92:3829–3840
6. Druker BJ, Guilhot F, O'Brien SG, Gathmann I, Kantarjian H, Gattermann N, Deininger MW, Silver RT, Goldman JM, Stone RM et al (2006) Five-year follow-up of patients receiving imatinib for chronic myeloid leukemia. N Engl J Med 355:2408–2417
7. Shah NP, Nicoll JM, Nagar B, Gorre ME, Paquette RL, Kuriyan J, Sawyers CL (2002) Multiple BCR-ABL kinase domain mutations confer polyclonal resistance to the tyrosine kinase inhibitor imatinib (STI571) in chronic phase and blast crisis chronic myeloid leukemia. Cancer Cell 2:117–125
8. O'Hare T, Shakespeare WC, Zhu X, Eide CA, Rivera VM, Wang F, Adrian LT, Zhou T, Huang WS, Xu Q et al (2009) AP24534, a pan-BCR-ABL inhibitor for chronic myeloid leukemia, potently inhibits the T315I mutant and overcomes mutation-based resistance. Cancer Cell 16:401–412
9. Nazarian R, Shi H, Wang Q, Kong X, Koya RC, Lee H, Chen Z, Lee MK, Attar N, Sazegar H et al (2010) Melanomas acquire resistance to B-RAF(V600E) inhibition by RTK or N-RAS upregulation. Nature 468:973–977
10. Kobayashi S, Boggon TJ, Dayaram T, Janne PA, Kocher O, Meyerson M, Johnson BE, Eck MJ, Tenen DG, Halmos B (2005) EGFR mutation and resistance of non-small-cell lung cancer to gefitinib. N Engl J Med 352:786–792
11. Flaherty KT, Puzanov I, Kim KB, Ribas A, McArthur GA, Sosman JA, O'Dwyer PJ, Lee RJ, Grippo JF, Nolop K et al (2010) Inhibition of mutated, activated BRAF in metastatic melanoma. N Engl J Med 363:809–819

12. Poulikakos PI, Rosen N (2011) Mutant BRAF melanomas–dependence and resistance. Cancer Cell 19:11–15
13. Hanahan D, Coussens LM (2012) Accessories to the crime: functions of cells recruited to the tumor microenvironment. Cancer Cell 21:309–322
14. Meads MB, Gatenby RA, Dalton WS (2009) Environment-mediated drug resistance: a major contributor to minimal residual disease. Nat Rev Cancer 9:665–674
15. Pittenger MF, Mackay AM, Beck SC, Jaiswal RK, Douglas R, Mosca JD, Moorman MA, Simonetti DW, Craig S, Marshak DR (1999) Multilineage potential of adult human mesenchymal stem cells. Science 284:143–147
16. Bewry NN, Nair RR, Emmons MF, Boulware D, Pinilla-Ibarz J, Hazlehurst LA (2008) Stat3 contributes to resistance toward BCR-ABL inhibitors in a bone marrow microenvironment model of drug resistance. Mol Cancer Ther 7:3169–3175
17. Nair RR, Tolentino J, Hazlehurst LA (2010) The bone marrow microenvironment as a sanctuary for minimal residual disease in CML. Biochem Pharmacol 80(5):602–612
18. Nefedova Y, Landowski TH, Dalton WS (2003) Bone marrow stromal-derived soluble factors and direct cell contact contribute to de novo drug resistance of myeloma cells by distinct mechanisms. Leukemia 17:1175–1182
19. Lagneaux L, Delforge A, De Bruyn C, Bernier M, Bron D (1999) Adhesion to bone marrow stroma inhibits apoptosis of chronic lymphocytic leukemia cells. Leuk Lymphoma 35:445–453
20. Nair RR, Tolentino JH, Argilagos RF, Zhang L, Pinilla-Ibarz J, Hazlehurst LA (2012) Potentiation of Nilotinib-mediated cell death in the context of the bone marrow microenvironment requires a promiscuous JAK inhibitor in CML. Leuk Res 36:756–763
21. Hazlehurst LA, Valkov N, Wisner L, Storey JA, Boulware D, Sullivan DM, Dalton WS (2001) Reduction in drug-induced DNA double-strand breaks associated with beta1 integrin-mediated adhesion correlates with drug resistance in U937 cells. Blood 98:1897–1903
22. Hoyt DG, Rusnak JM, Mannix RJ, Modzelewski RA, Johnson CS, Lazo JS (1996) Integrin activation suppresses etoposide-induced DNA strand breakage in cultured murine tumor-derived endothelial cells. Cancer Res 56:4146–4149
23. Ihle JN (2001) The stat family in cytokine signaling. Curr Opin Cell Biol 13:211–217
24. Herrington J, Smit LS, Schwartz J, Carter-Su C (2000) The role of STAT proteins in growth hormone signaling. Oncogene 19:2585–2597
25. Reddy EP, Korapati A, Chaturvedi P, Rane S (2000) IL-3 signaling and the role of Src kinases, JAKs and STATs: a covert liaison unveiled. Oncogene 19:2532–2547
26. Marrero MB, Schieffer B, Paxton WG, Heerdt L, Berk BC, Delafontaine P, Bernstein KE (1995) Direct stimulation of Jak/STAT pathway by the angiotensin II AT1 receptor. Nature 375:247–250
27. Shain KH, Yarde DN, Meads MB, Huang M, Jove R, Hazlehurst LA, Dalton WS (2009) Beta1 integrin adhesion enhances IL-6-mediated STAT3 signaling in myeloma cells: implications for microenvironment influence on tumor survival and proliferation. Cancer Res 69:1009–1015
28. Epling-Burnette PK, Liu JH, Catlett-Falcone R, Turkson J, Oshiro M, Kothapalli R, Li Y, Wang JM, Yang-Yen HF, Karras J et al (2001) Inhibition of STAT3 signaling leads to apoptosis of leukemic large granular lymphocytes and decreased Mcl-1 expression. J Clin Invest 107:351–362
29. Gritsko T, Williams A, Turkson J, Kaneko S, Bowman T, Huang M, Nam S, Eweis I, Diaz N, Sullivan D et al (2006) Persistent activation of stat3 signaling induces survivin gene expression and confers resistance to apoptosis in human breast cancer cells. Clin Cancer Res 12:11–19
30. Cuevas P, Diaz-Gonzalez D, Sanchez I, Lozano RM, Gimenez-Gallego G, Dujovny M (2006) Dobesilate inhibits the activation of signal transducer and activator of transcription 3, and the expression of cyclin D1 and bcl-XL in glioma cells. Neurol Res 28:127–130
31. Bhattacharya S, Ray RM, Johnson LR (2005) STAT3-mediated transcription of Bcl-2, Mcl-1 and c-IAP2 prevents apoptosis in polyamine-depleted cells. Biochem J 392:335–344

32. Cox D, Brennan M, Moran N (2010) Integrins as therapeutic targets: lessons and opportunities. Nat Rev Drug Discov 9:804–820
33. Heinrich PC, Behrmann I, Muller-Newen G, Schaper F, Graeve L (1998) Interleukin-6-type cytokine signalling through the gp130/Jak/STAT pathway. Biochem J 334(Pt 2):297–314
34. Quesnelle KM, Boehm AL, Grandis JR (2007) STAT-mediated EGFR signaling in cancer. J Cell Biochem 102:311–319
35. Heinrich PC, Behrmann I, Haan S, Hermanns HM, Muller-Newen G, Schaper F (2003) Principles of interleukin (IL)-6-type cytokine signalling and its regulation. Biochem J 374:1–20
36. Shuai K, Stark GR, Kerr IM, Darnell JE Jr (1993) A single phosphotyrosine residue of Stat91 required for gene activation by interferon-gamma. Science 261:1744–1746
37. Shuai K, Horvath CM, Huang LH, Qureshi SA, Cowburn D, Darnell JE Jr (1994) Interferon activation of the transcription factor Stat91 involves dimerization through SH2-phosphotyrosyl peptide interactions. Cell 76:821–828
38. Ara T, Song L, Shimada H, Keshelava N, Russell HV, Metelitsa LS, Groshen SG, Seeger RC, DeClerck YA (2009) Interleukin-6 in the bone marrow microenvironment promotes the growth and survival of neuroblastoma cells. Cancer Res 69:329–337
39. Catlett-Falcone R, Landowski TH, Oshiro MM, Turkson J, Levitzki A, Savino R, Ciliberto G, Moscinski L, Fernandez-Luna JL, Nunez G et al (1999) Constitutive activation of Stat3 signaling confers resistance to apoptosis in human U266 myeloma cells. Immunity 10:105–115
40. Monaghan KA, Khong T, Burns CJ, Spencer A (2011) The novel JAK inhibitor CYT387 suppresses multiple signalling pathways, prevents proliferation and induces apoptosis in phenotypically diverse myeloma cells. Leukemia 25:1891–1899
41. Hunsucker SA, Magarotto V, Kuhn DJ, Kornblau SM, Wang M, Weber DM, Thomas SK, Shah JJ, Voorhees PM, Xie H et al (2011) Blockade of interleukin-6 signalling with siltuximab enhances melphalan cytotoxicity in preclinical models of multiple myeloma. Br J Haematol 152:579–592
42. Burger R, Le Gouill S, Tai YT, Shringarpure R, Tassone P, Neri P, Podar K, Catley L, Hideshima T, Chauhan D et al (2009) Janus kinase inhibitor INCB20 has antiproliferative and apoptotic effects on human myeloma cells in vitro and in vivo. Mol Cancer Ther 8:26–35
43. Li J, Favata M, Kelley JA, Caulder E, Thomas B, Wen X, Sparks RB, Arvanitis A, Rogers JD, Combs AP et al (2010) INCB16562, a JAK1/2 selective inhibitor, is efficacious against multiple myeloma cells and reverses the protective effects of cytokine and stromal cell support. Neoplasia 12:28–38
44. Honemann D, Chatterjee M, Savino R, Bommert K, Burger R, Gramatzki M, Dorken B, Bargou RC (2001) The IL-6 receptor antagonist SANT-7 overcomes bone marrow stromal cell-mediated drug resistance of multiple myeloma cells. Int J Cancer 93:674–680
45. Hu M, Yao J, Cai L, Bachman KE, van den Brule F, Velculescu V, Polyak K (2005) Distinct epigenetic changes in the stromal cells of breast cancers. Nat Genet 37:899–905
46. Wang W, Li Q, Yamada T, Matsumoto K, Matsumoto I, Oda M, Watanabe G, Kayano Y, Nishioka Y, Sone S et al (2009) Crosstalk to stromal fibroblasts induces resistance of lung cancer to epidermal growth factor receptor tyrosine kinase inhibitors. Clin Cancer Res 15:6630–6638
47. Martinez-Outschoorn UE, Goldberg A, Lin Z, Ko YH, Flomenberg N, Wang C, Pavlides S, Pestell RG, Howell A, Sotgia F et al (2011) Anti-estrogen resistance in breast cancer is induced by the tumor microenvironment and can be overcome by inhibiting mitochondrial function in epithelial cancer cells. Cancer Biol Ther 12:924–938
48. Bates GJ, Fox SB, Han C, Leek RD, Garcia JF, Harris AL, Banham AH (2006) Quantification of regulatory T cells enables the identification of high-risk breast cancer patients and those at risk of late relapse. J Clin Oncol 24:5373–5380
49. Sasada T, Kimura M, Yoshida Y, Kanai M, Takabayashi A (2003) CD4+CD25+ regulatory T cells in patients with gastrointestinal malignancies: possible involvement of regulatory T cells in disease progression. Cancer 98:1089–1099

50. Gabrilovich DI, Nagaraj S (2009) Myeloid-derived suppressor cells as regulators of the immune system. Nat Rev Immunol 9:162–174
51. Mazzoni A, Bronte V, Visintin A, Spitzer JH, Apolloni E, Serafini P, Zanovello P, Segal DM (2002) Myeloid suppressor lines inhibit T cell responses by an NO-dependent mechanism. J Immunol 168:689–695
52. Lu T, Ramakrishnan R, Altiok S, Youn JI, Cheng P, Celis E, Pisarev V, Sherman S, Sporn MB, Gabrilovich D (2011) Tumor-infiltrating myeloid cells induce tumor cell resistance to cytotoxic T cells in mice. J Clin Invest 121:4015–4029
53. Shojaei F, Wu X, Malik AK, Zhong C, Baldwin ME, Schanz S, Fuh G, Gerber HP, Ferrara N (2007) Tumor refractoriness to anti-VEGF treatment is mediated by CD11b+Gr1+ myeloid cells. Nat Biotechnol 25:911–920
54. Schmid MC, Avraamides CJ, Foubert P, Shaked Y, Kang SW, Kerbel RS, Varner JA (2011) Combined Blockade of Integrin-alpha4beta1 Plus Cytokines SDF-1alpha or IL-1beta Potently Inhibits Tumor Inflammation and Growth. Cancer Res 71:6965–6975
55. Du R, Lu KV, Petritsch C, Liu P, Ganss R, Passegue E, Song H, Vandenberg S, Johnson RS, Werb Z et al (2008) HIF1alpha induces the recruitment of bone marrow-derived vascular modulatory cells to regulate tumor angiogenesis and invasion. Cancer Cell 13:206–220
56. Alam N, Goel HL, Zarif MJ, Butterfield JE, Perkins HM, Sansoucy BG, Sawyer TK, Languino LR (2007) The integrin-growth factor receptor duet. J Cell Physiol 213:649–653
57. Bergers G, Hanahan D (2008) Modes of resistance to anti-angiogenic therapy. Nat Rev Cancer 8:592–603
58. Takahashi M, Tsuda T, Ikeda Y, Honke K, Taniguchi N (2004) Role of N-glycans in growth factor signaling. Glycoconj J 20:207–212
59. Waetzig GH, Chalaris A, Rosenstiel P, Suthaus J, Holland C, Karl N, Valles Uriarte L, Till A, Scheller J, Grotzinger J et al (2010) N-linked glycosylation is essential for the stability but not the signaling function of the interleukin-6 signal transducer glycoprotein 130. J Biol Chem 285:1781–1789
60. Barondes SH, Cooper DN, Gitt MA, Leffler H (1994) Galectins. Structure and function of a large family of animal lectins. J Biol Chem 269:20807–20810
61. Brewer CF, Miceli MC, Baum LG (2002) Clusters, bundles, arrays and lattices: novel mechanisms for lectin-saccharide-mediated cellular interactions. Curr Opin Struct Biol 12:616–623
62. Sacchettini JC, Baum LG, Brewer CF (2001) Multivalent protein-carbohydrate interactions. A new paradigm for supermolecular assembly and signal transduction. Biochemistry 40:3009–3015
63. Collins BE, Paulson JC (2004) Cell surface biology mediated by low affinity multivalent protein-glycan interactions. Curr Opin Chem Biol 8:617–625
64. Jin L, Hope KJ, Zhai Q, Smadja-Joffe F, Dick JE (2006) Targeting of CD44 eradicates human acute myeloid leukemic stem cells. Nat Med 12:1167–1174
65. Varki A, Kannagi R, Toole BP (2009) Glycosylation changes in cancer. In: Varki A, Cummings RD, Esko JD, Freeze HH, Stanley P, Bertozzi CR, Hart GW, Etzler ME (eds.) Essentials of glycobiology. Cold Spring Harbor, NY
66. Hirabayashi J, Hashidate T, Arata Y, Nishi N, Nakamura T, Hirashima M, Urashima T, Oka T, Futai M, Muller WE et al (2002) Oligosaccharide specificity of galectins: a search by frontal affinity chromatography. Biochim Biophys Acta 1572:232–254
67. Granovsky M, Fata J, Pawling J, Muller WJ, Khokha R, Dennis JW (2000) Suppression of tumor growth and metastasis in Mgat5-deficient mice. Nat Med 6:306–312
68. Partridge EA, Le Roy C, Di Guglielmo GM, Pawling J, Cheung P, Granovsky M, Nabi IR, Wrana JL, Dennis JW (2004) Regulation of cytokine receptors by Golgi N-glycan processing and endocytosis. Science 306:120–124
69. Lajoie P, Partridge EA, Guay G, Goetz JG, Pawling J, Lagana A, Joshi B, Dennis JW, Nabi IR (2007) Plasma membrane domain organization regulates EGFR signaling in tumor cells. J Cell Biol 179:341–356

70. Daniels MA, Hogquist KA, Jameson SC (2002) Sweet 'n' sour: the impact of differential glycosylation on T cell responses. Nat Immunol 3:903–910
71. Barondes SH, Castronovo V, Cooper DN, Cummings RD, Drickamer K, Feizi T, Gitt MA, Hirabayashi J, Hughes C, Kasai K et al (1994) Galectins: a family of animal beta-galactoside-binding lectins. Cell 76:597–598
72. Liao DI, Kapadia G, Ahmed H, Vasta GR, Herzberg O (1994) Structure of S-lectin, a developmentally regulated vertebrate beta-galactoside-binding protein. Proc Nat Acad Sci USA 91:1428–1432
73. Hirabayashi J, Kasai K (1993) The family of metazoan metal-independent beta-galactoside-binding lectins: structure, function and molecular evolution. Glycobiology 3:297–304
74. Houzelstein D, Goncalves IR, Fadden AJ, Sidhu SS, Cooper DN, Drickamer K, Leffler H, Poirier F (2004) Phylogenetic analysis of the vertebrate galectin family. Mol Biol Evol 21:1177–1187
75. Di Lella S, Sundblad V, Cerliani JP, Guardia CM, Estrin DA, Vasta GR, Rabinovich GA (2011) When galectins recognize glycans: from biochemistry to physiology and back again. Biochemistry 50:7842–7857
76. de Waard A, Hickman S, Kornfeld S (1976) Isolation and properties of beta-galactoside binding lectins of calf heart and lung. J Biol Chem 251:7581–7587
77. Paz A, Haklai R, Elad-Sfadia G, Ballan E, Kloog Y (2001) Galectin-1 binds oncogenic H-Ras to mediate Ras membrane anchorage and cell transformation. Oncogene 20:7486–7493
78. Ellerhorst J, Nguyen T, Cooper DN, Lotan D, Lotan R (1999) Differential expression of endogenous galectin-1 and galectin-3 in human prostate cancer cell lines and effects of overexpressing galectin-1 on cell phenotype. Int J Oncol 14:217–224
79. van den Brule F, Califice S, Garnier F, Fernandez PL, Berchuck A, Castronovo V (2003) Galectin-1 accumulation in the ovary carcinoma peritumoral stroma is induced by ovary carcinoma cells and affects both cancer cell proliferation and adhesion to laminin-1 and fibronectin. Lab Invest J Tech Methods Pathol 83:377–386
80. Clausse N, van den Brule F, Waltregny D, Garnier F, Castronovo V (1999) Galectin-1 expression in prostate tumor-associated capillary endothelial cells is increased by prostate carcinoma cells and modulates heterotypic cell-cell adhesion. Angiogenesis 3:317–325
81. Horiguchi N, Arimoto K, Mizutani A, Endo-Ichikawa Y, Nakada H, Taketani S (2003) Galectin-1 induces cell adhesion to the extracellular matrix and apoptosis of non-adherent human colon cancer Colo201 cells. J Biochem 134:869–874
82. Lippert E, Gunckel M, Brenmoehl J, Bataille F, Falk W, Scholmerich J, Obermeier F, Rogler G (2008) Regulation of galectin-3 function in mucosal fibroblasts: potential role in mucosal inflammation. Clin Exp Immunol 152:285–297
83. Park SH, Min HS, Kim B, Myung J, Paek SH (2008) Galectin-3: a useful biomarker for differential diagnosis of brain tumors. Neuropathol: Official J Jpn Soc Neuropathol 28:497–506
84. Jiang HR, Al Rasebi Z, Mensah-Brown E, Shahin A, Xu D, Goodyear CS, Fukada SY, Liu FT, Liew FY, Lukic ML (2009) Galectin-3 deficiency reduces the severity of experimental autoimmune encephalomyelitis. J Immunol 182:1167–1173
85. Jeon SB, Yoon HJ, Chang CY, Koh HS, Jeon SH, Park EJ (2010) Galectin-3 exerts cytokine-like regulatory actions through the JAK-STAT pathway. J Immunol 185:7037–7046
86. Wan SY, Zhang TF, Ding Y (2011) Galectin-3 enhances proliferation and angiogenesis of endothelial cells differentiated from bone marrow mesenchymal stem cells. Transpl Proc 43:3933–3938
87. Chauhan D, Li G, Podar K, Hideshima T, Neri P, He D, Mitsiades N, Richardson P, Chang Y, Schindler J et al (2005) A novel carbohydrate-based therapeutic GCS-100 overcomes bortezomib resistance and enhances dexamethasone-induced apoptosis in multiple myeloma cells. Cancer Res 65:8350–8358
88. Streetly MJ, Maharaj L, Joel S, Schey SA, Gribben JG, Cotter FE (2010) GCS-100, a novel galectin-3 antagonist, modulates MCL-1, NOXA, and cell cycle to induce myeloma cell death. Blood 115:3939–3948

89. Cheng YL, Huang WC, Chen CL, Tsai CC, Wang CY, Chiu WH, Chen YL, Lin YS, Chang CF, Lin CF (2011) Increased galectin-3 facilitates leukemia cell survival from apoptotic stimuli. Biochem Biophys Res Commun 412:334–340
90. Yamamoto-Sugitani M, Kuroda J, Ashihara E, Nagoshi H, Kobayashi T, Matsumoto Y, Sasaki N, Shimura Y, Kiyota M, Nakayama R et al (2011) Galectin-3 (Gal-3) induced by leukemia microenvironment promotes drug resistance and bone marrow lodgment in chronic myelogenous leukemia. Proc Nat Acad Sci USA 108:17468–17473
91. Kageshita T, Kashio Y, Yamauchi A, Seki M, Abedin MJ, Nishi N, Shoji H, Nakamura T, Ono T, Hirashima M (2002) Possible role of galectin-9 in cell aggregation and apoptosis of human melanoma cell lines and its clinical significance. Int J Cancer J Int du Cancer 99:809–816
92. Lu LH, Nakagawa R, Kashio Y, Ito A, Shoji H, Nishi N, Hirashima M, Yamauchi A, Nakamura T (2007) Characterization of galectin-9-induced death of Jurkat T cells. J Biochem 141:157–172
93. Kuroda J, Yamamoto M, Nagoshi H, Kobayashi T, Sasaki N, Shimura Y, Horiike S, Kimura S, Yamauchi A, Hirashima M et al (2010) Targeting activating transcription factor 3 by Galectin-9 induces apoptosis and overcomes various types of treatment resistance in chronic myelogenous leukemia. Mol Cancer Res: MCR 8:994–1001
94. Kobayashi T, Kuroda J, Ashihara E, Oomizu S, Terui Y, Taniyama A, Adachi S, Takagi T, Yamamoto M, Sasaki N et al. (2010) Galectin-9 exhibits anti-myeloma activity through JNK and p38 MAP kinase pathways. Leuk: Official J Leuk Soc Am Leuk Res Fund UK 24:843–850
95. Cipolleschi MG, Dello Sbarba P, Olivotto M (1993) The role of hypoxia in the maintenance of hematopoietic stem cells. Blood 82:2031–2037
96. Wohlkoenig C, Leithner K, Deutsch A, Hrzenjak A, Olschewski A, Olschewski H (2011) Hypoxia-induced cisplatin resistance is reversible and growth rate independent in lung cancer cells. Cancer Lett 308:134–143
97. Tanturli M, Giuntoli S, Barbetti V, Rovida E, Dello Sbarba P (2011) Hypoxia selects bortezomib-resistant stem cells of chronic myeloid leukemia. PLoS ONE 6:e17008
98. Giuntoli S, Rovida E, Barbetti V, Cipolleschi MG, Olivotto M, Dello Sbarba P (2006) Hypoxia suppresses BCR/Abl and selects imatinib-insensitive progenitors within clonal CML populations. Leukemia 20:1291–1293
99. Graeber TG, Osmanian C, Jacks T, Housman DE, Koch CJ, Lowe SW, Giaccia AJ (1996) Hypoxia-mediated selection of cells with diminished apoptotic potential in solid tumours. Nature 379:88–91
100. Bristow RG, Hill RP (2008) Hypoxia and metabolism. Hypoxia, DNA repair and genetic instability. Nat Rev Cancer 8:180–192
101. Tang N, Wang L, Esko J, Giordano FJ, Huang Y, Gerber HP, Ferrara N, Johnson RS (2004) Loss of HIF-1alpha in endothelial cells disrupts a hypoxia-driven VEGF autocrine loop necessary for tumorigenesis. Cancer Cell 6:485–495
102. Pennacchietti S, Michieli P, Galluzzo M, Mazzone M, Giordano S, Comoglio PM (2003) Hypoxia promotes invasive growth by transcriptional activation of the met protooncogene. Cancer Cell 3:347–361
103. Lukashev D, Klebanov B, Kojima H, Grinberg A, Ohta A, Berenfeld L, Wenger RH, Sitkovsky M (2006) Cutting edge: hypoxia-inducible factor 1alpha and its activation-inducible short isoform I.1 negatively regulate functions of CD4+ and CD8+ T lymphocytes. J Immunol 177:4962–4965
104. Masson N, Willam C, Maxwell PH, Pugh CW, Ratcliffe PJ (2001) Independent function of two destruction domains in hypoxia-inducible factor-alpha chains activated by prolyl hydroxylation. EMBO J 20:5197–5206
105. Jaakkola P, Mole DR, Tian YM, Wilson MI, Gielbert J, Gaskell SJ, Kriegsheim A, Hebestreit HF, Mukherji M, Schofield CJ et al (2001) Targeting of HIF-alpha to the von Hippel-Lindau ubiquitylation complex by O2-regulated prolyl hydroxylation. Science 292:468–472

106. Mole DR, Blancher C, Copley RR, Pollard PJ, Gleadle JM, Ragoussis J, Ratcliffe PJ (2009) Genome-wide association of hypoxia-inducible factor (HIF)-1alpha and HIF-2alpha DNA binding with expression profiling of hypoxia-inducible transcripts. J Biol Chem 284:16767–16775
107. Weiss GJ, Infante JR, Chiorean EG, Borad MJ, Bendell JC, Molina JR, Tibes R, Ramanathan RK, Lewandowski K, Jones SF et al (2011) Phase 1 study of the safety, tolerability, and pharmacokinetics of TH-302, a hypoxia-activated prodrug, in patients with advanced solid malignancies. Clin Cancer Res 17:2997–3004
108. Romero-Ramirez L, Cao H, Nelson D, Hammond E, Lee AH, Yoshida H, Mori K, Glimcher LH, Denko NC, Giaccia AJ et al (2004) XBP1 is essential for survival under hypoxic conditions and is required for tumor growth. Cancer Res 64:5943–5947
109. Papandreou I, Denko NC, Olson M, Van Melckebeke H, Lust S, Tam A, Solow-Cordero DE, Bouley DM, Offner F, Niwa M et al (2011) Identification of an Ire1alpha endonuclease specific inhibitor with cytotoxic activity against human multiple myeloma. Blood 117:1311–1314
110. Volkmann K, Lucas JL, Vuga D, Wang X, Brumm D, Stiles C, Kriebel D, Der-Sarkissian A, Krishnan K, Schweitzer C et al (2011) Potent and selective inhibitors of the inositol-requiring enzyme 1 endoribonuclease. J Biol Chem 286:12743–12755
111. Fels DR, Ye J, Segan AT, Kridel SJ, Spiotto M, Olson M, Koong AC, Koumenis C (2008) Preferential cytotoxicity of bortezomib toward hypoxic tumor cells via overactivation of endoplasmic reticulum stress pathways. Cancer Res 68:9323–9330
112. Damiano JS, Cress AE, Hazlehurst LA, Shtil AA, Dalton WS (1999) Cell adhesion mediated drug resistance (CAM-DR): role of integrins and resistance to apoptosis in human myeloma cell lines. Blood 93:1658–1667
113. Emmons MF, Gebhard AW, Nair RR, Baz R, McLaughlin M, Cress AE, Hazlehurst LA (2011) Acquisition of resistance towards HYD1 correlates with a reduction in cleaved alpha 4 integrin expression and a compromised CAM-DR phenotype. Mol Cancer Ther 10(12):2257–2266
114. Lu X, Mu E, Wei Y, Riethdorf S, Yang Q, Yuan M, Yan J, Hua Y, Tiede BJ, Haffty BG et al (2011) VCAM-1 promotes osteolytic expansion of indolent bone micrometastasis of breast cancer by engaging alpha4beta1-positive osteoclast progenitors. Cancer Cell 20:701–714
115. Abe M, Hiura K, Ozaki S, Kido S, Matsumoto T (2009) Vicious cycle between myeloma cell binding to bone marrow stromal cells via VLA-4-VCAM-1 adhesion and macrophage inflammatory protein-1alpha and MIP-1beta production. J Bone Miner Metab 27:16–23
116. Chen Q, Lin TH, Der CJ, Juliano RL (1996) Integrin-mediated activation of MEK and mitogen-activated protein kinase is independent of Ras (corrected). J Biol Chem 271:18122–18127
117. Lee JW, Juliano RL (2002) The alpha5beta1 integrin selectively enhances epidermal growth factor signaling to the phosphatidylinositol-3-kinase/Akt pathway in intestinal epithelial cells. Biochim Biophys Acta 1542:23
118. Bill HM, Knudsen B, Moores SL, Muthuswamy SK, Rao VR, Brugge JS, Miranti CK (2004) Epidermal growth factor receptor-dependent regulation of integrin-mediated signaling and cell cycle entry in epithelial cells. Mol Cell Biol 24:8586–8599
119. Xiao T, Takagi J, Coller BS, Wang JH, Springer TA (2004) Structural basis for allostery in integrins and binding to fibrinogen-mimetic therapeutics. Nature 432:59–67
120. Xiong JP, Stehle T, Diefenbach B, Zhang R, Dunker R, Scott DL, Joachimiak A, Goodman SL, Arnaout MA (2001) Crystal structure of the extracellular segment of integrin alpha Vbeta3. Science 294:339–345
121. Rosano C, Rocco M (2010) Solution properties of full-length integrin alpha(IIb)beta3 refined models suggest environment-dependent induction of alternative bent/extended resting states. FEBS J 277:3190–3202
122. Lu C, Takagi J, Springer TA (2001) Association of the membrane proximal regions of the alpha and beta subunit cytoplasmic domains constrains an integrin in the inactive state. J Biol Chem 276:14642–14648

123. Luo BH, Carman CV, Springer TA (2007) Structural basis of integrin regulation and signaling. Annu Rev Immunol 25:619–647
124. Takagi J, Erickson HP, Springer TA (2001) C-terminal opening mimics 'inside-out' activation of integrin alpha5beta1. Nat Struct Biol 8:412–416
125. Calderwood DA, Zent R, Grant R, Rees DJ, Hynes RO, Ginsberg MH (1999) The Talin head domain binds to integrin beta subunit cytoplasmic tails and regulates integrin activation. J Biol Chem 274:28071–28074
126. Anthis NJ, Haling JR, Oxley CL, Memo M, Wegener KL, Lim CJ, Ginsberg MH, Campbell ID (2009) Beta integrin tyrosine phosphorylation is a conserved mechanism for regulating talin-induced integrin activation. J Biol Chem 284:36700–36710
127. Critchley DR, Gingras AR (2008) Talin at a glance. J Cell Sci 121:1345–1347
128. Arai A, Nosaka Y, Kanda E, Yamamoto K, Miyasaka N, Miura O (2001) Rap1 is activated by erythropoietin or interleukin-3 and is involved in regulation of beta1 integrin-mediated hematopoietic cell adhesion. J Biol Chem 276:10453–10462
129. Hazlehurst LA, Enkemann SA, Beam CA, Argilagos RF, Painter J, Shain KH, Saporta S, Boulware D, Moscinski L, Alsina M et al (2003) Genotypic and phenotypic comparisons of de novo and acquired melphalan resistance in an isogenic multiple myeloma cell line model. Cancer Res 63:7900–7906
130. Rice GP, Hartung HP, Calabresi PA (2005) Anti-alpha4 integrin therapy for multiple sclerosis: mechanisms and rationale. Neurology 64:1336–1342
131. Targan SR, Feagan BG, Fedorak RN, Lashner BA, Panaccione R, Present DH, Spehlmann ME, Rutgeerts PJ, Tulassay Z, Volfova M et al (2007) Natalizumab for the treatment of active Crohn's disease: results of the ENCORE Trial. Gastroenterology 132:1672–1683
132. Matsunaga T, Takemoto N, Sato T, Takimoto R, Tanaka I, Fujimi A, Akiyama T, Kuroda H, Kawano Y, Kobune M et al (2003) Interaction between leukemic-cell VLA-4 and stromal fibronectin is a decisive factor for minimal residual disease of acute myelogenous leukemia. Nat Med 9:1158–1165
133. Olson DL, Burkly LC, Leone DR, Dolinski BM, Lobb RR (2005) Anti-alpha4 integrin monoclonal antibody inhibits multiple myeloma growth in a murine model. Mol Cancer Ther 4:91–99
134. Pierschbacher MD, Ruoslahti E (1984) Cell attachment activity of fibronectin can be duplicated by small synthetic fragments of the molecule. Nature 309:30–33
135. Gould RJ, Polokoff MA, Friedman PA, Huang TF, Holt JC, Cook JJ, Niewiarowski S (1990) Disintegrins: a family of integrin inhibitory proteins from viper venoms. Proc Soc Exp Biol Med 195:168–171
136. Lu X, Lu D, Scully MF, Kakkar VV (2006) Integrins in drug targeting-RGD templates in toxins. Curr Pharm Des 12:2749–2769
137. DeRoock IB, Pennington ME, Sroka TC, Lam KS, Bowden GT, Bair EL, Cress AE (2001) Synthetic peptides inhibit adhesion of human tumor cells to extracellular matrix proteins. Cancer Res 61:3308–3313
138. Pennington ME, Lam KS, Cress AE (1996) The use of a combinatorial library method to isolate human tumor cell adhesion peptides. Mol Divers 2:19–28
139. Sroka TC, Pennington ME, Cress AE (2006) Synthetic D-amino acid peptide inhibits tumor cell motility on laminin-5. Carcinogenesis 27(9):1748–1757
140. Nair RR, Emmons MF, Cress AE, Argilagos RF, Lam K, Kerr WT, Wang HG, Dalton WS, Hazlehurst LA (2009) HYD1-induced increase in reactive oxygen species leads to autophagy and necrotic cell death in multiple myeloma cells. Mol Cancer Ther 8:2441–2451
141. Mas-Moruno C, Rechenmacher F, Kessler H (2010) Cilengitide: the first anti-angiogenic small molecule drug candidate design, synthesis and clinical evaluation. Anticancer Agents Med Chem 10:753–768
142. Doedens L, Opperer F, Cai M, Beck JG, Dedek M, Palmer E, Hruby VJ, Kessler H (2010) Multiple N-methylation of MT-II backbone amide bonds leads to melanocortin receptor subtype hMC1R selectivity: pharmacological and conformational studies. J Am Chem Soc 132:8115–8128

143. Reardon DA, Fink KL, Mikkelsen T, Cloughesy TF, O'Neill A, Plotkin S, Glantz M, Ravin P, Raizer JJ, Rich KM et al (2008) Randomized phase II study of cilengitide, an integrin-targeting arginine-glycine-aspartic acid peptide, in recurrent glioblastoma multiforme. J Clin Oncol 26:5610–5617
144. Yamada S, Bu XY, Khankaldyyan V, Gonzales-Gomez I, McComb JG, Laug WE (2006) Effect of the angiogenesis inhibitor Cilengitide (EMD 121974) on glioblastoma growth in nude mice. Neurosurgery 59:1304–1312 (discussion 1312)
145. Basoni C, Reuzeau E, Croft D, Genot E, Kramer IM (2006) CD44 and TGFbeta1 synergise to induce expression of a functional NADPH oxidase in promyelocytic cells. Biochem Biophys Res Commun 343:609–616
146. Khaldoyanidi S, Karakhanova S, Sleeman J, Herrlich P, Ponta H (2002) CD44 variant-specific antibodies trigger hemopoiesis by selective release of cytokines from bone marrow macrophages. Blood 99:3955–3961
147. Mielgo A, Brondani V, Landmann L, Glaser-Ruhm A, Erb P, Stupack D, Gunthert U (2007) The CD44 standard/ezrin complex regulates Fas-mediated apoptosis in Jurkat cells. Apoptosis 12:2051–2061
148. Okada T, Hawley RG, Kodaka M, Okuno H (1999) Significance of VLA-4-VCAM-1 interaction and CD44 for transendothelial invasion in a bone marrow metastatic myeloma model. Clin Exp Metastasis 17:623–629
149. van der Windt GJ, Schouten M, Zeerleder S, Florquin S, van der Poll T CD44 is Protective during hyperoxia-induced lung injury. Am J Respir Cell Mol Biol 44(3):377–383
150. Pure E, Assoian RK (2009) Rheostatic signaling by CD44 and hyaluronan. Cell Signal 21:651–655
151. Gunthert U, Hofmann M, Rudy W, Reber S, Zoller M, Haussmann I, Matzku S, Wenzel A, Ponta H, Herrlich P (1991) A new variant of glycoprotein CD44 confers metastatic potential to rat carcinoma cells. Cell 65:13–24
152. Weber GF, Bronson RT, Ilagan J, Cantor H, Schmits R, Mak TW (2002) Absence of the CD44 gene prevents sarcoma metastasis. Cancer Res 62:2281–2286
153. Asosingh K, Gunthert U, Bakkus MH, De Raeve H, Goes E, Van Riet I, Van Camp B, Vanderkerken K (2000) In vivo induction of insulin-like growth factor-I receptor and CD44v6 confers homing and adhesion to murine multiple myeloma cells. Cancer Res 60:3096–3104
154. Caers J, Gunthert U, De Raeve H, Van Valckenborgh E, Menu E, Van Riet I, Van Camp B, Vanderkerken K (2006) The involvement of osteopontin and its receptors in multiple myeloma cell survival, migration and invasion in the murine 5T33MM model. Br J Haematol 132:469–477
155. Masellis-Smith A, Belch AR, Mant MJ, Pilarski LM (1997) Adhesion of multiple myeloma peripheral blood B cells to bone marrow fibroblasts: a requirement for CD44 and alpha-4beta7. Cancer Res 57:930–936
156. Cichy J, Bals R, Potempa J, Mani A, Pure E (2002) Proteinase-mediated release of epithelial cell-associated CD44. Extracellular CD44 complexes with components of cellular matrices. J Biol Chem 277:44440–44447
157. Cichy J, Kulig P, Pure E (2005) Regulation of the release and function of tumor cell-derived soluble CD44. Biochim Biophys Acta 1745:59–64
158. Astier A, Manie SN, Avraham H, Hirai H, Law SF, Zhang Y, Golemis EA, Fu Y, Druker BJ, Haghayeghi N et al (1997) The related adhesion focal tyrosine kinase differentially phosphorylates p130Cas and the Cas-like protein, p105HEF1. J Biol Chem 272:19719–19724
159. Eisterer W, Bechter O, Soderberg O, Nilsson K, Terol M, Greil R, Thaler J, Herold M, Finke L, Gunthert U et al (2004) Elevated levels of soluble CD44 are associated with advanced disease and in vitro proliferation of neoplastic lymphocytes in B-cell chronic lymphocytic leukaemia. Leuk Res 28:1043–1051
160. Molica S, Vitelli G, Levato D, Giannarelli D, Gandolfo GM (2001) Elevated serum levels of soluble CD44 can identify a subgroup of patients with early B-cell chronic lymphocytic leukemia who are at high risk of disease progression. Cancer 92:713–719

161. Aruffo A, Stamenkovic I, Melnick M, Underhill CB, Seed B (1990) CD44 is the principal cell surface receptor for hyaluronate. Cell 61:1303–1313
162. Goodison S, Urquidi V, Tarin D (1999) CD44 cell adhesion molecules. Mol Pathol 52:189–196
163. Meyer MJ, Fleming JM, Lin AF, Hussnain SA, Ginsburg E, Vonderhaar BK (2010) CD44posCD49fhiCD133/2hi defines xenograft-initiating cells in estrogen receptor-negative breast cancer. Cancer Res 70:4624–4633
164. Stauder R, Van Driel M, Schwarzler C, Thaler J, Lokhorst HM, Kreuser ED, Bloem AC, Gunthert U, Eisterer W (1996) Different CD44 splicing patterns define prognostic subgroups in multiple myeloma. Blood 88:3101–3108
165. Van Driel M, Gunthert U, van Kessel AC, Joling P, Stauder R, Lokhorst HM, Bloem AC (2002) CD44 variant isoforms are involved in plasma cell adhesion to bone marrow stromal cells. Leukemia 16:135–143
166. Bourguignon LY, Gilad E, Brightman A, Diedrich F, Singleton P (2006) Hyaluronan-CD44 interaction with leukemia-associated RhoGEF and epidermal growth factor receptor promotes Rho/Ras co-activation, phospholipase C epsilon-Ca2+ signaling, and cytoskeleton modification in head and neck squamous cell carcinoma cells. J Biol Chem 281:14026–14040
167. Bourguignon LY, Singleton PA, Zhu H, Zhou B (2002) Hyaluronan promotes signaling interaction between CD44 and the transforming growth factor beta receptor I in metastatic breast tumor cells. J Biol Chem 277:39703–39712
168. Lee JL, Wang MJ, Sudhir PR, Chen JY (2008) CD44 engagement promotes matrix-derived survival through the CD44-SRC-integrin axis in lipid rafts. Mol Cell Biol 28:5710–5723
169. Redondo-Munoz J, Ugarte-Berzal E, Garcia-Marco JA, del Cerro MH, Van den Steen PE, Opdenakker G, Terol MJ, Garcia-Pardo A (2008) Alpha4beta1 integrin and 190-kDa CD44v constitute a cell surface docking complex for gelatinase B/MMP-9 in chronic leukemic but not in normal B cells. Blood 112:169–178
170. Nandi A, Estess P, Siegelman M (2004) Bimolecular complex between rolling and firm adhesion receptors required for cell arrest; CD44 association with VLA-4 in T cell extravasation. Immunity 20:455–465
171. Krause DS, Lazarides K, von Andrian UH, Van Etten RA (2006) Requirement for CD44 in homing and engraftment of BCR-ABL-expressing leukemic stem cells. Nat Med 12:1175–1180
172. Thorne RF, Legg JW, Isacke CM (2004) The role of the CD44 transmembrane and cytoplasmic domains in co-ordinating adhesive and signalling events. J Cell Sci 117:373–380
173. Lesley J, Hascall VC, Tammi M, Hyman R (2000) Hyaluronan binding by cell surface CD44. J Biol Chem 275:26967–26975
174. Banerji S, Wright AJ, Noble M, Mahoney DJ, Campbell ID, Day AJ, Jackson DG (2007) Structures of the Cd44-hyaluronan complex provide insight into a fundamental carbohydrate-protein interaction. Nat Struct Mol Biol 14:234–239
175. Ogino S, Nishida N, Umemoto R, Suzuki M, Takeda M, Terasawa H, Kitayama J, Matsumoto M, Hayasaka H, Miyasaka M et al (2010) Two-state conformations in the hyaluronan-binding domain regulate CD44 adhesiveness under flow condition. Structure 18:649–656
176. Wolny PM, Banerji S, Gounou C, Brisson AR, Day AJ, Jackson DG, Richter RP (2010) Analysis of CD44-hyaluronan interactions in an artificial membrane system: insights into the distinct binding properties of high and low molecular weight hyaluronan. J Biol Chem 285:30170–30180
177. Ilangumaran S, Borisch B, Hoessli DC (1999) Signal transduction via CD44: role of plasma membrane microdomains. Leuk Lymphoma 35:455–469
178. Marhaba R, Freyschmidt-Paul P, Zoller M (2006) In vivo CD44-CD49d complex formation in autoimmune disease has consequences on T cell activation and apoptosis resistance. Eur J Immunol 36:3017–3032
179. Verfaillie CM, Benis A, Iida J, McGlave PB, McCarthy JB (1994) Adhesion of committed human hematopoietic progenitors to synthetic peptides from the C-terminal heparin-binding domain of fibronectin: cooperation between the integrin alpha 4 beta 1 and the CD44 adhesion receptor. Blood 84:1802–1811

180. van der Voort R, Taher TE, Wielenga VJ, Spaargaren M, Prevo R, Smit L, David G, Hartmann G, Gherardi E, Pals ST (1999) Heparan sulfate-modified CD44 promotes hepatocyte growth factor/scatter factor-induced signal transduction through the receptor tyrosine kinase c-Met. J Biol Chem 274:6499–6506
181. Ruffell B, Johnson P (2008) Hyaluronan induces cell death in activated T cells through CD44. J Immunol 181:7044–7054
182. Tremmel M, Matzke A, Albrecht I, Laib AM, Olaku V, Ballmer-Hofer K, Christofori G, Heroult M, Augustin HG, Ponta H et al (2009) A CD44v6 peptide reveals a role of CD44 in VEGFR-2 signaling and angiogenesis. Blood 114:5236–5244
183. Piotrowicz RS, Damaj BB, Hachicha M, Incardona F, Howell SB, Finlayson M (2011) A6 peptide activates CD44 adhesive activity, induces FAK and MEK phosphorylation, and inhibits the migration and metastasis of CD44-expressing cells. Mol Cancer Ther 10:2072–2082
184. Riechelmann H, Sauter A, Golze W, Hanft G, Schroen C, Hoermann K, Erhardt T, Gronau S (2008) Phase I trial with the CD44v6-targeting immunoconjugate bivatuzumab mertansine in head and neck squamous cell carcinoma. Oral Oncol 44:823–829
185. Tilghman RW, Parsons JT (2008) Focal adhesion kinase as a regulator of cell tension in the progression of cancer. Semin Cancer Biol 18:45–52
186. Grigera PR, Jeffery ED, Martin KH, Shabanowitz J, Hunt DF, Parsons JT (2005) FAK phosphorylation sites mapped by mass spectrometry. J Cell Sci 118:4931–4935
187. Harte MT, Hildebrand JD, Burnham MR, Bouton AH, Parsons JT (1996) p130Cas, a substrate associated with v-Src and v-Crk, localizes to focal adhesions and binds to focal adhesion kinase. J Biol Chem 271:13649–13655
188. Hildebrand JD, Schaller MD, Parsons JT (1995) Paxillin, a tyrosine phosphorylated focal adhesion-associated protein binds to the carboxyl terminal domain of focal adhesion kinase. Mol Biol Cell 6:637–647
189. Schaller MD, Parsons JT (1995) pp125FAK-dependent tyrosine phosphorylation of paxillin creates a high-affinity binding site for Crk. Mol Cell Biol 15:2635–2645
190. Avraham H, Park SY, Schinkmann K, Avraham S (2000) RAFTK/Pyk2-mediated cellular signalling. Cell Signal 12:123–133
191. Ilic D, Furuta Y, Kanazawa S, Takeda N, Sobue K, Nakatsuji N, Nomura S, Fujimoto J, Okada M, Yamamoto T (1995) Reduced cell motility and enhanced focal adhesion contact formation in cells from FAK-deficient mice. Nature 377:539–544
192. Guinamard R, Okigaki M, Schlessinger J, Ravetch JV (2000) Absence of marginal zone B cells in Pyk-2-deficient mice defines their role in the humoral response. Nat Immunol 1:31–36
193. Buckbinder L, Crawford DT, Qi H, Ke HZ, Olson LM, Long KR, Bonnette PC, Baumann AP, Hambor JE, Grasser WA 3rd et al (2007) Proline-rich tyrosine kinase 2 regulates osteoprogenitor cells and bone formation, and offers an anabolic treatment approach for osteoporosis. Proc Natl Acad Sci USA 104:10619–10624
194. Roberts WG, Ung E, Whalen P, Cooper B, Hulford C, Autry C, Richter D, Emerson E, Lin J, Kath J et al (2008) Antitumor activity and pharmacology of a selective focal adhesion kinase inhibitor, PF-562,271. Cancer Res 68:1935–1944
195. Bagi CM, Roberts GW, Andresen CJ (2008) Dual focal adhesion kinase/Pyk2 inhibitor has positive effects on bone tumors: implications for bone metastases. Cancer 112:2313–2321
196. Infante JR, Camidge DR, Mileshkin LR, Chen EX, Hicks RJ, Rischin D, Fingert H, Pierce KJ, Xu H, Roberts WG et al. (2012) Safety, pharmacokinetic, and pharmacodynamic phase I dose-escalation trial of PF-00562271, an inhibitor of focal adhesion kinase, in advanced solid tumors. J Clin Oncol 30(13):1527–1533
197. Wei LH, Kuo ML, Chen CA, Chou CH, Lai KB, Lee CN, Hsieh CY (2003) Interleukin-6 promotes cervical tumor growth by VEGF-dependent angiogenesis via a STAT3 pathway. Oncogene 22:1517–1527
198. Hung SC, Pochampally RR, Chen SC, Hsu SC, Prockop DJ (2007) Angiogenic effects of human multipotent stromal cell conditioned medium activate the PI3 K-Akt pathway in hypoxic endothelial cells to inhibit apoptosis, increase survival, and stimulate angiogenesis. Stem Cells 25:2363–2370

199. Bid HK, Oswald D, Li C, London CA, Lin J, Houghton PJ (2012) Anti-angiogenic activity of a small molecule STAT3 inhibitor LLL12. PLoS ONE 7:e35513
200. Haynesworth SE, Baber MA, Caplan AI (1996) Cytokine expression by human marrow-derived mesenchymal progenitor cells in vitro: effects of dexamethasone and IL-1 alpha. J Cell Physiol 166:585–592
201. Casanova EA, Shakhova O, Patel SS, Asner IN, Pelczar P, Weber FA, Graf U, Sommer L, Burki K, Cinelli P (2011) Pramel7 mediates LIF/STAT3-dependent self-renewal in embryonic stem cells. Stem Cells 29:474–485
202. Nasef A, Mazurier C, Bouchet S, Francois S, Chapel A, Thierry D, Gorin NC, Fouillard L (2008) Leukemia inhibitory factor: Role in human mesenchymal stem cells mediated immunosuppression. Cell Immunol 253:16–22
203. Majumdar MK, Thiede MA, Mosca JD, Moorman M, Gerson SL (1998) Phenotypic and functional comparison of cultures of marrow-derived mesenchymal stem cells (MSCs) and stromal cells. J Cell Physiol 176:57–66
204. Ahr B, Denizot M, Robert-Hebmann V, Brelot A, Biard-Piechaczyk M (2005) Identification of the cytoplasmic domains of CXCR4 involved in Jak2 and STAT3 phosphorylation. J Biol Chem 280:6692–6700
205. Hattermann K, Mentlein R, Held-Feindt J (2012) CXCL12 mediates apoptosis resistance in rat C6 glioma cells. Oncol Rep 27:1348–1352
206. Hartmann TN, Burger JA, Glodek A, Fujii N, Burger M (2005) CXCR4 chemokine receptor and integrin signaling co-operate in mediating adhesion and chemoresistance in small cell lung cancer (SCLC) cells. Oncogene 24:4462–4471
207. Lis R, Touboul C, Mirshahi P, Ali F, Mathew S, Nolan DJ, Maleki M, Abdalla SA, Raynaud CM, Querleu D et al (2011) Tumor associated mesenchymal stem cells protects ovarian cancer cells from hyperthermia through CXCL12. Int J Cancer J Int du Cancer 128:715–725
208. Welte G, Alt E, Devarajan E, Krishnappa S, Jotzu C, Song YH (2011) Interleukin-8 derived from local tissue-resident stromal cells promotes tumor cell invasion. Mol Carcinog doi:10.1002/mc.20854
209. Neiva KG, Zhang Z, Miyazawa M, Warner KA, Karl E, Nor JE (2009) Cross talk initiated by endothelial cells enhances migration and inhibits anoikis of squamous cell carcinoma cells through STAT3/Akt/ERK signaling. Neoplasia 11:583–593
210. Mellado M, Rodriguez-Frade JM, Aragay A, del Real G, Martin AM, Vila-Coro AJ, Serrano A, Mayor F Jr, Martinez AC (1998) The chemokine monocyte chemotactic protein 1 triggers Janus kinase 2 activation and tyrosine phosphorylation of the CCR2B receptor. J Immunol 161:805–813
211. Fujimoto H, Sangai T, Ishii G, Ikehara A, Nagashima T, Miyazaki M, Ochiai A (2009) Stromal MCP-1 in mammary tumors induces tumor-associated macrophage infiltration and contributes to tumor progression. Int J Cancer J Int du Cancer 125:1276–1284
212. Tsuyada A, Chow A, Wu J, Somlo G, Chu P, Loera S, Luu T, Li X, Wu X, Ye W et al. (2012) CCL2 mediates crosstalk between cancer cells and stromal fibroblasts that regulates breast cancer stem cells. Cancer Res 72(11):2768–79
213. Faderl S, Pal A, Bornmann W, Albitar M, Maxwell D, Van Q, Peng Z, Harris D, Liu Z, Hazan-Halevy I et al (2009) Kit inhibitor APcK110 induces apoptosis and inhibits proliferation of acute myeloid leukemia cells. Cancer Res 69:3910–3917
214. Chaix A, Lopez S, Voisset E, Gros L, Dubreuil P, De Sepulveda P (2011) Mechanisms of STAT protein activation by oncogenic KIT mutants in neoplastic mast cells. J Biol Chem 286:5956–5966
215. Hasegawa T, Suzuki K, Sakamoto C, Ohta K, Nishiki S, Hino M, Tatsumi N, Kitagawa S (2003) Expression of the inhibitor of apoptosis (IAP) family members in human neutrophils: up-regulation of cIAP2 by granulocyte colony-stimulating factor and overexpression of cIAP2 in chronic neutrophilic leukemia. Blood 101:1164–1171
216. Sakamoto C, Suzuki K, Hato F, Akahori M, Hasegawa T, Hino M, Kitagawa S (2003) Antiapoptotic effect of granulocyte colony-stimulating factor, granulocyte-macrophage colony-stimulating factor, and cyclic AMP on human neutrophils: protein synthesis-dependent and protein synthesis-independent mechanisms and the role of the Janus kinase-STAT pathway. Int J Hematol 77:60–70

217. Kucerova L, Matuskova M, Hlubinova K, Altanerova V, Altaner C (2010) Tumor cell behaviour modulation by mesenchymal stromal cells. Mol Cancer 9:129
218. Gu L, Chiang KY, Zhu N, Findley HW, Zhou M (2007) Contribution of STAT3 to the activation of survivin by GM-CSF in CD34+ cell lines. Exp Hematol 35:957–966
219. Hofer EL, Labovsky V, La Russa V, Vallone VF, Honegger AE, Belloc CG, Wen HC, Bordenave RH, Bullorsky EO, Feldman L et al (2010) Mesenchymal stromal cells, colony-forming unit fibroblasts, from bone marrow of untreated advanced breast and lung cancer patients suppress fibroblast colony formation from healthy marrow. Stem Cells Dev 19:359–370
220. Syed ZA, Yin W, Hughes K, Gill JN, Shi R, Clifford JL (2011) HGF/c-met/Stat3 signaling during skin tumor cell invasion: indications for a positive feedback loop. BMC Cancer 11:180
221. Patel ZS, Grugan KD, Rustgi AK, Cucinotta FA, Huff JL (2012) Ionizing radiation enhances esophageal epithelial cell migration and invasion through a paracrine mechanism involving stromal-derived hepatocyte growth factor. Radiat Res 177:200–208
222. Di Nicola M, Carlo-Stella C, Magni M, Milanesi M, Longoni PD, Matteucci P, Grisanti S, Gianni AM (2002) Human bone marrow stromal cells suppress T-lymphocyte proliferation induced by cellular or nonspecific mitogenic stimuli. Blood 99:3838–3843
223. Carmo CR, Lyons-Lewis J, Seckl MJ, Costa-Pereira AP (2011) A novel requirement for Janus kinases as mediators of drug resistance induced by fibroblast growth factor-2 in human cancer cells. PLoS ONE 6:e19861
224. Udayakumar TS, Nagle RB, Bowden GT (2004) Fibroblast growth factor-1 transcriptionally induces membrane type-1 matrix metalloproteinase expression in prostate carcinoma cell line. Prostate 58:66–75
225. Nishimori H, Ehata S, Suzuki HI, Katsuno Y, Miyazono K (2012) Prostate cancer cells and bone stromal cells mutually interact with each other through bone morphogenetic protein-mediated signals. J Biol Chem 287(24):20037–20046
226. Di Maggio N, Mehrkens A, Papadimitropoulos A, Schaeren S, Heberer M, Banfi A, Martin I (2012) FGF-2 Maintains a niche-dependent population of self-renewing highly potent non-adherent mesenchymal progenitors through FGFR2c. Stem Cells 30(7):1455–1464
227. Dorff TB, Goldman B, Pinski JK, Mack PC, Lara PN, Jr Van Veldhuizen PJ, Jr Quinn DI, Vogelzang NJ, Thompson IM, Jr Hussain MH (2010) Clinical and correlative results of SWOG S0354: a phase II trial of CNTO328 (siltuximab), a monoclonal antibody against interleukin-6, in chemotherapy-pretreated patients with castration-resistant prostate cancer. Clin Cancer Res: Official J Am Assoc Cancer Res 16:3028–3034
228. Puchalski T, Prabhakar U, Jiao Q, Berns B, Davis HM (2010) Pharmacokinetic and pharmacodynamic modeling of an anti-interleukin-6 chimeric monoclonal antibody (siltuximab) in patients with metastatic renal cell carcinoma. Clinical Cancer Res: Official J Am Assoc Cancer Res 16:1652–1661
229. Hedvat M, Huszar D, Herrmann A, Gozgit JM, Schroeder A, Sheehy A, Buettner R, Proia D, Kowolik CM, Xin H et al (2009) The JAK2 inhibitor AZD1480 potently blocks Stat3 signaling and oncogenesis in solid tumors. Cancer Cell 16:487–497
230. Ioannidis S, Lamb ML, Wang T, Almeida L, Block MH, Davies AM, Peng B, Su M, Zhang HJ, Hoffmann E et al (2011) Discovery of 5-chloro-N2-[(1S)-1-(5-fluoropyrimidin-2-yl) ethyl]-N4-(5-methyl-1H-pyrazol-3-yl)pyrimidine-2,4-diamine (AZD1480) as a novel inhibitor of the Jak/Stat pathway. J Med Chem 54:262–276
231. Ramakrishnan V, Kimlinger T, Haug J, Timm M, Wellik L, Halling T, Pardanani A, Tefferi A, Rajkumar SV, Kumar S (2010) TG101209, a novel JAK2 inhibitor, has significant in vitro activity in multiple myeloma and displays preferential cytotoxicity for CD45+ myeloma cells. Am J Hematol 85:675–686
232. Hart S, Goh KC, Novotny-Diermayr V, Hu CY, Hentze H, Tan YC, Madan B, Amalini C, Loh YK, Ong LC et al. (2011) SB1518, a novel macrocyclic pyrimidine-based JAK2 inhibitor for the treatment of myeloid and lymphoid malignancies. Leuk: Official J Leuk Soc Am Leuk Res Fund UK 25:1751–1759

233. Diaz T, Navarro A, Ferrer G, Gel B, Gaya A, Artells R, Bellosillo B, Garcia-Garcia M, Serrano S, Martinez A et al (2011) Lestaurtinib inhibition of the Jak/STAT signaling pathway in hodgkin lymphoma inhibits proliferation and induces apoptosis. PLoS ONE 6:e18856
234. Santos FP, Kantarjian HM, Jain N, Manshouri T, Thomas DA, Garcia-Manero G, Kennedy D, Estrov Z, Cortes J, Verstovsek S (2010) Phase 2 study of CEP-701, an orally available JAK2 inhibitor, in patients with primary or post-polycythemia vera/essential thrombocythemia myelofibrosis. Blood 115:1131–1136
235. Kim BH, Oh SR, Yin CH, Lee S, Kim EA, Kim MS, Sandoval C, Jayabose S, Bach EA, Lee HK et al (2010) MS-1020 is a novel small molecule that selectively inhibits JAK3 activity. Br J Haematol 148:132–143
236. Quintas-Cardama A, Vaddi K, Liu P, Manshouri T, Li J, Scherle PA, Caulder E, Wen X, Li Y, Waeltz P et al (2010) Preclinical characterization of the selective JAK1/2 inhibitor INCB018424: therapeutic implications for the treatment of myeloproliferative neoplasms. Blood 115:3109–3117
237. Liu PC, Caulder E, Li J, Waeltz P, Margulis A, Wynn R, Becker-Pasha M, Li Y, Crowgey E, Hollis G et al (2009) Combined inhibition of Janus kinase 1/2 for the treatment of JAK2V617F-driven neoplasms: selective effects on mutant cells and improvements in measures of disease severity. Clin Cancer Res: Official J Am Assoc Cancer Res 15:6891–6900
238. Monaghan KA, Khong T, Burns CJ, Spencer A (2011) The novel JAK inhibitor CYT387 suppresses multiple signalling pathways, prevents proliferation and induces apoptosis in phenotypically diverse myeloma cells. Leuk: Official J Leuk Soc Am Leuk Res Fund UK 25:1891–1899
239. Zhang H, Zhang D, Luan X, Xie G, Pan X (2010) Inhibition of the signal transducers and activators of transcription (STAT) 3 signalling pathway by AG490 in laryngeal carcinoma cells. J Int Med Res 38:1673–1681
240. Meydan N, Grunberger T, Dadi H, Shahar M, Arpaia E, Lapidot Z, Leeder JS, Freedman M, Cohen A, Gazit A et al (1996) Inhibition of acute lymphoblastic leukaemia by a Jak-2 inhibitor. Nature 379:645–648
241. Huang C, Yang G, Jiang T, Huang K, Cao J, Qiu Z (2010) Effects of IL-6 and AG490 on regulation of Stat3 signaling pathway and invasion of human pancreatic cancer cells in vitro. J Exp Clin Cancer Res CR 29:51
242. Grandage VL, Everington T, Linch DC, Khwaja A (2006) Go6976 is a potent inhibitor of the JAK 2 and FLT3 tyrosine kinases with significant activity in primary acute myeloid leukaemia cells. Br J Haematol 135:303–316
243. Goh KC, Novotny-Diermayr V, Hart S, Ong LC, Loh YK, Cheong A, Tan YC, Hu C, Jayaraman R, William AD et al (2012) TG02, a novel oral multi-kinase inhibitor of CDKs, JAK2 and FLT3 with potent anti-leukemic properties. Leuk: Official J Leuk Soc Am Leuk Res Fund UK 26:236–243
244. Blechacz BR, Smoot RL, Bronk SF, Werneburg NW, Sirica AE, Gores GJ (2009) Sorafenib inhibits signal transducer and activator of transcription-3 signaling in cholangiocarcinoma cells by activating the phosphatase shatterproof 2. Hepatology 50:1861–1870
245. Xiong H, Du W, Zhang YJ, Hong J, Su WY, Tang JT, Wang YC, Lu R, Fang JY (2012) Trichostatin A, a histone deacetylase inhibitor, suppresses JAK2/STAT3 signaling via inducing the promoter-associated histone acetylation of SOCS1 and SOCS3 in human colorectal cancer cells. Mol Carcinog 51:174–184
246. Kim BH, Yin CH, Guo Q, Bach EA, Lee H, Sandoval C, Jayabose S, Ulaczyk-Lesanko A, Hall DG, Baeg GH (2008) A small-molecule compound identified through a cell-based screening inhibits JAK/STAT pathway signaling in human cancer cells. Mol Cancer Ther 7:2672–2680
247. Sai K, Wang S, Balasubramaniyan V, Conrad C, Lang FF, Aldape K, Szymanski S, Fokt I, Dasgupta A, Madden T et al (2012) Induction of cell-cycle arrest and apoptosis in glioblastoma stem-like cells by WP1193, a novel small molecule inhibitor of the JAK2/STAT3 pathway. J Neurooncol 107:487–501

248. Seavey MM, Lu LD, Stump KL, Wallace NH, Hockeimer W, O'Kane TM, Ruggeri BA, Dobrzanski P (2012) Therapeutic efficacy of CEP-33779, a novel selective JAK2 inhibitor, in a mouse model of colitis-induced colorectal cancer. Mol Cancer Ther 11:984–993
249. Michaud-Levesque J, Bousquet-Gagnon N, Beliveau R (2012) Quercetin abrogates IL-6/STAT3 signaling and inhibits glioblastoma cell line growth and migration. Exp Cell Res 318:925–935

Chapter 4
The Role of Autophagy in Drug Resistance and Potential for Therapeutic Targeting

Reshma Rangwala and Ravi Amaravadi

Abstract Autophagy is a cellular survival mechanism influenced by a wide variety of intracellular and extracellular stresses including energy and oxygen deprivation, signaling aberrancies, ER stress, DNA damage, systemic cancer therapies, and radiotherapies. There is growing evidence that it may potentiate cancer survival. A number of clinical trials have been launched using autophagy inhibition in combination with standard cancer therapeutics. The information gleaned from these as well as ongoing in vitro and in vivo studies will allow us to better understand the role of autophagy in cancer.

4.1 Introduction

Cancer cells are faced with multiple metabolic and therapeutic stresses and rely on intracellular stress response systems for survival. One such stress response that can be activated by nutrient deprivation, hypoxia, intracellular pathogens, chemotherapy, targeted therapies, and/or radiotherapy is autophagy. Autophagy is a term derived from the Greek word, "to eat" (-phag, y) "oneself" (auto). The term was first coined to describe vesicular structures noted on electron microscopy (EM) that contained cytoplasmic contents. Subsequent studies demonstrated that these contents consisted of organelles and proteins that were in various stages of catabolism [1, 2]. Autophagy is the only mechanism that eukaryotic cells possess that allows for the bulk degradation of intracellular organelles, which are often damaged and serve as a liability in stressed cells. Autophagy consists of multiple related vesicular trafficking programs within the cancer cell and is coordinated by a complex interplay between dedicated enzymes, cellular membranes, cytoskeleton, and motor proteins.

Initially, autophagy was described as type II programmed cell death, and "self-eating", and, if persistent, can result in exhaustion of all intracellular resources,

R. Rangwala · R. Amaravadi (✉)
Abramson Cancer Center and Department of Medicine, Perelman School of Medicine, University of Pennsylvania, 16 Penn Tower, 3400 Spruce Street, Philadelphia, PA 19104, USA
e-mail: Ravi.amaravadi@uphs.upenn.edu

D. E. Johnson (ed.), *Cell Death Signaling in Cancer Biology and Treatment*,
Cell Death in Biology and Diseases, DOI: 10.1007/978-1-4614-5847-0_4,

X.-M. Yin and Z. Dong (Series eds.), *Cell Death in Biology and Diseases*

and terminal starvation. However, unlike apoptosis and necrosis, autophagy is reversible once initiated and occurs at a basal level in all eukaryotic cells. In cancer cells, autophagy can be induced to even higher levels than in surrounding normal tissue. Autophagic catabolism serves a functional role in the stressed cell by producing energy sources that can be used by the cell to generate ATP and building blocks that can be recycled to fuel further growth [3–6]. These characteristics may explain how autophagy serves as a key survival mechanism in cancer cells. Autophagy is controlled by growth factor–kinase signaling, the ER stress response, the DNA damage response, and the immune system: systems that are modulated by cancer therapies.

This chapter will first define autophagy, catalog the molecular machinery required for functional autophagy, and detail the current tools used to measure autophagy and to characterize the known molecular links between autophagy and signaling pathways targeted by cancer therapies. The remainder of this chapter will review the current understanding of how metabolic and therapeutic stresses can induce autophagy, the evidence that supports autophagy's role as a tumor suppressor mechanism, and the evidence that supports autophagy's role as a tumor survival mechanism. Finally, this chapter will survey the current efforts to block autophagy in cancer therapy and possible interactions between autophagy and immune system.

4.2 Defining Autophagy

Three forms of autophagy have been identified, each defined on the basis by which the lysosome obtains the material targeted for recycling. In "macroautophagy," (hereafter referred to as autophagy) a double-membrane structure defined as the autophagosome, or autophagic vesicle, envelopes the cargo and then fuses with the lysosome. In "microautophagy," an invaginated lysosomal membrane engulfs the cargo [7]. In "chaperone-mediated autophagy," a chaperone protein delivers protein cargo directly to the LAMP2 receptor on lysosomal membranes [8]. Unlike the other two forms of autophagy, chaperone-mediated autophagy has been characterized in higher eukaryotes but not in yeast.

In addition to these described forms of autophagy, there is an increasing awareness of the distinct roles of "basal autophagy" and "stress-induced autophagy". Basal autophagy likely plays an essential homeostatic role in removing and recycling damaged parts, cellular metabolism and may also play a role in atypical protein secretion processes [9]. Basal autophagy may be regulated entirely at the posttranslational level in order to maintain the integrity of cellular constituents, while stress-induced autophagy may involve the regulatory input from pathways described below.

While autophagy has been generally considered a non-selective degradative process, selective autophagy has been described and includes mitophagy (mitochondria), ribophagy (ribosomes), pexophagy (peroxisomes), and reticulophagy (endoplasmic reticulum) [10–13]. The role of these multiple forms of autophagy in benign and malignant conditions is currently being explored.

Despite the emerging appreciation for the heterogeneity of autophagy, the components of autophagy are evolutionarily conserved and have been found to play key roles in physiological and pathologic conditions [14], including cancer [15]. In yeast, autophagy together with the CVT pathway, maintains survival when extracellular nutrients are sparse [16]. Much of our understanding of autophagy comes from studying deletion mutants in yeast (see Table 4.1). Key studies performed in *Caenorhabditis elegans* [17] and drosophila [18] have added to our understanding of autophagy's role in multicellular organisms.

Mechanistically, autophagy is initiated by the enclosure of cytosolic proteins and organelles. Closure of the phagophore isolates the cytosolic contents resulting in the formation of double-membrane-bound autophagosomes [15]. The later stages of autophagy are defined by the fusion of the autophagosomes with the lysosomes to form autolysosomes. The engulfed contents are digested by acid hydrolases, and the resultant breakdown products, including amino acids, sugars, and lipids, are shuttled back to the cytoplasm via lysosomal membrane permeases for reuse [15, 19].

4.3 The Core Machinery Required for Autophagy

A complex set of autophagy proteins and autophagy-associated proteins regulate the formation of the autophagosome, its subsequent fusion with the lysosome, and the degradation and recycling of cellular components [4–6] including class III phosphatidylinositol-3-kinase (PI3K). The core autophagy protein complexes consist largely of ATG genes. Currently, there are a number of mammalian ATG genes that have been characterized, with active, ongoing research to identify others. These genes include positive and negative regulators of autophagy and have many functions including kinase activity, ubiquitination, lipidation, conjugation, recycling, and release [20–22]. Many of these genes are part of "core" complexes that contribute to the autophagic process. This process can be further classified into five steps: induction, vesicle nucleation, vesicle expansion, cargo recruitment, and autolysosome formation (see Table 4.1).

4.3.1 The Five Steps in the Autophagic Process

4.3.1.1 Step 1: Induction

In yeast, induction is initiated by the activation of the Atg1 complex, which includes Atg1, Atg13, and Atg17-Atg31-Atg29 subcomplex [23]. The mammalian Atg1 complex, also known as the ULK complex, is composed of the mammalian

Atg1 homolog Unc-51-like kinases 1 or 2 (ULK1 and ULK2, respectively), the mammalian autophagy-related 13 homolog (Atg13), which is a putative counterpart of yeast Atg17, RB1-inducible coiled-coil 1 (RB1CC1, also known as FIP200), and Atg101, an Atg13-binding protein. ULK1 is phosphorylated at specific serine residues by both mTORC1 [24, 25] and AMPK1 (AMP-activated protein kinase) (see below), resulting in the disinhibition of the ULK complex activity. It is still unclear what the specific substrate of the ULK complex is that gives rise to vesicle nucleation. In addition, there is some evidence that autophagy can be activated in the absence of ULK1 and ULK2 in response to certain stimuli [26].

4.3.1.2 Step 2: Vesicle Nucleation

Vesicle nucleation is the initial process by which proteins and lipids are recruited for autophagosome formation. Studies indicate that the lipid membrane component of the autophagosome may be derived in whole or from parts of the mitochondria, the endoplasmic reticulum, the plasma membrane or the nuclear membrane [27–29]. As further evidence of this, Atg9 has been shown to traffic between the *trans*-Golgi network, endosomes, and autophagosome precursors [30], implying that the membrane may be derived from the above precursors. Finally, electron tomography has demonstrated connections between the ER and the autophagosomal membranes indicating that the ER cisternae associate with the developing autophagosomes [31, 32].

Vesicle nucleation starts with the recruitment of Atg proteins to the phagophore assembly site (PAS). While the exact mechanism of this step is unclear, activation of a phosphatidylinositol 3-kinase (PtdIns3K) complex is necessary. Generation of phosphoinositide signals on the surface of source membranes is accomplished by various protein complexes that include the class III phosphoinositide 3-kinase (PI3K), Vps34, and Beclin1 [33], which further complexes with UVRAG and Bif-1 and thus activates autophagy [34]. Association with Rubicon inhibits trafficking of autophagic vesicles [35]. These findings implicate Vps34/Beclin1 complexes in early and late roles in autophagic flux.

4.3.1.3 Step 3: Vesicle Expansion

Formation of the autophagosome is a de novo process. Membranes at the PAS expand and subsequently close to isolate the cytosolic cargo. The two ubiquitin-like conjugation systems, Atg8 and Atg12, are involved in vesicle expansion and completion. Atg8 is conjugated to the lipid phosphatidylethanolamine (PE) to form Atg8-PE, also known as LC3, while Atg12 is conjugated to Atg5. MAP1LC3 (LC3), the mammalian Atg8 homolog, is a key molecule used in assessing the autophagic process. Four mammalian Atg8 homologs have been identified: microtubule-associated protein 1 light chain 3 (LC3/MAP1–LC3/LC3B), $GABA_A$ receptor-associated

Table 4.1 Yeast autophagy genes and mammalian homologs and their respective functions

Yeast	Mammal	Function	References
Atg1 complex		*Induction*	
Atg1	Ulk1 and Ulk 2	Serine/threonine protein kinase	[24, 25]
Atg13	Atg13	Unknown	[24, 25]
	Atg101	Atg13-binding protein	[15]
Atg17	FIP200 (RB1CC1)	Modulates response of autophagy	[24]
Atg24 (Snx4)		PtdIns(3)P-binding protein	[200]
PtdIns3K complex		*Vesicle nucleation*	
Atg6 (Vps30)	Beclin1	Involved in activating autophagy	[33]
Atg14	Atg14L	Component of the class III PtdIns3K complex	[34, 35]
Vps34	PIK3C3	Kinase	[33]
Vps15	PIK3R4	Component of the class III PtdIns3K complex	[33]
Vps38	UVRAG	Multiple; involved in activation and maturation steps of autophagy	[34]
Atg18	WIPI proteins	Binds PtdIns(3)P and PtdIns(3,5)P_2	[201]
Atg21	WIPI proteins	Binds PtdIns(3)P and PtdIns(3,5)P_2	[202]
Atg8	MAPLC3, GABARAP, GATE-16, mATG8L	Ubiquitin-like conjugation system, involved in vesicle expansion	[36, 53]
Atg3	Atg3	Ubiquitin-conjugating enzyme (E2) analog, conjugates Atg8 to/LC3 to PE	[40]
Atg4	Atg4A Atg4B Atg4C Atg4D	Cysteine protease that processes Atg8/LC3, also removes PE from Atg8/LC3	[203, 204]
Atg5	Atg5	Component of the Atg12-Atg5-Atg16 complex, serves as a E3 ligase for Atg8/LC3 conjugation	[205]
Atg7	Atg7	Ubiquitin-activating (E1) enzyme homes, activates Atg8/LC3 and Atg12	[206]
Atg12	Atg12	Ubiquitin-like protein that modifies an internal lysine of Atg5	[205]
Atg16	Atg16L1 Atg16L2	Component of the Atg12-Atg5-Atg16 complex	[207]

(Continued)

Table 4.1 (Continued)

Yeast	Mammal	Function	References
Atg9	mAtg9A	Transmembrane protein, serves as a lipid carrier	[30]
Ypt1	Rab1	Required for correct localization of Atg8 to the PAS; in mammals, Rab1 is also required for autophagosome formation	[208]
Ypt7	Rab7	Small GTP-binding protein, facilitates transport from early to late endosomes and from late endosomes to lysosomes, facilitates clearance of autophagic compartments	[209]
Sec18	NSF	ATPase responsible for SNARE disassembly	[42]

Atg Autophagy related, *ULK* Unc51-like kinase, *FIP*200 focal adhesion kinase (*FAK*) family–interacting protein of 200 kDa, *RB*1*CC*1 retinoblastoma 1-inducible coiled-coil 1, *Beclin*1 Bcl-2 interacting myosin/moesin-like coiled-coil protein 1, *Vps* vacuolar protein sorting, *UVRAG* UV irradiation resistance-associated gene, *WIPI* WD repeat protein interacting with phosphoinositides, *Barkor* Beclin1-associated autophagy-related key regulator, *GATE*-16 Golgi-associated ATPase enhancer of 16 kDa/GABARAPL2, *GABARAP* gamma-aminobutyric acid receptor-associated protein, *LC*3 microtubule-associated protein 1 light chain 3, *TSC*1/2 tuberous sclerosis complex ½, *Ypt* yeast protein, *PE* phosphatidylethanolamine

protein (GABARAP), Golgi-associated ATPase enhancer of 16 kDa (GATE-16), and mAtg8L [36]. During the formation of the autophagosome, LC3 is conjugated to PE [36, 37]. Cleavage of proLC3 by the cysteine protease and redox sensor, Atg4 [38], leads to the formation of cytoplasmic LC3-I. A ubiquitin-like protein conjugation cascade involving an E1-like enzyme (Atg7) and E2-like enzyme (Atg3) is responsible for conjugation of LC3 to PE. On gel electrophoresis, this lipidated form migrates differently than the cytosolic form of LC3 (LC3-I) and is referred to as LC3-II. Once AV are formed and marked by lipidated LC3, LC3 is cleaved from PE by Atg4 and LC3 is recycled, a process which has been shown to be necessary for effective autophagy [36, 39–41]. There are emerging data that SNARE proteins are involved in the expansion of lipid membrane during autophagosome formation by recruiting key autophagy components to the site of construction [42].

4.3.1.4 Step 4: Cargo Recruitment

LC3 not only plays a direct role in vesicle nucleation, but also plays a role in the recruitment of cargo into the developing autophagic vesicle. p62/SQSTM1 has been shown to recruit aggregated proteins to autophagic vesicles [43], while NBR and NIX are responsible for the recruitments of organelles [44]. NBR and p62 contain ubiquitin-binding domains and LC3-binding domains. These domains allow for the tight sequestration of cargo by the surrounding LC3-containing membranes, while limiting the amount of cytosol included [45]. The adaptor protein NIX recruits mitochondria in a similar fashion to LC3-containing membranes [46]. Evidence indicates that active recruitment of cargo into the AV promotes AV closure [47].

4.3.1.5 Step 5: Fusion with Lysosome, Lysosomal Degradation, and Recycling

When the autophagosome is formed, it is transported on microtubules and fuses with the lysosome to form an autolysosome. Rab GTPases play a role in vesicle maturation and fusion with the lysosomes [48]. The endosomal sorting complexes required for transport (ESCRT) machinery [49] and multi-vesicular bodies [50] contribute to vesicle-lysosomal fusion and provide cross talk between the core autophagy machinery and the other components of endovesicular trafficking. Within the acidic environment of the autophagolysosome, the cytosolic cargo is degraded via pH-dependent hydrolases. The breakdown products of this degradation are released back into the cytosol by permeases [22]. AV components not exposed to lysosomal hydrolases are then recycled involving components of the outer membrane including Atg9, Atg2, Atg18, and Atg21 [51]. An alternative path of release occurs when the autophagosomes fuse with the plasma membrane and release their contents [52].

4.3.2 Redundancy, Non-Specificity, and the Non-Canonical Autophagy Program

There is tremendous redundancy in the autophagic machinery, including five human ULK homologs, six mammalian Atg8 homologs, and four mammalian Atg4 homologs, some of which are likely to be functional [53–57]. In addition, metabolic stress–induced autophagy can proceed in the absence of ULK1 and ULK2 [26]. Furthermore, while the components described in each of the steps above have been characterized in terms of discrete complexes critical to the autophagic process, the individual proteins have also been implicated in other protein–protein interactions [58], and these proteins have been implicated in processes irrespective of their role in autophagy. For example, Beclin1, a component of the mammalian class III PtdIns3K complex, has been implicated in angiogenesis [59], adaptation to stress, development, endocytosis, cytokinesis, immunity, tumorigenesis, aging, and cell death independent of autophagy [60].

Evidence for a non-canonical autophagy program comes from the study of MEFs deficient in Atg5 or Atg7. LC3-null vesicular structures formed in response to stress are thought to be functional autophagosomes derived from the *trans*-Golgi network and require the functional Rho-like GTPases, Rab9 [61]. These findings indicate that multiple layers exist in the autophagic machinery that provide for parallel "autophagy-like" pathways that may be recruited when canonical components are missing.

4.3.3 Unanswered Questions for the Autophagic Process

The significance of membrane composition and the fate of the membrane following autolysosome formation are unanswered questions. Furthermore, studies are needed to identify the mechanisms by which the non-selective versus selective autophagy pathways are discriminated, the mechanism(s) by which organelles and/or cytosolic debris are identified and marked for autophagy, and the regulatory components that limit degradation.

4.4 Measuring Autophagy in Cancer Cells

There are multiple in vitro tools to measure autophagy [62]. LC3-I is localized in the cytosol and LC3-II on the autophagosome surface; assessment of the ratios of LC3-I to LC3-II using protein-based assays can be employed as a surrogate for autophagosome induction and flux. Use of inhibitors to cathepsin B, H and L and pepstatin A can inhibit lysosomal function [63] and further define these processes. Lysosomal inhibition causes accumulation of autolysosomes and therefore LC3-II. This finding can be recapitulated by using other lysosomotropic agents like chloroquine derivatives or bafilomycin A1, an inhibitor of vacuolar H^+-ATPase. Because V-ATPase contributes to the acidification of other organelles, including endosomes, bafilomycin A1 may show multiple off-target effects.

In situ imaging of LC3 dynamics can be visualized by fusing LC3 to a green fluorescent probe (GFP). In the absence of autophagy induction, GFP-LC3 fluorescence is diffuse, while discrete puncta are formed upon induction or blockade of autophagic flux. It is important to note that in certain circumstances, GFP-LC3 can form puncta in cells independent of autophagy [64, 65] and that GFP fluorescence in lysosomes may occur even after degradation of the LC3 moiety. As such, this method may overestimate the number of autophagosomes. A recent iteration of the GFP-LC3 assay involves the mRFP-GFP-LC3 color change assay. This assay capitalizes on differential pH stability between GFP and mRFP, and differential pH of autophagosomes and lysosomes, the latter which has an acidic pH. In acidic environments, the fluorescence of mRFP is stable, while GFP is decreased. Merged mRFP-GFP-LC3 in autophagosomes is yellow, whereas it is red in autolysosomes [66]. Cells lacking Atg3 continue to have impaired but present autophagosome formation [67], indicating that LC3 lipidation is not the only process required for membrane maturation and should not be the only marker used to assess induction/flux.

Adaptor molecules such as p62/SQSTM1 and NBR1 bind to ubiquitin-labeled structures and recruit them inside autophagosomal. P62/SQSTM, which facilitates protein aggregate clearance by autophagy, is degraded by autophagy and can be found in cellular inclusion bodies that are presumably remnants of autophagic digestion [43, 68]. Inhibition of autophagy leads to the accumulation of p62/SQSTMI.

EM is often viewed as the gold standard for qualitatively assessing autophagy. Although EM measurements can be subjective, criteria have been defined for assessing autophagic vesicles [69]. Thus, meaningful interpretations of AV accumulation in peripheral blood cells and mouse and human tumor tissue can be made.

4.5 Molecular Links Between Autophagy and Signaling and Other Stress Pathways

4.5.1 PI3K/Akt/mTOR

The PI3K/AKT/mTOR signaling pathway is a major regulator of cell growth [70], metabolism [71], survival [3], and autophagy (Fig. 4.1). The mTOR kinase activity is divided among two protein complexes: mTORC1, which is responsible for control of protein translation and nutrient uptake and mTORC2, a regulator of AKT signaling and the cytoskeleton. mTORC1 is an upstream negative regulator of the ULK1-Atg13-FIP200-Atg101 kinase complex [10, 18]. Activation of mTORC1 complex is controlled by Rheb GTPase and the tuberous sclerosis complex (TSC) TSC1/2. Growth factor signaling through the PI3K/AKT signaling controls TSC1/2- dependent mTOR signaling.

Rapamycin, an inhibitor of mTOR, induces autophagy in an mTOR-dependent manner in rat hepatocytes while relieving the inhibitory effect by amino acids [72]. In response to nutrients, mTOR binds to the ULK1-Atg13-FIP200 complex and phosphorylates ULK1 and Atg13. Binding and subsequent phosphorylation of ULK1 by mTOR inhibit the kinase activity of ULK1 [24, 73]. Precise regulation of autophagic degradation and flux by mTORC1 kinase activity occurs as a result of the close proximity of these two pathways. mTORC1 is bound to the surface of lysosomes and likely autophagolysosomes through the regulator complex [74].

Other nodes of the PI3K/AKT/mTOR pathway are critical too. Overexpression of phosphatase and tensin homolog (PTEN), a negative regulator of PI3K-dependent lipid kinase activity, promotes autophagy [75], whereas targeted deletion of PTEN in mice strongly inhibits autophagy [76]. AKT inhibition promotes autophagy; constitutively, active AKT inhibits it [77]. Stabilization of TSC2, an inhibitor of mTOR signaling, promotes autophagy and suppresses tumorigenesis [78]. Stimulation of autophagy by mTOR has been described: mTOR and its downstream mediator S6 kinase 1 may positively regulate autophagy in 6-thioguanine-treated cells, possibly through the negative feedback inhibition of Akt [79].

4.5.2 MAPK Signaling

Abnormal activation of oncogenic Ras is frequently noted in cancers and activates autophagy. Expression of active Ras alone may induce a cellular senescence and is insufficient to transform cells [80, 81]. Human ovarian surface epithelial cells

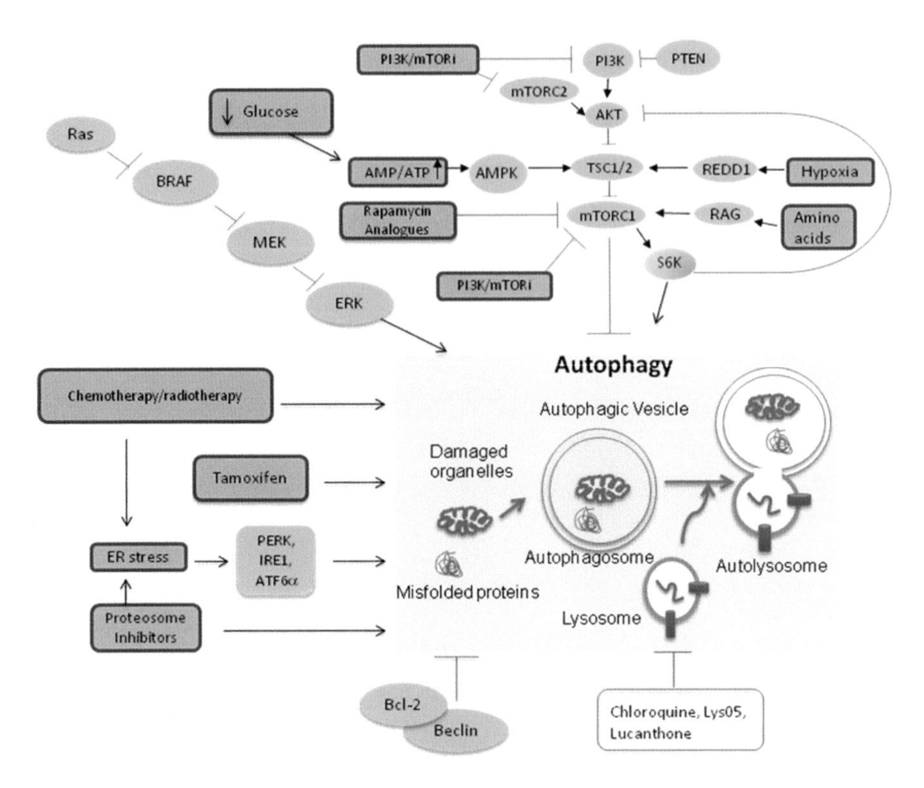

Fig. 4.1 Regulation of autophagy. Autophagy induction is influenced by a wide variety of intersecting signals internal and external to the cell comprising signaling transduction pathways; nutrient and energy deprivation; systemic cancer therapies including DNA-damaging agents, hormonal therapies, and proteasomal inhibitors; radiotherapy; and ER stress. Emerging cancer therapies including BRAF and PI3K/mTOR inhibition also induce autophagy

expressing $HRAS^{V12}$ undergo a caspase-independent cell death consistent with autophagy [82]. This is prevented by knockdown of Beclin1 or Atg5. This occurs because of upregulation of Beclin1 through the mitogen-activated protein kinase kinase (MEK)/extracellular signal-regulated kinase (ERK) cascade and by induction of Noxa (a BH3-only protein that displaces Beclin1 from Mcl-1). The unopposed Beclin1 promotes autophagy and leads to a type II cell death. However, HRAS-mediated autophagy is cytoprotective under certain conditions. BNIP3 (Bcl-2/adenovirus E1B 19-kDa-interacting protein 3) is a downstream effector of the Ras/Raf/ERK pathway and inducer of autophagy. Overexpression of BNIP3 by H-Ras(val12) competes with Beclin1 for binding to Bcl-2, induces autophagy, and reduces cell proliferation in the first 48 h. But overexpression of Ras for two weeks leads to enhanced cell proliferation and autophagy-mediated cell survival [83], indicating that autophagy's role in cell death or cell proliferation is context dependent and influenced by the degree of autophagy, stage of tumor, and oncogenic mediators.

Autophagy has also been shown to promote survival in Ras-transformed cells by induction of basal autophagy [84], potentially promoting the accumulation of abnormal mitochondria, depleting metabolites for the Krebs cycle, and decreasing oxygen consumption [85]. Data also indicate that hyperactivation of oncogenic BRAF induces autophagy; subsequent inhibition and/or depletion of its downstream effectors, MEK or ERK, inhibits it [86].

4.5.3 Bcl-2

Anti-apoptotic members of the B-cell lymphoma (Bcl-2) family are BH domain–containing proteins, which antagonize proapoptotic members including Bak and Bax. In addition to the regulation of apoptosis, Bcl-2 also inhibits autophagy [87]. Bcl-2 complexes with Beclin1, which serves as a platform for the recruitment of other proteins including UVRAG, Bif-1, and PI3KC3. Binding of Beclin1 to the anti-apoptotic protein Bcl-2 decreases Beclin1 association with Vps34 PtdIns3K activity, thus inhibiting autophagy [88]. The multiple resultant complexes are involved in different stages of autophagosome formation and maturation [89].

4.5.4 AMPK

The liver kinase B1 (LKB1)–AMP-activated protein kinase (AMPK) energy sensor pathway is a critical direct and indirect regulator of autophagy by effecting proliferation and apoptosis during metabolic stress [90]. Activation of this pathway further stimulates autophagy through multiple pathways including stabilization of p27 [91] and inhibition of mTOR signaling [92–94]. AMPK signaling also integrates the pro-autophagic effect of reactive oxygen species (ROS)-dependent activation of ataxia telangiectasia mutated (ATM) [94]. Furthermore, TAK1 (transforming growth factor—activated kinase 1) activates AMPK independently of LKB1, which leads to the induction of autophagy in TRAIL (tumor necrosis factor–related apoptosis-inducing ligand) treated epithelial cells [95]. Oxidative stress leads to a calmodulin-dependent activation of the LKB1–AMPK signaling pathway [96].

4.6 Induction of Autophagy by Metabolic and Therapeutic Stresses

Autophagy is required for and/or significantly impacts a wide variety of other cellular processes including but not limited to apoptosis [97, 98], necrosis [99], cell cycle control [100], immune modulation [101], angiogenesis [102], cell metabolism [103], protein and organelle turnover [104], and cell survival [104]. These internal and external processes are, in turn, mediated by nutrient, energy, oxygen, and hormonal demands. Processing these diverse cellular inputs is critical for maintaining cellular homeostasis. The primary role of autophagy is to protect cells under stress conditions, such as starvation and hypoxia. During periods of nutrient deprivation, autophagy degrades cytoplasmic materials to produce amino acids and fatty acids that can be used to synthesize new proteins or ATP [105]. When induction of autophagy exceeds homeostatic control, autophagic cell death, also known as type II programmed cell death, is induced [105–108]. Understanding how autophagic regulation is effected by glucose, oxygen, and amino acid stores may provide further insight that will ultimately allow us to leverage autophagy to our therapeutic advantage in treating disease states such as cancer.

4.6.1 The ER Stress Response and Autophagy

The ER stress response is a cellular homeostatic program initiated by an excess of unfolded or misfolded client proteins in the ER lumen. Protein kinase RNA-like endoplasmic reticulum kinase (PERK), activating transcription factor 6 (ATF6α), and inositol requiring transmembrane kinase and endonuclease 1 (IRE1) [109, 110] act as ER stress sensors. Accumulation of excessive unfolded proteins and calcium depletion in the ER results in their activation [111]. Previous work in yeast [112] and mammalian cells [113, 114] has established that ER stress can activate autophagy. The contribution of IRE1α and ATF6α in the activation of autophagy has been less well studied but may occur through the IRE1α-JNK/p38-p53-PUMA pathway [115]. The transcription factor, ATF6α, has unknown influences on the activation of cytoprotective autophagy.

4.6.2 DNA Damage Response and Autophagy

The DNA damage response is a cellular stress response to double-strand DNA breaks [116]. Although the mechanism that links DNA damage with autophagy is not completely understood, the DNA damage response results in alterations in other pathways including Bcl-2 family members [117], mTOR [118], and the ER stress response [119]. Activation of p53 is a common event during DNA damage. If the DNA damage is significant, p53 induces apoptosis. Nuclear p53 serves as a transcriptional factor and stimulates autophagy through upregulation of DRAM (damage-associated autophagy modulator), a lysosomal protein that may be critical for functional autophagy [120] and by promoting the dissociation of Beclin1-Bcl-2 and inhibiting mTOR signaling [121, 122].

4.6.3 Effect of Glucose Deprivation on Autophagy

Incubation of murine breast cancer cells in media containing less than 5 mM of glucose induces autophagy [123]. Glucose deprivation activates the AMP kinase (AMPK) and TSC2, which, in turn, inhibits the GTPase Rheb and its downstream effector mTORC1 [124, 125] by direct activation of the ULK1 complex [126]. Also, oncogenic Ras and Myc can shunt energy production from oxidative phosphorylation to glycolysis [127] as a result of PI3K/Akt/mTOR pathway activation. The loss of the tumor suppressor p53 promotes glycolysis through enhanced glucose transport [128], enhanced phosphoglycerate mutase expression [129], and suppression of the mitochondrial enzyme synthesis of cytochrome c oxidase 2 (SCO2) [130].

4.6.4 Hypoxia and Autophagy

Autophagy is induced in cells grown in hypoxic (1 % pO_2) conditions (normoxia corresponds to 21 % pO_2) [131]. Furthermore, hypoxic tumors show markers of increased autophagic activity [132] as well as increased generation of reactive oxidation species (ROS) [133] a direct inducer of autophagy [134, 135]. The effect of oxygen on the cell is mediated through the hypoxia-inducible factors (HIF), which function as heterodimers. Expression of these heterodimers is controlled by prolyl hydrolases. In the presence of oxygen, the HIF 1/2α subunits are targeted for degradation; in low oxygen states or in the presence of reactive oxidation species, HIF 1/2α subunits are stabilized and partner with a HIF-1β subunit, which then enter the nucleus and regulate hypoxia-induced gene expression [133]. Inhibition of autophagy sensitizes cells to hypoxia and ROS-induced cell death [132, 136].

4.6.5 Amino Acid Stores and Autophagy

Amino acid deprivation is one of the most potent inducers of autophagy [137]. mTORC1 regulates both protein and cytoplasmic amino acid levels by regulating the amino acid transporter expression and autophagy [138]. Furthermore, mTORC1 facilitates the transportation of leucine well as its degradation by autophagy and the ubiquitin-proteasomal system [139, 140]. Additional regulation occurs as a result of convergence of AMPK onto mTORC1. Ammonia, a glutaminolysis byproduct, stimulates autophagy by a mechanism dependent on Atg5 [141, 142].

4.6.6 DNA-Damaging Agents: DNA Alkylating Agents and Topoisomerase Inhibitors

Temozolomide, a DNA alkylating agent, is a potent inducer of autophagy [143] and mitochondrial-induced apoptosis [144]. Other DNA alkylators including the nitrogen mustards, methylnitrosourea (MNU), or N-methyl-N-nitroso-N$'$-nitroguanidine (MNNG) also induce autophagy. Reseau and colleagues demonstrated that pancreatic organ explants treated with either agent had an increased number of autophagic vesicles as compared to controls [145]. Autophagy induction has also been implicated as a resistance mechanism to topoisomerase I and topoisomerase II inhibitors [146].

4.6.7 Hormonal Therapies: Tamoxifen

Modulation of estrogen receptor signaling is a common therapeutic maneuver in the treatment for ER^+ (estrogen receptor) breast cancer; however, de novo and acquired resistance is common. It competes with endogenous estrogen in binding to the estrogen receptor and thus modulates the effects of ER complexes and downstream gene transcription. Recent studies indicate that tamoxifen-sensitive cells induce autophagy [147]. Although the exact mechanism is not known, tamoxifen binds with high affinity to microsomal antiestrogen-binding site (AEBS), a hetero-oligomeric complex involved in cholesterol metabolism. Binding of tamoxifen with AEBS induces autophagy through resultant sterol accumulation [148].

4.6.8 Proteasomal Inhibitors

Autophagy and the ubiquitin-proteasome system (UPS) are the two intracellular mechanisms by which proteins are degraded. Autophagy degrades long-lived, cytosolic proteins and damaged organelles; while shorter-lived proteins are targeted for catabolism by the UPS. Inhibition of the UPS by proteasomal inhibitors, like bortezomib, induces autophagy by multiple pathways including activation of HDAC6, activation of IRE1-JNK, stabilization of ATF4, inhibition of mTOR signaling, and decreased proteasomal degradation of LC3. Not surprisingly, these pathways show commonality with the regulatory mechanisms of ER stress. The induction of autophagy has also been shown to mitigate the anti-tumor effects of proteasome inhibition, including in multiple myeloma [149].

4.6.9 Histone Deacetylase Inhibitors

Acetylation and deacetylation of lysine residues are common epigenetic modulations. Aberrant deacetylation of non-histone proteins has also been implicated in a variety of pathologic states, including cancer [150]. Given its role in oncogenesis and tumor maintenance, HDAC inhibitors such as vorinostat (rINN) or suberoylanilide hydroxamic acid (SAHA) have been used clinically in the treatment for cancers including cutaneous T-cell lymphomas. Growing evidence indicates that the HDAC inhibitors induce autophagy dependent and independent of other intersecting pathways including mTOR, tyrosine kinase induction as a result of BCR-ABL fusion, and estrogen receptor–activated signaling [151–154] (see Fig. 4.1).

4.7 Autophagy and the Immune System: Role as Both Inducer and Repressor

Mutation in autophagy genes increases susceptibility to infection by intracellular pathogens. Autophagic machinery interfaces with pathways including immune responses and inflammation and involves direct interactions between autophagic proteins and immune signaling molecules [155]. Cross talk between these two seemingly disparate components has been identified in which autophagic proteins and immune and inflammatory response can both induce and suppress each other.

In mice, knockout of *Atg*5 in macrophages and neutrophils increases susceptibility to infection with *Listeria monocytogenes* and the protozoan *Toxoplasma gondii* [156]. Association between infection and tumorigenesis has been well established especially in the cases of viruses. These include HPV which has been associated with squamous cell carcinoma of the head and neck, esophagus, and cervix; HHV-6 which has been associated with Kaposi's sarcoma; hepatitis B and C have been implicated in hepatocellular carcinoma; and EBV which has been implicated in aggressive non-Hodgkin's lymphoma. The association between these infections and tumorigenesis may involve a viral strategy that blocks host autophagy, including IFN-inducible RNA-activated eIF2α and AKT/TOR [157]. In addition, several viral proteins target the core autophagy protein Beclin1 and its association with herpes simplex virus 1 (HSV-1) neurovirulence factor ICP34.5, the oncogenic herpesvirus-encoded viral BCL2-like proteins, the HIV accessory protein Nef, and the influenza virus matrix protein 2 [157]. Tumorigenesis initiated by these viruses is likely due to more than one mechanism; one can propose a multi-pronged approach by which the virus affects the host autophagic process and directly or indirectly induces tumorigenesis [158]. Michaud and colleagues report that autophagy is necessary to elicit an immune response to chemotherapy through the release of immunostimulatory extracellular ATP [159]. In contrast, Noman and colleagues found that in human lung cancer cells hypoxia-induced autophagy prevented T-cell-mediated cytotoxicity. Knockdown of autophagy genes restored sensitivity to cytotoxic T cells [131]. In an immunocompetent model of melanoma, the combination of genetic or pharmacological autophagy inhibition in combination with melanoma peptide vaccination resulted in synergistic tumor regression compared with either treatment alone [131]. These results suggest the complex interactions between tumor cell autophagy and the immune system may be context dependent.

4.8 Arguments for a Tumor Suppressor and Tumor Promoter Role for Autophagy in Cancer

4.8.1 Role as a Tumor Suppressor

Autophagy was initially considered a tumor suppressor mechanism. Beclin1 is a phylogenetically conserved protein essential for autophagy and homolog of the yeast gene apg6/vps30 [160]. Monoallelic deletion of *beclin1* on chromosome 17q21 occurs in 40–75 % of ovarian, breast, and prostate cancers [161]. Mice with monoallelic loss of *beclin1* show accelerated development of spontaneous malignant tumors [162, 163]. Similarly, a mouse with a mosaic deletion of *Atg*5 in all tissues was predisposed to the development of benign liver tumors, with no detection of tumors in any other organs [164]. Multiple groups have demonstrated that autophagy defects lead to increased DNA damage, genomic instability, and tumor progression [165–167]. Defects in autophagy can lead to the elevation of p62/SQSTMI, which has been shown to promote tumorigenesis through multiple mechanisms including alterations in NF-κB signaling [68]. There is also growing evidence that p62 binds to Keap1 with subsequent upregulation of the transcription factor NRF2 whose function is to coordinate antioxidant defense [168, 169]. Furthermore, persistent activation of NRF2 is critical for anchorage-independent growth of hepatocellular carcinoma cells in the context of p62/SQSTMI overexpression [170]. The tumor suppressor function of autophagy may also be attributed to its role in cellular senescence. Autophagy is activated during oncogene-induced senescence by oncogenic Ras. Inhibition of autophagy relieves this senescence [171]. Given that cancer is associated commonly with activating mutations in growth factor signaling pathways (such as PI3K/Akt/mTOR), overexpression of anti-apoptotic proteins (such as BCL2), it would be predicted that autophagy would be suppressed in most cancers [172–175]. However, recent evidence indicates that constitutive Ras activation actually provokes elevated levels of basal autophagy [84].

4.8.2 Role of Autophagy as a Tumor Promoter

The above data implicate autophagy as a guardian against genomic instability and resultant tumorigenesis and therefore support a role for autophagy as a tumor suppressor mechanism. However, the critical role of autophagy as a stress response in tumor cells, suggests that human tumors growing within the harsh conditions of the tumor microenvironment may in fact have high levels of autophagy. The earliest data supporting this view come from experiments characterizing autophagy in mouse models of Akt-driven solid tumors that are deficient in apoptosis [165]. Furthermore, these studies demonstrated that autophagy was induced in tumors grown in hypoxic environments.

Additional evidence to support this tumor potentiating role is the high levels of autophagy, as assessed by EM and LC3 immunohistochemistry, in both metastatic melanoma and pancreas cancer [85]. LC3B was elevated in 84 % of cases of 20 advanced cancer histologies, and elevated levels of autophagy in breast cancer and melanoma primary tumors correlated with lymph node metastases and poor survival [131].

Although monoallelic deletion of *beclin* has been found in a large subset of common tumors such as breast and prostate cancer, the retained allele of *beclin* in these tumors is always wild type. Beclin1 "deficient" cells usually have similar levels of autophagy induced by stresses as compared to *beclin* WT cells. Beclin1 is bound tightly in a complex involving cytosolic p53. In cells derived from $beclin^{+/-}$ mice, p53 levels were found to be lower than in cells derived from *beclin* WT mice. This finding suggests that p53 deficiency and not autophagy deficiency may explain why *beclin*-deficient mice develop malignancies, with the caveat that Liu and colleagues did not differentiate cytosolic from nuclear p53 [176]. Furthermore, no other mouse model of genetic autophagy deficiency has produced true spontaneous malignancies. For example, $FIP200^{-/-}$ mice are found to have impaired tumorigenesis [177]. In addition, inactivating somatic mutations have yet to be reported in any other autophagy genes, save for polymorphisms in the *ATG16L1* gene, which predispose patients to Crohn's disease [178] and not cancer. Finally, there is compelling evidence that Ras transformation produces a massive derangement in cancer cell metabolism that necessitates high levels of autophagy to avoid oncogene-induced senescence or cell death [84]. Therefore, autophagy may be essential for tumorigenesis and cancer maintenance and autophagy-deficient tumors may be rare.

4.8.3 Autophagy as Both Tumor Suppressor and Tumor Potentiator

How can these dual, opposing roles be reconciled? As detailed above, autophagy is regulated by multiple, intersecting pathways. Emerging evidence supports the concept that its tumor suppressor versus tumor potentiator role may be pathway specific and not necessarily mutually exclusive. Given the multiple mechanisms by which autophagy is regulated, mutations in any of these signaling nodes will affect cellular homeostasis and therefore the role of autophagy within that cellular environment. This, in turn, effects whether autophagy modulates pro-survival versus pro-death pathways. Both the degree and duration of autophagy has been proposed as an explanation as to the role in which autophagy functions. In *C. elegans*, physiological levels of autophagy promote survival during nutrient poor conditions, while either insufficient or excessive autophagy under these same nutrient poor conditions leads to cell death [179]. In addition, excessive autophagy likely compromises the integrity of the lysosomal membranes resulting in the release of cathepsins into the cytosol [180–182] and may underlie the mechanism of such autophagic inducing drugs like temozolomide [183]. These data implicate LMP as an effector pathway for autophagic-induced cell death, or type II cell death [184]. However, the same stressors that activate autophagic cell death in vitro can promote autophagic cell survival in vivo as a result of cytokines and components of the extracellular matrix that are responsible for this switch in cell fates [185]. Because of growing evidence in vivo that autophagy serves more as a survival mechanism than a pathway for cell death, it has become a target for novel therapeutic inhibitors.

4.9 Preclinical Evidence Supporting Autophagy Inhibition as a Therapeutic Strategy in Cancer

The cytotoxic effects of blocking autophagy with chloroquine in cells that are reliant on autophagy for survival was first demonstrated in apoptosis-defective cells exposed to the stress of growth factor withdrawal [186]. Chloroquine also augmented chemotherapy-induced tumor impairment in a model of Myc-induced lymphoma [3]. Since these papers, numerous investigators have demonstrated that combining chloroquine or siRNA against essential autophagy genes can augment the efficacy of many existing and emerging cancer therapies in many different disease models (see selected examples in Table 4.2). Because effective autophagy inhibition can be achieved in vivo with the anti-malarial drug chloroquine (CQ) [3], and there is extensive experience with CQ derivatives for the treatment for malaria [187], rheumatoid arthritis [188], and HIV [189], multiple trials have been launched across a wide variety of tumor types.

4.10 Clinical Trials Involving Hydroxychloroquine as a First-Generation Autophagy Inhibitor

A phase III trial in glioblastoma patients treated with radiation and carmustine with or without daily CQ found a median overall survival of 24 and 11 months in CQ- and placebo-treated patients, respectively [190]. While it was not adequately powered to detect a significant difference in survival, it established the safety of adding low dose CQ to DNA alkylators. Its long half-life and low potency may limit its efficacy as an autophagy inhibitor in patients [191]. To address these concerns, a phase I/II trial of HCQ with temozolomide and radiation for glioblastoma patients was launched and included pharmacodynamic (PD) and pharmacokinetic (PK) analyses. PD evidence of HCQ dose–dependent autophagy inhibition was observed using an EM assay on serial blood mononuclear cells [192]. The implications of autophagy as a mechanism by which tumor growth is potentiated have significant clinical implications: clinical effect observed with standard therapies may be improved upon by the addition of drugs that inhibit autophagy. To date, only the chloroquine

Table 4.2 Preclinical studies that demonstrate augmented anti-tumor activity with autophagy inhibition

Metabolic or therapeutic stress	Therapeutic class	In vivo malignancy	Reference
Growth factor limitation	Metabolic stress	Bax/bak-deficient cells	[186]
P53 activation or MNNG	DNA-damaging agent	Lymphoma	[3]
SAHA	Histone deacetylase inhibitor	CML	[151]
Akt-1	Akt inhibitor	Prostate	[210]
Imatinib	CKIT, Abl kinase inhibitor	CML	[211]
		GIST	[212]
Bortezomib	Proteasome inhibitor	Myeloma	[213]
Hypoxia	Metabolic stress	Colon	[214]
BEZ235	PI3 K/mTOR inhibitor	Glioma	[215]
Doxorubicin melphalan	DNA-damaging agent	Myeloma	[216]
Leucine deprivation	Metabolic Stress	Melanoma	[217]

derivatives (chloroquine and hydroxychloroquine) have been used in the clinical setting. Since 2007, more than 20 trials have been launched involving HCQ. A complete listing of ongoing and/or completed trials can be found at clinicaltrials.gov. The knowledge gained from the PD, PK, and predictive biomarkers in these studies will guide the development of more potent and specific autophagy inhibitors.

4.11 Emerging Inhibitors of Autophagy

Chemical autophagy inhibition has become an important tool to understand the role of autophagy in various disease states including cancer. A common concern is that most of the autophagy inhibitors described below are not pharmacological compounds and lack specificity (see Fig. 4.2).

1. Pan phosphatidylinositol 3-kinase (PtdIns3K) inhibitors, including wortmannin and LY294002: inhibit both class I and class III PtdIns3K. The implications of this lack of specificity are significant since the class III PtdIns3K product, phosphatidylinositol 3-phosphate (PtdIns(3)P) is essential for autophagy, whereas the class I PtdIns3K products, phosphatidylinositol (3,4)-bisphosphate (PtdIns(3,4,5)P_3) have inhibitory effects [193]. Because the pan PtdIns3K inhibitors inhibit both classes of PtdIns3K enzymes, both autophagy and S6 phosphorylation are downregulated. These two compounds serve as a proximal inhibitors of autophagy.
2. 3-methyladenine (3-MA): a pan PtdIns3K inhibitor. Unlike wortmannin and LY294002, 3-MA promotes autophagic flux under nutrient-rich conditions, but suppresses autophagy during periods of nutrient deprivation. These opposing effects are due to 3-MAs differing temporal effects on class I and class III PtdIns3K; 3-MA blocks class I PtdIns3K persistently, whereas its suppressive

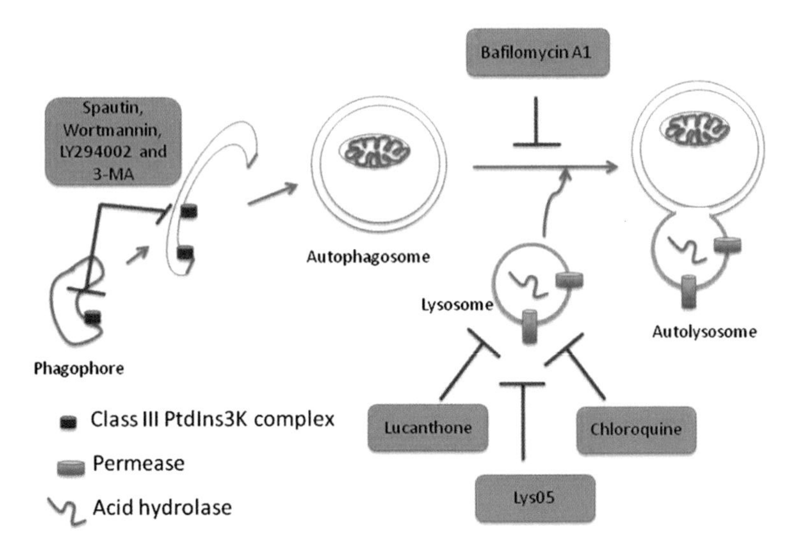

Fig. 4.2 Inhibitors of autophagy. Spautin, wortmannin, LY294002, and 3-MA are pan inhibitors of PtdIns3K. Lucanthone, Lys05, and chloroquine inhibit lysosomal function. Bafilomycin A1 prevents the maturation of autophagic vacuoles

effect on class III PtdIns3K is transient [194]. 3-MA serves as a proximal inhibitor of autophagy.

3. specific and potent autophagy inhibitor-1 (Spautin-1): a small molecule inhibitor of autophagy that promotes the degradation of Vps34/PtdIns3 kinase complexes by inhibiting two ubiquitin-specific peptidases, USP10 and USP13, which target Beclin1. Because USP10 also mediates the deubiquitination of p53, Beclin1 indirectly regulates p53 levels [176]. It serves as a proximal inhibitor of autophagy.
4. Bafilomycin A1: a vacuolar ATPase. This compound prevents maturation of autophagic vacuoles by inhibiting fusion between autophagosomes and lysosomes [195]. Bafilomycin A1 serves as a distal inhibitor of autophagy.
5. Lucanthone: an anti-schistosome agent that blocks topoisomerase II activity and inhibits AP endonuclease (APE1), a DNA base excision repair enzyme [196, 197]. It disrupts lysosomal function and serves as a distal inhibitor of autophagy. Autophagy inhibition by lucanthone induces apoptosis by a cathepsin D-mediated process [198].
6. Phenylethynesulfonamide (PES): a sulfonamide that binds to and inhibits heat shock protein (HSP)70 function, which leads to misfolding of critical lysosomal proteins. PES treatment of cell lines produces striking accumulation of autophagic vesicles similar to chloroquine treatment [199].
7. Chloroquine derivatives: an anti-malarial agent that also has been used as a lysosomal trafficking inhibitor. It inhibits lysosomal acidification and prevents the fusion of autophagosomes with lysosomes. It serves as a distal inhibitor of autophagy and is the only clinically applicable inhibitor of autophagy.

8. Lys05: a dimeric chloroquine found to be a 10-fold more potent autophagy inhibitor. This compound has single-agent anti-tumor activity in multiple xenograft models and, when given at doses above the maximal tolerated dose, reproduced an intestinal phenotype similar to patients and mice deficient in ATG16L1. These results suggest chloroquine derivatives can phenocopy a genetic autophagy deficiency (Mcafee et al. PNAS 2012 in press)

4.12 Summary and Future Directions

Providing for cellular homeostasis in the face of rapid proliferation can be a challenge as the cell integrates signals generated from intracellular alterations in glucose, oxygen, energy, and amino acid stores. An additional layer of complexity lies in the fact that the cell is also responsible for assimilating extracellular signals arising from the immune system, pathogens, radio-, targeted-, hormonal-, and/or chemotherapies. These diverse, sometimes diametrically opposed signals can each stimulate autophagy. In this capacity, autophagy can serve as a stress mechanism that can maintain survival despite limitations in essential nutrients and has been described as such in both physiological and pathophysiological settings. Interestingly, though, unchecked autophagy can also lead to cell death. Predicting these diametrically opposed roles is at the heart of leveraging autophagy as a novel target for cancer therapy. Currently, the therapeutic strategy targeting autophagy as a means to enhance clinical cancer benefit is by inhibiting it. An abundance of preclinical evidence exists supporting autophagy as a mechanism by which a cancerous cell protects itself from reactive oxygen and reactive nitrogen species accumulation, genetic instability, hypoxia, and diminished glucose stores. Multiple clinical trials have been launched utilizing chloroquine and the chloroquine derivative, hydroxychloroquine. Despite their extensive clinical use for the treatment for malaria, rheumatoid arthritis, and HIV, their oncology use may be limited by their narrow therapeutic window, their long half-life, and accumulating toxicities. Development of new autophagy inhibitors is being pursued and will likely advance this target as a cancer therapy adjunct.

Hand in hand with the development of new autophagy inhibitors is our ability to prospectively identify cancer patients who will benefit from autophagy inhibition. Currently, data from ongoing and completed trials that couple standard therapeutics with autophagy inhibition show mixed response. Further work needs to be pursued concerning pre-identification of sensitive tumor types, recognizing appropriate therapeutic regimens that synergistically act with autophagy inhibition to enhance clinical benefit, and development of more robust pharmacodynamic markers to assess autophagy inhibition in vivo.

A yet theoretical application for autophagy lies in the use of pharmacologic autophagic inducers in those cells or tumors in which autophagy stimulates cell death. The challenge for this application lies in the development of such drugs as well as our ability to prospectively identify those cells in which autophagy induces cell death, not cell survival.

Regardless of the above challenges, research into the metabolic drivers, identification of the delicate interplay between the immune system and autophagy as well as development of mouse models is being actively pursued. These insights will improve our understanding of the role of autophagy in normal, dysplastic, and cancerous cells and therefore will allow us to better target autophagy and ultimately improve cancer therapy.

References

1. Ashford TP, Porter KR (1962) Cytoplasmic components in hepatic cell lysosomes. J Cell Biol 12:198–202
2. Watanabe E et al (2009) Sepsis induces extensive autophagic vacuolization in hepatocytes: a clinical and laboratory-based study. Lab Invest 89(5):549–561
3. Amaravadi RK et al (2007) Autophagy inhibition enhances therapy-induced apoptosis in a Myc-induced model of lymphoma. J Clin Invest 117(2):326–336
4. Klionsky DJ (2007) Autophagy: from phenomenology to molecular understanding in less than a decade. Nat Rev Mol Cell Biol 8(11):931–937
5. Kroemer G, Jaattela M (2005) Lysosomes and autophagy in cell death control. Nat Rev Cancer 5(11):886–897
6. Levine B, Deretic V (2007) Unveiling the roles of autophagy in innate and adaptive immunity. Nat Rev Immunol 7(10):767–777
7. Muller O et al (2000) Autophagic tubes: vacuolar invaginations involved in lateral membrane sorting and inverse vesicle budding. J Cell Biol 151(3):519–528
8. Bandyopadhyay U et al (2008) The chaperone-mediated autophagy receptor organizes in dynamic protein complexes at the lysosomal membrane. Mol Cell Biol 28(18):5747–5763
9. Chen HY, White E (2011) Role of autophagy in cancer prevention. Cancer Prev Res (Phila) 4(7):973–983
10. He C, Klionsky DJ (2009) Regulation mechanisms and signaling pathways of autophagy. Annu Rev Genet 43:67–93
11. Gu Y, Wang C, Cohen A (2004) Effect of IGF-1 on the balance between autophagy of dysfunctional mitochondria and apoptosis. FEBS Lett 577(3):357–360
12. Bjorkoy G et al (2005) p62/SQSTM1 forms protein aggregates degraded by autophagy and has a protective effect on huntingtin-induced cell death. J Cell Biol 171(4):603–614
13. Wild P, Dikic I (2010) Mitochondria get a Parkin' ticket. Nat Cell Biol 12(2):104–106
14. Di Bartolomeo S, Nazio F, Cecconi F (2010) The role of autophagy during development in higher eukaryotes. Traffic 11(10):1280–1289
15. Mizushima N et al (2008) Autophagy fights disease through cellular self-digestion. Nature 451(7182):1069–1075
16. Wang CW, Klionsky DJ (2003) The molecular mechanism of autophagy. Mol Med 9(3–4):65–76
17. Kovacs AL, Zhang H (2010) Role of autophagy in *Caenorhabditis elegans*. FEBS Lett 584(7):1335–1341
18. Chang YY, Neufeld TP (2010) Autophagy takes flight in drosophila. FEBS Lett 584(7):1342–1349
19. Rong Y et al (2011) Spinster is required for autophagic lysosome reformation and mTOR reactivation following starvation. Proc Natl Acad Sci USA 108(19):7826–7831
20. Teter SA et al (2001) Degradation of lipid vesicles in the yeast vacuole requires function of Cvt17, a putative lipase. J Biol Chem 276(3):2083–2087
21. Epple UD et al (2001) Aut5/Cvt17p, a putative lipase essential for disintegration of autophagic bodies inside the vacuole. J Bacteriol 183(20):5942–5955

22. Yang Z et al (2006) Atg22 recycles amino acids to link the degradative and recycling functions of autophagy. Mol Biol Cell 17(12):5094–5104
23. Kabeya Y et al (2005) Atg17 functions in cooperation with Atg1 and Atg13 in yeast autophagy. Mol Biol Cell 16(5):2544–2553
24. Jung CH et al (2009) ULK-Atg13-FIP200 complexes mediate mTOR signaling to the autophagy machinery. Mol Biol Cell 20(7):1992–2003
25. Mizushima N (2010) The role of the Atg1/ULK1 complex in autophagy regulation. Curr Opin Cell Biol 22(2):132–9
26. Cheong H et al (2011) Ammonia-induced autophagy is independent of ULK1/ULK2 kinases. Proc Natl Acad Sci USA 108(27):11121–11126
27. Hailey DW et al (2010) Mitochondria supply membranes for autophagosome biogenesis during starvation. Cell 141(4):656–667
28. Ravikumar B et al (2010) Plasma membrane contributes to the formation of pre-autophagosomal structures. Nat Cell Biol 12(8):747–757
29. English L et al (2009) Autophagy enhances the presentation of endogenous viral antigens on MHC class I molecules during HSV-1 infection. Nat Immunol 10(5):480–487
30. Webber JL, Tooze SA (2010) New insights into the function of Atg9. FEBS Lett 584(7):1319–1326
31. Hayashi-Nishino M et al (2009) A subdomain of the endoplasmic reticulum forms a cradle for autophagosome formation. Nat Cell Biol 11(12):1433–1437
32. Yla-Anttila P et al (2009) 3D tomography reveals connections between the phagophore and endoplasmic reticulum. Autophagy 5(8):1180–1185
33. Funderburk SF, Wang QJ, Yue Z (2010) The Beclin 1-VPS34 complex–at the crossroads of autophagy and beyond. Trends Cell Biol 20(6):355–2
34. Itakura E et al (2008) Beclin 1 forms two distinct phosphatidylinositol 3-kinase complexes with mammalian Atg14 and UVRAG. Mol Biol Cell 19(12):5360–5372
35. Matsunaga K et al (2009) Two Beclin 1-binding proteins, Atg14L and Rubicon, reciprocally regulate autophagy at different stages. Nat Cell Biol 11(4):385–396
36. Kabeya Y et al (2000) LC3, a mammalian homologue of yeast Apg8p, is localized in autophagosome membranes after processing. EMBO J 19(21):5720–5728
37. Ichimura Y et al (2000) A ubiquitin-like system mediates protein lipidation. Nature 408(6811):488–492
38. Yoshimura K et al (2006) Effects of RNA interference of Atg4B on the limited proteolysis of LC3 in PC12 cells and expression of Atg4B in various rat tissues. Autophagy 2(3):200–208
39. Satoo K et al (2009) The structure of Atg4B-LC3 complex reveals the mechanism of LC3 processing and delipidation during autophagy. EMBO J 28(9):1341–1350
40. Tanida I et al (2002) Human Apg3p/Aut1p homologue is an authentic E2 enzyme for multiple substrates, GATE-16, GABARAP, and MAP-LC3, and facilitates the conjugation of hApg12p to hApg5p. J Biol Chem 277(16):13739–13744
41. Tanida I et al (2001) The human homolog of *Saccharomyces cerevisiae* Apg7p is a protein-activating enzyme for multiple substrates including human Apg12p, GATE-16, GABARAP, and MAP LC3. J Biol Chem 276(3):1701–1706
42. Nair U et al (2011) SNARE proteins are required for macroautophagy. Cell 146(2):290–302
43. Pankiv S et al (2007) p62/SQSTM1 binds directly to Atg8/LC3 to facilitate degradation of ubiquitinated protein aggregates by autophagy. J Biol Chem 282(33):24131–24145
44. Schweers RL et al (2007) NIX is required for programmed mitochondrial clearance during reticulocyte maturation. Proc Natl Acad Sci USA 104(49):19500–19505
45. Kim PK et al (2008) Ubiquitin signals autophagic degradation of cytosolic proteins and peroxisomes. Proc Natl Acad Sci USA 105(52):20567–20574
46. Novak I et al (2010) Nix is a selective autophagy receptor for mitochondrial clearance. EMBO Rep 11(1):45–51
47. Sandoval H et al (2008) Essential role for nix in autophagic maturation of erythroid cells. Nature 454(7201):232–235

48. Jager S et al (2004) Role for Rab7 in maturation of late autophagic vacuoles. J Cell Sci 117(Pt 20):4837–4848
49. Djeddi A et al (2012) Induction of autophagy in ESCRT mutants is an adaptive response for cell survival in *C. elegans*. J Cell Sci 125(Pt 3):685–694
50. Fader CM, Colombo MI (2009) Autophagy and multivesicular bodies: two closely related partners. Cell Death Differ 16(1):70–78
51. Young AR et al (2006) Starvation and ULK1-dependent cycling of mammalian Atg9 between the TGN and endosomes. J Cell Sci 119(Pt 18):3888–3900
52. Manjithaya R et al (2010) Unconventional secretion of *Pichia pastoris* Acb1 is dependent on GRASP protein, peroxisomal functions, and autophagosome formation. J Cell Biol **188**(4):537–46
53. Kabeya Y et al (2004) LC3, GABARAP and GATE16 localize to autophagosomal membrane depending on form-II formation. J Cell Sci 117(Pt 13):2805–2812
54. Tanida I et al (2003) GATE-16 and GABARAP are authentic modifiers mediated by Apg7 and Apg3. Biochem Biophys Res Commun 300(3):637–644
55. Tanida I, Ueno T, Kominami E (2004) LC3 conjugation system in mammalian autophagy. Int J Biochem Cell Biol 36(12):2503–2518
56. Weidberg H et al (2010) LC3 and GATE-16/GABARAP subfamilies are both essential yet act differently in autophagosome biogenesis. EMBO J 29(11):1792–1802
57. Marino G et al (2003) Human autophagins, a family of cysteine proteinases potentially implicated in cell degradation by autophagy. J Biol Chem 278(6):3671–3678
58. Behrends C et al (2010) Network organization of the human autophagy system. Nature 466(7302):68–76
59. Lee SJ et al (2011) Beclin 1 deficiency is associated with increased hypoxia-induced angiogenesis. Autophagy 7(8):829–839
60. Wirawan E et al (2012) Beclin1: a role in membrane dynamics and beyond. Autophagy 8(1):6–17
61. Nishida Y et al (2009) Discovery of Atg5/Atg7-independent alternative macroautophagy. Nature 461(7264):654–658
62. Klionsky DJ et al (2008) Guidelines for the use and interpretation of assays for monitoring autophagy in higher eukaryotes. Autophagy 4(2):151–175
63. Tanida I et al (2005) Lysosomal turnover, but not a cellular level, of endogenous LC3 is a marker for autophagy. Autophagy 1(2):84–91
64. Ciechomska IA, Tolkovsky AM (2007) Non-autophagic GFP-LC3 puncta induced by saponin and other detergents. Autophagy 3(6):586–590
65. Kuma A, Matsui M, Mizushima N (2007) LC3, an autophagosome marker, can be incorporated into protein aggregates independent of autophagy: caution in the interpretation of LC3 localization. Autophagy 3(4):323–328
66. Kimura S, Noda T, Yoshimori T (2007) Dissection of the autophagosome maturation process by a novel reporter protein, tandem fluorescent-tagged LC3. Autophagy 3(5):452–460
67. Sou YS et al (2008) The Atg8 conjugation system is indispensable for proper development of autophagic isolation membranes in mice. Mol Biol Cell 19(11):4762–4775
68. Mathew R et al (2009) Autophagy suppresses tumorigenesis through elimination of p62. Cell 137(6):1062–1075
69. Ma X et al (2011) Measurements of tumor cell autophagy predict invasiveness, resistance to chemotherapy, and survival in melanoma. Clin Cancer Res 17(10):3478–3489
70. Mendoza MC, Er EE, Blenis J (2011) The Ras-ERK and PI3 K-mTOR pathways: cross-talk and compensation. Trends Biochem Sci 36(6):320–328
71. Steelman LS et al (2011) Roles of the Raf/MEK/ERK and PI3 K/PTEN/Akt/mTOR pathways in controlling growth and sensitivity to therapy-implications for cancer and aging. Aging (Albany NY) 3(3):192–222
72. Blommaart EF et al (1995) Phosphorylation of ribosomal protein S6 is inhibitory for autophagy in isolated rat hepatocytes. J Biol Chem 270(5):2320–2326
73. Hosokawa N et al (2009) Nutrient-dependent mTORC1 association with the ULK1-Atg13-FIP200 complex required for autophagy. Mol Biol Cell 20(7):1981–1991

74. Sancak Y et al (2010) Ragulator-Rag complex targets mTORC1 to the lysosomal surface and is necessary for its activation by amino acids. Cell 141(2):290–303
75. Arico S et al (2001) The tumor suppressor PTEN positively regulates macroautophagy by inhibiting the phosphatidylinositol 3-kinase/protein kinase B pathway. J Biol Chem 276(38):35243–35246
76. Ueno T et al (2008) Loss of Pten, a tumor suppressor, causes the strong inhibition of autophagy without affecting LC3 lipidation. Autophagy 4(5):692–700
77. Laane E et al (2009) Cell death induced by dexamethasone in lymphoid leukemia is mediated through initiation of autophagy. Cell Death Differ 16(7):1018–1029
78. Kuo HP et al (2010) ARD1 stabilization of TSC2 suppresses tumorigenesis through the mTOR signaling pathway. Sci Signal 3(108):ra9
79. Zeng X, Kinsella TJ (2008) Mammalian target of rapamycin and S6 kinase 1 positively regulate 6-thioguanine-induced autophagy. Cancer Res 68(7):2384–2390
80. Yaswen P, Campisi J (2007) Oncogene-induced senescence pathways weave an intricate tapestry. Cell 128(2):233–234
81. Serrano M et al (1997) Oncogenic ras provokes premature cell senescence associated with accumulation of p53 and p16INK4a. Cell 88(5):593–602
82. Elgendy M et al (2011) Oncogenic Ras-induced expression of Noxa and Beclin-1 promotes autophagic cell death and limits clonogenic survival. Mol Cell 42(1):23–35
83. Wu SY et al (2011) Ras-related tumorigenesis is suppressed by BNIP3-mediated autophagy through inhibition of cell proliferation. Neoplasia 13(12):1171–1182
84. Guo JY et al (2011) Activated Ras requires autophagy to maintain oxidative metabolism and tumorigenesis. Genes Dev 25(5):460–470
85. Yang S et al (2011) Pancreatic cancers require autophagy for tumor growth. Genes Dev 25(7):717–729
86. Maddodi N et al (2010) Induction of autophagy and inhibition of melanoma growth in vitro and in vivo by hyperactivation of oncogenic BRAF. J Invest Dermatol 130(6):1657–1667
87. Zhou W et al (2011) Small interfering RNA targeting mcl-1 enhances proteasome inhibitor-induced apoptosis in various solid malignant tumors. BMC Cancer 11:485
88. Pattingre S et al (2005) Bcl-2 antiapoptotic proteins inhibit Beclin 1-dependent autophagy. Cell 122(6):927–939
89. Ku B et al (2008) An insight into the mechanistic role of Beclin 1 and its inhibition by prosurvival Bcl-2 family proteins. Autophagy 4(4):519–520
90. Hoyer-Hansen M, Jaattela M (2007) AMP-activated protein kinase: a universal regulator of autophagy? Autophagy 3(4):381–383
91. Liang J et al (2007) The energy sensing LKB1-AMPK pathway regulates p27(kip1) phosphorylation mediating the decision to enter autophagy or apoptosis. Nat Cell Biol 9(2):218–224
92. Egan D et al (2011) The autophagy initiating kinase ULK1 is regulated via opposing phosphorylation by AMPK and mTOR. Autophagy 7(6):643–644
93. Kim J, Guan KL (2011) Regulation of the autophagy initiating kinase ULK1 by nutrients: Roles of mTORC1 and AMPK. Cell Cycle 10(9):1337–1338
94. Alexander A, Walker CL (2010) Differential localization of ATM is correlated with activation of distinct downstream signaling pathways. Cell Cycle 9(18):3685–3686
95. Herrero-Martin G et al (2009) TAK1 activates AMPK-dependent cytoprotective autophagy in TRAIL-treated epithelial cells. EMBO J 28(6):677–685
96. Tzatsos A, Tsichlis PN (2007) Energy depletion inhibits phosphatidylinositol 3-kinase/Akt signaling and induces apoptosis via AMP-activated protein kinase-dependent phosphorylation of IRS-1 at Ser-794. J Biol Chem 282(25):18069–18082
97. Xu J et al (2012) MicroRNAs in autophagy and their emerging roles in crosstalk with apoptosis. Autophagy 8(6):873–882
98. Baldwin AS (2012) Regulation of cell death and autophagy by IKK and NF-kappaB: critical mechanisms in immune function and cancer. Immunol Rev 246(1):327–345
99. Jin Y et al (2012) Autophagic proteins: New facets of the oxygen paradox. Autophagy **8**(3):426–428

100. Wang RC, Levine B (2010) Autophagy in cellular growth control. FEBS Lett 584(7):1417–1426
101. Chaturvedi A, Pierce SK (2009) Autophagy in immune cell regulation and dysregulation. Curr Allergy Asthma Rep 9(5):341–346
102. Wang L et al (2012) The roles of integrin beta4 in vascular endothelial cells. J Cell Physiol 227(2):474–478
103. Lee J, Giordano S, Zhang J (2012) Autophagy, mitochondria and oxidative stress: cross-talk and redox signalling. Biochem J 441(2):523–540
104. Yang Z, Klionsky DJ (2010) Eaten alive: a history of macroautophagy. Nat Cell Biol 12(9):814–822
105. Levine B, Yuan J (2005) Autophagy in cell death: an innocent convict? J Clin Invest 115(10):2679–2688
106. Platini F et al (2010) Understanding autophagy in cell death control. Curr Pharm Des 16(1):101–113
107. Maiuri MC et al (2007) Self-eating and self-killing: crosstalk between autophagy and apoptosis. Nat Rev Mol Cell Biol 8(9):741–752
108. Chen Y, Azad MB, Gibson SB (2010) Methods for detecting autophagy and determining autophagy-induced cell death. Can J Physiol Pharmacol 88(3):285–295
109. Malhotra JD, Kaufman RJ (2007) The endoplasmic reticulum and the unfolded protein response. Semin Cell Dev Biol 18(6):716–731
110. Ron D, Walter P (2007) Signal integration in the endoplasmic reticulum unfolded protein response. Nat Rev Mol Cell Biol 8(7):519–529
111. Bertolotti A et al (2000) Dynamic interaction of BiP and ER stress transducers in the unfolded- protein response. Nat Cell Biol 2(6):326–332
112. Bernales S, Schuck S, Walter P (2007) ER-phagy: selective autophagy of the endoplasmic reticulum. Autophagy 3(3):285–287
113. Yorimitsu T et al (2006) Endoplasmic reticulum stress triggers autophagy. J Biol Chem 281(40):30299–30304
114. Kruse KB, Brodsky JL, McCracken AA (2006) Autophagy: an ER protein quality control process. Autophagy 2(2):135–137
115. Younce CW, Kolattukudy PE (2010) MCP-1 causes cardiomyoblast death via autophagy resulting from ER stress caused by oxidative stress generated by inducing a novel zinc-finger protein, MCPIP. Biochem J 426(1):43–53
116. Rodriguez-Rocha H et al (2011) DNA damage and autophagy. Mutat Res 711(1–2):158–166
117. Rieber M, Rieber MS (2008) Sensitization to radiation-induced DNA damage accelerates loss of bcl-2 and increases apoptosis and autophagy. Cancer Biol Ther 7(10):1561–1566
118. Feng Z et al (2005) The coordinate regulation of the p53 and mTOR pathways in cells. Proc Natl Acad Sci USA 102(23):8204–8209
119. Malzer E et al (2010) Impaired tissue growth is mediated by checkpoint kinase 1 (CHK1) in the integrated stress response. J Cell Sci 123(Pt 17):2892–2900
120. Crighton D et al (2006) DRAM, a p53-induced modulator of autophagy, is critical for apoptosis. Cell 126(1):121–134
121. Lorin S et al (2010) Evidence for the interplay between JNK and p53-DRAM signalling pathways in the regulation of autophagy. Autophagy 6(1):153–154
122. Feng Z, Levine AJ (2010) The regulation of energy metabolism and the IGF-1/mTOR pathways by the p53 protein. Trends Cell Biol 20(7):427–434
123. Wu H et al (2011) Central role of lactic acidosis in cancer cell resistance to glucose deprivation-induced cell death. J Pathol 227(2):189–199
124. Takagi H et al (2007) AMPK mediates autophagy during myocardial ischemia in vivo. Autophagy 3(4):405–407
125. Mihaylova MM, Shaw RJ (2011) The AMPK signalling pathway coordinates cell growth, autophagy and metabolism. Nat Cell Biol 13(9):1016–1023
126. Egan DF et al (2011) Phosphorylation of ULK1 (hATG1) by AMP-activated protein kinase connects energy sensing to mitophagy. Science 331(6016):456–461

127. Persons DA et al (1989) Increased expression of glycolysis-associated genes in oncogene-transformed and growth-accelerated states. Mol Carcinog 2(2):88–94
128. Kawauchi K et al (2008) p53 regulates glucose metabolism through an IKK-NF-kappaB pathway and inhibits cell transformation. Nat Cell Biol 10(5):611–618
129. Kondoh H et al (2005) Glycolytic enzymes can modulate cellular life span. Cancer Res 65(1):177–185
130. Matoba S et al (2006) p53 regulates mitochondrial respiration. Science 312(5780):1650–1653
131. Noman MZ et al (2011) Blocking hypoxia-induced autophagy in tumors restores cytotoxic T-cell activity and promotes regression. Cancer Res 71(18):5976–5986
132. Rouschop KM et al (2010) The unfolded protein response protects human tumor cells during hypoxia through regulation of the autophagy genes MAP1LC3B and ATG5. J Clin Invest 120(1):127–141
133. Semenza GL (2011) Hypoxia-inducible factor 1: regulator of mitochondrial metabolism and mediator of ischemic preconditioning. Biochim Biophys Acta 1813(7):1263–1268
134. Djavaheri-Mergny M et al (2006) NF-kappaB activation represses tumor necrosis factor-alpha-induced autophagy. J Biol Chem 281(41):30373–30382
135. Chen JL et al (2008) Novel roles for protein kinase Cdelta-dependent signaling pathways in acute hypoxic stress-induced autophagy. J Biol Chem 283(49):34432–34444
136. Huang J, Brumell JH (2009) NADPH oxidases contribute to autophagy regulation. Autophagy 5(6):887–889
137. Mortimore GE, Schworer CM (1977) Induction of autophagy by amino-acid deprivation in perfused rat liver. Nature 270(5633):174–176
138. Dodd KM, Tee AR (2012) Leucine and mTORC1: a complex relationship. Am J Physiol Endocrinol Metab 302(11):E1329–1342
139. Liu XM et al (2004) Platelet-derived growth factor stimulates LAT1 gene expression in vascular smooth muscle: role in cell growth. FASEB J 18(6):768–770
140. Kashiwagi H et al (2009) Regulatory mechanisms of SNAT2, an amino acid transporter, in L6 rat skeletal muscle cells by insulin, osmotic shock and amino acid deprivation. Amino Acids 36(2):219–230
141. Williams GS, Molinelli EJ, Smith GD (2008) Modeling local and global intracellular calcium responses mediated by diffusely distributed inositol 1,4,5-trisphosphate receptors. J Theor Biol 253(1):170–188
142. Eng CH et al (2010) Ammonia derived from glutaminolysis is a diffusible regulator of autophagy. Sci Signal 3(119):ra31
143. Kanzawa T et al (2004) Role of autophagy in temozolomide-induced cytotoxicity for malignant glioma cells. Cell Death Differ 11(4):448–457
144. Repnik U, Turk B (2010) Lysosomal-mitochondrial cross-talk during cell death. Mitochondrion 10(6):662–669
145. Resau JH et al (1985) Studies on the mechanisms of altered exocrine acinar cell differentiation and ductal metaplasia following nitrosamine exposure using hamster pancreatic explant organ culture. Carcinogenesis 6(1):29–35
146. Bae H, Guan JL (2011) Suppression of autophagy by FIP200 deletion impairs DNA damage repair and increases cell death upon treatments with anticancer agents. Mol Cancer Res 9(9):1232–1241
147. Gonzalez-Malerva L et al (2011) High-throughput ectopic expression screen for tamoxifen resistance identifies an atypical kinase that blocks autophagy. Proc Natl Acad Sci USA 108(5):2058–2063
148. de Medina P, Silvente-Poirot S, Poirot M (2009) Tamoxifen and AEBS ligands induced apoptosis and autophagy in breast cancer cells through the stimulation of sterol accumulation. Autophagy 5(7):1066–1067
149. Wu WK et al (2010) Macroautophagy modulates cellular response to proteasome inhibitors in cancer therapy. Drug Resist Updat 13(3):87–92
150. Ropero S, Esteller M (2007) The role of histone deacetylases (HDACs) in human cancer. Mol Oncol 1(1):19–25

151. Carew JS et al (2007) Targeting autophagy augments the anticancer activity of the histone deacetylase inhibitor SAHA to overcome Bcr-Abl-mediated drug resistance. Blood 110(1):313–322
152. Park JH et al (2011) A new synthetic HDAC inhibitor, MHY218, induces apoptosis or autophagy-related cell death in tamoxifen-resistant MCF-7 breast cancer cells. Invest New Drugs 30(5):1887–1898
153. Yamamoto S et al (2008) Suberoylanilide hydroxamic acid (SAHA) induces apoptosis or autophagy-associated cell death in chondrosarcoma cell lines. Anticancer Res 28(3A)1585–91
154. Hrzenjak A et al (2008) SAHA induces caspase-independent, autophagic cell death of endometrial stromal sarcoma cells by influencing the mTOR pathway. J Pathol 216(4):495–504
155. Saitoh T, Akira S (2010) Regulation of innate immune responses by autophagy-related proteins. J Cell Biol 189(6):925–935
156. Zhao Z et al (2008) Autophagosome-independent essential function for the autophagy protein Atg5 in cellular immunity to intracellular pathogens. Cell Host Microbe 4(5):458–469
157. Dreux M, Chisari FV (2010) Viruses and the autophagy machinery. Cell Cycle 9(7):1295–1307
158. Amaravadi RK et al (2011) Principles and current strategies for targeting autophagy for cancer treatment. Clin Cancer Res 17(4):654–666
159. Michaud M et al (2011) Autophagy-dependent anticancer immune responses induced by chemotherapeutic agents in mice. Science 334(6062):1573–1577
160. Liang XH et al (1999) Induction of autophagy and inhibition of tumorigenesis by beclin 1. Nature 402(6762):672–676
161. Aita VM et al (1999) Cloning and genomic organization of beclin 1, a candidate tumor suppressor gene on chromosome 17q21. Genomics 59(1):59–65
162. Qu X et al (2003) Promotion of tumorigenesis by heterozygous disruption of the beclin 1 autophagy gene. J Clin Invest 112(12):1809–1820
163. Yue Z et al (2003) Beclin 1, an autophagy gene essential for early embryonic development, is a haploinsufficient tumor suppressor. Proc Natl Acad Sci USA 100(25):15077–15082
164. Takamura A et al (2011) Autophagy-deficient mice develop multiple liver tumors. Genes Dev 25(8):795–800
165. Degenhardt K et al (2006) Autophagy promotes tumor cell survival and restricts necrosis, inflammation, and tumorigenesis. Cancer Cell 10(1):51–64
166. Karantza-Wadsworth V et al (2007) Autophagy mitigates metabolic stress and genome damage in mammary tumorigenesis. Genes Dev 21(13):1621–1635
167. Mathew R, White E (2007) Why sick cells produce tumors: the protective role of autophagy. Autophagy 3(5):502–505
168. Komatsu M et al (2010) The selective autophagy substrate p62 activates the stress responsive transcription factor Nrf2 through inactivation of Keap1. Nat Cell Biol 12(3):213–223
169. Lau A et al (2010) A noncanonical mechanism of Nrf2 activation by autophagy deficiency: direct interaction between Keap1 and p62. Mol Cell Biol 30(13):3275–3285
170. Inami Y et al (2011) Persistent activation of Nrf2 through p62 in hepatocellular carcinoma cells. J Cell Biol 193(2):275–284
171. Young AR et al (2009) Autophagy mediates the mitotic senescence transition. Genes Dev 23(7):798–803
172. Sinha S, Levine B (2008) The autophagy effector Beclin 1: a novel BH3-only protein. Oncogene 27(Suppl 1):S137–S148
173. Guertin DA, Sabatini DM (2007) Defining the role of mTOR in cancer. Cancer Cell 12(1):9–22
174. Diaz-Troya S et al (2008) The role of TOR in autophagy regulation from yeast to plants and mammals. Autophagy 4(7):851–865
175. Maiuri MC et al (2007) BH3-only proteins and BH3 mimetics induce autophagy by competitively disrupting the interaction between Beclin 1 and Bcl-2/Bcl-X(L). Autophagy 3(4):374–376
176. Liu J et al (2011) Beclin1 controls the levels of p53 by regulating the deubiquitination activity of USP10 and USP13. Cell 147(1):223–234

177. Wei H et al (2011) Suppression of autophagy by FIP200 deletion inhibits mammary tumorigenesis. Genes Dev 25(14):1510–1527
178. Zhang HF et al (2009) ATG16L1 T300A polymorphism and Crohn's disease susceptibility: evidence from 13,022 cases and 17,532 controls. Hum Genet 125(5–6):627–631
179. Kang C, Avery L (2008) To be or not to be, the level of autophagy is the question: dual roles of autophagy in the survival response to starvation. Autophagy 4(1):82–84
180. Park MA et al (2008) PERK-dependent regulation of HSP70 expression and the regulation of autophagy. Autophagy 4(3):364–367
181. Hsu KF et al (2009) Cathepsin L mediates resveratrol-induced autophagy and apoptotic cell death in cervical cancer cells. Autophagy 5(4):451–460
182. Bhoopathi P et al (2010) Cathepsin B facilitates autophagy-mediated apoptosis in SPARC overexpressed primitive neuroectodermal tumor cells. Cell Death Differ 17(10):1529–1539
183. Mathieu V et al (2007) Galectin-1 knockdown increases sensitivity to temozolomide in a B16F10 mouse metastatic melanoma model. J Invest Dermatol 127(10):2399–2410
184. Debnath J, Baehrecke EH, Kroemer G (2005) Does autophagy contribute to cell death? Autophagy 1(2):66–74
185. Lu Z et al (2008) The tumor suppressor gene ARHI regulates autophagy and tumor dormancy in human ovarian cancer cells. J Clin Invest 118(12):3917–3929
186. Lum JJ et al (2005) Growth factor regulation of autophagy and cell survival in the absence of apoptosis. Cell 120(2):237–248
187. O'Neill PM et al (1998) 4-Aminoquinolines–past, present, and future: a chemical perspective. Pharmacol Ther 77(1):29–58
188. Kremer JM (2001) Rational use of new and existing disease-modifying agents in rheumatoid arthritis. Ann Intern Med 134(8):695–706
189. Romanelli F, Smith KM, Hoven AD (2004) Chloroquine and hydroxychloroquine as inhibitors of human immunodeficiency virus (HIV-1) activity. Curr Pharm Des 10(21):2643–2648
190. Sotelo J, Briceno E, Lopez-Gonzalez MA (2006) Adding chloroquine to conventional treatment for glioblastoma multiforme: a randomized, double-blind, placebo-controlled trial. Ann Intern Med 144(5):337–343
191. Carmichael SJ, Charles B, Tett SE (2003) Population pharmacokinetics of hydroxychloroquine in patients with rheumatoid arthritis. Ther Drug Monit 25(6):671–681
192. Rosenfeld MRGS, Brem S, Mikkelson T, Wang D, Piao S, Davis L, O'Dwyer PJ, Amaravadi RK (2010) Pharmacokinetic analysis and pharmacodynamic evidence of autophagy inhibition in patients with newly diagnosed glioblastoma treated on a phase I trial of hydroxychloroquine in combination with adjuvant temozolomide and radiation (ABTC 0603). J Clin Oncol 28(15s):3086
193. Petiot A et al (2000) Distinct classes of phosphatidylinositol 3'-kinases are involved in signaling pathways that control macroautophagy in HT-29 cells. J Biol Chem 275(2):992–998
194. Wu YT et al (2010) Dual role of 3-methyladenine in modulation of autophagy via different temporal patterns of inhibition on class I and III phosphoinositide 3-kinase. J Biol Chem 285(14):10850–10861
195. Werner G et al (1984) Metabolic products of microorganisms. 224. Bafilomycins, a new group of macrolide antibiotics. Production, isolation, chemical structure and biological activity. J Antibiot (Tokyo) 37(2):110–117
196. Dassonneville L, Bailly C (1999) Stimulation of topoisomerase II-mediated DNA cleavage by an indazole analogue of lucanthone. Biochem Pharmacol 58(8):1307–1312
197. Luo M, Kelley MR (2004) Inhibition of the human apurinic/apyrimidinic endonuclease (APE1) repair activity and sensitization of breast cancer cells to DNA alkylating agents with lucanthone. Anticancer Res 24(4):2127–2134
198. Carew JS et al (2011) Lucanthone is a novel inhibitor of autophagy that induces cathepsin D-mediated apoptosis. J Biol Chem 286(8):6602–6613
199. Leu JI et al (2011) HSP70 inhibition by the small-molecule 2-phenylethynesulfonamide impairs protein clearance pathways in tumor cells. Mol Cancer Res 9(7):936–947

200. Nice DC et al (2002) Cooperative binding of the cytoplasm to vacuole targeting pathway proteins, Cvt13 and Cvt20, to phosphatidylinositol 3-phosphate at the pre-autophagosomal structure is required for selective autophagy. J Biol Chem 277(33):30198–30207
201. Guan J et al (2001) Cvt18/Gsa12 is required for cytoplasm-to-vacuole transport, pexophagy, and autophagy in *Saccharomyces cerevisiae* and *Pichia pastoris*. Mol Biol Cell 12(12):3821–3838
202. Stromhaug PE et al (2004) Atg21 is a phosphoinositide binding protein required for efficient lipidation and localization of Atg8 during uptake of aminopeptidase I by selective autophagy. Mol Biol Cell 15(8):3553–3566
203. Tanida I et al (2006) Atg8L/Apg8L is the fourth mammalian modifier of mammalian Atg8 conjugation mediated by human Atg4B, Atg7 and Atg3. FEBS J 273(11):2553–2562
204. Li M et al (2011) Kinetics comparisons of mammalian Atg4 homologues indicate selective preferences toward diverse Atg8 substrates. J Biol Chem 286(9):7327–7338
205. Mizushima N et al (1998) A protein conjugation system essential for autophagy. Nature 395(6700):395–398
206. Tanida I et al (1999) Apg7p/Cvt2p: A novel protein-activating enzyme essential for autophagy. Mol Biol Cell 10(5):1367–1379
207. Mizushima N, Noda T, Ohsumi Y (1999) Apg16p is required for the function of the Apg12p-Apg5p conjugate in the yeast autophagy pathway. EMBO J 18(14):3888–3896
208. Lynch-Day MA et al (2010) Trs85 directs a Ypt1 GEF, TRAPPIII, to the phagophore to promote autophagy. Proc Natl Acad Sci USA 107(17):7811–7816
209. Balderhaar HJ et al (2010) The Rab GTPase Ypt7 is linked to retromer-mediated receptor recycling and fusion at the yeast late endosome. J Cell Sci 123(Pt 23):4085–4094
210. Degtyarev M et al (2008) Akt inhibition promotes autophagy and sensitizes PTEN-null tumors to lysosomotropic agents. J Cell Biol 183(1):101–116
211. Bellodi C et al (2009) Targeting autophagy potentiates tyrosine kinase inhibitor-induced cell death in Philadelphia chromosome-positive cells, including primary CML stem cells. J Clin Invest 119(5):1109–1123
212. Gupta A et al (2010) Autophagy inhibition and antimalarials promote cell death in gastrointestinal stromal tumor (GIST). Proc Natl Acad Sci USA 107(32):14333–14338
213. Ding WX et al (2009) Oncogenic transformation confers a selective susceptibility to the combined suppression of the proteasome and autophagy. Mol Cancer Ther 8(7):2036–2045
214. Rouschop KM et al (2009) Autophagy is required during cycling hypoxia to lower production of reactive oxygen species. Radiother Oncol 92(3):411–416
215. Fan QW et al (2010) Akt and autophagy cooperate to promote survival of drug-resistant glioma. Sci Signal 3(147):81
216. Pan Y et al (2011) Targeting autophagy augments in vitro and in vivo antimyeloma activity of DNA-damaging chemotherapy. Clin Cancer Res 17(10):3248–3258
217. Sheen JH et al (2011) Defective regulation of autophagy upon leucine deprivation reveals a targetable liability of human melanoma cells in vitro and in vivo. Cancer Cell 19(5):613–628

Chapter 5
MicroRNAs in Cell Death and Cancer

Jong Kook Park and Thomas D. Schmittgen

Abstract MicroRNAs (miRNAs) are small, non-coding RNAs (ncRNAs) that post transcriptionally regulate protein levels by binding to the 3′ UTR of the mRNA. miRNAs are differentially expressed in many solid tumors and often create a unique signature for each tumor type. This chapter explains the function of miRNAs in cancer based on their potential target genes.

5.1 Introduction

Approximately 97 % of the human genome is non-coding sequences. Over the past two decades, it has been considered that non-coding RNAs (ncRNAs) are important to fundamental biologic processes, contributing significantly to pathophysiological cell mechanisms. A growing number of non-coding transcripts have been found to play important roles in several gene regulation mechanisms [1–3]. ncRNAs are traditionally classified into long ncRNAs and shorter species of ncRNAs. Three main categories of small ncRNAs are short-interfering RNAs (siRNAs), microRNAs (miRNAs), and piwi-interacting RNAs (piRNAs). It is increasingly difficult to discern the boundaries between the various small ncRNAs, however, the double-stranded structure of precursors is the feature of both siRNA and miRNA [4]. piRNAs utilize their function in the germline [5]. miRNAs are single-stranded 21–25 nucleotide ncRNAs that regulate gene expression posttranscriptionally by base-pairing with the 3′ untranslated region (3′ UTR) of target messenger RNA (mRNA) [6]. Since the discovery of the first miRNA, Lin-4, in *Caenorhabditis elegans* by Victor Ambros and Gary Ruvkun [7, 8], over 1,500 native miRNAs have been identified in vertebrates.

Given that multiple mRNAs can be targeted by a single miRNA and that individual mRNA can be modulated by numerous miRNAs, interactions of miRNA with mRNA provide significant insights into highly complicated cell signaling

J. K. Park · T. D. Schmittgen (✉)
Division of Pharmaceutics, College of Pharmacy,
Ohio State University, Columbus, OH, USA
e-mail: schmittgen.2@osu.edu

D. E. Johnson (ed.), *Cell Death Signaling in Cancer Biology and Treatment*,
Cell Death in Biology and Diseases, DOI: 10.1007/978-1-4614-5847-0_5,

X.-M. Yin and Z. Dong (Series eds.), *Cell Death in Biology and Diseases*

networks. Indeed, alterations of miRNA expression contribute to the pathogenesis of various human malignancies such as cancer, cardiovascular, neurodegenerative, and autoimmune diseases [9–12].

5.2 Location and Biogenesis of miRNAs

Genome location analysis of miRNAs has suggested that over 70 % of miRNAs are transcribed from intronic regions of both protein-coding and long non-protein-coding genes [13, 14]. Expression of these miRNAs shows the remarkable coincidence with the expression of their host gene transcripts. Otherwise, miRNAs are located in intergenic regions and apparently possess independent transcription units. Long primary transcripts containing multiple miRNAs are commonly located as clusters of polycistronic units. The first step of miRNA biogenesis is the transcription of the nascent primary miRNA precursors (pri-miRNAs), which is mainly mediated by RNA polymerase II [15] (Fig. 5.1). The pri-miRNAs have a hairpin structure containing the mature miRNA sequences. Pri-miRNAs also have a 5' cap structure and poly-A tails, similar to mRNAs [16]. Binding of DiGeorge syndrome critical region gene 8 (DGCR8) to the pri-miRNAs recruits the class 2 RNase III enzyme, Drosha, following the formation of a multi-protein complex called the microprocessor [17, 18]. Co-factors including the DEAD box RNA helicases p68 (DDX5), p72 (DDX17), and heterogeneous nuclear ribonucleoproteins (hnRNPs) also compose the microprocessor complex [18]. DGCR8 contains two double-stranded RNA-binding domains (dsRBDs) and is able to bind to a single-stranded portion of the pri-miRNAs for the appropriate processing [19]. Cleavage of the pri-miRNAs by Drosha produces ~60–70-nt precursor miRNAs (pre-miRNAs) (Fig. 5.1). Following the initial cleavage by Drosha, the pre-miRNAs are then recognized by the nuclear export protein, Exportin-5, and actively transported to the cytoplasm in a Ran-GTP-dependent manner [20]. Recently, it was proposed that other factors such as nuclear export receptor Exportin 1 (XPO1), cap-binding complex (CBC), and arsenite resistance protein 2 (ARS2) act a part of the pri-miRNAs processing [21, 22]. Once inside the cytoplasm, the endoribonuclease Dicer cleaves the terminal loop of the pre-miRNA. A partner protein, HIV-1 transactivating response (TAR) RNA-binding protein (TRBP) in humans (R2D2 and loquacious (Loqs), in *Drosophila*), assists the cleavage resulting in 22–23-nt mature miRNA duplexes (miRNA/miRNA*) (Fig. 5.1) [23–25]. Following Dicer processing, the guide strand is preferentially incorporated into miRNA-induced silencing complex (miRISC) and guides Argonaute (AGO) proteins to target genes (Fig. 5.1). miRNA*, the passenger strand, is commonly degraded [26, 27].

5.3 miRNA Decay

Expression of miRNAs is coordinated by transcription rates, processing, and decay. Treatment of RNA polymerase II inhibitors or interruption of miRNA processing demonstrated that miRNAs are highly stable and have prolonged half-lives

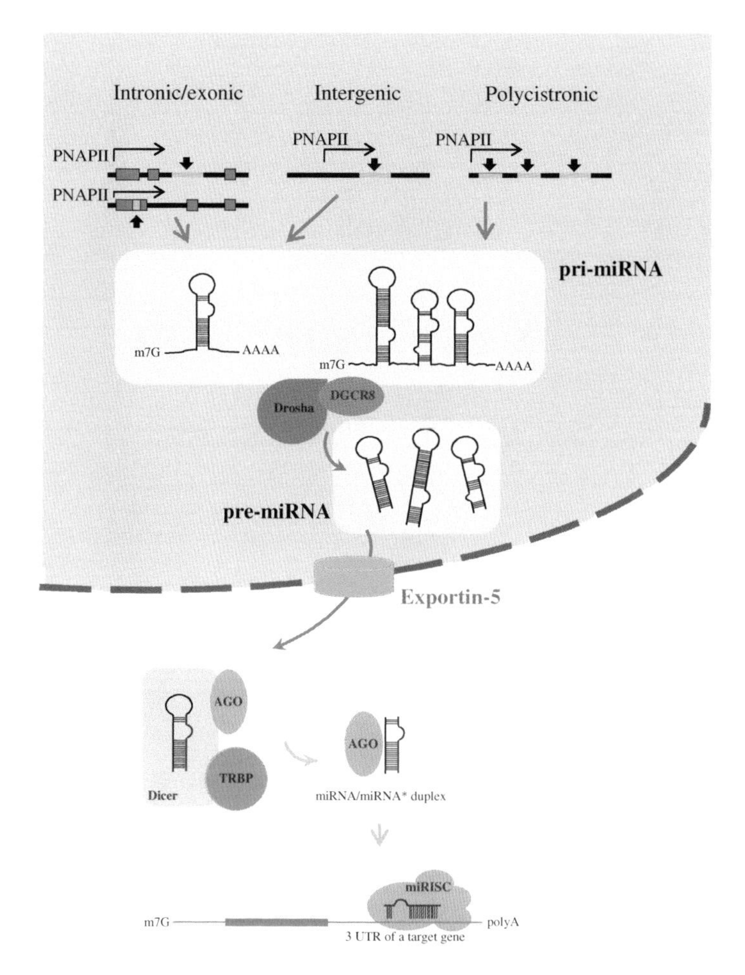

Fig. 5.1 Biogenesis of miRNAs. Primary transcripts (pri-miRNAs) are encoded from individual miRNA genes. Pri-miRNAs are processed into hairpin-structured pre-miRNAs by Drosha and a co-factor, DGCR8, in the nucleus. Pre-miRNAs are exported from the nucleus via exportin-5

from several hours to days in cell lines and organs [28]. Since miRNAs can act as on and off switches and play critical roles in developmental transitions, it is required that expression of miRNAs is actively regulated in cells. In mouse retinal neurons, turnover of several miRNAs due to rapid decay can be prevented by blocking the action potential or glutamate receptors. This suggests that turnover-mediated transition of miRNA expression plays an important role for neuronal function [29]. miR-29b, which is primarily confined in the nucleus, is an example of sequence-specific decomposition of miRNA [30]. miR-29b rapidly decays due to the uracils at position 9–11. Uracil-rich sequence affects the dynamics of miRNA decay depending on the sequence context [31]. m169, a transcript of murine cytomegalovirus (MCMV), inhibits the function of miR-27 through degradation. Its expression kinetics correlate with miR-27 degradation which implies that miRNA degradation mediated by viral RNA may be a manipulation strategy of virus against host cells [32].

Moreover, recent progress demonstrated the effect of enzymes on miRNA stability. PAP-associated domain containing 4 (PAPD4, alternative synonyms are GLD-2 and TUTase2) is the cytoplasmic poly (A) RNA polymerase. A single adenosine monophosphate monomer is added to the 3′ end of mature miR-122 by PAPD4 leading to the stabilization of miR-122. Mature miR-122 levels are especially lower in liver tissues of PAPD4-null mice than in those of heterozygous mice [33]. Stability of miRNAs is also affected by incorporation into AGO proteins. 5′–3′ exoribonuclease 2 (XRN-2) is an exoribonuclease enzyme which catalyzes the degradation of single-stranded mature miRNAs. Complexation of miRNAs and target genes with AGO proteins prevents miRNA release resulting in the protection of vulnerable miRNA substrates from degradation. Therefore, expression levels of AGO proteins affect the abundance of mature miRNA. Suppression of AGO proteins leads to diminished expression of mature miRNAs [34].

5.4 Principles of miRNA–mRNA Interactions and miRNA's Functions

A principal function of miRNAs is the inhibition of protein synthesis of protein-coding genes through suppression of translation and/or mRNA degradation. AGO family proteins are crucial elements of miRISC and directly associate with miRNAs. AGO proteins orientate miRNAs for association with their target mRNAs. There are four AGO proteins (AGO1–AGO4) in mammals. Among them, AGO2 has an RNase H-like P-element-induced wimpy testis (PIWI) domain which can endonucleolytically cleave target mRNAs at the miRNA–mRNA duplex region [35].

miRNAs interact with the 3′ UTR of their target mRNAs by means of Watson–Crick complementary for direct posttranscriptional repression. The most rigorous condition for target recognition is perfect complementary and uninterrupted base-pairing of miRNA nucleotides 2–8 (seed sequence) with target gene sequences. A residue on the opposite side at position 1 of the miRNA enhances the site efficiency even though there is no miRNA–mRNA association. In addition, additional complementarity 3′ of miRNAs stabilizes the miRNA–mRNA interactions. Furthermore, multiple targeting sites of the identical or distinct miRNAs are typically claimed for potent repression of gene expression [36]. Processing bodies (P-bodies) consist of various proteins which are necessary for miRNA-mediated gene silencing. GW182 (also known as trinucleotide repeat containing 6A), AGO proteins, and RNA helicases are enriched in P-bodies, and disruption of P-bodies can result from lack of GW182 proteins [37]. There is a positive correlation between miRNA-mediated gene silencing and the cumulation of target genes in P-bodies. For example, cationic amino acid transporter-1 (CAT-1) mRNA is localized to P-bodies when it is repressed by miR-122. Upon stress, the interaction of CAT1 with HuR results in the release of CAT-1 from P-bodies followed by re-initiation of translation activation [38]. Disappearance of P-bodies in developing mouse oocytes shows the coincidence with global suppression of miRNA functions even though there are a plentiful amount of miRNA [39, 40].

In addition to the inhibitory function of protein synthesis, miRNA can have a decoy activity. Posttranscriptional CCAAT/enhancer-binding protein (C/EBP), alpha (CEBPα) suppression is mediated by interaction with a heterogeneous ribonucleoprotein hnRNP-E2 (also known as poly(rC) binding protein 2). miR-328 interacts with hnRNP-E2 independently of seed sequence, resulting in the release of CEBPα mRNA from the translation inhibitory effect of hnRNP-E2 [41]. Furthermore, miRNAs can also paradoxically activate the target mRNA's translation. Under serum deprivation conditions, both AGO and fragile X mental retardation–related protein 1 (FXR1) are assembled into an AU-rich element (ARE) in tumor necrosis factor alpha (TNFα) mRNA, which is mediated by base-pairing of miR-369-3 with ARE to activate translation [42]. In addition, let-7 and the synthetic miR-cxcr4 repress the translation of their target genes in proliferating cells, but increase the translation of target genes in serum starved cells, all of which propose that miRNA's functions can be shifted depending on cell proliferation status [42].

5.5 miRNAs in Cancer

It is generally accepted that tumors are initially developed by genetic and epigenetic alterations of protein-coding oncogenes and tumor suppressors. On account of structured endeavors to identify additional alterations contributing to human malignancy, the importance of miRNAs has been realized in the pathogenesis of cancer over the past several years. Extensive profiling studies have shown that various miRNAs are differentially expressed in distinct types of human cancer [43–50]. Furthermore, comprehensive repression of miRNA biogenesis by targeting DGCR8, Drosha, or Dicer1 enhances cellular transformation and tumorigenesis, which distinctly signifies the role of miRNAs in cancer-related signaling pathways [51]. Deregulation of miRNAs can result from various causes such as genomic alterations (i.e., regional gain and loss at chromosomal loci), modification of transcription factors, epigenetic mechanisms, and abnormal processing of miRNAs. Undoubtedly, a number of miRNAs have been identified to have a promising role in the pathophysiology of many types of tumors. Overexpressed miRNAs in cancer can serve as oncogenes by targeting protein-coding tumor suppressor genes. On the contrary, miRNAs have been suggested as tumor suppressors based on their reduced expression, deletion status, and interaction with protein-coding oncogenes (Tables 5.1 and 5.2).

Table 5.1 Tumor-suppressive miRNAs

microRNA	Expression pattern	Identified targets	Regulation factors	References
miR-15a/16-1	Downregulated in prostate cancer, pituitary adenomas, multiple myeloma, and CLL	BCL2, MCL1, CCND1, WNT3A, FGF2, FGFR1, MAP3KIP3	Deletion	[52–55, 57–59]
miR-26a-1, miR-26a-2	Reduced in HCC and nasopharyngeal carcinoma	EZH2, CDK6, CCNE1, CCNE2, CCND2, BDNF, GSK3B	Transcribed by C/EBP alpha	[60, 63, 65–68]
miR-34a, miR-34b/34c	Loss in breast, lung, colon, pancreatic cancer, HCC, and neuroblastoma	CCND1, CCNE2, CDK4, CDK6, c-MYC, MET, E2F3, SIRT1, BCL2, CD44	Deletion, hypermethylation, transactivated by TP53	[61, 69, 70]
miR-122	Downregulated in HCC	ADAM10, SRF, IGF1R, ADAM17, CUTL1, SMARCD1	Controlled by HNF1A, HNF3A, HNF3B, and HNF4A. Transcriptionally activated by C/EBP alpha	[62, 71–73, 77, 78]
miR-143/145	Downregulated in colon, lung, cervical, and prostate cancer	DNMT3A, RREB1, MLL–AF4, ERK5, OCT4, SOX2, KLF4, EGFR, NUDT1, c-MYC, MMP11, ADAM17, CCNA2	Repressed by Ras activation. Epigenetically repressed in ALL	[79, 80, 82–85, 87, 88]

5.5.1 Tumor-Suppressive miRNAs

Association of miRNAs with human cancer was first suggested in 2002. Both miR-15a and miR-16-1 are located on chromosome 13q14 and are frequently deleted or downregulated in chronic lymphocytic leukemia (CLL) [52]. Expression of both miRNAs is also downregulated in pituitary adenomas, prostate carcinoma, multiple myeloma, and diffuse B-cell lymphomas [53–55]. The tumor-suppressive function of these two miRNAs is corroborated by their specific target genes. Anti-apoptotic proteins, B-cell lymphoma 2 (BCL2) and myeloid cell leukemia-1 (MCL1) are critical cellular oncogenes which determine proliferation, differentiation, and tumorigenesis [56]. BCL2 expression is inversely proportional to miR-15a

Table 5.2 Oncogenic miRNAs

microRNA	Expression pattern	Identified targets	Regulation factors	References
hsa-miR-21	Upregulated in most cancer	PDCD4, TPM1, SPRY2, TIMP3, RECK, MASPIN, PTEN	Induced by Ras/ERK, AP-1, NFκB, deltaEF1, and STAT3. Suppressed by FOXO3A and NFIB	[44, 90, 91, 100, 101]
miR-221/222	Overexpressed in lung, breast, prostate cancer, and HCC	CDKN1B, CDKN1C, PTEN, TIMP3, TRPS1, PUMA	Induced by NFκB, c-Jun, and FOSL1. Negatively regulated by ER in breast cancer. Modulated by EGFR and MET	[102, 103, 105–108, 110, 111]
miR-17-92	Upregulated in colon, lung, breast cancer, and B-cell lymphomas	ZBTB4, E2F1, PTEN, BCL2L11, CDKN1A, DKK3, TGFBR2, SMAD4, TSP1, CTGF	Transactivated by E2F, MYCN, and c-Myc. Negatively regulated by p53 under hypoxic condition	[113–119]
miR-106b-25	Increased in pancreas, prostate, colon cancer, and HCC	CDKN1A, BCL2L11, ITGB8, PTEN, E2F1, E2F3, BCL2L11	Activated by E2F1 and E2F3	[120–125]
miR-130/301	Upregulated in breast, colon, and pancreatic cancer	HOXA5, GAX, TP53INP1, RUNX3, CDKN1A, FOXF2, BBC3, PTEN, COL2A1, NKRF	Transcribed by NFκB	[126–131]
miR-155	Upregulated in thyroid, breast, colon, cervical, lung cancer, and PDAC	SHIP1, TP53INP1, HGAL, SOCS1, PPP2CA, FOXO3A	Epigenetically repressed by BRCA1. Transcriptionally activated by NFκB and MYB	[132–142]

and miR-16-1 in CLL, and both miRNAs post transcriptionally repress BCL2 [57]. Transcriptome and proteome analysis in MEG-01 cells showed that ectopic expression of miR-15a and miR-16-1 results in upregulation or downregulation of numerous genes including MCL1 [58]. In prostate cancer cells, reconstitution of miR-15a and miR-16-1 expression using lentiviral vectors induces apoptosis and growth arrest, whereas knockdown of miR-15a and miR-16-1 leads to tumorigenesis of untransformed prostate cells [53]. Other than BCL-2, cyclin D1 (CCND1) and wingless-type MMTV integration site family, member 3A (WNT3A) both of which elevate tumorigenic potentials are directly regulated by the miR-15 and miR-16-1 cluster [53]. Fibroblast growth factor 2 (FGF2) and fibroblast growth factor receptor 1 (FGFR1) are also repressed by direct interactions with miR-15a and miR-16-1 in prostate cancer–associated fibroblasts resulting in promotion of cell growth and migration [59].

Supporting the prominent role of miRNAs in cancers, miR-26a, the miR-34 family, and miR-122 are identified to be commonly downregulated in hepatocellular carcinoma (HCC) [60–62]. Two distinct host mRNA genes, carboxy-terminal domain, RNA polymerase II, polypeptide A small phosphatase-like (CTDSPL) and carboxy-terminal domain, RNA polymerase II, polypeptide A small phosphatase 2 (CTDSP2) encode pri-miR-26a-1 and pri-miR-26a-2, respectively, generating identical mature miR-26a. Transcription of pri-miR-26a-1 is directly activated by C/EBPα [63]. C/EBPα impedes tumor cell growth through stabilization of cyclin-dependent kinase inhibitor 1A (CDKN1A, p21, Cip1) and is commonly downregulated in HCC [64]. miR-26a has tumor-suppressive biologic effects by targeting several protein-coding oncogenes. For example, miR-26a regulates cyclin D2, cyclin E1, and cyclin E2, all of which regulate cell cycle transition from G1 to S phase [65, 66]. Additional targets of miR-26a include brain-derived neurotrophic factor (BDNF) and enhancer of zeste homolog 2 (EZH2) [67, 68].

Recent work also suggests the connection between the miR-34 family and tumor suppressor p53. Primary transcripts of mature miR-34a and miR-34b/c are activated by p53 and further processed into mature forms. Expression of the miR-34 family is downregulated in human cancers due to frequent functional p53 deficiency (reviewed in [69]). Deletion of miR-34 family members has also been observed in several types of cancer [69]. Validated target genes of the miR-34 family are associated with apoptosis, senescence, migration, and cell cycle mechanisms [69, 70]. This further supports the p53 network that multiple transcriptional targets of p53 act in concert with miR-34 family to suppress tumor formation.

Several miRNAs are expressed in a tissue-specific manner, implying their involvement in tissue specification. miR-122 is a liver-specific miRNA whose expression is controlled by C/EBPα, hepatocyte nuclear factor 1α, 3α, 3β, 4α, and REV-ERBα. miR-122 is highly abundant in hepatocytes, but undetectable in other tissues [71–73]. miR-122 modulates hepatitis C virus accumulation and cholesterol metabolic pathways in liver cells [74, 75]. miR-122 is frequently downregulated in HCC patients with poor prognosis, suggesting that miR-122 is an important regulator of HCC pathogenesis [76]. Indeed, reconstruction of miR-122 expression in HCC cells enhances the effects of sorafenib and doxorubicin and reduces tumorigenic properties such as angiogenesis and metastasis in orthotopic transplantation HCC models [77, 78].

Additional miRNAs also target multiple protein-coding oncogenes. Expression of the miR-143 and miR-145 cluster is broadly downregulated in several types of cancer including colon, lung, cervical, and prostate cancer [79–82]. Expression of miR-143 by promoter hypermethylation is also repressed in acute lymphoblastic leukemia (ALL) [83]. In addition, miR-145 is weakly expressed in human embryonic stem cells and regulates the pluripotency factors, OCT4, SOX2, and KLF4 [84]. miR-145 also targets EGFR and nucleoside diphosphate linked moiety X-type motif 1 (NUDT1) resulting in the inhibition of cell proliferation of human lung adenocarcinoma cells [85]. Overexpression of miR-143 also induces apoptosis and impairs the growth of a xenograft model of human colon carcinoma [86]. DNA methyltranferase 3A (DNMT3A) is targeted by miR-143, and overexpression of miR-143 in colon cells halts tumor growth and colony formation [87]. KRAS and Ras-responsive element-binding protein 1 (RREB1) are direct targets of the miR-143/145 cluster. Constitutively, active KRAS induces RREB1 and restrains the transcription of miR-143/145. This suggests that cellular transformation can be initiated by KRAS along with activation of other oncogenic signals related with the miR-143/145 cluster [88]. Depending on the cellular context, the miR-143/145 cluster acts as oncogenic miRNAs. It was reported that expression of miR-143 is elevated in hepatitis B virus (HBV)-related HCC [89]. Nuclear factor kappa B (NFκB) transcribes miR-143 and facilitates invasion and metastasis of HCC in vivo via targeting fibronectin type III domain containing 3B (FNDC3B). Knockdown of miR-143 expression in vivo also leads to inhibition of metastasis of HBV-HCC in transgenic mice [89].

5.5.2 Oncogenic miRNAs

One of the first miRNAs identified as oncogenic in the human genome is miR-21. miR-21 was first identified in glioblastoma [90] and is upregulated in most types of human malignancies [44]. A term "oncomiR addiction" markedly describes the momentous function of miR-21 on all stages of tumor development. Overexpression of miR-21 in a genetically engineered mouse model facilitates a pre-B malignant lymphoid-like phenotype [91]. Switching off miR-21 expression leads to tumor regression and recovery from symptoms of lymphoma, clearly illustrating the role of miR-21 in tumor initiation as well as maintenance of established tumors. Furthermore, a survival advantage followed by the suppression of miR-21 also demonstrates the clinical impact of targeting one aberrantly expressed miRNA [91]. Posttranscriptional regulation of miR-21 is mediated by the receptor-mediated SMADs (R-SMADs). Interactions of R-SMADs with primary transcripts of miR-21 (pri-miR-21) promote Drosha processing by stabilizing the binding of pri-miR-21 with Drosha. TGF-β and BMP4 enhance the expression of mature miR-21 through the complex formation of R-SMADs, pri-miR-21, and DEAD (Asp-Glu-Ala-Asp) box polypeptide 5 (DDX5, RNA helicase, p68, a component of microprocessor complex) [92]. Moreover, miR-21 is located at a TMEM49 intronic region, and several pathways modulate the expression of miR-21 at the transcription level. Akt2 transcriptionally activates miR-21 induction through binding of NFκB, CREB, and CBP/p300 to the miR-21 promoter regions under conditions of hypoxia [93]. AP-1 and deltaEF1 increase miR-21 expression by binding to the miR-21 promoter. Treatment of BMP-6 transcriptionally inhibits miR-21 expression via suppression of AP-1 and deltaEF1 in breast tumor cells [94]. In addition, signal transducer and activator of transcription 3 (STAT3) also directly binds to the miR-21 promoter and regulates miR-21 levels [94]. Interferon (IFN) increases miR-21 through co-regulation of STAT3 and NFκB. Apoptosis induced by IFN can be suppressed due to an increased amount of mature miR-21 [95]. Enhanced expression of miR-21 resulted from activation of EGFR signaling and neurotensin treatment in lung cancer and colon tumors, respectively [96, 97]. In contrast, suppression of miR-21 promoter activity is induced by the binding of FOXO3A to its promoter region. FOXO3A is a trigger of apoptosis of tumor cells via upregulation of several pro-apoptotic genes such as Bim and PUMA. Fas ligand (FasL) is a target of miR-21, and upregulation of FasL by FOXO3A partly results from miR-21 depletion [98]. Nuclear factor I/B (NFIB) also binds to the miR-21 promoter and acts as a negative regulator of miR-21 transcription [99]. The potential oncogenic activity of miR-21 has been further supported by numerous validated target genes (reviewed in [100] and summarized in Table 5.2). All of these are closely related with growth, apoptosis, invasion, and metastasis of tumor cells. Overexpression of miR-21 therefore leads to facilitation of growth, evasion of apoptosis, potent metastasis, and resistance toward anticancer agents [100, 101].

Deregulation of other oncogenic miRNAs including miR-221 and miR-222 has also been observed in human malignancies. miR-221 and miR-222 are clustered in an intergenic region on chromosome Xp11.3. They have the same seed sequence, thus share common predicted target genes. They are upregulated in lung, breast, prostate cancer, and HCC [102–104]. Contribution of the miR-221/222 cluster to tumorigenesis of lung cancer and HCC has been suggested. Expression of miR-221/222 is modulated by EGFR and MET receptors in lung cancer, and they contribute to gefitinib resistance and tumorigenesis of non-small-cell lung cancer (NSCLC). Overexpression of miR-221 facilitates the growth of tumorigenic murine hepatic progenitor cells in vivo, suggesting the important role of miR-221 in hepatocarcinogenesis [105]. Oncogenic activities of miR-221/222 are also related with their transcription modulators. NFκB and c-Jun contribute to the oncogenesis of prostate carcinoma and glioblastoma and trigger the expression of miR-221/222 [106]. A downstream transcription factor of EMT-promoting Ras-ERK pathway, FOSL1, stimulates the expression of miR-221/222 in breast cancer. FOSL1 is present in basal-like (BL) sub-type breast cancers; therefore, upregulation of miR-221/222 by FOSL1 in breast cancer contributes to aggressiveness of BL sub-type partly by targeting tricho-rhino-phalangeal syndrome type 1 (TRPS1) which represses the GATA family at the transcriptional level and decreases ZEB2 expression [107]. miR-221/222 are negatively modulated by the binding of ER alpha together with nuclear receptor co-repressor (NCoR) and silencing mediator of retinoic acid and thyroid hormone receptor (SMRT) [108]. A number of evaluated target genes of miR-221/222 such as CDKN1B (p27, Kip1), PTEN, PUMA, and TIMP3 also substantiate the functional significance of miR-221/222 in tumor proliferation, apoptosis, invasion, and metastasis [102, 109–111]. This combined knowledge suggests the therapeutic potential of miR-221/222 as a treatment for cancer.

Of all of the miRNA clusters, the miR-17~92 polycistronic cluster is one of the most studied. This oncogenic cluster is located on chromosome 13q31.3 and is transcribed from a long non-coding miR-17~92 cluster host gene (MIR17HG). Six different mature miRNAs (miR-17, miR-18a, miR-19a, miR-20a, miR-19b, and miR-92a) are produced from a single primary transcript. Based on homology of their seed sequence, six mature miRNAs can be sorted into four groups, that is, miR-17/miR-20a (AAAGUG), miR-18a (AAGGUG), miR-19a/miR-19b (GUGCAA), and miR-92a (AUUGCA). Mature miRNAs from this cluster are overexpressed in breast, lung, pancreas, colon, prostate, stomach cancer, chronic myeloid leukemia (CML), and B-cell lymphomas (reviewed in [112]). The miR-17~92 cluster gene is transcriptionally activated by MYC [113] and MYCN [114] and suppressed by p53 [115]. Overexpression of the miR-17~92 cluster gene facilitates the onset and progression of c-Myc-induced lymphoma, indicating an orchestrated interaction between c-Myc and the miR-17~92 cluster for tumorigenesis [113]. In addition, stimulating angiogenesis and tumor growth by overexpression of miR-17~92 is derived from the weakened TGF beta signals by repressing related target genes, TGFBR2 and SMAD4 [116]. Functional studies related with their targets have been also successful, suggesting that they are genuine oncogenic miRNAs [112, 117, 118]. The effects of individual miRNAs of the miR-17~92 cluster in human cancer still remain to be uncovered, but overexpression of miR-17~92 generally enhances cell cycle progression and inhibits apoptosis of tumor cells, for example, through repression of E2F1, PTEN, and BCL2L11 [119].

The intronic miR-106b-25 cluster is a paralog of miR-17~92 polycistronic gene cluster. It is transcribed from intron 13 of its host gene, minichromosome maintenance protein 7 (MCM7), and contains mature miR-106b, miR-93, and miR-25 [120]. Inhibition of a miR-106b~25 cluster effectively inhibits proliferation and anchorage-independent growth of HCC cells [121]. A miR-106b~25 cluster is a target of E2F1 and E2F3. In turn, E2F1 is repressed by miR-106b and miR-93. Also, the pro-apoptotic BCL2L11 gene is also targeted by miR-25, all of which imply that the miR-106b~25 cluster can blunt the effects of overexpressed E2F which leads to cell death [122, 123]. Recently, overexpression of a single miR-93 enhances tumor growth and angiogenesis in human astrocytoma U87 cells by targeting integrin, beta 8 (ITGB8) [124]. Expression of the oncogenic miR-106b~25 cluster is upregulated in prostate cancer, and all individual miRNAs of this cluster directly target and repress PTEN expression, suggesting the association of a miR-106b~25 cluster with prostate tumorigenesis [125].

The miR-130/301 seed family is also known as an oncogenic miRNA. Four members of this family are miR-130a, miR-130b, miR-301a, and miR-301b. Their 5’ seed sequences are identical and thus would be expected to regulate the identical target genes. Consistent with the effects of NFκB on regulating oncogenic miR-21 and miR-221/222 expression, miR-301a transcription is also promoted by NFκB. In turn, miR-301a directly suppresses NFκB-repressing factor (NKRF), which implies a positive feedback mechanism to elevate NFκB activity in tumor cells [126]. miR-130a regulates the angiogenic phenotype of vascular endothelial cells by direct modulation of anti-angiogenic factors, GAX and HOXA5 [127]. Experimentally confirmed target genes of miR-130b are tumor protein p53-inducible nuclear protein 1 (TP53INP1), runt-related transcription factor 3 (RUNX3), and CDKN1A (p21, Cip1), indicating the functional contribution to cell growth and metastasis [128–130]. Upregulation of miR-301 is considered as a poor prognosis factor of invasive ductal breast cancer. Downregulation of miR-301 expression in breast tumor cells results in reduced cell proliferation and migration. Nodal or distant relapses of breast cancer are considered to be related with target genes of miR-301, namely, FOXF2, BBC3, PTEN, and COL2A1 [131].

Double-stranded fold-back motif of B-cell integration cluster (BIC) RNA is the precursor of miR-155. BIC/miR-155 is greatly overexpressed in Hodgkin's lymphoma and diffuse large-cell B-cell lymphoma [132]. Explicit evidence of involvement of miR-155 in B-cell malignancy is that ALL is generated in a transgenic mouse overexpressing miR-155 specifically in B-cell lineage [133]. Deregulation of miR-155 expression has been shown to be related with several other kinds of cancer. miR-155 is upregulated in thyroid carcinoma, breast cancer, colon cancer, cervical cancer, pancreatic ductal adenocarcinoma (PDAC), and lung cancer (reviewed in [134]). miR-155 is one of the most promising bona fide oncogenic miRNAs suppressing apoptosis via efficient repression of several tumor suppressor genes [135–139]. It was also reported that Akt activity is enhanced through downregulation of phosphatase protein phosphatase 2A catalytic subunit alpha (PPP2A) mediated by miR-155 [97]. Transcription of miR-155 is activated by NFκB and MYB transcription factors [140, 141]. Recently, it was demonstrated that BRCA1 associated with HDAC2 epigenetically restrains the expression of miR-155 through deacetylation of histones on the miR-155 promoter [142].

In conclusion, since the discovery of Lin-4 in *C. elegans*, much progress has been made in characterizing miRNAs, in particular, their role in gene regulation. Better understanding of miRNA involvement in the regulation of the cancerous networks may lead to successful miRNA-based cancer therapy in the near future.

References

1. Holley CL, Topkara VK (2011) An introduction to small non-coding RNAs: miRNA and snoRNA. Cardiovasc Drugs Ther/Sponsored Int Soc Cardiovasc Pharmacother 25(2):151–159
2. Mattick JS, Makunin IV (2006) Non-coding RNA. Hum Mol Genet 15(1):R17–R29
3. Huarte M, Guttman M, Feldser D, Garber M, Koziol MJ, Kenzelmann-Broz D, Khalil AM, Zuk O, Amit I, Rabani M, Attardi LD, Regev A, Lander ES, Jacks T, Rinn JL (2010) A large intergenic noncoding RNA induced by p53 mediates global gene repression in the p53 response. Cell 142(3):409–419

4. Carthew RW, Sontheimer EJ (2009) Origins and mechanisms of miRNAs and siRNAs. Cell 136(4):642–655
5. Malone CD, Hannon GJ (2009) Small RNAs as guardians of the genome. Cell 136(4):656–668
6. Ambros V (2004) The functions of animal microRNAs. Nature 431(7006):350–355
7. Lee RC, Feinbaum RL, Ambros V (1993) The *C. elegans* heterochronic gene lin-4 encodes small RNAs with antisense complementarity to lin-14. Cell 75(5):843–854
8. Wightman B, Ha I, Ruvkun G (1993) Posttranscriptional regulation of the heterochronic gene lin-14 by lin-4 mediates temporal pattern formation in *C. elegans*. Cell 75(5):855–862
9. Wiemer EA (2007) The role of microRNAs in cancer: no small matter. Eur J Cancer 43(10):1529–1544
10. Zhao Y, Samal E, Srivastava D (2005) Serum response factor regulates a muscle-specific microRNA that targets Hand2 during cardiogenesis. Nature 436(7048):214–220
11. Schaefer A, O'Carroll D, Tan CL, Hillman D, Sugimori M, Llinas R, Greengard P (2007) Cerebellar neurodegeneration in the absence of microRNAs. J Exp Med 204(7):1553–1558
12. Taganov KD, Boldin MP, Chang KJ, Baltimore D (2006) NF-kappaB-dependent induction of microRNA miR-146, an inhibitor targeted to signaling proteins of innate immune responses. Proc Nat Acad Sci USA 103(33):12481–12486
13. Rodriguez A, Griffiths-Jones S, Ashurst JL, Bradley A (2004) Identification of mammalian microRNA host genes and transcription units. Genome Res 14(10A):1902–1910
14. Berezikov E, Chung WJ, Willis J, Cuppen E, Lai EC (2007) Mammalian mirtron genes. Mol Cell 28(2):328–336
15. Lee Y, Kim M, Han J, Yeom KH, Lee S, Baek SH, Kim VN (2004) MicroRNA genes are transcribed by RNA polymerase II. EMBO J 23(20):4051–4060
16. Cai X, Hagedorn CH, Cullen BR (2004) Human microRNAs are processed from capped, polyadenylated transcripts that can also function as mRNAs. RNA 10(12):1957–1966
17. Denli AM, Tops BB, Plasterk RH, Ketting RF, Hannon GJ (2004) Processing of primary microRNAs by the microprocessor complex. Nature 432(7014):231–235
18. Gregory RI, Yan KP, Amuthan G, Chendrimada T, Doratotaj B, Cooch N, Shiekhattar R (2004) The microprocessor complex mediates the genesis of microRNAs. Nature 432(7014):235–240
19. Senturia R, Faller M, Yin S, Loo JA, Cascio D, Sawaya MR, Hwang D, Clubb RT, Guo F (2010) Structure of the dimerization domain of DiGeorge critical region 8. Protein Sci publ Protein Soc 19(7):1354–1365
20. Yi R, Qin Y, Macara IG, Cullen BR (2003) Exportin-5 mediates the nuclear export of pre-microRNAs and short hairpin RNAs. Genes Dev 17(24):3011–3016
21. Bussing I, Yang JS, Lai EC, Grosshans H (2010) The nuclear export receptor XPO-1 supports primary miRNA processing in *C. elegans* and *Drosophila*. EMBO J 29(11):1830–1839
22. Gruber JJ, Zatechka DS, Sabin LR, Yong J, Lum JJ, Kong M, Zong WX, Zhang Z, Lau CK, Rawlings J, Cherry S, Ihle JN, Dreyfuss G, Thompson CB (2009) Ars2 links the nuclear cap-binding complex to RNA interference and cell proliferation. Cell 138(2):328–339
23. Haase AD, Jaskiewicz L, Zhang H, Laine S, Sack R, Gatignol A, Filipowicz W (2005) TRBP, a regulator of cellular PKR and HIV-1 virus expression, interacts with Dicer and functions in RNA silencing. EMBO Rep 6(10):961–967
24. Tsutsumi A, Kawamata T, Izumi N, Seitz H, Tomari Y (2011) Recognition of the pre-miRNA structure by *Drosophila* dicer-1. Nat Struct Mol Biol 18(10):1153–1158
25. Saito K, Ishizuka A, Siomi H, Siomi MC (2005) Processing of pre-microRNAs by the dicer-1-loquacious complex in *Drosophila* cells. PLoS Biol 3(7):e235
26. Matranga C, Tomari Y, Shin C, Bartel DP, Zamore PD (2005) Passenger-strand cleavage facilitates assembly of siRNA into Ago2-containing RNAi enzyme complexes. Cell 123(4):607–620
27. Okamura K, Ishizuka A, Siomi H, Siomi MC (2004) Distinct roles for argonaute proteins in small RNA-directed RNA cleavage pathways. Genes Dev 18(14):1655–1666

28. Gantier MP, McCoy CE, Rusinova I, Saulep D, Wang D, Xu D, Irving AT, Behlke MA, Hertzog PJ, Mackay F, Williams BR (2011) Analysis of microRNA turnover in mammalian cells following dicer1 ablation. Nucleic Acids Res 39(13):5692–5703
29. Krol J, Busskamp V, Markiewicz I, Stadler MB, Ribi S, Richter J, Duebel J, Bicker S, Fehling HJ, Schubeler D, Oertner TG, Schratt G, Bibel M, Roska B, Filipowicz W (2010) Characterizing light-regulated retinal microRNAs reveals rapid turnover as a common property of neuronal microRNAs. Cell 141(4):618–631
30. Hwang HW, Wentzel EA, Mendell JT (2007) A hexanucleotide element directs microRNA nuclear import. Science 315(5808):97–100
31. Zhang Z, Zou J, Wang GK, Zhang JT, Huang S, Qin YW, Jing Q (2011) Uracils at nucleotide position 9–11 are required for the rapid turnover of miR-29 family. Nucleic Acids Res 39(10):4387–4395
32. Libri V, Helwak A, Miesen P, Santhakumar D, Borger JG, Kudla G, Grey F, Tollervey D, Buck AH (2012) Murine cytomegalovirus encodes a miR-27 inhibitor disguised as a target. Proc Nat Acad Sci USA 109(1):279–284
33. Katoh T, Sakaguchi Y, Miyauchi K, Suzuki T, Kashiwabara S, Baba T, Suzuki T (2009) Selective stabilization of mammalian microRNAs by 3' adenylation mediated by the cytoplasmic poly(A) polymerase GLD-2. Genes Dev 23(4):433–438
34. Chatterjee S, Fasler M, Bussing I, Grosshans H (2011) Target-mediated protection of endogenous microRNAs in *C. elegans*. Dev Cell 20(3):388–396
35. Hammell CM (2008) The microRNA-argonaute complex: a platform for mRNA modulation. RNA Biol 5(3):123–127
36. Wang Y, Li Y, Ma Z, Yang W, Ai C (2010) Mechanism of microRNA-target interaction: molecular dynamics simulations and thermodynamics analysis. PLoS Comput Biol 6(7):e1000866
37. Liu J, Valencia-Sanchez MA, Hannon GJ, Parker R (2005) MicroRNA-dependent localization of targeted mRNAs to mammalian P-bodies. Nat Cell Biol 7(7):719–723
38. Bhattacharyya SN, Habermacher R, Martine U, Closs EI, Filipowicz W (2006) Stress-induced reversal of microRNA repression and mRNA P-body localization in human cells. Cold Spring Harb Symp Quant Biol 71:513–521
39. Ma J, Flemr M, Stein P, Berninger P, Malik R, Zavolan M, Svoboda P, Schultz RM (2010) MicroRNA activity is suppressed in mouse oocytes. Current Biol CB 20(3):265–270
40. Suh N, Baehner L, Moltzahn F, Melton C, Shenoy A, Chen J, Blelloch R (2010) MicroRNA function is globally suppressed in mouse oocytes and early embryos. Current Biol CB 20(3):271–277
41. Eiring AM, Harb JG, Neviani P, Garton C, Oaks JJ, Spizzo R, Liu S, Schwind S, Santhanam R, Hickey CJ, Becker H, Chandler JC, Andino R, Cortes J, Hokland P, Huettner CS, Bhatia R, Roy DC, Liebhaber SA, Caligiuri MA, Marcucci G, Garzon R, Croce CM, Calin GA, Perrotti D (2010) miR-328 functions as an RNA decoy to modulate hnRNP E2 regulation of mRNA translation in leukemic blasts. Cell 140(5):652–665
42. Vasudevan S, Tong Y, Steitz JA (2007) Switching from repression to activation: microRNAs can up-regulate translation. Science 318(5858):1931–1934
43. Lu J, Getz G, Miska EA, Alvarez-Saavedra E, Lamb J, Peck D, Sweet-Cordero A, Ebert BL, Mak RH, Ferrando AA, Downing JR, Jacks T, Horvitz HR, Golub TR (2005) MicroRNA expression profiles classify human cancers. Nature 435(7043):834–838
44. Volinia S, Calin GA, Liu CG, Ambs S, Cimmino A, Petrocca F, Visone R, Iorio M, Roldo C, Ferracin M, Prueitt RL, Yanaihara N, Lanza G, Scarpa A, Vecchione A, Negrini M, Harris CC, Croce CM (2006) A microRNA expression signature of human solid tumors defines cancer gene targets. Proc Nat Acad Sci USA 103(7):2257–2261
45. Budhu A, Jia HL, Forgues M, Liu CG, Goldstein D, Lam A, Zanetti KA, Ye QH, Qin LX, Croce CM, Tang ZY, Wang XW (2008) Identification of metastasis-related microRNAs in hepatocellular carcinoma. Hepatology 47(3):897–907

46. Murakami Y, Yasuda T, Saigo K, Urashima T, Toyoda H, Okanoue T, Shimotohno K (2006) Comprehensive analysis of microRNA expression patterns in hepatocellular carcinoma and non-tumorous tissues. Oncogene 25(17):2537–2545
47. Subramanian S, Lui WO, Lee CH, Espinosa I, Nielsen TO, Heinrich MC, Corless CL, Fire AZ, van de Rijn M (2008) MicroRNA expression signature of human sarcomas. Oncogene 27(14):2015–2026
48. Navarro A, Gaya A, Martinez A, Urbano-Ispizua A, Pons A, Balague O, Gel B, Abrisqueta P, Lopez-Guillermo A, Artells R, Montserrat E, Monzo M (2008) MicroRNA expression profiling in classic Hodgkin lymphoma. Blood 111(5):2825–2832
49. Dyrskjot L, Ostenfeld MS, Bramsen JB, Silahtaroglu AN, Lamy P, Ramanathan R, Fristrup N, Jensen JL, Andersen CL, Zieger K, Kauppinen S, Ulhoi BP, Kjems J, Borre M, Orntoft TF (2009) Genomic profiling of microRNAs in bladder cancer: miR-129 is associated with poor outcome and promotes cell death in vitro. Cancer Res 69(11):4851–4860
50. Calin GA, Ferracin M, Cimmino A, Di Leva G, Shimizu M, Wojcik SE, Iorio MV, Visone R, Sever NI, Fabbri M, Iuliano R, Palumbo T, Pichiorri F, Roldo C, Garzon R, Sevignani C, Rassenti L, Alder H, Volinia S, Liu CG, Kipps TJ, Negrini M, Croce CM (2005) A microRNA signature associated with prognosis and progression in chronic lymphocytic leukemia. N Engl J Med 353(17):1793–1801
51. Kumar MS, Lu J, Mercer KL, Golub TR, Jacks T (2007) Impaired microRNA processing enhances cellular transformation and tumorigenesis. Nat Genet 39(5):673–677
52. Calin GA, Dumitru CD, Shimizu M, Bichi R, Zupo S, Noch E, Aldler H, Rattan S, Keating M, Rai K, Rassenti L, Kipps T, Negrini M, Bullrich F, Croce CM (2002) Frequent deletions and down-regulation of micro- RNA genes miR15 and miR16 at 13q14 in chronic lymphocytic leukemia. Proc Nat Acad Sci USA 99(24):15524–15529
53. Bonci D, Coppola V, Musumeci M, Addario A, Giuffrida R, Memeo L, D'Urso L, Pagliuca A, Biffoni M, Labbaye C, Bartucci M, Muto G, Peschle C, De Maria R (2008) The miR-15a-miR-16-1 cluster controls prostate cancer by targeting multiple oncogenic activities. Nat Med 14(11):1271–1277
54. Bottoni A, Piccin D, Tagliati F, Luchin A, Zatelli MC, Degli Uberti EC (2005) miR-15a and miR-16-1 down-regulation in pituitary adenomas. J Cell Physiol 204(1):280–285
55. Roccaro AM, Sacco A, Thompson B, Leleu X, Azab AK, Azab F, Runnels J, Jia X, Ngo HT, Melhem MR, Lin CP, Ribatti D, Rollins BJ, Witzig TE, Anderson KC, Ghobrial IM (2009) MicroRNAs 15a and 16 regulate tumor proliferation in multiple myeloma. Blood 113(26):6669–6680
56. Youle RJ, Strasser A (2008) The BCL-2 protein family: opposing activities that mediate cell death. Nat Rev Mol Cell Biol 9(1):47–59
57. Cimmino A, Calin GA, Fabbri M, Iorio MV, Ferracin M, Shimizu M, Wojcik SE, Aqeilan RI, Zupo S, Dono M, Rassenti L, Alder H, Volinia S, Liu CG, Kipps TJ, Negrini M, Croce CM (2005) miR-15 and miR-16 induce apoptosis by targeting BCL2. Proc Nat Acad Sci USA 102(39):13944–13949
58. Calin GA, Cimmino A, Fabbri M, Ferracin M, Wojcik SE, Shimizu M, Taccioli C, Zanesi N, Garzon R, Aqeilan RI, Alder H, Volinia S, Rassenti L, Liu X, Liu CG, Kipps TJ, Negrini M, Croce CM (2008) MiR-15a and miR-16-1 cluster functions in human leukemia. Proc Nat Acad Sci USA 105(13):5166–5171
59. Musumeci M, Coppola V, Addario A, Patrizii M, Maugeri-Sacca M, Memeo L, Colarossi C, Francescangeli F, Biffoni M, Collura D, Giacobbe A, D'Urso L, Falchi M, Venneri MA, Muto G, De Maria R, Bonci D (2011) Control of tumor and microenvironment cross-talk by miR-15a and miR-16 in prostate cancer. Oncogene 30(41):4231–4242
60. Chen L, Zheng J, Zhang Y, Yang L, Wang J, Ni J, Cui D, Yu C, Cai Z (2011) Tumor-specific expression of microRNA-26a suppresses human hepatocellular carcinoma growth via cyclin-dependent and -independent pathways. Mol Ther J Am Soc Gene Ther 19(8):1521–1528

61. Tryndyak VP, Ross SA, Beland FA, Pogribny IP (2009) Down-regulation of the microRNAs miR-34a, miR-127, and miR-200b in rat liver during hepatocarcinogenesis induced by a methyl-deficient diet. Mol Carcinog 48(6):479–487
62. Tsai WC, Hsu PW, Lai TC, Chau GY, Lin CW, Chen CM, Lin CD, Liao YL, Wang JL, Chau YP, Hsu MT, Hsiao M, Huang HD, Tsou AP (2009) MicroRNA-122, a tumor suppressor microRNA that regulates intrahepatic metastasis of hepatocellular carcinoma. Hepatology 49(5):1571–1582
63. Mohamed JS, Lopez MA, Boriek AM (2010) Mechanical stretch up-regulates microRNA-26a and induces human airway smooth muscle hypertrophy by suppressing glycogen synthase kinase-3beta. J Biol Chem 285(38):29336–29347
64. Harris TE, Albrecht JH, Nakanishi M, Darlington GJ (2001) CCAAT/enhancer-binding protein-alpha cooperates with p21 to inhibit cyclin-dependent kinase-2 activity and induces growth arrest independent of DNA binding. J Biol Chem 276(31):29200–29209
65. Kota J, Chivukula RR, O'Donnell KA, Wentzel EA, Montgomery CL, Hwang HW, Chang TC, Vivekanandan P, Torbenson M, Clark KR, Mendell JR, Mendell JT (2009) Therapeutic microRNA delivery suppresses tumorigenesis in a murine liver cancer model. Cell 137(6):1005–1017
66. Zhu Y, Lu Y, Zhang Q, Liu JJ, Li TJ, Yang JR, Zeng C, Zhuang SM (2011) MicroRNA-26a/b and their host genes cooperate to inhibit the G1/S transition by activating the pRb protein. Nucleic Acids Res 40:4615–4625
67. Caputo V, Sinibaldi L, Fiorentino A, Parisi C, Catalanotto C, Pasini A, Cogoni C, Pizzuti A (2011) Brain derived neurotrophic factor (BDNF) expression is regulated by microRNAs miR-26a and miR-26b allele-specific binding. PLoS ONE 6(12):e28656
68. Lu J, He ML, Wang L, Chen Y, Liu X, Dong Q, Chen YC, Peng Y, Yao KT, Kung HF, Li XP (2011) MiR-26a inhibits cell growth and tumorigenesis of nasopharyngeal carcinoma through repression of EZH2. Cancer Res 71(1):225–233
69. Hermeking H (2010) The miR-34 family in cancer and apoptosis. Cell Death Differ 17(2):193–199
70. Liu C, Kelnar K, Liu B, Chen X, Calhoun-Davis T, Li H, Patrawala L, Yan H, Jeter C, Honorio S, Wiggins JF, Bader AG, Fagin R, Brown D, Tang DG (2011) The microRNA miR-34a inhibits prostate cancer stem cells and metastasis by directly repressing CD44. Nat Med 17(2):211–215
71. Zeng C, Wang R, Li D, Lin XJ, Wei QK, Yuan Y, Wang Q, Chen W, Zhuang SM (2010) A novel GSK-3 beta-C/EBP alpha-miR-122-insulin-like growth factor 1 receptor regulatory circuitry in human hepatocellular carcinoma. Hepatology 52(5):1702–1712
72. Xu H, He JH, Xiao ZD, Zhang QQ, Chen YQ, Zhou H, Qu LH (2010) Liver-enriched transcription factors regulate microRNA-122 that targets CUTL1 during liver development. Hepatology 52(4):1431–1442
73. Gatfield D, Le Martelot G, Vejnar CE, Gerlach D, Schaad O, Fleury-Olela F, Ruskeepaa AL, Oresic M, Esau CC, Zdobnov EM, Schibler U (2009) Integration of microRNA miR-122 in hepatic circadian gene expression. Genes Dev 23(11):1313–1326
74. Lanford RE, Hildebrandt-Eriksen ES, Petri A, Persson R, Lindow M, Munk ME, Kauppinen S, Orum H (2010) Therapeutic silencing of microRNA-122 in primates with chronic hepatitis C virus infection. Science 327(5962):198–201
75. Esau C, Davis S, Murray SF, Yu XX, Pandey SK, Pear M, Watts L, Booten SL, Graham M, McKay R, Subramaniam A, Propp S, Lollo BA, Freier S, Bennett CF, Bhanot S, Monia BP (2006) miR-122 regulation of lipid metabolism revealed by in vivo antisense targeting. Cell Metab 3(2):87–98
76. Jopling C (2012) Liver-specific microRNA-122: biogenesis and function. RNA Biol 9(2):137–142
77. Bai S, Nasser MW, Wang B, Hsu SH, Datta J, Kutay H, Yadav A, Nuovo G, Kumar P, Ghoshal K (2009) MicroRNA-122 inhibits tumorigenic properties of hepatocellular carcinoma cells and sensitizes these cells to sorafenib. J Biol Chem 284(46):32015–32027

78. Fornari F, Gramantieri L, Giovannini C, Veronese A, Ferracin M, Sabbioni S, Calin GA, Grazi GL, Croce CM, Tavolari S, Chieco P, Negrini M, Bolondi L (2009) MiR-122/cyclin G1 interaction modulates p53 activity and affects doxorubicin sensitivity of human hepatocarcinoma cells. Cancer Res 69(14):5761–5767
79. Akao Y, Nakagawa Y, Hirata I, Iio A, Itoh T, Kojima K, Nakashima R, Kitade Y, Naoe T (2010) Role of anti-oncomirs miR-143 and -145 in human colorectal tumors. Cancer Gene Ther 17(6):398–408
80. Clape C, Fritz V, Henriquet C, Apparailly F, Fernandez PL, Iborra F, Avances C, Villalba M, Culine S, Fajas L (2009) miR-143 interferes with ERK5 signaling, and abrogates prostate cancer progression in mice. PLoS ONE 4(10):e7542
81. Wijnhoven BP, Hussey DJ, Watson DI, Tsykin A, Smith CM, Michael MZ, South Australian Oesophageal Research G (2010) MicroRNA profiling of Barrett's oesophagus and oesophageal adenocarcinoma. Br J Surg 97(6):853–861
82. Sachdeva M, Mo YY (2010) miR-145-mediated suppression of cell growth, invasion and metastasis. Am J Transl Res 2(2):170–180
83. Dou L, Zheng D, Li J, Li Y, Gao L, Wang L, Yu L (2012) Methylation-mediated repression of microRNA-143 enhances MLL-AF4 oncogene expression. Oncogene 31(4):507–517
84. Xu N, Papagiannakopoulos T, Pan G, Thomson JA, Kosik KS (2009) MicroRNA-145 regulates OCT4, SOX2, and KLF4 and represses pluripotency in human embryonic stem cells. Cell 137(4):647–658
85. Cho WC, Chow AS, Au JS (2011) MiR-145 inhibits cell proliferation of human lung adenocarcinoma by targeting EGFR and NUDT1. RNA Biol 8(1):125–131
86. Borralho PM, Simoes AE, Gomes SE, Lima RT, Carvalho T, Ferreira DM, Vasconcelos MH, Castro RE, Rodrigues CM (2011) miR-143 overexpression impairs growth of human colon carcinoma xenografts in mice with induction of apoptosis and inhibition of proliferation. PLoS ONE 6(8):e23787
87. Ng EK, Tsang WP, Ng SS, Jin HC, Yu J, Li JJ, Rocken C, Ebert MP, Kwok TT, Sung JJ (2009) MicroRNA-143 targets DNA methyltransferases 3A in colorectal cancer. Br J Cancer 101(4):699–706
88. Kent OA, Chivukula RR, Mullendore M, Wentzel EA, Feldmann G, Lee KH, Liu S, Leach SD, Maitra A, Mendell JT (2010) Repression of the miR-143/145 cluster by oncogenic Ras initiates a tumor-promoting feed-forward pathway. Genes Dev 24(24):2754–2759
89. Zhang X, Liu S, Hu T, Liu S, He Y, Sun S (2009) Up-regulated microRNA-143 transcribed by nuclear factor kappa B enhances hepatocarcinoma metastasis by repressing fibronectin expression. Hepatology 50(2):490–499
90. Chan JA, Krichevsky AM, Kosik KS (2005) MicroRNA-21 is an antiapoptotic factor in human glioblastoma cells. Cancer Res 65(14):6029–6033
91. Medina PP, Nolde M, Slack FJ (2010) OncomiR addiction in an in vivo model of microRNA-21-induced pre-B-cell lymphoma. Nature 467(7311):86–90
92. Davis BN, Hilyard AC, Lagna G, Hata A (2008) SMAD proteins control DROSHA-mediated microRNA maturation. Nature 454(7200):56–61
93. Polytarchou C, Iliopoulos D, Hatziapostolou M, Kottakis F, Maroulakou I, Struhl K, Tsichlis PN (2011) Akt2 regulates all Akt isoforms and promotes resistance to hypoxia through induction of miR-21 upon oxygen deprivation. Cancer Res 71(13):4720–4731
94. Du J, Yang S, An D, Hu F, Yuan W, Zhai C, Zhu T (2009) BMP-6 inhibits microRNA-21 expression in breast cancer through repressing deltaEF1 and AP-1. Cell Res 19(4):487–496
95. Yang CH, Yue J, Fan M, Pfeffer LM (2010) IFN induces miR-21 through a signal transducer and activator of transcription 3-dependent pathway as a suppressive negative feedback on IFN-induced apoptosis. Cancer Res 70(20):8108–8116
96. Seike M, Goto A, Okano T, Bowman ED, Schetter AJ, Horikawa I, Mathe EA, Jen J, Yang P, Sugimura H, Gemma A, Kudoh S, Croce CM, Harris CC (2009) MiR-21 is an EGFR-regulated anti-apoptotic factor in lung cancer in never-smokers. Proc Nat Acad Sci USA 106(29):12085–12090

97. Bakirtzi K, Hatziapostolou M, Karagiannides I, Polytarchou C, Jaeger S, Iliopoulos D, Pothoulakis C (2011) Neurotensin signaling activates microRNAs-21 and -155 and Akt, promotes tumor growth in mice, and is increased in human colon tumors. Gastroenterology 141(5):1749–1761, e1741
98. Wang K, Li PF (2010) Foxo3a regulates apoptosis by negatively targeting miR-21. J Biol Chem 285(22):16958–16966
99. Fujita S, Ito T, Mizutani T, Minoguchi S, Yamamichi N, Sakurai K, Iba H (2008) miR-21 gene expression triggered by AP-1 is sustained through a double-negative feedback mechanism. J Mol Biol 378(3):492–504
100. Krichevsky AM, Gabriely G (2009) miR-21: a small multi-faceted RNA. J Cell Mol Med 13(1):39–53
101. Zhu S, Wu H, Wu F, Nie D, Sheng S, Mo YY (2008) MicroRNA-21 targets tumor suppressor genes in invasion and metastasis. Cell Res 18(3):350–359
102. Garofalo M, Di Leva G, Romano G, Nuovo G, Suh SS, Ngankeu A, Taccioli C, Pichiorri F, Alder H, Secchiero P, Gasparini P, Gonelli A, Costinean S, Acunzo M, Condorelli G, Croce CM (2009) miR-221&222 regulate TRAIL resistance and enhance tumorigenicity through PTEN and TIMP3 downregulation. Cancer Cell 16(6):498–509
103. Shah MY, Calin GA (2011) MicroRNAs miR-221 and miR-222: a new level of regulation in aggressive breast cancer. Genome Med 3(8):56
104. Fornari F, Gramantieri L, Ferracin M, Veronese A, Sabbioni S, Calin GA, Grazi GL, Giovannini C, Croce CM, Bolondi L, Negrini M (2008) MiR-221 controls CDKN1C/p57 and CDKN1B/p27 expression in human hepatocellular carcinoma. Oncogene 27(43):5651–5661
105. Garofalo M, Romano G, Di Leva G, Nuovo G, Jeon YJ, Ngankeu A, Sun J, Lovat F, Alder H, Condorelli G, Engelman JA, Ono M, Rho JK, Cascione L, Volinia S, Nephew KP, Croce CM (2011) EGFR and MET receptor tyrosine kinase-altered microRNA expression induces tumorigenesis and gefitinib resistance in lung cancers. Nat Med 18(1):74–82
106. Galardi S, Mercatelli N, Farace MG, Ciafre SA (2011) NF-kB and c-Jun induce the expression of the oncogenic miR-221 and miR-222 in prostate carcinoma and glioblastoma cells. Nucleic Acids Res 39(9):3892–3902
107. Stinson S, Lackner MR, Adai AT, Yu N, Kim HJ, O'Brien C, Spoerke J, Jhunjhunwala S, Boyd Z, Januario T, Newman RJ, Yue P, Bourgon R, Modrusan Z, Stern HM, Warming S, de Sauvage FJ, Amler L, Yeh RF, Dornan D (2011) TRPS1 targeting by miR-221/222 promotes the epithelial-to-mesenchymal transition in breast cancer. Sci Signal 4(177):ra41
108. Di Leva G, Gasparini P, Piovan C, Ngankeu A, Garofalo M, Taccioli C, Iorio MV, Li M, Volinia S, Alder H, Nakamura T, Nuovo G, Liu Y, Nephew KP, Croce CM (2010) MicroRNA cluster 221–222 and estrogen receptor alpha interactions in breast cancer. J Natl Cancer Inst 102(10):706–721
109. le Sage C, Nagel R, Egan DA, Schrier M, Mesman E, Mangiola A, Anile C, Maira G, Mercatelli N, Ciafre SA, Farace MG, Agami R (2007) Regulation of the p27(Kip1) tumor suppressor by miR-221 and miR-222 promotes cancer cell proliferation. EMBO J 26(15):3699–3708
110. Zhang CZ, Zhang JX, Zhang AL, Shi ZD, Han L, Jia ZF, Yang WD, Wang GX, Jiang T, You YP, Pu PY, Cheng JQ, Kang CS (2010) MiR-221 and miR-222 target PUMA to induce cell survival in glioblastoma. Mol Cancer 9:229
111. Lu Y, Roy S, Nuovo G, Ramaswamy B, Miller T, Shapiro C, Jacob ST, Majumder S (2011) Anti-microRNA-222 (anti-miR-222) and -181B suppress growth of tamoxifen-resistant xenografts in mouse by targeting TIMP3 protein and modulating mitogenic signal. J Biol Chem 286(49):42292–42302
112. Olive V, Jiang I, He L (2010) mir-17-92, a cluster of miRNAs in the midst of the cancer network. Int J Biochem Cell Biol 42(8):1348–1354
113. Mu P, Han YC, Betel D, Yao E, Squatrito M, Ogrodowski P, de Stanchina E, D'Andrea A, Sander C, Ventura A (2009) Genetic dissection of the miR-17~92 cluster of microRNAs in Myc-induced B-cell lymphomas. Genes Dev 23(24):2806–2811

114. De Brouwer S, Mestdagh P, Lambertz I, Pattyn F, De Paepe A, Westermann F, Schroeder C, Schulte JH, Schramm A, De Preter K, Vandesompele J, Speleman F (2011) Dickkopf-3 is regulated by the MYCN-induced miR-17-92 cluster in neuroblastoma. Int J Cancer J Int Du Cancer 130(11):2591–2598
115. Yan HL, Xue G, Mei Q, Wang YZ, Ding FX, Liu MF, Lu MH, Tang Y, Yu HY, Sun SH (2009) Repression of the miR-17-92 cluster by p53 has an important function in hypoxia-induced apoptosis. EMBO J 28(18):2719–2732
116. Dews M, Fox JL, Hultine S, Sundaram P, Wang W, Liu YY, Furth E, Enders GH, El-Deiry W, Schelter JM, Cleary MA, Thomas-Tikhonenko A (2010) The myc-miR-17~92 axis blunts TGF{beta} signaling and production of multiple TGF{beta}-dependent antiangiogenic factors. Cancer Res 70(20):8233–8246
117. Hong L, Lai M, Chen M, Xie C, Liao R, Kang YJ, Xiao C, Hu WY, Han J, Sun P (2010) The miR-17-92 cluster of microRNAs confers tumorigenicity by inhibiting oncogene-induced senescence. Cancer Res 70(21):8547–8557
118. Kim K, Chadalapaka G, Lee SO, Yamada D, Sastre-Garau X, Defossez PA, Park YY, Lee JS, Safe S (2012) Identification of oncogenic microRNA-17-92/ZBTB4/specificity protein axis in breast cancer. Oncogene 31(8):1034–1044
119. van Haaften G, Agami R (2010) Tumorigenicity of the miR-17-92 cluster distilled. Genes Dev 24(1):1–4
120. Petrocca F, Vecchione A, Croce CM (2008) Emerging role of miR-106b-25/miR-17-92 clusters in the control of transforming growth factor beta signaling. Cancer Res 68(20):8191–8194
121. Li Y, Tan W, Neo TW, Aung MO, Wasser S, Lim SG, Tan TM (2009) Role of the miR-106b-25 microRNA cluster in hepatocellular carcinoma. Cancer Sci 100(7):1234–1242
122. Bueno MJ, Gomez de Cedron M, Laresgoiti U, Fernandez-Piqueras J, Malumbres M (2010) Multiple E2F-induced microRNAs prevent replicative stress in response to mitogenic signaling. Mol Cell Biol 30(12):2983–2995
123. Kan T, Sato F, Ito T, Matsumura N, David S, Cheng Y, Agarwal R, Paun BC, Jin Z, Olaru AV, Selaru FM, Hamilton JP, Yang J, Abraham JM, Mori Y, Meltzer SJ (2009) The miR-106b-25 polycistron, activated by genomic amplification, functions as an oncogene by suppressing p21 and Bim. Gastroenterology 136(5):1689–1700
124. Fang L, Deng Z, Shatseva T, Yang J, Peng C, Du WW, Yee AJ, Ang LC, He C, Shan SW, Yang BB (2011) MicroRNA miR-93 promotes tumor growth and angiogenesis by targeting integrin-beta8. Oncogene 30(7):806–821
125. Poliseno L, Salmena L, Riccardi L, Fornari A, Song MS, Hobbs RM, Sportoletti P, Varmeh S, Egia A, Fedele G, Rameh L, Loda M, Pandolfi PP (2010) Identification of the miR-106b~25 microRNA cluster as a proto-oncogenic PTEN-targeting intron that cooperates with its host gene MCM7 in transformation. Sci Signal 3(117):ra29
126. Lu Z, Li Y, Takwi A, Li B, Zhang J, Conklin DJ, Young KH, Martin R, Li Y (2011) miR-301a as an NF-kappaB activator in pancreatic cancer cells. EMBO J 30(1):57–67
127. Chen Y, Gorski DH (2008) Regulation of angiogenesis through a microRNA (miR-130a) that down-regulates antiangiogenic homeobox genes GAX and HOXA5. Blood 111(3).1217–1226
128. Yeung ML, Yasunaga J, Bennasser Y, Dusetti N, Harris D, Ahmad N, Matsuoka M, Jeang KT (2008) Roles for microRNAs, miR-93 and miR-130b, and tumor protein 53-induced nuclear protein 1 tumor suppressor in cell growth dysregulation by human T-cell lymphotrophic virus 1. Cancer Res 68(21):8976–8985
129. Lai KW, Koh KX, Loh M, Tada K, Subramaniam MM, Lim XY, Vaithilingam A, Salto-Tellez M, Iacopetta B, Ito Y, Soong R, Singapore Gastric Cancer C (2010) MicroRNA-130b regulates the tumour suppressor RUNX3 in gastric cancer. Eur J Cancer 46(8):1456–1463
130. Borgdorff V, Lleonart ME, Bishop CL, Fessart D, Bergin AH, Overhoff MG, Beach DH (2010) Multiple microRNAs rescue from Ras-induced senescence by inhibiting p21(Waf1/Cip1). Oncogene 29(15):2262–2271

131. Shi W, Gerster K, Alajez NM, Tsang J, Waldron L, Pintilie M, Hui AB, Sykes J, P'ng C, Miller N, McCready D, Fyles A, Liu FF (2011) MicroRNA-301 mediates proliferation and invasion in human breast cancer. Cancer Res 71(8):2926–2937
132. Kluiver J, Poppema S, de Jong D, Blokzijl T, Harms G, Jacobs S, Kroesen BJ, van den Berg A (2005) BIC and miR-155 are highly expressed in Hodgkin, primary mediastinal and diffuse large B cell lymphomas. J Pathol 207(2):243–249
133. Costinean S, Zanesi N, Pekarsky Y, Tili E, Volinia S, Heerema N, Croce CM (2006) Pre-B cell proliferation and lymphoblastic leukemia/high-grade lymphoma in E(mu)-miR155 transgenic mice. Proc Nat Acad Sci USA 103(18):7024–7029
134. Faraoni I, Antonetti FR, Cardone J, Bonmassar E (2009) miR-155 gene: a typical multifunctional microRNA. Biochim Biophys Acta 1792(6):497–505
135. Gironella M, Seux M, Xie MJ, Cano C, Tomasini R, Gommeaux J, Garcia S, Nowak J, Yeung ML, Jeang KT, Chaix A, Fazli L, Motoo Y, Wang Q, Rocchi P, Russo A, Gleave M, Dagorn JC, Iovanna JL, Carrier A, Pebusque MJ, Dusetti NJ (2007) Tumor protein 53-induced nuclear protein 1 expression is repressed by miR-155, and its restoration inhibits pancreatic tumor development. Proc Nat Acad Sci USA 104(41):16170–16175
136. Kong W, He L, Coppola M, Guo J, Esposito NN, Coppola D, Cheng JQ (2010) MicroRNA-155 regulates cell survival, growth, and chemosensitivity by targeting FOXO3a in breast cancer. J Biol Chem 285(23):17869–17879
137. Lee DW, Futami M, Carroll M, Feng Y, Wang Z, Fernandez M, Whichard Z, Chen Y, Kornblau S, Shpall EJ, Bueso-Ramos CE, Corey SJ (2012) Loss of SHIP-1 protein expression in high-risk myelodysplastic syndromes is associated with miR-210 and miR-155. Oncogene 31(37):4085–4094
138. Dagan LN, Jiang X, Bhatt S, Cubedo E, Rajewsky K, Lossos IS (2012) miR-155 regulates HGAL expression and increases lymphoma cell motility. Blood 119(2):513–520
139. Jiang S, Zhang HW, Lu MH, He XH, Li Y, Gu H, Liu MF, Wang ED (2010) MicroRNA-155 functions as an OncomiR in breast cancer by targeting the suppressor of cytokine signaling 1 gene. Cancer Res 70(8):3119–3127
140. Vargova K, Curik N, Burda P, Basova P, Kulvait V, Pospisil V, Savvulidi F, Kokavec J, Necas E, Berkova A, Obrtlikova P, Karban J, Mraz M, Pospisilova S, Mayer J, Trneny M, Zavadil J, Stopka T (2011) MYB transcriptionally regulates the miR-155 host gene in chronic lymphocytic leukemia. Blood 117(14):3816–3825
141. Wang B, Majumder S, Nuovo G, Kutay H, Volinia S, Patel T, Schmittgen TD, Croce C, Ghoshal K, Jacob ST (2009) Role of microRNA-155 at early stages of hepatocarcinogenesis induced by choline-deficient and amino acid-defined diet in C57BL/6 mice. Hepatology 50(4):1152–1161
142. Chang S, Wang RH, Akagi K, Kim KA, Martin BK, Cavallone L, Kathleen Cuningham Foundation Consortium for Research into Familial Breast C, Haines DC, Basik M, Mai P, Poggi E, Isaacs C, Looi LM, Mun KS, Greene MH, Byers SW, Teo SH, Deng CX, Sharan SK (2011) Tumor suppressor BRCA1 epigenetically controls oncogenic microRNA-155. Nat Med 17(10):1275–1282

Chapter 6
Targeting DNA Repair Pathways for Cancer Therapy

Conchita Vens and Robert W. Sobol

Abstract DNA repair pathways maintain the integrity of the genome, reducing the onset of cancer, disease, and aging. The majority of anticancer therapeutics (radiation and chemotherapy) function as genotoxins, eliciting genomic DNA damage in an attempt to induce cell death in the tumor. However, cellular DNA repair proteins counteract the effectiveness of these therapeutic genotoxins by repairing and removing the cell death-inducing DNA lesions, implicating DNA repair proteins as prime targets for improving response to currently available anticancer regimens. To trigger a tumor-specific cell death response (with minimal normal cell toxicity), the level of genomic DNA damage must therefore surpass the DNA repair capacity of the tumor without overwhelming the DNA repair potential of normal tissue. Interestingly, cancer-specific DNA repair defects offer novel approaches for tumor-selective therapy. This has become highly relevant as it is suggested that most cancer cells are likely to be defective in some aspect of DNA repair. Herein, we describe the molecular pathways that participate in the repair of DNA damage induced by radiation- and chemotherapeutics and discuss strategies that are being developed to target DNA repair for cancer treatment and highlight key DNA repair inhibitors that can enhance response. Further, we present novel therapeutic strategies being considered to exploit inherent weaknesses in tumor cells such as defects in one or more DNA repair pathways or related processes that may provide the opportunity to selectively increase tumor-specific cell death.

C. Vens
Division of Biological Stress Response, The Netherlands Cancer Institute, NL, Amsterdam, The Netherlands

R. W. Sobol (✉)
Department of Pharmacology and Chemical Biology, University of Pittsburgh Cancer Institute, Hillman Cancer Center; University of Pittsburgh School of Medicine, Pittsburgh, PA, USA
e-mail: rws9@pitt.edu

R. W. Sobol
Department of Human Genetics, University of Pittsburgh School of Public Health, Pittsburgh, PA, USA

D. E. Johnson (ed.), *Cell Death Signaling in Cancer Biology and Treatment*,
Cell Death in Biology and Diseases, DOI: 10.1007/978-1-4614-5847-0_6,

X.-M. Yin and Z. Dong (Series eds.), *Cell Death in Biology and Diseases*

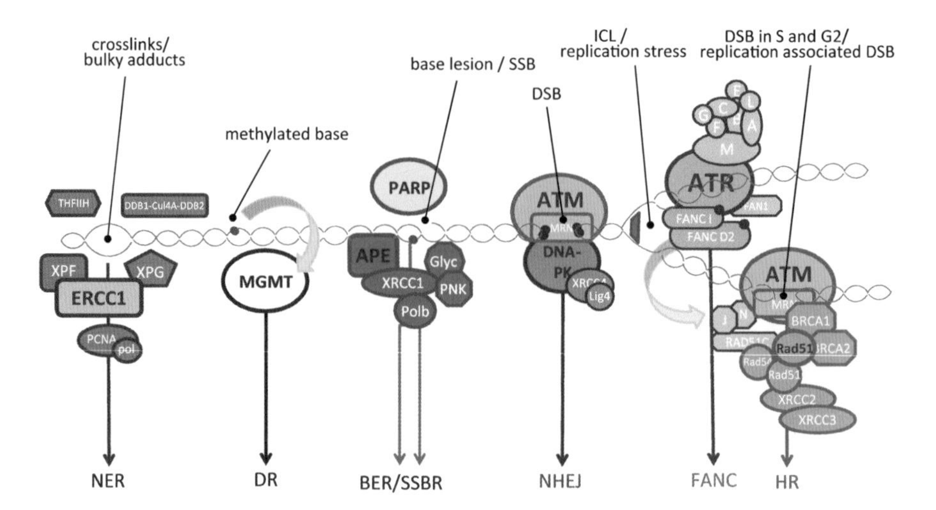

Fig. 6.1 Schematic representation of cellular DNA damage repair. This figure depicts cellular repair processes that deal with chemo- and radiotherapy-induced DNA lesions. Prevalent DNA repair targets for cancer treatment are highlighted in bold

6.1 Role of DNA Repair Pathways for Cancer Treatment

Human cells must repair tens of thousands of DNA lesions per day [1]. If they are not repaired, these lesions lead to mutations or genome aberrations that threaten cell survival and genomic integrity. To combat these threats, cells have evolved multiple DNA repair and DNA damage response (DDR) mechanisms that signal the presence of lesions and promote their repair or regulate cellular processes in response to the DNA damage (Fig. 6.1) [2]. Defects in these repair and response pathways can promote tumorigenesis and, indeed, are common in human cancers [3, 4]. On the other hand, current therapy options for cancer patients exploit the DNA-damaging properties of certain drugs and agents. The success of radiation exposure during radiotherapy and the success of most chemotherapy agents rely on the destructive nature that these agents have on cellular DNA, ultimately resulting in death and hopefully eradication of the tumor cells. Hence, DNA damage and repair mechanisms play a crucial role in determining treatment outcome. On a cellular level, resistance to treatment is profoundly determined by the capacity of the cancer cell to respond to and repair the individual DNA lesions that are induced by the chemotherapeutic agents or radiation.

Our increasing knowledge of these processes has led to the development of new concepts that target and exploit the cancer cell DDR [5]. Counteracting resistance to chemotherapy by targeting the appropriate DNA repair pathway is a promising strategy in cancer treatment. Another is to exploit the DNA repair and response defects that are present in cancer cells thereby specifically targeting tumor cells while sparing healthy cells from a high load of unrepaired DNA damage [6, 7]. Together, these strategies might provide promising avenues in the conquest against cancer.

Here, we will briefly describe essential cellular DNA repair and response mechanisms and illustrate novel concepts and promising strategies to exploit and target DNA repair for cancer treatment.

6.1.1 DNA Damage Response

The initial DDR of a cell involves the recognition of the DNA damage followed by the propagation of a series of signals ranging from alterations in RNA or protein expression and modification of protein function or stability through post-translation modification, among other signals. The cell's defense to genotoxic lesions is triggered and accomplished by a series of events that mediate and regulate proliferation, cell death, or DNA repair crucial to its survival [2, 4]. The initial steps for an appropriate response require detection of the lesion, signaling of its presence and promotion of repair. Cells act upon DNA damage not only by promoting and executing repair but also respond by halting the cell cycle or by promoting cell death mechanisms in order to prevent propagation of the damage. The DDR therefore has an impact on transcription, cellular metabolism, cell cycle regulators, as well as cell death, via apoptosis and senescence.

One of the most prominent members of the DNA damage signaling pathway that links DNA damage with cell cycle checkpoints is the protein ataxia telangiectasia-mutated (ATM) [8], a protein kinase that is recruited to DNA double-strand breaks (DSB) such as those induced by ionizing radiation. The formation of DSBs triggers the activation of ATM and activated ATM then phosphorylates a wide range of downstream substrate proteins thereby signaling the presence of the damage throughout the cell to facilitate repair [9]. Initial activation of ATM is promoted by its autophosphorylation that initiates a signaling cascade of further phosphorylation events that constitute the DDR [10]. With excessive unrepaired DNA damage present, this cellular response can culminate in an apoptotic response in which p53 is central but not necessarily always required.

Another key DDR signaling component is the ataxia telangiectasia-RAD3-related (ATR) kinase that gets activated after replication stress-induced DNA damage. Replication stress, caused, for example, by exposure to hydroxyurea (HU), results in the formation of large stretches of single-stranded DNA coated with replication protein A (RPA) that triggers activation of ATR. Similar to processes in ATM-mediated signaling, ATR signaling is promoted by regulatory proteins. ATRIP and TopBP1, together with RAD17-mediated 9-1-1 (Rad9-Rad1-Hus1) clamp loading, "sense" the damage and trigger the activation of ATR.

Further downstream of these initial events, the cell cycle checkpoint-regulating protein kinases CHK1 and CHK2 are among the most important targets (substrates) of ATM and ATR. Supported by the activation of p53 and mediated via multiple paths, this signaling cascade ultimately results in the reduction of cyclin-dependent kinase activity that drives cell cycle progression. The halt in cell cycle progression is thought to allow time for repair and, most importantly, if not successfully repaired, to prevent propagation of the DNA damage. Cell cycle checkpoints are in place at the border from G1 to S, within S and at the G2/M border. The prevalent blocks and their extent depend on the damaging agent and the number and type of lesion. Another important downstream target of ATM and ATR is p53, an essential player in the induction of apoptosis upon DNA damage. Thus, DDR mechanisms have a crucial role in the protection against genome instability and chemo/radiotherapy response.

ATM/ATR signaling also enhances repair by recruiting repair factors to the site of the lesion and activating DNA repair proteins through phosphorylation or indirectly, by modulating acetylation, ubiquitylation, SUMOylation, or DNA repair gene transcription. These kinases also influence chromatin structure through phosphorylation of the histone H2A variant (γH2AX). Thereby, they facilitate recruitment of DDR factors and expedite DNA repair while amplifying DSB signaling that is crucial to cellular survival following exposure to DNA-damaging agents.

The role of the DDR is broad with respect to the type of cancer therapeutic. DDR activity is involved upon exposure of a whole range of chemotherapeutic agents and upon radiation. Indeed, DDR and cell cycle blocks are induced by radiation, topoisomerase I and II poisons, anthracyclines, alkylating drugs including platinum analogues and antitumor antibiotics. Interference in DDR by the use of inhibitors will likely affect the response and cellular survival in most cancer treatment options.

6.1.2 Direct-Reversal Repair and Mismatch Repair

One of the first DNA repair proteins to be considered as a viable target for improving chemotherapy was O^6-alkylguanine-DNA alkyltransferase (MGMT or AGT), a protein encoded by the O^6-methylguanine-DNA methyltransferase gene (*MGMT*) [11]. In depth, discussion on the function of MGMT and its role in cancer and chemotherapy can be found in many excellent reviews [12–14]. MGMT falls within the category of direct-reversal (DR) DNA repair proteins that also include the AlkB family of proteins [13, 15]. Unlike most other DNA repair pathways that correct lesions by removing the base containing the lesion [base excision repair (BER), see below], removing a short oligonucleotide containing the lesion [nucleotide excision repair (NER), see below] or removing long tracts of DNA (mismatch repair, MMR) followed by a DNA synthesis step (repair-directed DNA synthesis), DR proteins such as MGMT or the AlkB proteins reverse the damage to the DNA base directly, and the mechanism of repair does not involve a DNA synthesis step. This section will focus on the role of MGMT in DNA lesion repair and the subsequent role that the MMR proteins play in the cellular response when MGMT is unable to repair the O^6-alkylguanine lesion. Further discussion on the mechanism of action of AlkB proteins can be found elsewhere [13].

Like many DNA repair proteins, MGMT repairs lesions from both carcinogenic compounds and from chemotherapeutic agents. As such, MGMT acts to suppress cancer formation by removing lesions induced by carcinogens such as methylnitrosourea (MNU), the tobacco smoke lung carcinogen 4-(methylnitrosamino)-1-(3-pyridyl)-1-butanone (NKK), and the colon carcinogen azoxymethane. Conversely, many chemotherapeutic agents, including Temozolomide (Temodar, TMZ), dacarbazine, streptozotocin, procarbazine, BCNU (camustine), CCNU (lomustine), and gliadel trigger cell death by inducing the formation of an alkyl lesion (methyl- or chloroethyl-) on the O^6 position of guanine bases in DNA [12–14, 16]. Upon MGMT binding to the DNA containing the alkyl lesion, the O^6-alkyl group is transferred from the guanine base onto a Cysteine (Cys) residue (amino acid residue Cys145 in humans) in the MGMT protein [12, 14]. Upon transfer of the alkyl group to MGMT, the protein undergoes a conformational change that both releases the protein from the repaired DNA and promotes the ubiquitylation and subsequent proteasome-mediated degradation. This suicide mechanism of MGMT has been taken advantage of clinically, as will be described below regarding the development and evaluation of MGMT inhibitors (see Sect. 6.3.2).

The chemotherapeutic agents mentioned above induce the formation of methyl or chloroethyl adducts on the O^6 position of guanine bases in DNA. If not repaired, the majority of the chloroethyl lesions are converted to G-C interstrand DNA cross-links. Such lesions are primarily repaired by a concerted effort of the NER, HR, and fanconi anemia (FA) pathways (see below). In general, interstrand DNA cross-links are highly genotoxic, inducing cell death. Conversely, if the methyl lesion (O^6-MeG) is not removed by MGMT, during cellular replication, the mispairing of O^6-MeG with thymine leads to the formation of a O^6-MeG:T mispair, a substrate for MMR. Repair mediated by the MMR pathway facilitates the removal of the DNA strand containing the newly synthesized "T" base. However, re-synthesis of the DNA in the process of MMR will regenerate the O^6-MeG:T mispair, perpetuating the O^6-MeG lesion and the presence of the mispair. As such, in the absence of MGMT-mediated repair, O^6-MeG is suggested to initiate a futile cycle of MMR or alternately to trigger ATR protein kinase activation through the action of several MMR proteins [17], leading to apoptosis and cell death [18–20]. Details on the MMR pathway can be found elsewhere [21]. However, for the purpose of this discussion, we should consider the MMR pathway as an essential sensor to trigger cell death from chemotherapeutic agents that induce the O^6-MeG lesion. Briefly, recognition of the O^6-MeG:T mispair by the MMR protein MSH2 induces recruitment and activation of ATR and subsequently CHK1 and CHK2 to activate an apoptotic response [22, 23]. In fact, much of the resistance to agents such as TMZ observed clinically is due to high expression of MGMT (and subsequent repair of the lesion) or loss of MMR (therefore preventing the initiation of apoptotic signaling) [24–26]. Currently, TMZ along with radiation and surgery are the standard of care for glioblastoma multiforme (GBM), the most common and aggressive primary brain tumor [27]. Median survival is less than two years [28–30], and unfortunately, almost all patients eventually recur with the disease and the large majority of recurrent tumors are resistant to chemotherapy [31, 32]. Inhibition of MGMT-mediated repair has been taken advantage of experimentally and in many clinical trials [33] since MGMT can be inhibited with the O^6-MeG analogue O^6-benzylguanine [34] (Sect. 6.3.2). Improved prognosis has also been reported in tumors with loss of *MGMT* expression due to promoter methylation [35] whereas poor prognosis is observed when *MGMT* expression levels are high or MMR capacity is compromised. Hence, elevated expression of *MGMT* and/or a non-functional MMR pathway contribute much of the observed resistance to TMZ in many tumor cell lines and in clinical trials.

6.1.3 Base Excision Repair

As suggested in its namesake, the BER pathway is the primary mechanism to remove and repair base lesions. A special sub-pathway of BER, single-strand break repair (SSBR), is also essential for the repair of single-strand DNA breaks [13, 36, 37]. The types of base lesions repaired by the proteins of the BER pathway range from base deamination products (e.g., conversion of C to U or 5 meC to T) to oxidative modification of bases (8-oxo-7,8-dihydro-2′- deoxyguanosine; 8-oxodG), alkylation products such as N7-meG and N3-meA that are induced by chemotherapeutic agents such as TMZ [38] and many others [36, 37]. These and many other lesions are induced in genomic (and mitochondrial) DNA by a multitude of anticancer treatments including radiation, monofunctional alkylators such as TMZ, cisplatin, and 5FU, among others. There are over 20 proteins brought to bear to facilitate the complete process of BER [36]. Repair is initiated following recognition of the base lesion by one of the eleven DNA glycosylases in humans. This group of proteins is further subdivided into two classes: Bifunctional and monofunctional DNA glycosylases. A β-bifunctional glycosylase such as OGG1 excises the modified base and hydrolyzes the DNA backbone (via a β-elimination step) 3′ to the incised base, leaving a 3′ unsaturated aldehyde (after β-elimination) and a 5′ phosphate at the termini of the repair gap. Alternatively, a β,δ-bifunctional glycosylase such as NEIL1 hydrolyzes the glycosidic bond to release the lesion and then cleave the DNA backbone 3′ to the resulting apurinic/apyrimidinic (AP or abasic) site via β-elimination and 5′ to the abasic site via δ-elimination. More detail on the mechanism of these bifunctional DNA glycosylases and the specific BER proteins involved in processing the resulting repair gaps can be found elsewhere [36, 37].

For the purpose of describing the complete BER pathway, we will focus on the initiation of BER by monofunctional DNA glycosylases, with an emphasis on the methylpurine DNA glycosylase (MPG) (also called AAG or ANPG). MPG is the primary glycosylase for the repair of the chemotherapy-induced DNA lesions such as N7-meG and N3-meA. These lesions are removed by hydrolysis of the glycosidic bond, producing an abasic site, a substrate for AP-Endonuclease 1 (APE1). Given the highly toxic nature of the intermediates in BER [13], it has been suggested that the product of each BER reaction "hands off" the toxic BER intermediate to the next enzyme in the pathway likening the complete reaction to the hand-off of a baton in a relay race [39]. Such a process or hand-off mechanism has the advantage of eliminating or avoiding the accumulation of free BER intermediates that are prone to induce cell death [13]. Once formed by the glycosylase, the resulting abasic site is then handed off to APE1 to be hydrolyzed on the 5′ end. The resulting single-nucleotide repair gap contains a 3′OH and a 5′deoxyribose-phosphate (5′dRP) moiety at the margins. It has been suggested that this BER intermediate (a single-strand break with a 5′dRP moiety) recruits poly(ADP)ribose polymerase (PARP)1 to the lesion site. Recruitment then triggers activation of PARP1. Activated PARP1 polymerizes NAD^+ to yield the polymer poly (ADP) ribose (PAR), an essential posttranslational modification. The first protein to be modified by PAR is PARP1 itself (auto-modification). Subsequently, it has been observed that XRCC1 and many other proteins are modified [40]. Once modified, activated PARP1 then facilitates chromatin relaxation (likely to provide access to the lesions for repair) [41, 42] and recruitment of the remaining BER proteins required to complete repair, including XRCC1, DNA Ligase III, and DNA polymerase β (Polβ). Whereas XRCC1 is a scaffold protein, Polβ carries out two essential enzymatic functions in BER. First, the repair gap is tailored by the 5′dRP lyase activity of Polβ. Next, Polβ fills the single-nucleotide gap, preparing the strand for ligation by either DNA ligase I (LigI) or a complex of DNA ligase III (LigIII), and XRCC1 [36].

Although some BER substrates (base lesions) induced by chemotherapeutic agents are cytotoxic [43], most are found to be mutagenic. However, essentially, every intermediate throughout the BER pathway (abasic sites, 5′dRP lesions, and single-strand DNA breaks) is toxic [13] and as such, there has been considerable interest in developing BER inhibitors to enhance the accumulation of the cytotoxic repair intermediates following chemotherapy or radiation treatment. This is discussed further in the sections below.

6.1.4 Nucleotide Excision Repair

Another multi-protein, highly complex DNA repair pathway is the NER pathway. NER plays an important role in the repair of DNA lesions induced by many genotoxins and chemotherapeutics including DNA cross-linking agents such as chloroethylating agents (see Sect. 6.1.2), cisplatin, carboplatin, and lesions induced by photodynamic therapy (PTD). Put simply, NER facilitates the removal of bulky DNA adducts that grossly distort the DNA double helix and those that cause a block to transcription. Molecular details on the proteins involved in NER can be found in several excellent reviews [44–47]. Overall, the pathway consists of two complementary sub-pathways that have some overlap. The two sub-pathways are distinct regarding the lesion recognition step but converge and utilize the same proteins to remove the oligonucleotide containing the lesion and for the steps involving new DNA synthesis.

The global genomic NER (GG-NER) pathway surveys the entire genome for DNA helix distorting lesions whereas the transcription-coupled repair NER

(TC-NER) pathway is recruited to facilitate removal of DNA lesions that block the elongating RNA polymerase and stall transcription. The GG-NER pathway utilizes the DDB1/DDB2 heterodimer, part of the DDB1-Cul4A-DDB2 E3 ubiquitin ligase, to facilitate lesion recognition and repair [48]. As such, targeting the proteasome (Sect. 6.4.2) or deubiquitinating enzymes (DUBs) (Sect. 6.5.4) would therefore indirectly impact NER function. The TC-NER pathway partners with the TFIIH transcription complex to recognize and repair lesions that halt transcription. Upon lesion recognition, XPG mediates cleavage of the DNA strand containing the lesion on the 3′ side of the lesion, and subsequently, the ERCC1/XPF heterodimer hydrolyzes the DNA strand containing the lesion on the 5' side. Replication factors then facilitate DNA synthesis and ligation. Of all the proteins in this pathway, ERCC1 has emerged as a valuable biomarker of response to chemotherapeutic agents that induce DNA damage repaired by NER (e.g., cisplatin) and is under consideration as a drug target [49, 50]. Currently, biomarker measurements have included both mRNA and protein analysis. However, it is not yet clear whether protein levels of ERCC1 are a valid biomarker [51, 52].

6.1.5 Non-Homologous-End-Joining

One of the most cytotoxic lesions is a DNA double-strand break (DSB). If not repaired, DSBs lead to chromosome breaks, loss of genetic material, and gross genomic rearrangements. Whereas tolerance to the presence of DSBs might vary in different cell types and cellular states, only a few DSBs will cause cell death or prevent clonogenicity in most cells including cancer cells [53]. These lesions are induced by multiple agents such as ionizing radiation, bleomycin, and topoisomerase II inhibitors, but can also be induced indirectly at replication forks when converting DNA single-strand breaks (SSBs) induced by topoisomerase I inhibitors (camptothecin).

Two major cellular pathways deal with the repair of DNA DSBs. The use of homologous DNA for repair distinguishes those repair pathways. As indicated by the name, the non-homologous-end-joining (NHEJ) repair pathway does not require any homologous sequences. Proteins of the NHEJ pathway can repair the two ends in a DSB by simple end joining while the homologous recombination (HR) repair pathway (see Sect. 6.1.6) requires homologous DNA stretches as templates for DNA synthesis and repair.

After the initial recognition of the DSB that is held in place and stabilized by the binding of the MRN complex (MRE11, RAD50, NBS) and promoted by ATM (see above), DSB repair is executed by the DNA-dependent protein kinase (DNA-PK). DNA-PK is comprised of the catalytic subunit DNA-PKcs and the relatively small Ku proteins (Ku70/80). They promote the simple ligation of the two broken DNA ends. Damage-induced DSBs, in particular after ionizing radiation, are rarely re-ligateable, and some end processing might be required that is accomplished by other enzymes such as Artemis. The DNA ligase IV-XRCC4 complex finally re-ligates the two ends. DNA PK–independent DSB end-joining activity has been observed in cells that have impaired NHEJ activity, the so-called alternative or B-NHEJ pathway. PARP and Ligase III activity appears to be implicated in this cellular DSB repair option [54, 55].

The role of ATM seems to be of particular importance in the repair of a certain proportion of DSBs, namely those in heterochromatic regions of the genome [56, 57]. These, judging from the repair kinetics, require more time to repair but influence survival substantially as indicated by the hypersensitivity to radiation of cells with impaired ATM function.

Based on the cytotoxic nature of DSBs, cells impaired in any step of the NHEJ process are highly sensitive to ionizing radiation. Genetic defects in or inhibition of NHEJ also profoundly affects survival of cells by other DNA-damaging agents that cause DSBs (directly or indirectly) such as DNA cross-linkers, bleomycin, and topoisomerase inhibitors.

6.1.6 Homologous Recombination DSB Repair and the Fanconi Pathway

The HR repair pathway, in contrast to NHEJ (see above), requires homologous DNA stretches as templates for DNA synthesis and repair [44, 58]. To assure accurate repair, HR tends to use the sister chromatid as a template, restricting this pathway to the S and G2 phase of the cell cycle. Together with the FA pathway, it has a crucial role in the surveillance of replication fork progression. As anticipated by its requirement for a homologous template, HR mainly determines survival of S- and G2-phase cells. HR's involvement upon radiation is not only required at directly induced DSBs but also at secondarily induced DSBs that result from replication attempts on nicked DNA [59]. In agreement with such a function, HR has been shown to determine radiosensitivity in a cell cycle phase-dependent manner. HR and the FA pathway are also crucial in resolving blocked replication forks [60]. Such blocks are, for example, caused by interstrand cross-links (ICLs) that tether the two DNA strands together and prevent separation during replication. Survival upon other cancer therapeutic agents that induce replication-blocking lesions such as alkylating agents and topoisomerase inhibitors is strongly determined by the functionality of HR. These blocks must be repaired or bypassed to allow cells to survive.

A complex multi-member process assures HR-driven repair. In brief, an ordered assembly of nucleoprotein filaments of RAD52, RAD51, and RAD54 upon RPA coating of the resected DNA promotes and catalyzes homologous DNA pairing. The extent of the resection at the break site is mediated by the MRN complex and appears to partly define the use of HR instead of NHEJ. Strand exchange is assisted by the RAD51 paralogs RAD51B, RAD51C, RAD51D, XRCC2, and XRCC3. In concert, these proteins direct and provide the recombinase activity, that is, crucial to resolve the complex-branched structures that arise in this process. Notably, the products of the breast cancer susceptibility genes BRCA1 and BRCA2 are involved in the FA and HR repair pathways, assisting DSB and cross-link repair.

To allow the resolution of blocked replication fork structures, in particular following exposure to DNA cross-linking agents, another replication-associated repair process is required, the FA pathway. Its members were discovered while analyzing FA patients, victims of a human genetic disease that is characterized, among other features [61], by extreme cellular sensitivity to drugs that produce ICLs. Subsequently, their role and actions in cellular cross-link repair was revealed.

The products of at least 15 genes have been currently implicated in this pathway [61–63]. This pathway constitutes a major signaling cascade upon replication fork stalling: the FA "core complex" consists of at least 8 FA elements (FANCA, FANCB, FANCC, FANCE, FANCF, FANCG, FANCL, and FANCM) and acts by realizing the mono-ubiquitylation of the FA ID complex (FANCD2 and FANCI). This activation of the ID complex allows chromatin binding and is thought to facilitate DNA repair, in particular HR. The FA-mediated recruitment of the RAD51 recombinase and the BRCA1-FANCJ helicase activity allows re-establishment of the replication fork. Resolution of stalled replication forks appears to be also supported by the translocase activity of FANCM that can remodel branched DNA structures. The FA core complex is regulated by ATR and cell cycle checkpoint elements (CHK1) allowing the activation of the pathway [64]. Importantly, any mutation upstream of the FA pathway that will disrupt the mono-ubiquitylation of FANCD2 will result in the cellular ICL hypersensitivity phenotype.

As illustrated above, upon exposure to DNA-damaging agents, it is the multitude of cellular repair capacities that ultimately determines survival. HR and FA have been shown to determine the cellular sensitivity to a wide range of cancer therapeutics. HR-defective cells are hypersensitive to cross-linkers, IR, topoisomerase inhibitors, and alkylators as they induce DSBs and replication stalling.

6.2 Strategies Targeting DNA Repair for Cancer Treatment

The requirement for DNA repair and genome maintenance in response to radiation and genotoxic chemotherapeutics implicates DNA repair proteins as prime targets for improving response to currently available anticancer regimens. In addition, frequent cancer-specific DNA repair and DDR defects offer tumor-selective therapy options. Thus, strategies targeting DNA repair pathways represent promising new avenues to improve outcome in cancer treatment.

Targeting DNA repair pathways in cancer treatment has been proposed in several settings (Fig. 6.2). Most evidently, inhibition of cellular DNA repair will cause increased sensitivity to chemotherapeutic agents or radiotherapy [65]. As illustrated above, cellular death upon exposure to most chemotherapeutic

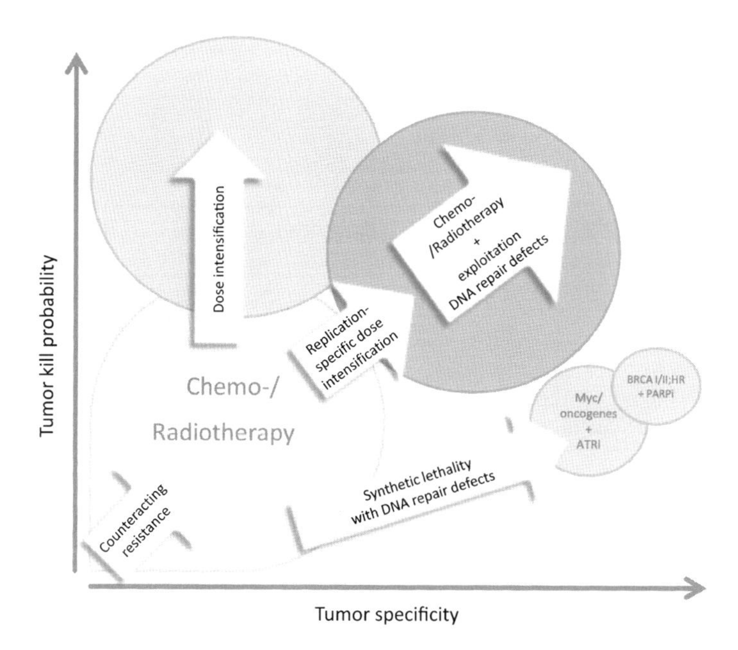

Fig. 6.2 Strategies to improve cancer treatment by targeting DNA damage response and repair. The overall goal is to increase tumor cell kill (*y-axis*) while sparing normal tissue by increasing tumor specificity (*x-axis*) of the cancer treatment. While some strategies will achieve increased cell death and thereby an increased probability to control the tumor (for example by applying DNA-PK inhibitors in combination with radiotherapy or by re-sensitizing chemotherapy-resistant tumors), others (such as those based on the exploitation of tumor-specific defects) will increase tumor specificity of cancer therapies. However, in the manner in which current chemo- and radiotherapy regimens are largely applied, each of the targeted approaches will be only beneficial in a small fraction of patients with tumors that harbor the respective DNA repair defects. Nevertheless, considering the wide range of combination possibilities and the high number of targets and exploitation opportunities, together they represent a promising avenue in cancer treatment

agents and ionizing radiation partly depends on the repair capacity for the respective lesions. HR, NHEJ, BER, and DR processes are responsible for the resistance to these agents. Hence, inhibitors to DNA repair elements might be useful as dose intensifiers, augmenting the cell-killing properties of many if not all therapeutic agents. This can be particularly useful in a setting in which cancer cells obtained resistance to certain agents due to an improved repair capacity. Thus, counteracting marked cancer cell resistance is one strategy in which DNA repair inhibitors are proposed to act as dose intensifiers. By lowering the tolerance and inhibiting alternative repair routes while augmenting cell kill, the application of "intelligent" dose intensification by DNA repair inhibition can also prevent the development of chemotherapy resistance.

Dose intensification will not be, in general, however, well tolerated, since chemotherapeutic drugs and radiotherapy doses are often administered at maximum-tolerated levels. Non-cancerous cells, with few exceptions, are as much exposed to the chemotherapeutic agents as are the cancer cells. Indeed, it is the normal tissue response that defines the dose level and use of dose intensifiers. Some tumor properties could, however, provide a tumor-specific effect when targeting DNA repair. Further, in line with the rationale of most classical cancer therapeutics, the proliferative nature of tumors is exploitable. For example, targeting replication-associated repair pathways such as HR with novel targeted agents could be beneficial in radiotherapy regimens in which the healthy cells in the irradiated area are nonproliferative. A gain could also be expected if chemotherapy dose limits are not defined by the proliferative cells, allowing dose intensification by targeting replication-associated DNA repair. Other tumor-specific properties such as exposure to hypoxic conditions or the altered metabolic status in cancer cells can offer opportunities to achieve tumor-specific dose intensification and will be discussed in the following paragraphs.

Another implication of DNA repair-targeting strategies has, however, become highly relevant. It is suggested that most cancer cells are likely to be defective in some aspect of DNA repair. Considering the multitude of repair options of healthy cells, DNA repair defects in cancer cells can be exploited by targeting the remaining repair processes. The combined lethal effect of two genetic variations that are otherwise non-lethal is termed "synthetic lethality" [66]. In compliance with the synthetic lethality concept, cancer cells defective in the primary repair pathway are viable but rely heavily on secondary backup repair for survival. As this is not only restricted to repair of endogenously produced lesions, this concept also applies to cells exposed to exogenous damage by exacerbating the effects of chemo- and radiotherapy in the defective cancer cells only (also may be called synthetic sickness). Despite mutations and genetic defects, the differential expression of DNA repair proteins or the altered engagement of DNA repair sub-pathways can be a base of tumor-specific activities. Hence, these tumor-specific DDR and repair defects offer promising novel approaches to tumor-selective therapy.

Investigation of the functionality of the individual DNA repair pathways in cancer cells and the knowledge on which pathways are implicated upon the inhibition of DNA repair drug targets are necessary to combine these cancer therapy options in an intelligent manner while focusing on a differential effect in the cancer versus normal cells.

6.3 DNA Repair Targets

The recognition that DNA repair processes are prime targets for chemo- and radiosensitization has driven the development of specific inhibitors to elements of DDR, NHEJ, HR, DR, and BER. The more recent discovery of tumor-specific targeting opportunities by the inhibition of DNA repair processes fueled such attempts and has yielded a multitude of DNA repair inhibitors [5]. Some of these novel agents are currently being evaluated in the clinic while others are being tested preclinically. Novel DNA repair targets have been identified,

Table 6.1 Targets in DNA damage response and repair

DNA Repair protein target	DNA Repair pathway involved	Compounds (DNA repair inhibitors)	Context	Strategy	Development stage	Reference
PARP1/2	BER	Olaparib	Monotherapy	BRCA1 or BRCA2 deficiency	Clinical validation	[248]
PARP1/2	BER	Olaparib	Chemosensitizer	Combined with cisplatin and gemcitabine	Clinical validation	[249]
PARP1/2	BER	Olaparib	Radiation sensitizer	Combined with radiation	Preclinical	[250]
PARP1/2	BER	Veliparib	Monotherapy	BRCA1 or BRCA2 deficiency	Preclinical	[251]
PARP1/2	BER	Veliparib	Chemosensitizer	Combined with cyclophosphamide, carboplatin, temozolomide or topotecan	Clinical validation and preclinical	[134, 251–255]
PARP1/2	BER	Veliparib	Radiation sensitizer	Combined with radiation	Preclinical	[254, 256–258]
MGMT	DR	O^6-benzylguanine	Chemosensitizer	Combined with carmustine	Clinical validation	[259]
MGMT	DR	O^6-benzylguanine	Chemosensitizer	Combined with temozolimide	Clinical validation	[260]
MGMT	DR	Lomeguatrib	Chemosensitizer	Combined with temozolimide	Clinical validation	[261, 262]
CHK1	DDR	AZD7762	Radiation sensitizer	Combined with radiation	Preclinical	[263]
CHK1	DDR	AZD7762	Chemosensitizer	Combined with irinotecan	Preclinical	[115]
CHK1	DDR	SCH900776	Chemosensitizer	Combined with SN38	Preclinical	[264]

(continued)

Table 6.1 (continued)

DNA Repair protein target	DNA Repair pathway involved	Compounds (DNA repair inhibitors)	Context	Strategy	Development stage	Reference
CHK1	DDR	PF-0477736	Monotherapy	Myc-driven cancers	Preclinical	[265]
None (Abasic site in DNA)	BER	TRC102 (methoxyamine)	Chemosensitizer	Combined with pemetrexed	Preclinical	[142]
None (Abasic site in DNA)	BER	TRC102 (methoxyamine)	Chemosensitizer	Combined with temozolimide	Preclinical	[133–135]
None (Abasic site in DNA)	BER	TRC102 (methoxyamine)	Radio- and chemosensitizer	Combined with iododeoxyuridine + radiation	Preclinical	[136]
None (Abasic site in DNA)	BER	TRC102 (methoxyamine)	Chemosensitizer	Combined with BCNU	Preclinical	[137]
None (Abasic site in DNA)	BER	TRC102 (methoxyamine)	Chemosensitizer	Combined with manumycin A	Preclinical	[138]
None (Abasic site in DNA)	BER	TRC102 (methoxyamine)	Chemosensitizer	Combined with fludarabine	Preclinical	[139]
None (Abasic site in DNA)	BER	TRC102 (methoxyamine)	Radiation sensitizer	Combined with radiation	Preclinical	[140, 141]
ATR	DDR	NU6027	Chemosensitizer	Combined with PARP inhibitors	Preclinical	[151]
ATR	DDR	NVP-BEZ235	Monotherapy	Cyclin-E overexpressing cells	Compound screen	[152]
ATM	DDR	KU55933	Radiation sensitizer	Combined with radiation	Compound screen and preclinical	[266]
ATM	DDR	KU60019	Radiation sensitizer	Combined with radiation	Preclinical	[167]
ATM	DDR	CP466722	Radiation sensitizer	Combined with radiation	Preclinical	[153]

(continued)

Table 6.1 (continued)

DNA Repair protein target	DNA Repair pathway involved	Compounds (DNA repair inhibitors)	Context	Strategy	Development stage	Reference
DNA-PK	DDR	NU7026	Radiation sensitizer	Combined with radiation in EGFRvIII-expressing tumors	Preclinical	[164]
DNA-PK	DDR	NU7441	Radio- and Chemosensitizer	Combined with doxorubicin or radiation	Compound screen and preclinical	[267]
DNA-PK	DDR	DT01 (dbait, DRIIM)	Radiation sensitizer	Metastatic melanoma with relapsed cutaneous tumors	Compound screen and clinical validation	[158]
DNA-PK	DDR	DT01 (dbait, DRIIM)	Chemosensitizer	Combined with 5-fluorouracil or irinotecan	Preclinical	[268]
DNA-PK	DDR	HNI-38	Radiation sensitizer	Combined with radiation	Preclinical	[269]
DNA-PK	DDR	KU-0060648; A dual inhibitor of DNA-PK and PI-3 K	Chemosensitizer	Combined with etoposide	Preclinical	[159]
NAMPT	NAD^+ biosynthesis	FK866 (Apo866)	Chemosensitizer	Combined with various chemotherapeutics	Preclinical	[190–192]
NAMPT	NAD^+ biosynthesis	GMX1778	Chemosensitizer	Combined with various chemotherapeutics	Preclinical	[183]
NAMPT	NAD^+ biosynthesis	CB30865	Chemosensitizer	Combined with various chemotherapeutics	Preclinical	[184]
NAMPT	NAD^+ biosynthesis	CHS-828	Chemosensitizer	Combined with various chemotherapeutics	Preclinical	[185]

and compounds that specifically inhibit their activity are sought in order to apply tumor-specific anticancer strategies.

We will list currently explored DNA repair targets and some of the most advanced compounds according to their developmental stage (Table 6.1). Rationales and applied strategies will be discussed while pointing to opportunities on combinations and other DNA repair targets.

6.3.1 Poly(ADP)Ribose Polymerase

One of the most advanced and applied DNA repair target inhibitors to date are the PARP inhibitors [67–69]. Since the discovery that cells with defects in the BRCA genes are selectively killed by the inhibition of PARP, PARP inhibitors have rapidly made their way into the clinic [70, 71]. As tumors from carriers of mutations in the breast cancer susceptibility genes BRCA1 and BRCA2 are almost exclusively composed of such BRCA-defected cells while normal cells of these carriers still carry a functional allele, hence are HR proficient, these PARP inhibitors achieve tumor-specific kill with little normal cell toxicity. The proposed mechanism that causes such selectivity points to the dependence of BER-inhibited cells on HR due to secondarily induced cytotoxic DSBs (Fig. 6.3a) [60]. Indeed, early synthetic lethality screens in yeast indicated such an opportunity revealing a crucial link between BER and HR for cellular survival [72]. Other hypotheses assume a direct role of PARP inhibitors in replication fork stalling [73]. Most compounds with PARP inhibitory activity target PAR generation by blocking the catalytic activity of the enzyme. In principle, these compounds compete with NAD^+ for the PARP catalytic site and are therefore not necessarily specific to PARP1 and could impact the activity of the other PARP isoforms [74]. Several PARP inhibitors are in clinical development. To date, the leading compounds Olaparib (AZD2281, AstraZeneca; originally developed by KuDos) and Veliparib (ABT-888, Abbott) are probably the two most extensively studied in the clinic whereas at least one compound, namely Iniparib (BSI-201; Sanofi-Aventis), has been reported to lack effective PARP inhibitory activity [75, 76].

Although registration of these drugs is still awaiting approval, several studies have shown their beneficial application [77, 78]. One obstacle could be that these PARP inhibitors were expected to act in a fraction of tumors, those exhibiting HR defects due to BRCA1 & BRCA2 mutations only. It should be noted that those early clinical trials revealed that not all BRCA mutation carriers benefit, indicating that a certain degree and type of HR defect is required to be exploitable with PARP inhibition. The impact of individual BRCA mutations with respect to HR functionality and/or PARP inhibitor sensitivity could be variable [79]. The status and propensity to use the remaining DSB repair mechanism NHEJ, for example via 53BP1 channeling, also influences the extent of PARP inhibitor toxicity [60, 80]. In addition, other general drug resistance mechanisms such as increased compound rejection by

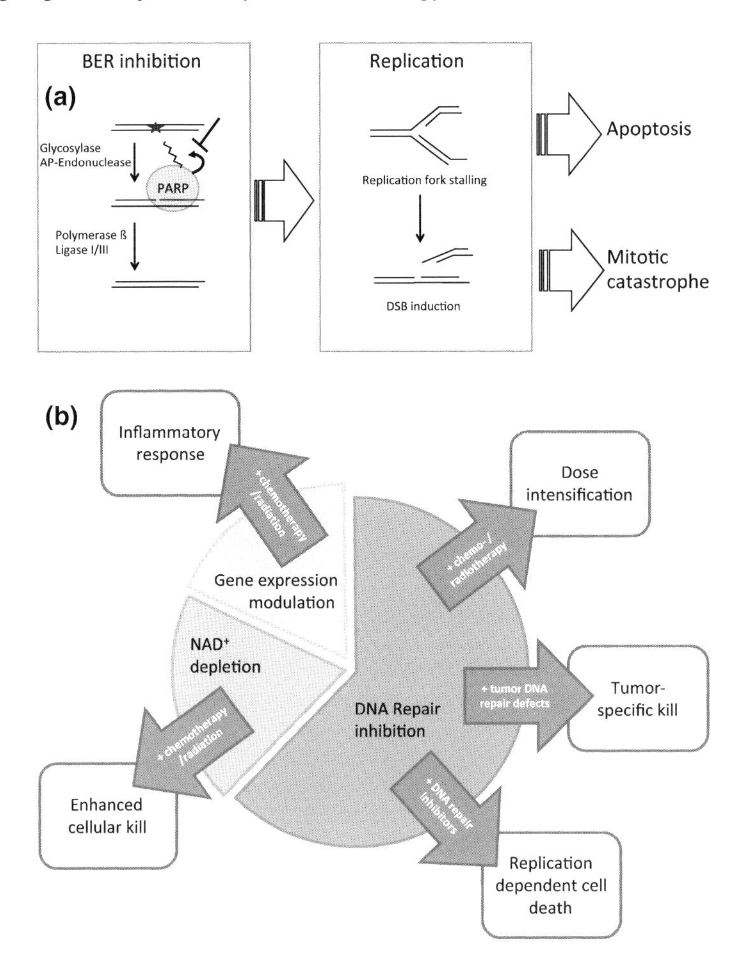

Fig. 6.3 Cellular functions of PARP and opportunities for cancer treatment. **a** The mode of action of *PARP* inhibitors is based primarily on inhibition of the poly(ADP)ribosylation activity of the enzyme *PARP*1. As a result, BER or SSBR cannot be executed. In addition, the lack of poly(ADP)ribosylation is likely to negatively impact chromatin and lesion accessibility. Auto-poly(ADP)ribosylation of *PARP*1 is thought to promote its repulsion from DNA. A failure to recruit downstream BER elements or an increase in lesion shielding by trapping *PARP*1 on the DNA further inhibits BER. BER intermediates accumulated upon chemo- and radiotherapy, however, will then cause replication problems that in turn will induce *DSB*s. Those will ultimately lead to cell death via apoptosis and/or mitotic catastrophe in particular if not repaired by HR. **b** PARP activity has several cellular roles that can be taken advantage of in cancer treatment. PARP promotes DNA repair and its inhibition, when combined with chemotherapy or with tumor-specific defects, enhances tumor cell kill. *NAD*$^+$ *depletion*, as a consequence of chemo- or radiotherapy-induced PARP activity, also induces cell kill. Agents depleting cellular NAD$^+$ levels could indirectly inhibit PARP. Conversely, PARP inhibitors can have the effect of lowering chemotherapy-induced *NAD*$^+$ *depletion*, thereby altering the mode of cell death. Lastly PARP also has regulatory functions regarding DNA damage–induced gene expression that is often connected to an inflammatory or fibrotic response. PARP inhibitors have been reported to exhibit anti-inflammatory properties

drug transporters, decreased tumor perfusion, cellular metabolism, or simply pharmacogenetics of a certain compound might also be responsible for a lack of benefit.

However, new studies indicate that other defects that supposedly result in impaired HR or DSB repair, such as when ATM is mutated, result in PARP inhibitor-mediated selective kill [81, 82]. Mantle cell lymphoma harbors ATM defects and preclinical studies demonstrate sensitivity of these tumors to PARP inhibition [83, 84]. Other genetic analyses indicated frequent ATM mutations in human cancer. These studies therefore warrant the application of PARP inhibitors for cancer treatment in a much larger patient population. In agreement with a synthetic lethal interaction with HR-driven processes, cells with an impaired FA pathway are hypersensitive to PARP inhibition, thus further enlarging the potential patient population that might benefit from such therapies [81]. Some studies indicated an impact of PTEN deletion on HR via a reduction of Rad51 or Rad51 paralogs, a condition that could be exploited by PARP inhibition [85, 86]. However, this could only be partly confirmed in a separate study [87]. HR impairment was observed when cells experience hypoxia [88, 89] which sensitized them to PARP inhibition. PARP inhibitor sensitivity in hypoxic cells or those with defects in FA, ATM, or PTEN is, however, generally less pronounced than when BRCA-defective. Yet, a benefit, in particular when combined with radio- or chemotherapy, can be anticipated since PARP inhibitors appear to enlarge the therapeutic window and could spare healthy cells from chemo- to radiosensitization.

Historically, prior to the discovery of specific killing of tumors with defective HR, PARP inhibitors have been actively studied in preclinical and clinical investigations to potentiate the cytotoxic effects of chemo- and radiotherapy (Fig. 6.3a). PARP inhibitors, due to their DNA repair-inhibiting properties, are radiation sensitizers and are highly effective in sensitizing cells to chemotherapeutic agents such as DNA alkylators (e.g., TMZ) and topoisomerase I inhibitors (irinotecan and topotecan). Combining PARP inhibitors with chemo-/radiotherapy could be beneficial in a large patient population with the added advantage of single-agent activity in a fraction of patients that happen to have tumors with exploitable DNA repair defects. As noted above, in such dose-intensification strategies, the benefit of tumor-specific killing needs to be evaluated against normal tissue effects that are connected to the use of DNA repair inhibitors. To exemplify, since the radiosensitizing effect of PARP inhibitors is most effective in proliferating cells, PARP inhibitors have been proposed in clinical radiotherapy settings in which normal tissue toxicity within the radiation field is not defined by a highly proliferating (stem) cell fraction such as in the treatment of lung cancer and glioblastoma [90, 91]. Radiation-induced lung toxicity is strongly determined by inflammatory and fibrotic processes. Therefore, PARP inhibition might impact lung toxicity in a beneficial way by altering the inflammatory DDR [92, 93].

Despite crude chemosensitization and synthetic lethal activity, other strategies exploit the BER inhibitory properties of PARP inhibitors. These strategies aim to prevent the quick development of resistance to alkylators such as temozolomide (TMZ). Resistance to TMZ is associated with MGMT expression and MMR defects (see above). PARP inhibitors, however, are able to (re-) sensitize

MMR-defective cells to the anti-tumor effect of TMZ [94–98] thereby counteracting the resistance to TMZ.

Other PARP inhibitor combinations have been pursued. Assuming that BRCA defects do limit HR functions but do not fully impair HR, BER inhibition and consequently increased engagement of HR might be the basis for the observed synergistic cytotoxicity of Olaparib and Cisplatin [99, 100]. Depending on the cytotoxic agents and genetic background, BRCA-defected cells die by apoptosis or mitotic catastrophe upon PARP inhibition.

DNA damage–induced and PARP-activation-mediated consumption of NAD^+ has been implicated in the increase in genotoxin-induced cell death [74, 101]. The PARP1 and PARP2 proteins [102] act as sensors of DNA damage such as DNA single-strand breaks and become hyperactivated, consuming NAD^+ as a substrate to synthesize PAR [103]. Consumption of NAD^+ after DNA damage leads to ATP depletion, likely due to continued re-synthesis of NAD^+ as well as ongoing cellular utilization of NAD^+ and ATP for metabolic functions [74, 101]. Following up on the observation that cell death due to BER inhibition and the accumulation of BER intermediates results in PARP hyperactivation [103], it was shown that the combination of BER and NAD^+ biosynthesis inhibition significantly sensitizes glioma cells to TMZ. Dual targeting of these two interacting pathways (DNA repair and NAD^+ biosynthesis) may prove to be an effective treatment combination for patients with resistant and recurrent GBM. Thus, in summary, several distinct roles of PARP including the promotion of DNA repair and its impact on damage-induced gene expression as well as the relationship to NAD^+ levels can be exploited for cancer therapy (Fig. 6.3b).

6.3.2 *O^6-Methylguanine-DNA Methyltransferase*

As described in Sect. 6.1 above, MGMT is the sole protein responsible for the repair of O^6-alkylguanine lesions (formed by chemotherapeutic agents such as TMZ). Tumors with elevated MGMT expression are resistant to TMZ and related chemotherapeutic agents, and so, an active area of investigation has been the development of MGMT inhibitors. If MGMT is inhibited (or MGMT is not expressed), the tumor becomes highly sensitive to the agent (provided the tumor cell is proficient in MMR—see Sect. 6.1.2). There have been multiple methodologies proposed to inhibit or overcome resistance mediated by MGMT expression in the tumor as well as to prevent sensitivity of normal tissue (primarily hematopoietic cells) [12]. As we alluded in Sect. 6.1.2, the mechanism of action of MGMT in the repair or de-alkylation of guanine suggested O^6-benzylguanine as an ideal inhibitor [104]. This inhibitor, also called BG, is an analogue of the O^6-alkylguanine base and contains a benzyl ring instead of an alkyl group. The BG compound readily reacts with MGMT, and the benzyl moiety is transferred to Cys145 as shown (Fig. 6.4), releasing free guanine and rendering MGMT inactive. In some cases, this has been shown to trigger ubiquitylation and proteasome-mediated destruction

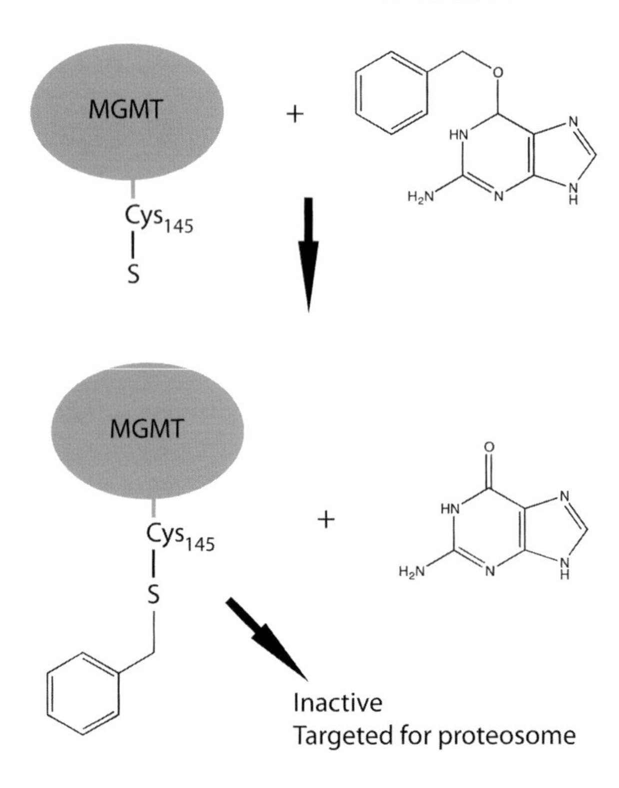

Fig. 6.4 Schematic representation depicting the mechanism of action of the MGMT inhibitor O^6-benzylguanine (*BG*). The benzyl moiety of BG is transferred to the *Cys*145 residue in *MGMT*, releasing free guanine, rendering *MGMT* inactive. In some cases, this has been shown to trigger ubiquitylation and proteasome-mediated destruction of *MGMT*

of MGMT [105]. Further details on BG and related analogues to inhibit or regulate MGMT can be found elsewhere [106]. A second inhibitor with greater inhibitory activity that has been widely tested is the alkylguanine analogue 6-[4-bromo-2-thienyl]methoxypurin-2-amine [107], also called lomeguatrib [108, 109].

A significant challenge with all or most chemosensitizers is the observed increase in normal cell/tissue sensitivity or cell death. This has been addressed with regard to MGMT by the use of an ex vivo gene therapy approach to express a mutant of MGMT in the bone marrow or hematopoietic stem cells (cells are modified ex vivo and re-delivered to the patient), as described [12]. This is feasible since the G156A or P140K mutants of MGMT are 60-fold and 500-fold more resistant to BG than the wild-type protein, respectively [110].

6.3.3 Cell Cycle Checkpoints

DNA damage activates checkpoints to arrest proliferation. Normal cells have intact G1, S, and G2 checkpoints that are mediated by the ATM/CHK2 and ATR/CHK1 pathways. Owing to mutations in the p53 or pRB tumor suppressor genes, cancer cells, however, lack a G1 checkpoint and as a result, rely on G2 checkpoints to prevent cell division and propagation of the damage. Thus, the S/G2 checkpoint is an attractive target for cancer-specific sensitization to DNA-damaging agents [111]. In order to exploit the cancer-specific defects, inhibitors have been developed that abrogate the G2 checkpoint. CHK1 has been a prime target for such attempts, as activated CHK1 mediates the arrest by phosphorylating Cdc25A and Cdc25C leading to their degradation and inactivation that otherwise promote S-phase progression and entry into mitosis. Loss of the G2 cell cycle checkpoint, despite the presence of unrepaired damage, is thought to provoke mitotic catastrophe and ultimately cell kill. Consistent with this idea, loss of intra-S or G2/M checkpoints increases the cytotoxicity of DNA-damaging agents such as ionizing radiation and cisplatin. CHK1 knockdown sensitizes cells to 5-fluorouracil, doxorubicin, and etoposide. Despite the proposed mechanism that CHK1 inhibition causes sensitization in p53-defective cancer cells due to the G2 block abrogation in a G1 block-deficient background, the data to support this are contradictory [112–114]. Xeno-transplant studies on human triple-negative breast cancer demonstrated the benefit of combining irinotecan with CHK1 inhibitors, inducing checkpoint bypass and apoptosis. A role of p53 was supported by the gain of CHK1 sensitization after p53 knockdown in the resistant tumors [115]. A more complicated rationale, however, argues that a series of DDR defects in tumors should be considered and could be exploited by the inhibition of CHK1. These are based on the secondary effects of CHK1 inactivation on replication and are discussed below.

A large battery of CHK1 inhibitors is available to support cancer treatment in combination with radio- and chemotherapy [116]. Older CHK1 inhibitors such as UCN01 were not very selective but did demonstrate potent chemosensitization to cisplatin and camptothecin. More potent and specific CHK1 inhibitors were developed; however, concomitant CHK2 inhibition to some degree is common to most CHK1 inhibitors. In general, cells appear to depend on a functional G2 checkpoint when exposed to agents that cause replication stress. Hence, potentiation is greatest to cross-linkers, topoisomerase I poisons and nucleoside analogues such as gemcitabine. Consistent with replication stress hypersensitivity when G2 arrest is abrogated, PARP inhibitors also cause problems. PARP inhibition can cause a G2 checkpoint dependence and the combination of PARP inhibitors with CHK1 inhibitors is synthetic lethal [117]. These cellular sensitivity features are the basis for the clinical trials testing the combination of older (UCN01) or later generation compounds (for example AZD7762, AstraZeneca; PF477736, Pfizer; LY2606368, Eli Lilly and SCH900776, Schering Plough) with cisplatin, topotecan, and gemcitabine. For a more detailed review, see [116]. Unfortunately, safety requirements as assessed in these initial studies were not met in at least one compound (AZD7762) [118]. Similar to chemotherapy regimens, CHK1 inhibitors prevent ionizing radiation (IR) induced S and G2 arrest and demonstrated some potential to radiosensitize in a p53-dependent manner. CHK1 is upregulated in Myc-overexpressing lymphomas, and single-agent activity is expected since CHK1 inhibition is cytotoxic to these cells.

Interestingly, recent data indicate that CHK1 activity might prevent replication-induced DNA damage or is implicated in DNA repair [119, 120]. A conversion of halted replication forks (induced by the chemotherapeutics) into persistent and cytotoxic DSBs has been postulated. The CDK-driven unscheduled initiation of replication origins can be accounted for with regard to this DSB induction and explains chemosensitization beyond G2/M checkpoint abrogation [120]. The notion that CHK1 inhibition promotes replication stress merits their evaluation in synthetic lethal strategies exploiting tumor-specific DNA repair defects [121].

6.3.4 AP-Endonuclease 1

Targeted knockout (KO) of AP(apurinic/apyrimidinic)-endonuclease 1 (Ape1 or Ref1) in mice is lethal [122], and deletion of the *Ape1* gene in mouse cells induces apoptosis within 24 h [123]. Interestingly, granzyme A(GzmA)-mediated cell death is enhanced by GzmA-mediated cleavage of Ape1. It is suggested that the proteolysis of APE1 enhances GzmA-mediated cell death by promoting apoptosis [124]. In human cells, RNA interference–mediated depletion of APE1 suppresses cell and tumor growth [125]. These and many other studies therefore support the development of APE1 inhibitors to enhance radiation and chemotherapy response [126, 127].

The first such compound to be tested clinically that impacts APE1 function in BER is methoxyamine (TRC102; Tracon Pharmaceuticals). Methoxyamine hydrochloride (MX) was first suggested to be a mutator, inducing the formation of 5,6-dihydro-6-methoxyaminecytosine residues in DNA [128]. It was subsequently determined that MX traps aldehyde groups, forming a stable intermediate [129]. The AP site in DNA is not a chemically unique species but exists as an equilibrium mixture of the ring-closed cyclic hemiacetals and open-chain aldehyde, and hydrate forms [130]. The transient open-chain aldehyde form is reactive with aldehyde-specific reagents such as methoxyamine, allowing the trapping or quantitative measurement of AP sites in DNA [131]. It was subsequently determined that the reaction between MX and the open-chain aldehyde form of an AP site blocks repair of DNA base lesions by BER. The trapped AP site is a highly cytotoxic intermediate and sensitizes cells to the cytotoxic effects of alkylating agents [132–134] and other DNA-damaging agents that may give rise to spontaneous or enzyme-mediated AP sites, such as TMZ [135], iododeoxyuridine + radiation [136], BCNU [137], manumycin A [138], fludarabine [139], radiation [140, 141], and pemetrexed [142]. Interestingly, elevated expression of the downstream BER protein DNA polymerase β (Polβ) reverses MX sensitization of alkylating agents, suggesting that the cleaved open-chain aldehyde may be the preferred MX substrate [134]. MX is in the clinic under the brand name TRC102 and is undergoing clinical evaluation in Phase II trials for solid tumors as a sensitizer to Temodar® (TMZ) or Alimta® (Pemetrexed) and in hematologic malignancies with Fludara® (Fludarabine).

In parallel, there are a number of groups developing direct APE1 active site inhibitors [143–146], and an overview of the development of APE1 inhibitors has just been reported [147]. As might be expected, many of these APE1 inhibitors are themselves cytotoxic and enhance the cytotoxicity of DNA-damaging agents such as TMZ [144, 145] or other alkylating agents [143]. Although Ape1 KO or GzmA-mediated Ape1 cleavage in mouse cells induces apoptosis, the cell death mechanism(s) induced by these recently developed APE1 inhibitors has yet to be resolved. Interestingly, the cell death mechanism triggered by some of these APE1 inhibitors may involve the accumulation of DNA DSBs since it was observed that the compounds are more cytotoxic in cells deficient in the HR proteins BRCA1 or BRCA2 [148].

6.3.5 Ataxia Telangiectasia-RAD3 Related

Similar strategies and rationales that apply to the CHK1 target also apply to ATR. A direct link connects ATR with CHK1 [121]. After initial Rad17 binding, Claspin-loaded single-strand DNA mediates the activation of CHK1 by ATR upon replication stress. ATR is essential in the surveillance of replication stress, especially when associated with exposed single-stranded DNA mostly connected to replication problems. Stalled replication forks can collapse which results in the formation of DSBs, a signal that mainly, but not exclusively, triggers ATM whereas the single-stranded DNA-induced damage response appears to be ATM independent. Endogenous and exogenous damage induces replication stress that requires a proficient ATR/CHK1 response for survival. One prevalent rationale to inhibit ATR for cancer therapy is to exacerbate the levels of replication stress that might be augmented in cancer cells due to inherited defects in the DDR and DNA repair pathways. Cancer cells are exposed to a higher load of replication stress compared to normal cells and will suffer most from targeting ATR. In addition, since targeting replication-associated processes, any selective killing property will be augmented in highly proliferative cells. Such a strategy is supported by two observations: (1) the activated DDR found in early stages of tumorigenesis [149, 150] indicates an increased load of "endogenous" DNA damage and (2) the discovery that a wide variety of oncogenes generate such damage.

Based on this proposed endogenous damage (replication stress)-induced killing mechanism, such ATR inhibitors have been proposed to act as single agents. However, combination strategies similar to those for the CHK1 inhibitor are envisioned. As noted before, the response of normal cells should be carefully taken into consideration. A few compounds have been pursued by industry. The ATR inhibitor NU6027 appears to impair HR and enhances the cytotoxicity of PARP inhibitors [151], a theme that is consistent to other DDR inhibitors. One compound (NVP-BEZ235) has been recently discovered and found to trigger preferential cell kill in cyclin-E overexpressing cells [152] and is awaiting entry into clinical trials for cancer therapy.

6.3.6 Ataxia Telangiectasia-Mutated

ATM inhibitors are less advanced in their clinical development than PARP or CHK1 inhibitors. Since involved in DSB repair, NHEJ, and HR, ATM inhibition results in reduced cellular DSB repair activity and cell cycle checkpoint defects. As a result, ATM inhibitors are highly potent radiosensitizers while exhibiting little toxicity on their own. Therefore, they have been proposed to be applied in this context [153]. Several ATM-inhibiting compounds have been identified and pursued in preclinical studies: KU55933 and KU60019 (KuDos/AstraZeneca) and CP466722 (Pfizer).

Combination strategies for ATM inhibitors have been proposed to enhance the cytotoxicity of PARP inhibitors (see above). Interestingly, p53 disruption in normal cells sensitized those cells to the combination of PARP and ATM inhibitors [83]. Due to its crucial role in DSB repair, the inhibition of ATM will radio- and chemosensitize most cells, including normal cells, with no evident DNA repair defects. However, synergistic cytotoxicity can be observed under certain conditions that could be exploited for tumor-targeted strategies. Cells with BER defects for example rely on secondary DSB repair pathways such as HR for survival upon damaging agents. Similar to the PARP inhibitor/HR-defect synthetic lethal interaction, the conversion of unrepaired SSBs to DSBs could be a mechanism that underlies such dependence. BER defects of different kinds have been reported in tumors, and it is suggested that inhibition of DSB repair processes will be beneficial in such a setting [154].

6.3.7 DNA-Dependent Protein Kinase

DNA-PK is crucial to DSB repair as well as cellular survival following radiation and topoisomerase II poisons, and so, DNA-PK has been a long sought target for drug development [155]. Kinase activities are relatively easy to target, and drugs inhibiting DNA-PK activity have been available for some time. One should note that a considerable fraction of first generation tyrosine kinase inhibitors do also inhibit DNA-PK to some extent. The radiosensitization phenotype by some of these agents should therefore be re-evaluated with respect to its origin (see below). Some examples of currently pursued inhibitors are NU7026 and NU7441 [156, 157]. Other targeting strategies apply short modified DNA molecules that are supposed to interfere in DNA-PK signaling (DT01, DNA therapeutics) or short peptides that resemble Ku80 (HNI-38) and disrupt the interaction with DNA-PK [158]. The dual DNA-PK and PI3-K inhibitor KU-0060648 takes advantage of the inhibition of two cellular processes central to promoting survival upon chemo- and radiotherapeutic insults. Surprisingly, this inhibitor enhanced etoposide-induced xenograft tumor growth delay substantially without exacerbating etoposide toxicity in mice indicating that the combination provided some tumor specificity [159].

Most DNA-PK-inhibiting agents are very potent radiosensitizers, enhancing clonogenic cell kill markedly [160]. Radiosensitization by one compound, the DNA-PK inhibitor BEZ235, has been shown to be associated with an accelerated p53-dependent senescence [161]. Although proposed as radiation dose intensifiers, DNA-PK inhibitors, similar to ATM inhibitors (though often to a greater extent), will sensitize normal tissue to radiation at least as much as it does sensitize the tumors. However, some DNA repair defects inherent to tumors might provide some therapeutic gain when applying low inhibitor doses that could spare normal tissue. Opportunities exist for example in the treatment of ATM-deficient tumors since DNA-PK inhibition induces tumor cell kill in ATM-deficient cells [162].

Considering that disruption of the catalytic activity of DNA-PK confers severe immunodeficiency, DNA-PK inhibitors will have to be evaluated carefully for toxicities in preclinical studies before entering clinical trial [163]. Tumor-targeted delivery strategies could, however, make use of these highly potent radiation dose intensifiers. Other strategies exploiting the differential expression or use of DNA-PK and Ku-dependent DNA repair in some tissues with respect to the tumors could apply these inhibitors successfully. An example for this is the sensitization of EGFRvIII-overexpressing cells with DNA-PK inhibitors [164].

6.4 Indirect DNA Repair Modulators

6.4.1 Signaling Pathways, EGFR and the PI3K/AKT Pathway

Targeting the cellular signaling pathways via EGFR, PI3K/AKT, or MAPK can modulate the DNA repair status of cells. The effects are multiple and involve different pathways. Cellular signaling influences DNA repair in multiple ways including changes in the expression levels of crucial DNA repair proteins such as RAD51 or the regulation of the activation or translocation of enzymes and kinases such as DNA-PK. For example, the inhibition of the MAPK pathway can lead to reduced DSB repair by both homologous and non-homologous pathways [165, 166]. Links between the AKT, MAPK, and EGFR pathways and DNA repair have been found, particularly with DSB repair by NHEJ or HR [166–171]. Hence, targeting these pathways can result in the concomitant inhibition of DNA repair.

Based on knowledge that the PI3K pathway promotes DNA repair and survival, inhibitors of AKT have been evaluated as chemosensitizers for alkylating agents [172, 173]. AKT inhibitors such as LY294002 and wortmannin radiosensitized and caused enhanced sensitivity to alkylating agents such as TMZ, or the cross-linkers cisplatin in various human tumor cell lines.

The suppression of ATR/CHK1 checkpoints has been observed upon treatment with wortmannin (a fairly non-specific PI3K inhibitor). For more details, see [174] and references within. Note that an influence on Rad51 expression or on NHEJ via DNA-PK has also been observed.

Co-targeting PARP and the PI3K pathway has been shown to synergistically decrease growth and further induce apoptosis [175]. The molecular mechanism of this combination effect is not completely clear but could originate from secondary effects on DNA repair pathways and the DDR from the inhibition of cellular signaling in human cancer cells. Further detail on AKT inhibitors is discussed in more detail in Chap. 13.

6.4.2 Proteasome Inhibitors

Proteasome-mediated destruction and removal of key DNA repair and DDR proteins are an essential aspect of the cellular response to genotoxins [4, 176]. Failure to remove critical DNA repair and DDR proteins in a timely fashion effectively halts the process, triggering a cascade of events leading to cell death [177]. As the endpoint in ubiquitin-mediated protein turnover, targeting the proteasome is a high-profile target to enhance DNA damage–induced cell death [178, 179]. Further detail on the proteasome pathway and proteasome inhibitors is discussed in more detail in Chap. 12.

6.4.3 NAD^+ Biosynthesis Inhibitors

The metabolite NAD^+ is an essential substrate for all PARPs [74]. Since PARP1, as well as PARP2 and PARP3, is a critical protein involved in DNA repair and the DDR [36, 103], we have included NAD^+ biosynthesis inhibitors as an indirect DNA repair modulator. All of the NAD^+ biosynthesis inhibitors developed to date target NAMPT, a pivotal and rate-limiting enzyme in the salvage pathway of NAD^+ biosynthesis [180]. NAMPT catalyzes the synthesis of nicotinamide mononucleotide (NMN) from nicotinamide to 5-phosphoribosyl-1-pyrophosphate [181]. The resulting NMN is then converted to NAD^+ by one of the three isoforms of NMNAT (1, 2 or 3), located in the nucleus, cytosol, or mitochondria, respectively [182]. The most studied inhibitor of NAMPT is FK866 (also referred to as Apo866) although several other NAMPT inhibitors have been reported including GMX1778 [183], CB30865 [184], and CHS-828 [185], and one group is actively screening for NAMPT inhibitors [186]. FK866 binds at the interface of the NAMPT dimer [187] and effectively inhibits NAMPT and depletes cellular NAD^+ levels within 24 h, inducing apoptosis [188] although the overall level of cell death may be offset by FK866-mediated induction of autophagy [189]. Further detail on the shift of cell death triggered by NAD^+ biosynthesis inhibitors is discussed in Chap. 2.

As an NAMPT inhibitor, FK866 does have a chemo-potentiating effect [190–192] that is more pronounced when combined with a second DNA repair inhibitor [133] or when combined with TRAIL [193]. Interestingly, it has been suggested that some tumors are deficient in the NAPRT1-mediated NAD^+ biosynthesis pathway that is responsible for generating NAD^+ from nicotinic acid (NA) [183, 194]. In sum, these studies suggest that NAMPT is a potentially valuable target to impact both metabolism and DNA repair and may provide a type of synthetic lethality based on DNA repair and NAD^+ biosynthesis status [133, 195].

6.5 Promising Future Targets and Opportunities

6.5.1 ERCC1

An emerging DNA repair target protein is ERCC1, encoded by the excision repair cross-complementing rodent repair deficiency, complementation group 1 gene (ERCC1) [49, 50]. As an essential NER protein, elevated ERCC1 expression enhances the repair of cisplatin-induced genomic DNA damage [15] and decreased expression is suggested to improve response to platinum-based therapies [196]. As described above (Sect. 6.1.4), ERCC1 exists as an obligate heterodimer with the protein XPF to yield an essential NER structure-specific nuclease ERCC1/XPF [15]. The ERCC1/XPF heterodimer is recruited to function in NER by interaction with the NER protein XPA [197] although multiple interaction sites are found to facilitate recruitment of ERCC1/XPF to the NER complex [198]. In this regard, it has been suggested that the function of ERCC1 in NER can be disrupted by interfering with the interaction between ERCC1 and XPA [199]. Alternatively, since heterodimer formation (ERCC1/XPF) is needed to maintain the stability of ERCC1 [200], it might be feasible to interrupt the heterodimer complex and force ERCC1 proteolysis [201]. Finally, one might also consider deregulating expression of ERCC1 through modulation of the RAS kinase or ERK1/2 pathways [202]. Using the Raf kinase and VEGF receptor inhibitor sorafenib reduces ERCC1 levels and effectively enhances the response to radiation and chemotherapy [203]. Of course this treatment would have multiple effects on gene regulation.

6.5.2 DNA Polymerases

There are 15 DNA polymerases in the human cell [204] and several have been suggested to be viable drug targets either because they facilitate repair of DNA damage or may be overexpressed in cancer [205]. DNA polymerase β (Polβ) is a member of the X family of DNA polymerases [204, 206] and is an essential BER protein, as detailed above. Polβ has two active sites and several critical functional domains that may be considered as targets to inhibit Polβ and BER. Further, the central and pivotal role of Polβ in BER implicates this protein as a prime target to enhance the response to chemotherapeutic agents or radiation. Although many inhibitors of Polβ have been developed and characterized very few, if any, show specificity and all are cytotoxic even in Polβ KO cells. Three recent reviews on Polβ inhibitors should be referred to for more detail [195, 207, 208].

Other polymerases have been also identified as potential targets. A siRNA screen and large-scale tumor sample analysis identified DNA polymerase theta (POLQ) as a target that determines radiosensitivity and is overexpressed in tumors. POLQ depletion radiosensitized these cells arguing for the development of POLQ inhibitors [209–211].

6.5.3 DNA Glycosylases

DNA glycosylases are the initiating enzymes in BER (see above). In general, DNA glycosylases probe the DNA helix for base lesions by a base-flipping mechanism, interrogating each base as it is flipped out of the major groove as the enzyme moves along the DNA helix. Identification of the lesion then promotes hydrolysis [212]. As discussed above, radiation and chemotherapeutic agents induce the formation of BER substrates and many of these base lesions block replication and are therefore cytotoxic. For example, the methylated base 3-methyladenine, induced by the chemotherapeutic agent TMZ and repaired by MPG, triggers lesion-induced apoptosis as well as sister chromatid exchange, chromatid and chromosome gaps and breaks, and S-phase arrest [213]. As such, it has been hypothesized that inhibition of MPG would enhance response to TMZ and other alkylators. Similarly, UNG, the primary glycosylase responsible for the removal of deoxyuracil, is reported to govern the efficacy of pemetrexed [142], and therefore, UNG might be considered a potential target to enhance the response to this chemotherapeutic agent. Finally, it was recently demonstrated that the DNA glycosylase OGG1 is in a complex with the BER protein PARP1. The OGG1–PARP1 complex promotes PARP1 activity whereas loss of OGG1 suppressed PARP1 activity [214]. From this study, it might be suggested that an inhibitor designed to disrupt the OGG1–PARP1 interaction may function as a pseudo- or indirect PARP1 inhibitor. However, there are multiple challenges to developing effective DNA glycosylase inhibitors such as the large overlap in enzyme substrate specificity [36, 37] and the requirement for some glycosylases in normal cell survival [215, 216] and immunoglobulin class-switch recombination [217]. To date, we know of no DNA glycosylase inhibitors with the exception of Ugi, a virally encoded UNG inhibitor [218].

6.5.4 DUBs

As with most posttranslational modifications, both the synthesis and removal of the ubiquitin modification is a highly regulated process [219, 220]. Recently, several studies have revealed the potential for targeting DUBs to enhance radiation and chemotherapeutic response [221, 222]. Toward that end, several groups have reported the development of DUB inhibitors that are effective in preclinical studies [223–226].

6.5.5 HR/FA

HR and FA strongly determine cellular survival upon cross-linkers, topoisomerase poisons, alkylators, and radiation. These pathways are mostly engaged in a proliferation-dependent manner with less influence on damage repair and response in quiescent cells. Thus, targeting HR and FA can be an effective means to potentiate radio- and chemotherapy [227].

Novel agents targeting RAD51, albeit indirectly, are now emerging in clinical trials. MP470 (Supergen) is a tyrosine kinase inhibitor with RAD51 protein–suppressing properties. More specific inhibitors are sought in high-throughput screening efforts of which some provided some compounds with convincing specificity and potency [228].

The value of HR inhibition can be deduced from experiments in which elements of the HR or FA pathway such as RAD51 or BRCA2 have been downregulated, supporting the potential of HR/FA-targeting strategies. Downregulation of Rad51 or BRCA2 sensitized glioma cells to the cytotoxic effects of TMZ or nimustine in cells with low MGMT levels [229]. The marked resistance of gliomas to chemo- or radiotherapy regimens that might be partly defined by HR or increased Rad51 levels can be defined by targeting RAD51, indicating a similar strategy option to counteract resistance by HR inhibition [230].

FA- and HR-defected cells have been found to lack the increased radio-resistance that is observed in hypoxic areas of the tumors [231]. Inhibitors targeted to HR or FA will not only preferentially sensitize proliferating cancer cells but also hypoxic tumors to radiation. Owing to less-effective DSB induction under low oxygen concentration, radiotherapy often fails in patients with hypoxic tumors. Tumor specificity of HR-targeting drugs will be increased when applied to BER-defected cancer cells as illustrated above. BRCA2-knockdown radiosensitizes cells that express a dominant-negative Polβ mutant to a greater extent than isogenic controls [232]. Similarly, NHEJ-defected cancer cells are expected to rely on HR for survival upon genotoxic threats. Together, these interactions and data support strategies applying HR- or FA-inhibiting drugs for tumor-specific sensitization.

6.5.6 Poly(ADP)ribose Glycohydrolase

Poly(ADP)ribose glycohydrolase (PARG) [233] is an enzyme involved in BER and is the major factor responsible for removing the PAR molecules synthesized by PARP1 and PARP2, among other PARPs [234]. As seen with PARP1, PARG is also a substrate for caspase-3 [235], supporting the significance of PAR metabolism in cell death processes. Mouse KO studies or RNA interference studies to knockdown PARG in human cells have suggested that blocking PAR degradation is an effective means to modulate chemotherapy and radiation response [134, 236, 237]. Pre-clinical evaluations using early phase small-molecule inhibitors such as gallotannin, tannic acid, and related small molecules have implicated PARG in genotoxin or chemotherapy sensitivity [238–240]. The inhibitor GPI-16552 (N-bis-(3-phenyl-propyl)9-oxo-fluorene-2,7-diamide) has shown early success in cell and mouse models as a chemotherapy sensitizer [97] but has mostly been effective to reduce inflammation [241–245]. Recent drug discovery efforts seem promising and have yielded PARG inhibitors with increased specificity and cell permeability [246, 247].

6.6 Summary and Concluding Remarks

Inducing tumor-specific or selective cell death is a significant challenge in the treatment of cancer. Although radiation and most chemotherapeutic treatments are designed to induce significant genotoxic damage to the tumor cells, a multitude of DNA repair and DDR gene products respond to the presence of genotoxic stress (DNA damage) by orchestrating a massively complex cellular response to survive and repair the genome insult. Most DNA repair pathways focus on specific damage types (e.g., DSBs), but there is considerable overlap and backup capacity when evaluating the overall involvement of DR, BER, MMR, NER, HR, and NHEJ proteins in response to these agents (radiation and chemotherapy). It has been observed that some cancer cells can survive the loss of key DNA repair proteins (e.g., BRCA1/2). In fact, we have come to realize that many cancer cell types are defective in one or more DNA repair or DDR genes/proteins, precipitating the need to identify the key stress nodes in these defective cells in response to radiation and classical chemotherapy treatments. With this in mind, we have described the molecular pathways that participate in the repair of DNA damage induced by radiation and chemotherapeutics. Further, we present novel therapeutic strategies being considered to exploit inherent weaknesses in tumor cells such as defects in one or more DNA repair pathways that may provide the opportunity to selectively increase tumor-specific cell death.

References

1. Lindahl T (1993) Instability and decay of the primary structure of DNA. Nature 362(6422):709–715
2. Jackson SP, Bartek J (2009) The DNA-damage response in human biology and disease. Nature 461(7267):1071–1078. Epub 2009/10/23
3. Hanahan D, Weinberg RA (2011) Hallmarks of cancer: the next generation. Cell 144(5):646–674. Epub 2011/03/08
4. Harper JW, Elledge SJ (2007) The DNA damage response: ten years after. Mol Cell 28(5):739–745. Epub 2007/12/18

5. Ljungman M (2009) Targeting the DNA damage response in cancer. Chem Rev 109(7):2929–2950. Epub 2009/06/24
6. Alberts B (2009) Redefining cancer research. Science 325(5946):1319. Epub 2009/09/12
7. Helleday T, Petermann E, Lundin C, Hodgson B, Sharma RA (2008) DNA repair pathways as targets for cancer therapy. Nat Rev Cancer 8(3):193–204
8. Lavin MF (2008) Ataxia-telangiectasia: from a rare disorder to a paradigm for cell signalling and cancer. Nat Rev Mol Cell Biol 9(10):759–769. Epub 2008/09/25
9. Matsuoka S, Ballif BA, Smogorzewska A, McDonald ER 3rd, Hurov KE, Luo J et al (2007) ATM and ATR substrate analysis reveals extensive protein networks responsive to DNA damage. Science 316(5828):1160–1166
10. Bakkenist CJ, Kastan MB (2003) DNA damage activates ATM through intermolecular autophosphorylation and dimer dissociation. Nature 421(6922):499–506
11. Wood RD, Mitchell M, Sgouros J, Lindahl T (2001) Human DNA repair genes. Science 291(5507):1284–1289
12. Gerson SL (2004) MGMT: its role in cancer aetiology and cancer therapeutics. Nat Rev Cancer 4(4):296–307
13. Fu D, Calvo JA, Samson LD (2012) Balancing repair and tolerance of DNA damage caused by alkylating agents. Nat Rev Cancer 12(2):104–120. Epub 2012/01/13
14. Kaina B, Christmann M, Naumann S, Roos WP (2007) MGMT: key node in the battle against genotoxicity, carcinogenicity and apoptosis induced by alkylating agents. DNA Repair (Amst) 6(8):1079–1099
15. Friedberg EC, Walker GC, Siede W, Wood RD, Schultz RA, Ellenberger T (2006) DNA repair and mutagenesis. 2nd Edn. ASM Press, Washington, p 1164
16. Wyatt MD, Pittman DL (2006) Methylating agents and DNA repair responses: methylated bases and sources of strand breaks. Chem Res Toxicol 19(12):1580–1594
17. Wang JY, Edelmann W (2006) Mismatch repair proteins as sensors of alkylation DNA damage. Cancer Cell 9(6):417–418
18. Friedman HS, Johnson SP, Dong Q, Schold SC, Rasheed BK, Bigner SH et al (1997) Methylator resistance mediated by mismatch repair deficiency in a glioblastoma multiforme xenograft. Cancer Res 57(14):2933–2936
19. Roos WP, Batista LF, Naumann SC, Wick W, Weller M, Menck CF et al (2007) Apoptosis in malignant glioma cells triggered by the temozolomide-induced DNA lesion O6-methylguanine. Oncogene 26(2):186–197
20. Caporali S, Falcinelli S, Starace G, Russo MT, Bonmassar E, Jiricny J et al (2004) DNA damage induced by temozolomide signals to both ATM and ATR: role of the mismatch repair system. Mol Pharmacol 66(3):478–491
21. Jiricny J (2006) The multifaceted mismatch-repair system. Nat Rev Mol Cell Biol 7(5):335–346
22. Pabla N, Ma Z, McIlhatton MA, Fishel R, Dong Z (2011) hMSH2 recruits ATR to DNA damage sites for activation during DNA damage-induced apoptosis. J Biol Chem 286(12):10411–10418. Epub 2011/02/03
23. Wang Y, Qin J. MSH2 and ATR form a signaling module and regulate two branches of the damage response to DNA methylation. Proceedings Nat Acad Sci 100(26):15387–15392
24. Hegi ME, Liu L, Herman JG, Stupp R, Wick W, Weller M et al (2008) Correlation of O6-methylguanine methyltransferase (MGMT) promoter methylation with clinical outcomes in glioblastoma and clinical strategies to modulate MGMT activity. J Clin Oncol 26(25):4189–4199
25. Sarkaria JN, Kitange GJ, James CD, Plummer R, Calvert H, Weller M et al (2008) Mechanisms of chemoresistance to alkylating agents in malignant glioma. Clin Cancer Res 14(10):2900–2908
26. Pollack IF, Hamilton RL, Sobol RW, Burnham J, Yates AJ, Holmes EJ et al (2006) MGMT expression strongly correlates with outcome in childhood malignant gliomas: results from the CCG-945 cohort. J Clin Oncol 24(21):3431–3437
27. Holland EC (2000) Glioblastoma multiforme: the terminator. Proc Natl Acad Sci USA 97(12):6242–6244. Epub 2000/06/07

28. Cohen MH, Johnson JR, Pazdur R (2005) Food and drug administration drug approval summary: temozolomide plus radiation therapy for the treatment of newly diagnosed glioblastoma multiforme. Clin Cancer Res 11(19 Pt 1):6767–6771
29. Stupp R, Hegi ME, Mason WP, van den Bent MJ, Taphoorn MJ, Janzer RC et al (2009) Effects of radiotherapy with concomitant and adjuvant temozolomide versus radiotherapy alone on survival in glioblastoma in a randomised phase III study: 5 year analysis of the EORTC-NCIC trial. Lancet Oncol 10(5):459–466
30. Stupp R, Mason WP, van den Bent MJ, Weller M, Fisher B, Taphoorn MJ et al (2005) Radiotherapy plus concomitant and adjuvant temozolomide for glioblastoma. N Engl J Med 352(10):987–996
31. Cahill DP, Levine KK, Betensky RA, Codd PJ, Romany CA, Reavie LB et al (2007) Loss of the mismatch repair protein MSH6 in human glioblastomas is associated with tumor progression during temozolomide treatment. Clin Cancer Res 13(7):2038–2045
32. Yip S, Miao J, Cahill DP, Iafrate AJ, Aldape K, Nutt CL et al (2009) MSH6 mutations arise in glioblastomas during temozolomide therapy and mediate temozolomide resistance. Clin Cancer Res 15(14):4622–4629
33. Friedman HS, Keir S, Pegg AE, Houghton PJ, Colvin OM, Moschel RC et al (2002) O^6-benzylguanine-mediated enhancement of chemotherapy. Mol Cancer Ther 1(11):943–948
34. Tserng KY, Ingalls ST, Boczko EM, Spiro TP, Li X, Majka S et al (2003) Pharmacokinetics of O^6-benzylguanine (NSC637037) and its metabolite, 8-oxo-O6-benzylguanine. J Clin Pharmacol 43(8):881–893
35. Esteller M, Garcia-Foncillas J, Andion E, Goodman SN, Hidalgo OF, Vanaclocha V et al (2000) Inactivation of the DNA-repair gene MGMT and the clinical response of gliomas to alkylating agents. N Engl J Med 343(19):1350–1354
36. Almeida KH, Sobol RW (2007) A unified view of base excision repair: lesion-dependent protein complexes regulated by post-translational modification. DNA Repair 6(6):695–711
37. Svilar D, Goellner EM, Almeida KH, Sobol RW (2011) Base excision repair and lesion-dependent sub-pathways for repair of oxidative DNA damage. Antioxid Redox Signal 14(12):2491–2507. Epub 2010/07/24
38. Sobol RW (2009) Temozolomide. In: Schwab M (ed) Encyclopedia of cancer, 2nd edn. Springer, Berlin Heidelberg, pp 2928–2933
39. Wilson SH, Kunkel TA (2000) Passing the baton in base excision repair. Nat Struct Biol 7(3):176–178
40. Gagne JP, Pic E, Isabelle M, Krietsch J, Ethier C, Paquet E et al (2012) Quantitative proteomics profiling of the poly(ADP-ribose)-related response to genotoxic stress. Nucleic Acids Res 40(16):7788–7805. Epub 2012/06/07
41. Gibson BA, Kraus WL (2012) New insights into the molecular and cellular functions of poly(ADP-ribose) and PARPs. Nat Rev Mol Cell Biol 13(7):411–424. Epub 2012/06/21
42. Sousa FG, Matuo R, Soares DG, Escargueil AE, Henriques JA, Larsen AK et al PARPs and the DNA damage response. Carcinogenesis 33(8):1433–1440. Epub 2012/03/21
43. Roos WP, Kaina B (2012) DNA damage-induced apoptosis: from specific DNA lesions to the DNA damage response and apoptosis. Cancer Lett In Press. Epub 2012/01/21
44. Hoeijmakers JH (2001) Genome maintenance mechanisms for preventing cancer. Nature 411(6835):366–374
45. De Laat WL, Jaspers NG, Hoeijmakers JH (1999) Molecular mechanism of nucleotide excision repair. Genes Dev 13(7):768–785. Epub 1999/04/10
46. Wood RD (1996) DNA repair in eukaryotes. Annu Rev Biochem 65:135–167. Epub 1996/01/01
47. Shuck SC, Short EA, Turchi JJ (2008) Eukaryotic nucleotide excision repair: from understanding mechanisms to influencing biology. Cell Res 18(1):64–72. Epub 2008/01/02
48. Kapetanaki MG, Guerrero-Santoro J, Bisi DC, Hsieh CL, Rapic-Otrin V, Levine AS (2006) The DDB1-CUL4ADDB2 ubiquitin ligase is deficient in xeroderma pigmentosum group E and targets histone H2A at UV-damaged DNA sites. Proc Natl Acad Sci USA 103(8):2588–2593

49. Barakat K, Gajewski M, Tuszynski JA (2012) DNA repair inhibitors: the next major step to improve cancer therapy. Curr Top Med Chem 12(12):1376–1390. Epub 2012/07/17
50. Postel-Vinay S, Vanhecke E, Olaussen KA, Lord CJ, Ashworth A, Soria JC (2012) The potential of exploiting DNA-repair defects for optimizing lung cancer treatment. Nat Rev Clin Oncol 9(3):144–155. Epub 2012/02/15
51. Bhagwat NR, Roginskaya VY, Acquafondata MB, Dhir R, Wood RD, Niedernhofer LJ (2009) Immunodetection of DNA repair endonuclease ERCC1-XPF in human tissue. Cancer Res 69(17):6831–6838. Epub 2009/09/03
52. Niedernhofer LJ, Bhagwat N, Wood RD. ERCC1 and non-small-cell lung cancer. N Engl J Med. 2007;356(24):2538-40; author reply 40-1. Epub 2007/06/15
53. Lobrich M, Shibata A, Beucher A, Fisher A, Ensminger M, Goodarzi AA et al (2010) GammaH2AX foci analysis for monitoring DNA double-strand break repair: strengths, limitations and optimization. Cell Cycle 9(4):662–669. Epub 2010/02/09
54. Audebert M, Salles B, Calsou P (2004) Involvement of poly(ADP-ribose) polymerase-1 and XRCC1/DNA ligase III in an alternative route for DNA double-strand breaks rejoining. J Biol Chem 279(53):55117–55126
55. Mladenov E, Iliakis G (2011) Induction and repair of DNA double strand breaks: the increasing spectrum of non-homologous end joining pathways. Mutat Res 711(1–2):61–72. Epub 2011/02/19
56. Goodarzi AA, Noon AT, Deckbar D, Ziv Y, Shiloh Y, Lobrich M et al (2008) ATM signaling facilitates repair of DNA double-strand breaks associated with heterochromatin. Mol Cell 31(2):167–177. Epub 2008/07/29
57. Murray JM, Stiff T, Jeggo PA (2012) DNA double-strand break repair within heterochromatic regions. Biochem Soc Trans 40(1):173–178. Epub 2012/01/21
58. Moynahan ME, Jasin M (2010) Mitotic homologous recombination maintains genomic stability and suppresses tumorigenesis. Nat Rev Mol Cell Biol 11(3):196–207. Epub 2010/02/24
59. Jeggo PA, Geuting V, Lobrich M (2011) The role of homologous recombination in radiation-induced double-strand break repair. Radiother Oncol 101(1):7–12. Epub 2011/07/09
60. Allen C, Ashley AK, Hromas R, Nickoloff JA (2011) More forks on the road to replication stress recovery. J Mol Cell Biol 3(1):4–12. Epub 2011/02/01
61. Kim H, D'Andrea AD (2012) Regulation of DNA cross-link repair by the Fanconi anemia/BRCA pathway. Genes Dev 26(13):1393–1408. Epub 2012/07/04
62. Crossan GP, Patel KJ (2012) The fanconi anaemia pathway orchestrates incisions at sites of crosslinked DNA. J Pathol 226(2):326–337. Epub 2011/10/01
63. Yuan F, Song L, Qian L, Hu JJ, Zhang Y (2010) Assembling an orchestra: Fanconi anemia pathway of DNA repair. Front Biosci 15:1131–1149. Epub 2010/06/03
64. Wang W (2008) A major switch for the Fanconi anemia DNA damage-response pathway. Nat Struct Mol Biol 15(11):1128–1130. Epub 2008/11/06
65. Begg AC, Stewart FA, Vens C (2011) Strategies to improve radiotherapy with targeted drugs. Nat Rev Cancer 11(4):239–253. Epub 2011/03/25
66. Kaelin WG Jr (2005) The concept of synthetic lethality in the context of anticancer therapy. Nat Rev Cancer 5(9):689–698
67. Javle M, Curtin NJ (2011) The role of PARP in DNA repair and its therapeutic exploitation. Br J Cancer 105(8):1114–1122. Epub 2011/10/13
68. Sandhu SK, Yap TA, De Bono JS (2010) Poly(ADP-ribose) polymerase inhibitors in cancer treatment: a clinical perspective. Eur J Cancer 46(1):9–20. Epub 2009/11/21
69. Peralta-Leal A, Rodriguez MI, Oliver FJ (2008) Poly(ADP-ribose)polymerase-1 (PARP-1) in carcinogenesis: potential role of PARP inhibitors in cancer treatment. Clin Transl Oncol 10(6):318–323
70. Bryant HE, Schultz N, Thomas HD, Parker KM, Flower D, Lopez E et al (2005) Specific killing of BRCA2-deficient tumours with inhibitors of poly(ADP-ribose) polymerase. Nature 434(7035):913–917

71. Farmer H, McCabe N, Lord CJ, Tutt AN, Johnson DA, Richardson TB et al (2005) Targeting the DNA repair defect in BRCA mutant cells as a therapeutic strategy. Nature 434(7035):917–921
72. Hendricks CA, Razlog M, Matsuguchi T, Goyal A, Brock AL, Engelward BP (2002) The *S.cerevisiae* Mag1 3-methyladenine DNA glycosylase modulates susceptibility to homologous recombination. DNA Repair 1:645–659
73. Shaheen M, Allen C, Nickoloff JA, Hromas R (2011) Synthetic lethality: exploiting the addiction of cancer to DNA repair. Blood 117(23):6074–6082. Epub 2011/03/29
74. Hassa PO, Haenni SS, Elser M, Hottiger MO (2006) Nuclear ADP-ribosylation reactions in mammalian cells: where are we today and where are we going? Microbiol Mol Biol Rev 70(3):789–829
75. Patel AG, De Lorenzo SB, Flatten KS, Poirier GG, Kaufmann SH (2012) Failure of iniparib to inhibit poly(ADP-Ribose) polymerase in vitro. Clin Cancer Res 18(6):1655–1662. Epub 2012/02/01
76. Liu X, Shi Y, Maag DX, Palma JP, Patterson MJ, Ellis PA et al (2012) Iniparib nonselectively modifies cysteine-containing proteins in tumor cells and is not a bona fide PARP inhibitor. Clin Cancer Res 18(2):510–523. Epub 2011/12/01
77. Audeh MW, Carmichael J, Penson RT, Friedlander M, Powell B, Bell-McGuinn KM et al (2010) Oral poly(ADP-ribose) polymerase inhibitor olaparib in patients with BRCA1 or BRCA2 mutations and recurrent ovarian cancer: a proof-of-concept trial. Lancet 376(9737):245–251. Epub 2010/07/09
78. Tutt A, Robson M, Garber JE, Domchek SM, Audeh MW, Weitzel JN et al (2010) Oral poly(ADP-ribose) polymerase inhibitor olaparib in patients with BRCA1 or BRCA2 mutations and advanced breast cancer: a proof-of-concept trial. Lancet 376(9737):235–244. Epub 2010/07/09
79. Vollebergh MA, Jonkers J, Linn SC (2012) Genomic instability in breast and ovarian cancers: translation into clinical predictive biomarkers. Cell Mol Life Sci 69(2):223–245. Epub 2011/09/17
80. Bunting SF, Callen E, Kozak ML, Kim JM, Wong N, Lopez-Contreras AJ et al (2012) BRCA1 functions independently of homologous recombination in DNA interstrand crosslink repair. Mol Cell 46(2):125–135. Epub 2012/03/27
81. McCabe N, Turner NC, Lord CJ, Kluzek K, Bialkowska A, Swift S et al (2006) Deficiency in the repair of DNA damage by homologous recombination and sensitivity to poly(ADP-Ribose) polymerase inhibition. Cancer Res 66(16):8109–8115
82. Yap TA, Sandhu SK, Carden CP, De Bono JS (2011) Poly(ADP-ribose) polymerase (PARP) inhibitors: exploiting a synthetic lethal strategy in the clinic. CA Cancer J Clin 61(1):31–49. Epub 2011/01/06
83. Williamson CT, Kubota E, Hamill JD, Klimowicz A, Ye R, Muzik H et al (2012) Enhanced cytotoxicity of PARP inhibition in mantle cell lymphoma harbouring mutations in both ATM and p53. EMBO Mol Med 4(6):515–527. Epub 2012/03/15
84. Williamson CT, Muzik H, Turhan AG, Zamo A, O'Connor MJ, Bebb DG et al ATM deficiency sensitizes mantle cell lymphoma cells to poly(ADP-ribose) polymerase-1 inhibitors. Mol Cancer Ther 9(2):347–57. Epub 2010/02/04
85. McEllin B, Camacho CV, Mukherjee B, Hahm B, Tomimatsu N, Bachoo RM et al (2010) PTEN loss compromises homologous recombination repair in astrocytes: implications for glioblastoma therapy with temozolomide or poly(ADP-ribose) polymerase inhibitors. Cancer Res 70(13):5457–5464. Epub 2010/06/10
86. Mendes-Pereira AM, Martin SA, Brough R, McCarthy A, Taylor JR, Kim JS et al (2009) Synthetic lethal targeting of PTEN mutant cells with PARP inhibitors. EMBO molecular medicine 1(6–7):315–322. Epub 2010/01/06
87. Fraser M, Zhao H, Luoto KR, Lundin C, Coackley C, Chan N et al (2011) PTEN deletion in prostate cancer cells does not associate with loss of RAD51 function: implications for radiotherapy and chemotherapy. Clin Cancer Res 18(4):1015–1027. Epub 2011/11/25

88. Chan N, Pires IM, Bencokova Z, Coackley C, Luoto KR, Bhogal N et al (2010) Contextual synthetic lethality of cancer cell kill based on the tumor microenvironment. Cancer Res 70(20):8045–8054. Epub 2010/10/07
89. Liu SK, Coackley C, Krause M, Jalali F, Chan N, Bristow RG (2008) A novel poly(ADP-ribose) polymerase inhibitor, ABT-888, radiosensitizes malignant human cell lines under hypoxia. Radiother Oncol 88(2):258–268. Epub 2008/05/06
90. Chalmers AJ (2010) Overcoming resistance of glioblastoma to conventional cytotoxic therapies by the addition of PARP inhibitors. Anti-Cancer Agents Med Chem 10(7):520–533. Epub 2010/10/01
91. Chalmers AJ, Lakshman M, Chan N, Bristow RG (2010) Poly(ADP-ribose) polymerase inhibition as a model for synthetic lethality in developing radiation oncology targets. Semin Radiat Oncol 20(4):274–281. Epub 2010/09/14
92. Aguilar-Quesada R, Munoz-Gamez JA, Martin-Oliva D, Peralta-Leal A, Quiles-Perez R, Rodriguez-Vargas JM et al (2007) Modulation of transcription by PARP-1: consequences in carcinogenesis and inflammation. Curr Med Chem 14(11):1179–1187
93. Giansanti V, Dona F, Tillhon M, Scovassi AI (2010) PARP inhibitors: new tools to protect from inflammation. Biochem pharmacol 80(12):1869–1877. Epub 2010/04/27
94. De la Lastra CA, Villegas I, Sanchez-Fidalgo S (2007) Poly(ADP-ribose) polymerase inhibitors: new pharmacological functions and potential clinical implications. Curr Pharm Des 13(9):933–962. Epub 2007/04/14
95. Wedge SR, Porteous JK, Newlands ES (1996) 3-Aminobenzamide and/or O6-benzylguanine evaluated as an adjuvant to temozolomide or BCNU treatment in cell lines of variable mismatch repair status and O6-alkylguanine-DNA alkyltransferase activity. Br J Cancer 74(7):1030–1036. Epub 1996/10/01
96. Tentori L, Turriziani M, Franco D, Serafino A, Levati L, Roy R et al (1999) Treatment with temozolomide and poly(ADP-ribose) polymerase inhibitors induces early apoptosis and increases base excision repair gene transcripts in leukemic cells resistant to triazene compounds. Leukemia 13(6):901–909
97. Tentori L, Leonetti C, Scarsella M, D'Amati G, Vergati M, Portarena I et al (2003) Systemic administration of GPI 15427, a novel poly(ADP-ribose) polymerase-1 inhibitor, increases the antitumor activity of temozolomide against intracranial melanoma, glioma, lymphoma. Clin Cancer Res 9(14):5370–5379
98. Curtin NJ, Wang LZ, Yiakouvaki A, Kyle S, Arris CA, Canan-Koch S et al (2004) Novel poly(ADP-ribose) polymerase-1 inhibitor, AG14361, restores sensitivity to temozolomide in mismatch repair-deficient cells. Clin Cancer Res 10(3):881–889
99. Rottenberg S, Jaspers JE, Kersbergen A, van der Burg E, Nygren AO, Zander SA et al (2008) High sensitivity of BRCA1-deficient mammary tumors to the PARP inhibitor AZD2281 alone and in combination with platinum drugs. Proc Natl Acad Sci USA 105(44):17079–17084. Epub 2008/10/31
100. Evers B, Drost R, Schut E, de Bruin M, van der Burg E, Derksen PW et al (2008) Selective inhibition of BRCA2-deficient mammary tumor cell growth by AZD2281 and cisplatin. Clin Cancer Res 14(12):3916–3925 Epub 2008/06/19
101. Berger NA (1985) Poly(ADP-ribose) in the cellular response to DNA damage. Radiat Res 101(1):4–15
102. Hottiger MO, Hassa PO, Luscher B, Schuler H, Koch-Nolte F (2010) Toward a unified nomenclature for mammalian ADP-ribosyltransferases. Trends Biochem Sci 35(4):208–219. Epub 2010/01/29
103. Tang J, Goellner EM, Wang XW, Trivedi RN, St. Croix CM, Jelezcova E et al (2010) Bioenergetic metabolites regulate base excision repair-dependent cell death in response to DNA damage. Mol Cancer Res 8(1):67–79. Epub 2010/01/14
104. Dolan ME, Pegg AE (1997) O6-Benzylguanine and its role in chemotherapy. Clin Cancer Res 3(6):837–847. Epub 1997/06/01
105. Xu-Welliver M, Pegg AE (2002) Degradation of the alkylated form of the DNA repair protein, O(6)-alkylguanine-DNA alkyltransferase. Carcinogenesis 23(5):823–30. Epub 2002/05/23

106. Pegg AE (2011) Multifaceted roles of alkyltransferase and related proteins in DNA repair, DNA damage, resistance to chemotherapy, and research tools. Chem Res Toxicol 24(5):618–639. Epub 2011/04/07
107. Middleton MR, Kelly J, Thatcher N, Donnelly DJ, McElhinney RS, McMurry TB et al (2000) O(6)-(4-bromothenyl)guanine improves the therapeutic index of temozolomide against A375 M melanoma xenografts. Int J Cancer 85(2):248–252. Epub 2000/01/11
108. Zhang J, Stevens MF, Bradshaw TD (2012) Temozolomide: mechanisms of action, repair and resistance. Curr Mol Pharmacol 5(1):102–114. Epub 2011/11/30
109. Khan O, Middleton MR (2007) The therapeutic potential of O6-alkylguanine DNA alkyltransferase inhibitors. Expert Opin Investig Drugs 16(10):1573–1584. Epub 2007/10/10
110. Davis BM, Roth JC, Liu L, Xu-Welliver M, Pegg AE, Gerson SL (1999) Characterization of the P140 K, PVP(138-140)MLK, and G156A O6-methylguanine-DNA methyltransferase mutants: implications for drug resistance gene therapy. Hum Gene Ther 10(17):2769–2778. Epub 1999/12/10
111. Chen T, Stephens PA, Middleton FK, Curtin NJ (2011) Targeting the S and G2 checkpoint to treat cancer. Drug Discovery Today 17(5–6):194–202. Epub 2011/12/24
112. Zenvirt S, Kravchenko-Balasha N, Levitzki A (2010) Status of p53 in human cancer cells does not predict efficacy of CHK1 kinase inhibitors combined with chemotherapeutic agents. Oncogene 29(46):6149–6159. Epub 2010/08/24
113. Cho SH, Toouli CD, Fujii GH, Crain C, Parry D (2004) Chk1 is essential for tumor cell viability following activation of the replication checkpoint. Cell Cycle 4(1):131–139. Epub 2004/11/13
114. Sorensen CS, Hansen LT, Dziegielewski J, Syljuasen RG, Lundin C, Bartek J et al (2005) The cell-cycle checkpoint kinase Chk1 is required for mammalian homologous recombination repair. Nat Cell Biol 7(2):195–201
115. Ma CX, Cai S, Li S, Ryan CE, Guo Z, Schaiff WT et al (2012) Targeting Chk1 in p53-deficient triple-negative breast cancer is therapeutically beneficial in human-in-mouse tumor models. J Clin Invest 122(4):1541–1552. Epub 2012/03/27
116. Ma CX, Janetka JW, Piwnica-Worms H (2010) Death by releasing the breaks: CHK1 inhibitors as cancer therapeutics. Trends Mol Med 17(2):88–96. Epub 2010/11/23
117. Mitchell C, Park M, Eulitt P, Yang C, Yacoub A, Dent P (2010) Poly(ADP-ribose) polymerase 1 modulates the lethality of CHK1 inhibitors in carcinoma cells. Mol Pharmacol 78(5):909–917. Epub 2010/08/11
118. Sausville EA, LoRusso P, Carducci MA, Barker PN, Agbo F, Oakes P et al (2011) Phase I dose-escalation study of AZD7762 in combination with gemcitabine (gem) in patients (pts) with advanced solid tumors. J Clin Oncol 29(suppl):abstr 3058
119. McNeely S, Conti C, Sheikh T, Patel H, Zabludoff S, Pommier Y et al (2010) Chk1 inhibition after replicative stress activates a double strand break response mediated by ATM and DNA-dependent protein kinase. Cell Cycle 9(5):995–1004. Epub 2010/02/18
120. Sorensen CS, Syljuasen RG (2012) Safeguarding genome integrity: the checkpoint kinases ATR, CHK1 and WEE1 restrain CDK activity during normal DNA replication. Nucleic Acids Res 40(2):477–486. Epub 2011/09/23
121. Toledo LI, Murga M, Fernandez-Capetillo O (2011) Targeting ATR and Chk1 kinases for cancer treatment: a new model for new (and old) drugs. Mol Oncol 5(4):368–373. Epub 2011/08/09
122. Xanthoudakis S, Smeyne RJ, Wallace JD, Curran T (1996) The redox/DNA repair protein, Ref-1, is essential for early embryonic development in mice. Proc Nat Acad Sci 93(17):8919–8923
123. Izumi T, Brown DB, Naidu CV, Bhakat KK, Macinnes MA, Saito H et al (2005) Two essential but distinct functions of the mammalian abasic endonuclease. Proc Natl Acad Sci USA 102(16):5739–5743
124. Fan Z, Beresford PJ, Zhang D, Xu Z, Novina CD, Yoshida A et al (2003) Cleaving the oxidative repair protein Ape1 enhances cell death mediated by granzyme A. Nat Immunol 4(2):145–153

125. Fishel ML, He Y, Reed AM, Chin-Sinex H, Hutchins GD, Mendonca MS et al (2008) Knockdown of the DNA repair and redox signaling protein Ape1/Ref-1 blocks ovarian cancer cell and tumor growth. DNA Repair (Amst) 7(2):177–186. Epub 2007/11/03
126. Bapat A, Fishel ML, Kelley MR (2009) Going ape as an approach to cancer therapeutics. Antioxid Redox Signal 11(3):651–668. Epub 2008/08/22
127. Abbotts R, Madhusudan S (2010) Human AP endonuclease 1 (APE1): from mechanistic insights to druggable target in cancer. Cancer Treat Rev 36(5):425–435. Epub 2010/01/09
128. Phillps JH, Brown DM, Grossman L (1996) The efficiency of induction of mutations by hydroxylamine. J Mol Biol 21(3):405–419. Epub 1966/11/28
129. Goldszer F, Tindell GL, Walle UK, Walle T (1981) Chemical trapping of labile aldehyde intermediates in the metabolism of propranolol and oxprenolol. Res Commun Chem Pathol Pharmacol 34(2):193–205. Epub 1981/11/01
130. Beger RD, Bolton PH (1998) Structures of apurinic and apyrimidinic sites in duplex DNAs. J Biol Chem 273(25):15565–15573. Epub 1998/06/23
131. Talpaert-Borle M, Liuzzi M (1983) Reaction of apurinic/apyrimidinic sites with [14C] methoxyamine. A method for the quantitative assay of AP sites in DNA. Biochim Biophys Acta 740(4):410–416. Epub 1983/09/09
132. Liu L, Taverna P, Whitacre CM, Chatterjee S, Gerson SL (1999) Pharmacologic disruption of base excision repair sensitizes mismatch repair-deficient and -proficient colon cancer cells to methylating agents. Clin Cancer Res 5(10):2908–2917
133. Goellner EM, Grimme B, Brown AR, Lin YC, Wang XH, Sugrue KF et al (2011) Overcoming temozolomide resistance in glioblastoma via dual inhibition of NAD + biosynthesis and base excision repair. Cancer Res 71(6):2308–2317. Epub 2011/03/17
134. Tang JB, Svilar D, Trivedi RN, Wang XH, Goellner EM, Moore B et al (2011) N-methylpurine DNA glycosylase and DNA polymerase beta modulate BER inhibitor potentiation of glioma cells to temozolomide. Neurooncology 13(5):471–486. Epub 2011/03/08
135. Taverna P, Liu L, Hwang HS, Hanson AJ, Kinsella TJ, Gerson SL (2001) Methoxyamine potentiates DNA single strand breaks and double strand breaks induced by temozolomide in colon cancer cells. Mutat Res 485(4):269–281
136. Taverna P, Hwang HS, Schupp JE, Radivoyevitch T, Session NN, Reddy G et al (2003) Inhibition of base excision repair potentiates iododeoxyuridine-induced cytotoxicity and radiosensitization. Cancer Res 63(4):838–846. Epub 2003/02/20
137. Liu L, Yan L, Donze JR, Gerson SL (2003) Blockage of abasic site repair enhances antitumor efficacy of 1,3-bis-(2-chloroethyl)-1-nitrosourea in colon tumor xenografts. Mol Cancer Ther 2(10):1061–1066. Epub 2003/10/28
138. She M, Pan I, Sun L, Yeung SC (2005) Enhancement of manumycin A-induced apoptosis by methoxyamine in myeloid leukemia cells. Leukemia 19(4):595–602. Epub 2005/03/04
139. Bulgar AD, Snell M, Donze JR, Kirkland EB, Li L, Yang S et al (2010) Targeting base excision repair suggests a new therapeutic strategy of fludarabine for the treatment of chronic lymphocytic leukemia. Leukemia 24(10):1795–1799. Epub 2010/09/03
140. Kinsella TJ (2009) Coordination of DNA mismatch repair and base excision repair processing of chemotherapy and radiation damage for targeting resistant cancers. Clin Cancer Res 15(6):1853–1859. Epub 2009/02/26
141. Vermeulen C, Verwijs-Janssen M, Cramers P, Begg AC, Vens C (2007) Role for DNA polymerase beta in response to ionizing radiation. DNA Repair (Amst) 6(2):202–212
142. Bulgar AD, Weeks LD, Miao Y, Yang S, Xu Y, Guo C et al (2012) Removal of uracil by uracil DNA glycosylase limits pemetrexed cytotoxicity: overriding the limit with methoxyamine to inhibit base excision repair. Cell Death Dis 3:e252. Epub 2012/01/13
143. Srinivasan A, Wang L, Cline CJ, Xie Z, Sobol RW, Xie XQ et al (2012) Identification and characterization of human apurinic/apyrimidinic endonuclease-1 inhibitors. Biochemistry 51(31):6246–6259. Epub 2012/07/14
144. Rai G, Vyjayanti VN, Dorjsuren D, Simeonov A, Jadhav A, Wilson DM 3rd et al (2012) Synthesis, biological evaluation, and structure-activity relationships of a novel class of apurinic/apyrimidinic endonuclease 1 inhibitors. J Med chem 55(7):3101–3112. Epub 2012/03/30

145. Bapat A, Glass LS, Luo M, Fishel ML, Long EC, Georgiadis MM et al (2010) Novel small-molecule inhibitor of apurinic/apyrimidinic endonuclease 1 blocks proliferation and reduces viability of glioblastoma cells. J Pharmacol Exp Ther 334(3):988–998. Epub 2010/05/28
146. Zawahir Z, Dayam R, Deng J, Pereira C, Neamati N (2009) Pharmacophore guided discovery of small-molecule human apurinic/apyrimidinic endonuclease 1 inhibitors. J Med Chem 52(1):20–32
147. Al-Safi RI, Odde S, Shabaik Y, Neamati N (2012) Small-molecule inhibitors of APE1 DNA repair function: an overview. Curr Mol Pharmacol 5(1):14–35. Epub 2011/11/30
148. Sultana R, McNeill DR, Abbotts R, Mohammed MZ, Zdzienicka MZ, Qutob H et al (2012) Synthetic lethal targeting of DNA double-strand break repair deficient cells by human apurinic/apyrimidinic endonuclease inhibitors. Int J Cancer 131(10):2433–2444. Epub 2012/03/02
149. Bartkova J, Horejsi Z, Koed K, Kramer A, Tort F, Zieger K et al (2005) DNA damage response as a candidate anti-cancer barrier in early human tumorigenesis. Nature 434(7035):864–870. Epub 2005/04/15
150. Gorgoulis VG, Vassiliou LV, Karakaidos P, Zacharatos P, Kotsinas A, Liloglou T et al (2005) Activation of the DNA damage checkpoint and genomic instability in human precancerous lesions. Nature 434(7035):907–913. Epub 2005/04/15
151. Peasland A, Wang LZ, Rowling E, Kyle S, Chen T, Hopkins A et al (2011) Identification and evaluation of a potent novel ATR inhibitor, NU6027, in breast and ovarian cancer cell lines. Br J Cancer 105(3):372–381. Epub 2011/07/07
152. Toledo LI, Murga M, Zur R, Soria R, Rodriguez A, Martinez S et al (2011) A cell-based screen identifies ATR inhibitors with synthetic lethal properties for cancer-associated mutations. Nat Struct Mol Biol 18(6):721–727. Epub 2011/05/10
153. Rainey MD, Charlton ME, Stanton RV, Kastan MB (2008) Transient inhibition of ATM kinase is sufficient to enhance cellular sensitivity to ionizing radiation. Cancer Res 68(18):7466–7474. Epub 2008/09/17
154. Vens C, Begg AC (2010) Targeting base excision repair as a sensitization strategy in radiotherapy. Semin Radiat Oncol 20(4):241–249. Epub 2010/09/14
155. O'Connor MJ, Martin NM, Smith GC (2007) Targeted cancer therapies based on the inhibition of DNA strand break repair. Oncogene 26(56):7816–7824. Epub 2007/12/11
156. Nutley BP, Smith NF, Hayes A, Kelland LR, Brunton L, Golding BT et al (2005) Preclinical pharmacokinetics and metabolism of a novel prototype DNA-PK inhibitor NU7026. Br J Cancer 93(9):1011–1018. Epub 2005/10/27
157. Zhao Y, Thomas HD, Batey MA, Cowell IG, Richardson CJ, Griffin RJ et al (2006) Preclinical evaluation of a potent novel DNA-dependent protein kinase inhibitor NU7441. Cancer Res 66(10):5354–5362. Epub 2006/05/19
158. Quanz M, Berthault N, Roulin C, Roy M, Herbette A, Agrario C et al (2009) Small-molecule drugs mimicking DNA damage: a new strategy for sensitizing tumors to radiotherapy. Clin Cancer Res 15(4):1308–1316. Epub 2009/02/05
159. Munck JM, Batey MA, Zhao Y, Jenkins H, Richardson CJ, Cano C et al (2012) Chemosensitization of cancer cells by KU-0060648, A Dual Inhibitor of DNA-PK and PI-3 K. Mol Cancer Ther. Epub 2012/05/12
160. Veuger SJ, Curtin NJ, Richardson CJ, Smith GC, Durkacz BW (2003) Radiosensitization and DNA repair inhibition by the combined use of novel inhibitors of DNA-dependent protein kinase and poly(ADP-ribose) polymerase-1. Cancer Res 63(18):6008–6015. Epub 2003/10/03
161. Azad A, Jackson S, Cullinane C, Natoli A, Neilsen PM, Callen DF et al (2011) Inhibition of DNA-dependent protein kinase induces accelerated senescence in irradiated human cancer cells. Mol Cancer Res 9(12):1696–1707. Epub 2011/10/20
162. Gurley KE, Kemp CJ (2001) Synthetic lethality between mutation in ATM and DNA-PK(cs) during murine embryogenesis. Curr Biol 11(3):191–194. Epub 2001/03/07
163. Taccioli GE, Amatucci AG, Beamish HJ, Gell D, Xiang XH, Torres Arzayus MI et al (1998) Targeted disruption of the catalytic subunit of the DNA-PK gene in mice confers severe combined immunodeficiency and radiosensitivity. Immunity 9(3):355–366. Epub 1998/10/13

164. Mukherjee B, McEllin B, Camacho CV, Tomimatsu N, Sirasanagandala S, Nannepaga S et al (2009) EGFRvIII and DNA double-strand break repair: a molecular mechanism for radioresistance in glioblastoma. Cancer Res 69(10):4252–4259. Epub 2009/05/14
165. Golding SE, Rosenberg E, Neill S, Dent P, Povirk LF, Valerie K (2007) Extracellular signal-related kinase positively regulates ataxia telangiectasia mutated, homologous recombination repair, and the DNA damage response. Cancer Res 67(3):1046–1053. Epub 2007/02/07
166. Kriegs M, Kasten-Pisula U, Rieckmann T, Holst K, Saker J, Dahm-Daphi J et al (2010) The epidermal growth factor receptor modulates DNA double-strand break repair by regulating non-homologous end-joining. DNA Repair (Amst) 9(8):889–897. Epub 2010/07/10
167. Golding SE, Rosenberg E, Valerie N, Hussaini I, Frigerio M, Cockcroft XF et al (2009) Improved ATM kinase inhibitor KU-60019 radiosensitizes glioma cells, compromises insulin, AKT and ERK prosurvival signaling, and inhibits migration and invasion. Mol Cancer Ther 8(10):2894–2902. Epub 2009/10/08
168. Toulany M, Kehlbach R, Florczak U, Sak A, Wang S, Chen J et al (2008) Targeting of AKT1 enhances radiation toxicity of human tumor cells by inhibiting DNA-PKcs-dependent DNA double-strand break repair. Mol Cancer Ther 7(7):1772–1781. Epub 2008/07/23
169. Zaidi SH, Huddart RA, Harrington KJ (2009) Novel targeted radiosensitisers in cancer treatment. Curr Drug Discov Technol 6(2):103–134. Epub 2009/06/13
170. Dittmann K, Mayer C, Fehrenbacher B, Schaller M, Raju U, Milas L et al (2005) Radiation-induced epidermal growth factor receptor nuclear import is linked to activation of DNA-dependent protein kinase. J Biol Chem 280(35):31182–31189. Epub 2005/07/08
171. Kirshner J, Jobling MF, Pajares MJ, Ravani SA, Glick AB, Lavin MJ et al (2006) Inhibition of transforming growth factor-beta1 signaling attenuates ataxia telangiectasia mutated activity in response to genotoxic stress. Cancer Res 66(22):10861–10869. Epub 2006/11/09
172. Sinnberg T, Lasithiotakis K, Niessner H, Schittek B, Flaherty KT, Kulms D et al (2009) Inhibition of PI3 K-AKT-mTOR signaling sensitizes melanoma cells to cisplatin and temozolomide. J Invest Dermatol 129(6):1500–1515. Epub 2008/12/17
173. Chen L, Han L, Shi Z, Zhang K, Liu Y, Zheng Y et al (2012) LY294002 enhances cytotoxicity of temozolomide in glioma by down-regulation of the PI3 K/Akt pathway. Mol Med Rep 5(2):575–579. Epub 2011/11/17
174. Pal J, Fulciniti M, Nanjappa P, Buon L, Tai YT, Tassone P et al (2012) Targeting PI3 K and RAD51 in Barrett's adenocarcinoma: impact on DNA damage checkpoints, expression profile and tumor growth. Cancer Genomics Proteomics 9(2):55–66. Epub 2012/03/09
175. Kimbung S, Biskup E, Johansson I, Aaltonen K, Ottosson-Wadlund A, Gruvberger-Saal S et al (2012) Co-targeting of the PI3 K pathway improves the response of BRCA1 deficient breast cancer cells to PARP1 inhibition. Cancer Lett 319(2):232–241. Epub 2012/01/24
176. Huang TT, D'Andrea AD (2006) Regulation of DNA repair by ubiquitylation. Nat Rev Mol Cell Biol 7(5):323–334 Epub 2006/04/25
177. Vucic D, Dixit VM, Wertz IE (2011) Ubiquitylation in apoptosis: a post-translational modification at the edge of life and death. Nat Rev Mol Cell Biol 12(7):439–452. Epub 2011/06/24
178. Frezza M, Schmitt S, Dou QP (2011) Targeting the ubiquitin-proteasome pathway: an emerging concept in cancer therapy. Curr Top Med Chem 11(23):2888–2905. Epub 2011/08/10
179. Ruschak AM, Slassi M, Kay LE, Schimmer AD (2011) Novel proteasome inhibitors to overcome bortezomib resistance. J Natl Cancer Inst 103(13):1007–1017. Epub 2011/05/25
180. Galli M, Van Gool F, Rongvaux A, Andris F, Leo O (2010) The nicotinamide phosphoribosyltransferase: a molecular link between metabolism, inflammation, and cancer. Cancer Res 70(1):8–11. Epub 2009/12/24
181. Wang T, Zhang X, Bheda P, Revollo JR, Imai S, Wolberger C (2006) Structure of Nampt/PBEF/visfatin, a mammalian NAD + biosynthetic enzyme. Nat Struct Mol Biol 13(7):661–662. Epub 2006/06/20

182. Berger F, Lau C, Dahlmann M, Ziegler M (2005) Subcellular compartmentation and differential catalytic properties of the three human nicotinamide mononucleotide adenylyltransferase isoforms. J Biol Chem 280(43):36334–36341
183. Watson M, Roulston A, Belec L, Billot X, Marcellus R, Bedard D et al (2009) The small molecule GMX1778 is a potent inhibitor of NAD + biosynthesis: strategy for enhanced therapy in nicotinic acid phosphoribosyltransferase 1-deficient tumors. Mol Cell Biol 29(21):5872–5888. Epub 2009/08/26
184. Fleischer TC, Murphy BR, Flick JS, Terry-Lorenzo RT, Gao ZH, Davis T et al Chemical proteomics identifies Nampt as the target of CB30865, an orphan cytotoxic compound. Chem Biol 17(6):659–664. Epub 2010/07/09
185. Olesen UH, Christensen MK, Bjorkling F, Jaattela M, Jensen PB, Sehested M et al (2008) Anticancer agent CHS-828 inhibits cellular synthesis of NAD. Biochem Biophys Res Commun 367(4):799–804
186. Zhang RY, Qin Y, Lv XQ, Wang P, Xu TY, Zhang L et al (2011) A fluorometric assay for high-throughput screening targeting nicotinamide phosphoribosyltransferase. Anal Biochem 412(1):18–25. Epub 2011/01/08
187. Khan JA, Tao X, Tong L (2006) Molecular basis for the inhibition of human NMPRTase, a novel target for anticancer agents. Nat Struct Mol Biol 13(7):582–588
188. Hasmann M, Schemainda I (2003) FK866, a highly specific noncompetitive inhibitor of nicotinamide phosphoribosyltransferase, represents a novel mechanism for induction of tumor cell apoptosis. Cancer Res 63(21):7436–7442
189. Billington RA, Genazzani AA, Travelli C, Condorelli F (2008) NAD depletion by FK866 induces autophagy. Autophagy 4(3):385–387. Epub 2008/01/30
190. Pogrebniak A, Schemainda I, Azzam K, Pelka-Fleischer R, Nussler V, Hasmann M (2006) Chemopotentiating effects of a novel NAD biosynthesis inhibitor, FK866, in combination with antineoplastic agents. Eur J Med Res 11(8):313–321
191. Bi TQ, Che XM, Liao XH, Zhang DJ, Long HL, Li HJ et al (2011) Overexpression of nampt in gastric cancer and chemopotentiating effects of the nampt inhibitor FK866 in combination with fluorouracil. Oncol Rep 26(5):1251–1257. Epub 2011/07/12
192. Travelli C, Drago V, Maldi E, Kaludercic N, Galli U, Boldorini R et al (2011) Reciprocal potentiation of the antitumoral activities of FK866, an inhibitor of nicotinamide phosphoribosyltransferase, and etoposide or cisplatin in neuroblastoma cells. J Pharmacol Exp Ther 338(3):829–840. Epub 2011/06/21
193. Zoppoli G, Cea M, Soncini D, Fruscione F, Rudner J, Moran E et al (2010) Potent synergistic interaction between the Nampt inhibitor APO866 and the apoptosis activator TRAIL in human leukemia cells. Exp Hematol 38(11):979–988. Epub 2010/08/11
194. Olesen UH, Thougaard AV, Jensen PB, Sehested M (2010) A preclinical study on the rescue of normal tissue by nicotinic acid in high-dose treatment with APO866, a specific nicotinamide phosphoribosyltransferase inhibitor. Mol Cancer Ther 9(6):1609–1617
195. Goellner EM, Svilar D, Almeida KH, Sobol RW (2012) Targeting DNA polymerase β for therapeutic intervention. Curr Mol Pharmacol 5(1):68–87. Epub 2011/11/30
196. Martin LP, Hamilton TC, Schilder RJ (2008) Platinum resistance: the role of DNA repair pathways. Clin Cancer Res 14(5):1291–1295. Epub 2008/03/05
197. Orelli B, McClendon TB, Tsodikov OV, Ellenberger T, Niedernhofer LJ, Scharer OD (2010) The XPA-binding domain of ERCC1 is required for nucleotide excision repair but not other DNA repair pathways. J Biol Chem 285(6):3705–3712. Epub 2009/11/27
198. Su Y, Orelli B, Madireddy A, Niedernhofer LJ, Scharer OD (2012) Multiple DNA binding domains mediate the function of the ERCC1-XPF protein in nucleotide excision repair. J Biol Chem 287(26):21846–21855. Epub 2012/05/02
199. Tsodikov OV, Ivanov D, Orelli B, Staresincic L, Shoshani I, Oberman R et al (2007) Structural basis for the recruitment of ERCC1-XPF to nucleotide excision repair complexes by XPA. Embo J 26(22):4768–4776. Epub 2007/10/20
200. Sijbers AM, van der Spek PJ, Odijk H, van den Berg J, van Duin M, Westerveld A et al (1996) Mutational analysis of the human nucleotide excision repair gene ERCC1. Nucleic Acids Res 24(17):3370–3380. Epub 1996/09/01

201. Barakat KH, Torin Huzil J, Luchko T, Jordheim L, Dumontet C, Tuszynski J (2009) Characterization of an inhibitory dynamic pharmacophore for the ERCC1-XPA interaction using a combined molecular dynamics and virtual screening approach. J Mol Graph Model 28(2):113–130. Epub 2009/05/29
202. Lee-Kwon W, Park D, Bernier M (1998) Involvement of the ras/extracellular signal-regulated kinase signalling pathway in the regulation of ERCC-1 mRNA levels by insulin. Biochem J 331 (Pt 2):591–597. Epub 1998/06/11
203. Yadav A, Kumar B, Teknos TN, Kumar P (2011) Sorafenib enhances the antitumor effects of chemoradiation treatment by downregulating ERCC-1 and XRCC-1 DNA repair proteins. Mol Cancer Ther 10(7):1241–1251. Epub 2011/05/10
204. Burgers PM, Koonin EV, Bruford E, Blanco L, Burtis KC, Christman MF et al (2001) Eukaryotic DNA polymerases: proposal for a revised nomenclature. J Biol Chem 276(47):43487–43490
205. Lange SS, Takata K, Wood RD (2011) DNA polymerases and cancer. Nat Rev Cancer 11(2):96–110. Epub 2011/01/25
206. Bebenek K, Kunkel TA (2004) Functions of DNA polymerases. Adv Protein Chem 69:137–65. Epub 2004/12/14
207. Wilson SH, Beard WA, Shock DD, Batra VK, Cavanaugh NA, Prasad R et al (2010) Base excision repair and design of small molecule inhibitors of human DNA polymerase beta. Cell Mol Life Sci 67(21):3633–3647. Epub 2010/09/17
208. Barakat KH, Gajewski MM, Tuszynski JA (2012) DNA polymerase beta (pol beta) inhibitors: a comprehensive overview. Drug discovery today 17(15–16):913–920. Epub 2012/05/09
209. Higgins GS, Prevo R, Lee YF, Helleday T, Muschel RJ, Taylor S et al (2010) A small interfering RNA screen of genes involved in DNA repair identifies tumor-specific radiosensitization by POLQ knockdown. Cancer Res 70(7):2984–2993. Epub 2010/03/18
210. Higgins GS, Harris AL, Prevo R, Helleday T, McKenna WG, Buffa FM (2010) Overexpression of POLQ confers a poor prognosis in early breast cancer patients. Oncotarget 1(3):175–184. Epub 2010/08/12
211. Lemee F, Bergoglio V, Fernandez-Vidal A, Machado-Silva A, Pillaire MJ, Bieth A et al (2010) DNA polymerase theta up-regulation is associated with poor survival in breast cancer, perturbs DNA replication, and promotes genetic instability. Proc Natl Acad Sci USA 107(30):13390–13395. Epub 2010/07/14
212. Slupphaug G, Mol CD, Kavli B, Arvai AS, Krokan HE, Tainer JA (1996) A nucleotide-flipping mechanism from the structure of human uracil-DNA glycosylase bound to DNA. Nature 384(6604):87–92. Epub 1996/11/07
213. Engelward BP, Allan JM, Dreslin AJ, Kelly JD, Wu MM, Gold B et al (1998) A chemical and genetic approach together define the biological consequences of 3-methyladenine lesions in the mammalian genome. J Biol Chem 273(9):5412–5418
214. Noren Hooten N, Kompaniez K, Barnes J, Lohani A, Evans MK (2011) Poly(ADP-ribose) polymerase 1 (PARP-1) binds to 8-oxoguanine-DNA glycosylase (OGG1). J Biol Chem 286(52):44679–44690. Epub 2011/11/08
215. Endres M, Biniszkiewicz D, Sobol RW, Harms C, Ahmadi M, Lipski A et al (2004) Increased postischemic brain injury in mice deficient in uracil-DNA glycosylase. J Clin Invest 113(12):1711–1721
216. Kruman II, Schwartz E, Kruman Y, Cutler RG, Zhu X, Greig NH et al (2004) Suppression of uracil-DNA glycosylase induces neuronal apoptosis. J Biol Chem 279(42):43952–43960
217. Imai K, Slupphaug G, Lee WI, Revy P, Nonoyama S, Catalan N et al (2003) Human uracil-DNA glycosylase deficiency associated with profoundly impaired immunoglobulin class-switch recombination. Nat Immunol 4(10):1023–1028. Epub 2003/09/06
218. Wang Z, Mosbaugh DW. Uracil-DNA glycosylase inhibitor gene of bacteriophage PBS2 encodes a binding protein specific for uracil-DNA glycosylase. J Biol Chem. 1989;264(2):1163-71. Epub 1989/01/15

219. Komander D, Clague MJ, Urbe S (2009) Breaking the chains: structure and function of the deubiquitinases. Nat Rev Mol Cell Biol 10(8):550–563. Epub 2009/07/25
220. Shabek N, Ciechanover A (2010) Degradation of ubiquitin: the fate of the cellular reaper. Cell Cycle 9(3):523–530. Epub 2010/01/29
221. Mattern MR, Wu J, Nicholson B (2012) Ubiquitin-based anticancer therapy: carpet bombing with proteasome inhibitors vs surgical strikes with E1, E2, E3, or DUB inhibitors. Biochim Biophys Acta 1823(11):2014–2021. Epub 2012/05/23
222. Opferman JT, Green DR (2010) DUB-le trouble for cell survival. Cancer Cell 17(2):117–119. Epub 2010/02/18
223. Altun M, Kramer HB, Willems LI, McDermott JL, Leach CA, Goldenberg SJ et al (2011) Activity-based chemical proteomics accelerates inhibitor development for deubiquitylating enzymes. Chem Biol 18(11):1401–1412. Epub 2011/11/29
224. Kapuria V, Peterson LF, Fang D, Bornmann WG, Talpaz M, Donato NJ (2010) Deubiquitinase inhibition by small-molecule WP1130 triggers aggresome formation and tumor cell apoptosis. Cancer Res 70(22):9265–9276. Epub 2010/11/04
225. Kramer HB, Nicholson B, Kessler BM, Altun M (2012) Detection of ubiquitin-proteasome enzymatic activities in cells: application of activity-based probes to inhibitor development. Biochim Biophys Acta 1823(11):2029–2037. Epub 2012/05/23
226. Seiberlich V, Goldbaum O, Zhukareva V, Richter-Landsberg C (2012) The small molecule inhibitor PR-619 of deubiquitinating enzymes affects the microtubule network and causes protein aggregate formation in neural cells: implications for neurodegenerative diseases. Biochim Biophys Acta 1823(11):2057–2068. Epub 2012/05/09
227. Chernikova SB, Game JC, Brown JM (2012) Inhibiting homologous recombination for cancer therapy. Cancer Biol Ther 13(2):61–68. Epub 2012/02/18
228. Huang F, Motlekar NA, Burgwin CM, Napper AD, Diamond SL, Mazin AV (2011) Identification of specific inhibitors of human RAD51 recombinase using high-throughput screening. ACS chemical biology 6(6):628–635. Epub 2011/03/25
229. Quiros S, Roos WP, Kaina B (2011) Rad51 and BRCA2–New molecular targets for sensitizing glioma cells to alkylating anticancer drugs. PLoS ONE 6(11):e27183. Epub 2011/11/11
230. Short SC, Giampieri S, Worku M, Alcaide-German M, Sioftanos G, Bourne S et al (2011) Rad51 inhibition is an effective means of targeting DNA repair in glioma models and CD133 + tumor-derived cells. Neurooncology 13(5):487–499. Epub 2011/03/03
231. Sprong D, Janssen HL, Vens C, Begg AC (2005) Resistance of hypoxic cells to ionizing radiation is influenced by homologous recombination status. Int J Radiat Oncol Biol Phys 64(2):562–572. Epub 2005/12/14
232. Neijenhuis S, Verwijs-Janssen M, van den Broek LJ, Begg AC, Vens C (2010) Targeted radiosensitization of cells expressing truncated DNA polymerase {beta}. Cancer Res 70(21):8706–8714. Epub 2010/10/28
233. Slade D, Dunstan MS, Barkauskaite E, Weston R, Lafite P, Dixon N et al (2011) The structure and catalytic mechanism of a poly(ADP-ribose) glycohydrolase. Nature 477(7366):616–620. Epub 2011/09/06
234. Brochu G, Duchaine C, Thibeault L, Lagueux J, Shah GM, Poirier GG (1994) Mode of action of poly(ADP-ribose) glycohydrolase. Biochim Biophys Acta 1219(2):342–350
235. Affar EB, Germain M, Winstall E, Vodenicharov M, Shah RG, Salvesen GS et al (2001) Caspase-3-mediated processing of poly(ADP-ribose) glycohydrolase during apoptosis. J Biol Chem 276(4):2935–42. Epub 2000/10/29
236. Li Q, Li M, Wang YL, Fauzee NJ, Yang Y, Pan J et al (2012) RNA interference of PARG could inhibit the metastatic potency of colon carcinoma cells via PI3-kinase/Akt pathway. Cell Physiol Biochem Int J Exp Cell Physiol Biochem Pharmacol 29(3–4):361–72. Epub 2012/04/18
237. Koh DW, Lawler AM, Poitras MF, Sasaki M, Wattler S, Nehls MC et al (2004) Failure to degrade poly(ADP-ribose) causes increased sensitivity to cytotoxicity and early embryonic lethality. Proc Natl Acad Sci USA 101(51):17699–17704

238. Ying W, Swanson RA (2000) The poly(ADP-ribose) glycohydrolase inhibitor gallotannin blocks oxidative astrocyte death. Neuroreport 11(7):1385–8. Epub 2000/06/07
239. Sun Y, Zhang T, Wang B, Li H, Li P (2012) Tannic acid, an inhibitor of poly(ADP-ribose) glycohydrolase, sensitizes ovarian carcinoma cells to cisplatin. Anti-cancer drugs 23(9):979–990. Epub 2012/07/13
240. Formentini L, Arapistas P, Pittelli M, Jacomelli M, Pitozzi V, Menichetti S et al (2008) Mono-galloyl glucose derivatives are potent poly(ADP-ribose) glycohydrolase (PARG) inhibitors and partially reduce PARP-1-dependent cell death. Br J Pharmacol 155(8):1235–1249. Epub 2008/09/23
241. Cuzzocrea S, Mazzon E, Genovese T, Crisafulli C, Min WK, Di Paola R et al (2007) Role of poly(ADP-ribose) glycohydrolase in the development of inflammatory bowel disease in mice. Free Radic Biol Med 42(1):90–105
242. Cuzzocrea S, Genovese T, Mazzon E, Crisafulli C, Min W, Di Paola R et al (2006) Poly(ADP-ribose) glycohydrolase activity mediates post-traumatic inflammatory reaction after experimental spinal cord trauma. J Pharmacol Exp Ther 319(1):127–138
243. Cuzzocrea S, Di Paola R, Mazzon E, Cortes U, Genovese T, Muia C et al (2005) PARG activity mediates intestinal injury induced by splanchnic artery occlusion and reperfusion. Faseb J 19(6):558–566
244. Falsig J, Christiansen SH, Feuerhahn S, Burkle A, Oei SL, Keil C et al (2004) Poly(ADP-ribose) glycohydrolase as a target for neuroprotective intervention: assessment of currently available pharmacological tools. Eur J Pharmacol 497(1):7–16
245. Lu XC, Massuda E, Lin Q, Li W, Li JH, Zhang J (2003) Post-treatment with a novel PARG inhibitor reduces infarct in cerebral ischemia in the rat. Brain Res 978(1–2):99–103. Epub 2003/07/02
246. Steffen JD, Coyle DL, Damodaran K, Beroza P, Jacobson MK (2011) Discovery and structure-activity relationships of modified salicylanilides as cell permeable inhibitors of poly(ADP-ribose) glycohydrolase (PARG). J Med Chem 54(15):5403–5413. Epub 2011/06/23
247. Okita N, Ashizawa D, Ohta R, Abe H, Tanuma S (2010) Discovery of novel poly(ADP-ribose) glycohydrolase inhibitors by a quantitative assay system using dot-blot with anti-poly(ADP-ribose). Biochem Biophys Res Commun 392(4):485–489. Epub 2010/01/19
248. Kaye SB, Lubinski J, Matulonis U, Ang JE, Gourley C, Karlan BY et al (2012) Phase II, open-label, randomized, multicenter study comparing the efficacy and safety of olaparib, a poly (ADP-ribose) polymerase inhibitor, and pegylated liposomal doxorubicin in patients with BRCA1 or BRCA2 mutations and recurrent ovarian cancer. J Clin Oncol 30(4):372–379. Epub 2011/12/29
249. Rajan A, Carter CA, Kelly RJ, Gutierrez M, Kummar S, Szabo E et al (2012) A phase I combination study of olaparib with cisplatin and gemcitabine in adults with solid tumors. Clin Cancer Res 18(8):2344–2351. Epub 2012/03/01
250. van Vuurden DG, Hulleman E, Meijer OL, Wedekind LE, Kool M, Witt H et al (2011) PARP inhibition sensitizes childhood high grade glioma, medulloblastoma and ependymoma to radiation. Oncotarget 2(12):984–996. Epub 2011/12/21
251. Clark CC, Weitzel JN, O'Connor TR (2012) Enhancement of synthetic lethality via combinations of ABT-888, a PARP inhibitor, and carboplatin in vitro and in vivo using BRCA1 and BRCA2 isogenic models. Mol Cancer Ther 11(9):1948–1958. Epub 2012/07/11
252. Kummar S, Chen A, Ji J, Zhang Y, Reid JM, Ames M et al (2011) Phase I study of PARP inhibitor ABT-888 in combination with topotecan in adults with refractory solid tumors and lymphomas. Cancer Res 71(17):5626–5634. Epub 2011/07/29
253. Palma JP, Wang YC, Rodriguez LE, Montgomery D, Ellis PA, Bukofzer G et al (2009) ABT-888 confers broad in vivo activity in combination with temozolomide in diverse tumors. Clin Cancer Res 15(23):7277–7290. Epub 2009/11/26
254. Donawho CK, Luo Y, Penning TD, Bauch JL, Bouska JJ, Bontcheva-Diaz VD et al (2007) ABT-888, an orally active poly(ADP-ribose) polymerase inhibitor that potentiates DNA-damaging agents in preclinical tumor models. Clin Cancer Res 13(9):2728–2737. Epub 2007/05/03

255. Kummar S, Ji J, Morgan R, Lenz HJ, Puhalla SL, Belani CP et al (2012) A phase I study of veliparib in combination with metronomic cyclophosphamide in adults with refractory solid tumors and lymphomas. Clin Cancer Res 18(6):1726–1734. Epub 2012/02/07
256. Barreto-Andrade JC, Efimova EV, Mauceri HJ, Beckett MA, Sutton HG, Darga TE et al (2011) Response of human prostate cancer cells and tumors to combining PARP inhibition with ionizing radiation. Mol Cancer Ther 10(7):1185–1193. Epub 2011/05/17
257. Efimova EV, Mauceri HJ, Golden DW, Labay E, Bindokas VP, Darga TE et al (2010) Poly(ADP-ribose) polymerase inhibitor induces accelerated senescence in irradiated breast cancer cells and tumors. Cancer Res 70(15):6277–6282. Epub 2010/07/09
258. Nowsheen S, Bonner JA, Yang ES (2011) The poly(ADP-Ribose) polymerase inhibitor ABT-888 reduces radiation-induced nuclear EGFR and augments head and neck tumor response to radiotherapy. Radiother Oncol 99(3):331–338. Epub 2011/07/02
259. Apisarnthanarax N, Wood GS, Stevens SR, Carlson S, Chan DV, Liu L et al (2012) Phase I clinical trial of O6-benzylguanine and topical carmustine in the treatment of cutaneous T-cell lymphoma, mycosis fungoides type. Arch Dermatol 148(5):613–620. Epub 2012/01/18
260. Warren KE, Gururangan S, Geyer JR, McLendon RE, Poussaint TY, Wallace D et al (2012) A phase II study of O6-benzylguanine and temozolomide in pediatric patients with recurrent or progressive high-grade gliomas and brainstem gliomas: a pediatric brain tumor consortium study. J Neurooncol 106(3):643–649. Epub 2011/10/05
261. Kefford RF, Thomas NP, Corrie PG, Palmer C, Abdi E, Kotasek D et al (2009) A phase I study of extended dosing with lomeguatrib with temozolomide in patients with advanced melanoma. Br J Cancer 100(8):1245–1249. Epub 2009/04/16
262. Watson AJ, Middleton MR, McGown G, Thorncroft M, Ranson M, Hersey P et al (2009) O(6)-methylguanine-DNA methyltransferase depletion and DNA damage in patients with melanoma treated with temozolomide alone or with lomeguatrib. Br J Cancer 100(8):1250–1256. Epub 2009/04/16
263. Ma Z, Yao G, Zhou B, Fan Y, Gao S, Feng X (2012) The Chk1 inhibitor AZD7762 sensitises p53 mutant breast cancer cells to radiation in vitro and in vivo. Mol Med Rep 6(4):897–903. Epub 2012/07/25
264. Montano R, Chung I, Garner KM, Parry D, Eastman A (2011) Preclinical development of the novel Chk1 inhibitor SCH900776 in combination with DNA-damaging agents and antimetabolites. Mol Cancer Ther 11(2):427–438. Epub 2011/12/29
265. Ferrao PT, Bukczynska EP, Johnstone RW, McArthur GA (2012) Efficacy of CHK inhibitors as single agents in MYC-driven lymphoma cells. Oncogene 31(13):1661–1672. Epub 2011/08/16
266. Hickson I, Zhao Y, Richardson CJ, Green SJ, Martin NM, Orr AI et al (2004) Identification and characterization of a novel and specific inhibitor of the ataxia-telangiectasia mutated kinase ATM. Cancer Res 64(24):9152–9159. Epub 2004/12/18
267. Leahy JJ, Golding BT, Griffin RJ, Hardcastle IR, Richardson C, Rigoreau L et al (2004) Identification of a highly potent and selective DNA-dependent protein kinase (DNA-PK) inhibitor (NU7441) by screening of chromenone libraries. Bioorg Med Chem Lett 14(24):6083–6087
268. Devun F, Bousquet G, Biau J, Herbette A, Roulin C, Berger F et al (2012) Preclinical study of the DNA repair inhibitor Dbait in combination with chemotherapy in colorectal cancer. J gastroenterol 47(3):266–275. Epub 2011/11/10
269. Kim CH, Park SJ, Lee SH (2002) A targeted inhibition of DNA-dependent protein kinase sensitizes breast cancer cells following ionizing radiation. J Pharmacol Exp Ther 303(2):753–759. Epub 2002/10/22

Chapter 7
Molecular Chaperones and How Addiction Matters in Cancer Therapy

Monica L. Guzman, Maeve A. Lowery, Tony Taldone, John Koren III, Erica DaGama Gomes and Gabriela Chiosis

Abstract Constitutive expression of HSP90 in normal cells is required for its evolutionarily conserved housekeeping function of folding and translocating cellular proteins to their proper cellular compartment. Under the stress of malignant transformation, cellular proteins and networks become perturbed requiring specific maintenance by a stress-modified HSP90. We here detail the many functions this oncogenic HSP90 takes on in cancer cells and discuss how the addiction of cancer-altered networks on HSP90 can be harvested therapeutically by small molecule HSP90 inhibitors.

7.1 HSPs as Buffers of Cellular Instability

To allow for normal functioning, cells use intricate molecular machineries comprised of thousands of proteins programmed to execute well-defined functions. Dysregulation in these pathways may lead to altered functions that confer a pathologic phenotype. In cancer, not one but multiple pathways and molecules lose their inherited state leading to cells characterized by many defects such as aberrant proliferation, cell cycle, invasive potential, and evasion of apoptosis [1]. While at the

M. L. Guzman
Division of Hematology and Oncology, Weill Cornell Medical College, NY, USA

M. A. Lowery
Department of Medicine, Gastrointestinal Service,
Memorial Sloan-Kettering Cancer Center, NY, USA

T. Taldone· J. Koren III· E. D. Gomes· G. Chiosis (✉)
Department of Medicine, Breast Cancer Service,
Memorial Sloan-Kettering Cancer Center, NY, USA
e-mail: chiosisg@mskcc.org

T. Taldone· J. Koren III· E. D. Gomes· G. Chiosis
Department of Molecular Pharmacology and Chemistry, Sloan-Kettering Institute, NY, USA

D. E. Johnson (ed.), *Cell Death Signaling in Cancer Biology and Treatment*,
Cell Death in Biology and Diseases, DOI: 10.1007/978-1-4614-5847-0_7,

X.-M. Yin and Z. Dong (Series eds.), *Cell Death in Biology and Diseases*

cellular level such dysregulation may be advantageous (i.e., increased survival), at the molecular level these changes occur at a cost of protein robustness that further translates into local energetic instability. To regain a pseudo-stable state, cells must co-opt chaperones, such as the heat shock proteins (HSPs), to bind such aberrant proteins and maintain them in a functional conformation. By these mechanisms, HSPs buffer the molecular instability and help maintain a functional cellular state under the transforming pressure [2]. As such, cancer cells often express high levels of several HSPs, which augment the aggressiveness of tumors and also allow cells to survive lethal conditions, including killing by therapies. In addition to conferring resistance to treatment, elevated HSP expression also facilitates cancer by inhibiting programmed cell death and by promoting autonomous growth [3–5].

It is not surprising that the cell uses the HSPs to buffer proteome instability. After all, the HSP chaperones are the evolutionarily evolved cellular machinery that maintains protein folding and functional conformation [6]. They are also one of the most abundant proteins with a long half-life and with a quick production mechanism. Indeed, HSPs were first identified as stress proteins that confer resistance to physical stresses, such as elevated temperatures and chemical insults in all cellular organisms. These stresses may lead to acute protein misfolding from whose potential cytotoxicity the cells become protected by the quick induction of HSPs [7].

While also expressed under normal conditions, HSPs are rapidly elevated after stress and confer a temperature-resistant phenotype. In cancer cells, cells that are under numerous acute and chronic stresses caused by internal factors such as proteome and genome instability and alterations, as well as environmental factors such as hypoxic conditions and nutrient deprivation, HSPs become vital for survival [8]. As we will discuss below, in cancer cells, HSPs are used to re-wire molecular pathways so that the cells can gain survival advantage under the transforming pressure.

7.2 HSP90 as an Important Cancer HSP

The major cancer HSPs are the HSP90 and HSP70 family of chaperones. In spite of several recent efforts to target HSP70, HSP90 remains to date the only one with a timely promise of delivering a new anticancer therapy [9, 10]. It is the most studied and perhaps understood cancer HSP family with four main known mammalian paralogs, the cytoplasmic heat shock protein 90 alpha and beta (HSP90α and β) [11], the endoplasmic reticulum (ER) glucose-regulated protein94 (GRP94) [12], and the mitochondrial tumor necrosis factor receptor–associated protein-1 (TRAP-1) [13]. Each is known to be highly expressed in cancer, but while HSP90α/β have been widely investigated, and important roles assigned to them in maintaining the functional conformation of a large number of aberrant malignancy-driving proteins, little is known on the roles of GRP94 and TRAP-1 [14, 15]. Because most small molecules used to investigate the role of HSP90 are pan-HSP90 inhibitors and few studies differentiate the role of HSP90α from that of HSP90β, we will here refer to HSP90 without delineating the role of individual paralogs.

HSP90 is a member of the GHKL family, proteins characterized by an atypical fold in their ATP-binding regulatory pocket, referred to as the Bergerat fold [16]. Its structure consists of three domains: mainly an amino-terminal region (N-domain) that contains the ATP pocket and co-chaperone-interacting motifs, a middle domain that provides docking sites for client proteins and co-chaperones and that participates in forming the active ATPase, and finally a carboxy-terminal domain that contains a dimerization motif, a second ligand-binding region and interaction sites for other co-chaperones [17, 18]. Dimerization of two HSP90 molecules through their C-domains is believed to be necessary for chaperone function.

HSP90 function is regulated by a plethora of co-chaperones, of which approximately 20 have been identified to date. A full list of these co-chaperones has recently been reviewed by Johnson [6]. Many of these co-chaperones compete for the same binding site on HSP90 or bind alternate HSP90 conformations, and thus, distinct HSP90-co-chaperone complexes likely exist at any time. Several of these co-chaperones have tetratricopeptide (TPR) domains and bind to the MEEVD acceptor site on HSP90. Others, such as SGT1 and GCUNC-45, bind to a sequence in the N-terminal domain of HSP90. Most have no effect on the ATPase activity of HSP90, while others stimulate (i.e., AHA1) or inhibit (i.e., p23, Cdc37) it. Co-chaperones also regulate the conformation of HSP90, such as p23 that is believed to stabilize the closed conformation, and control the chaperoning of client proteins. Cdc37, for example, pre-binds clients that are kinases and delivers them to HSP90, whereas HSP-organizing protein (HOP) presents HSP70-bound clients to HSP90.

HSP90 interacts with and regulates its client proteins in a cycle that is driven by the binding and hydrolysis of ATP. Through this catalytic cycle, HSP90 undergoes considerable structural changes, and this dynamic nature of HSP90 is key to its ability to function as a chaperone [19]. This conformational flux is constantly altered by the binding of regulatory nucleotides (i.e., ATP/ADP) and co-chaperones (i.e., HOP, Cdc37, p23, AHA1, and immunophilins) and regulated in part by a series of tyrosine phosphorylation events.

Because HSP90 activity is intrinsically linked to its conformation, which is in turn dependent on the binding and release of ATP/ADP, co-chaperones, and client proteins, regulating the HSP90 cycle offers several ways of interfering with HSP90 chaperoning function. Specifically, small molecules that bind to the regulatory ATP/ADP pocket, interfere with binding of its co-chaperones or with the conformational flexibility of HSP90, have all been shown to lead to HSP90-mediated activities in cancer cells [10]. To date, most advanced in development are those inhibitors that bind to the N-terminal regulatory pocket, with several in clinical evaluation in cancers [20, 21]. Section 7.5 will detail these efforts.

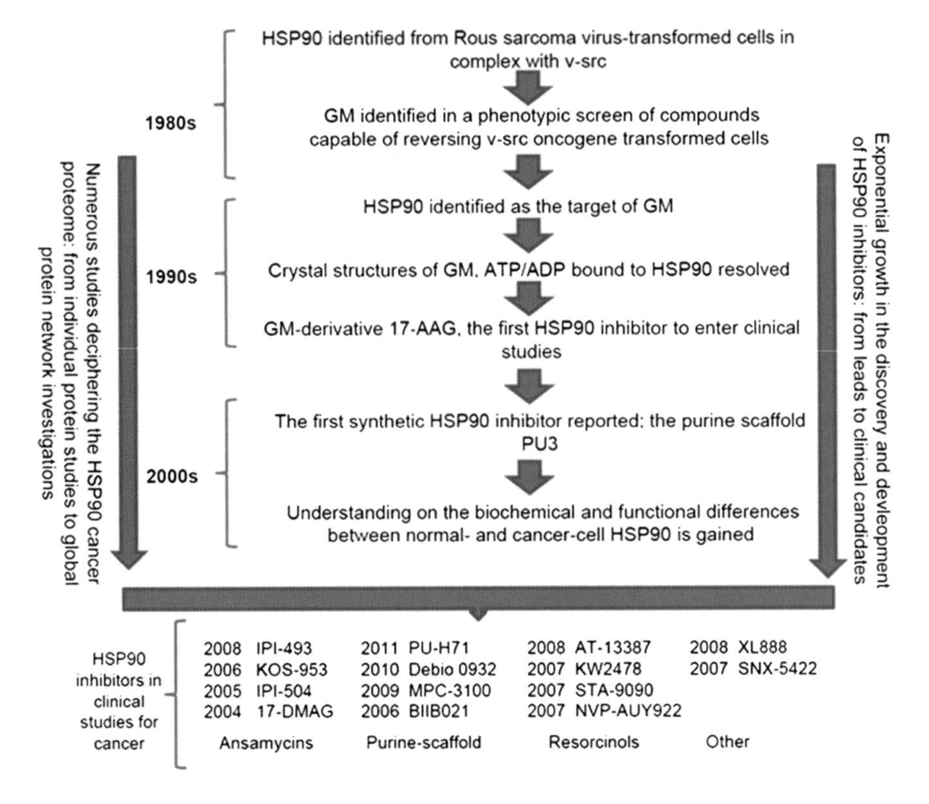

Fig. 7.1 Pivotal steps in the validation and translation of HSP90 as a target in cancer

7.3 Differences in Tumor and Normal Cell HSP90: A Rationale for Targeting HSP90 in Cancers

In fact, much of our understanding on the biology of HSP90 in cancer comes from studies with such small molecules [22]. Key milestones in these studies, leading to validation of HSP90 as a cancer chaperone, are outlined in Fig. 7.1. While the HSP90 protein was first discovered in the 1980s, it has drawn little interest as a potential target in cancer. After all, it is abundantly (~1–3 % of total cellular protein) and ubiquitously expressed in most if not all human cells and not particularly variable in expression between normal and cancer cells. Knockdown of even 50 % of HSP90 in cancer cells has little effect on their viability, while knockout of at least the HSP90β paralog is embryonically lethal [23]. Such findings match poorly with the belief that a good therapeutic target has to be crucial to the malignant phenotype and be of low expression in vital organs and tissues [24].

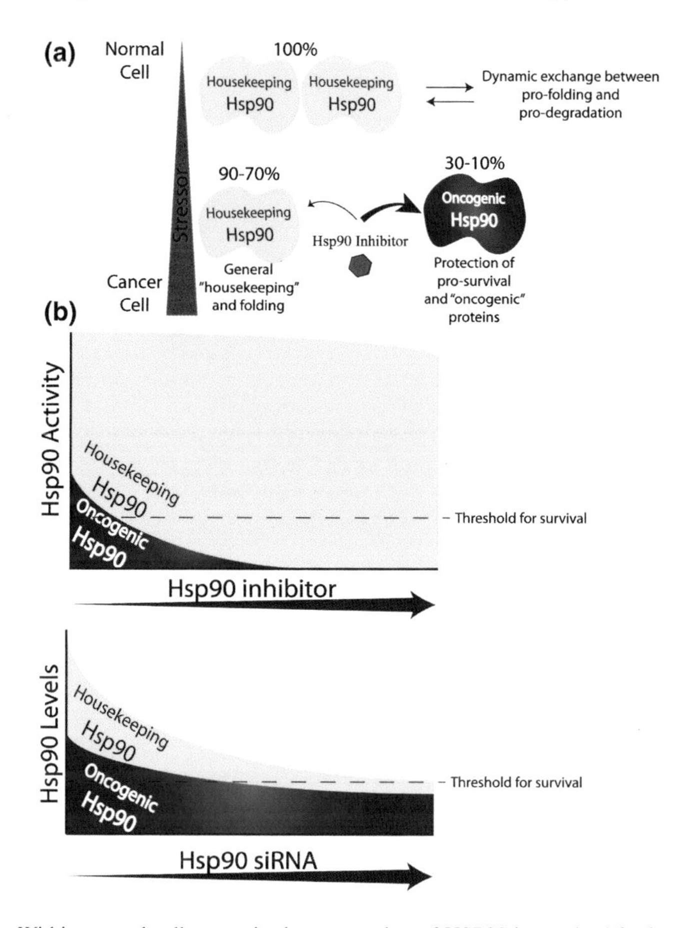

Fig. 7.2 **a** Within normal cells, constitutive expression of HSP90 is required for its evolutionarily conserved housekeeping function of folding and translocating cellular proteins to their proper cellular compartment ("housekeeping complex"). Upon malignant transformation, cellular proteins are perturbed through mutations, hyperactivity, and retention in incorrect cellular compartments or other means. The presence of these functionally altered proteins is required to initiate and maintain the malignant phenotype, and it is these oncogenic proteins that are specifically maintained by a subset of stress-modified HSP90 ("oncogenic complex"). Certain HSP90 inhibitors specifically bind to the "oncogenic complex". **b** By acting specifically on the "oncogenic HSP90", pharmacologic HSP90 inhibition is more efficient than genetic means at lowering HSP90 function below the threshold necessary for cancer cell survival

Interest on the target potential of HSP90 grew, however, after the serendipitous discovery of a natural product, geldanamycin (GM) [25]. Found in a screen searching for compounds able to revert the phenotype of cells transfected with the v-src oncogene, it was later demonstrated to do so by specifically binding to the N-terminal regulatory pocket of HSP90 and, by this, to inhibit its function. While surprising in light of the available genetic data, low concentrations of this natural product were active on many cancer cells and induced differentiation, reduced cell proliferation, and/or induced death, while being of no significant toxicity at such concentrations to normal cells [25]. Subsequent crystal structures of HSP90 in complex with GM and the regulatory nucleotides have uncovered the unique characteristics of this pocket and its high potential for druggability, further spurring the interest in HSP90 as a drug target [10, 11, 17].

The low toxicity seen with HSP90 inhibitors on normal cells remained unexplained for more than a decade. Then, Kamal et al. [26] proposed that, while no specific mutations differentiated HSP90 from normal and cancer cells, in cancer cells the chaperone was found entirely in complexes of high affinity and sensitivity to small molecule inhibitors. In normal cells, on the other hand, a dynamic complex of HSP90, of low affinity and decreased sensitivity to pharmacologic inhibition, was present. This mechanism provided a satisfactory explanation for the distinct sensitivity of normal and cancer cells to GM and other HSP90 inhibitors. It, however, fell short on explaining other observations, such as the little effect 50 % reduction in HSP90 levels had on cancer cells. An explanation came 8 years later when Moulick et al. [27] demonstrated that HSP90 in cancer cells was not in its entirety in the high-affinity form, but rather it was composed of a "housekeeping HSP90" species, of low affinity to small molecule inhibitors, such as the HSP90 found in normal cells, and also of a distinct HSP90, defined as the "oncogenic HSP90" species. The "oncogenic HSP90," of high affinity to certain small molecule inhibitors, comprised a functionally distinct HSP90 pool, enriched or expanded in cancer cells (Fig. 7.2a). The authors showed it to be a highly co-chaperone-dependent HSP90 that certain cancer cells use to maintain the altered proteins and protein networks that are needed to drive the malignant phenotype. Thus, while the tumor becomes addicted to survival on a network of HSP90-oncoproteins, these proteins become dependent on "oncogenic HSP90" for functioning and stability.

In this view, small molecules by their ability to interact specifically with the "oncogenic HSP90" will primarily and selectively affect these complexes and will act on the "housekeeping HSP90" only at higher or at saturating concentrations (Fig. 7.2b; HSP90 inhibitor). On the other hand, genetic means will equally reduce the expression of both "oncogenic" and "housekeeping" HSP90 pools, and thus, it is conceivable that more than 50 % reduction in HSP90 levels would be necessary to lower HSP90 to the threshold level required for cell survival (Fig. 7.2b; HSP90 siRNA).

With these concepts in mind, it is therefore important to understand that the pharmacologically targeted tumor HSP90 is not identical with the tumor HSP90, but rather encompasses a small portion of it (Fig. 7.2a; HSP90 inhibitor).

7.4 HSP90 as a Regulator of Important Cancer-Mediating Proteins

Now that we understand that HSP90 chaperones a significant portion of the altered cancer proteome, we will detail below the nature of such proteome. Significant research over the past 30 years has identified that HSP90 regulates many cancer proteins and protein networks involved in signaling, cell cycle and proliferation, DNA

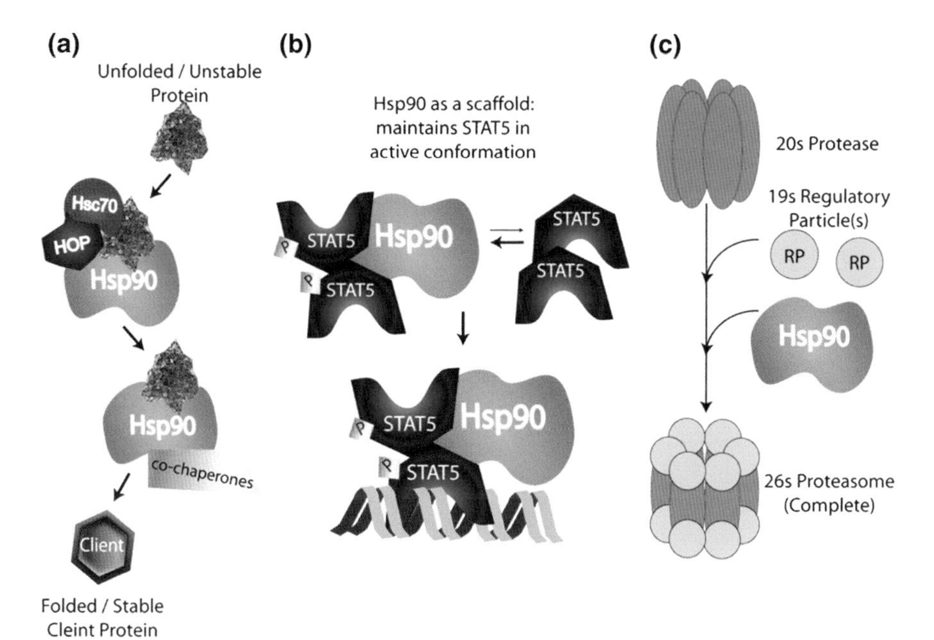

Fig. 7.3 Mechanisms by which HSP90 chaperones the cancer proteome. Cancer cells require HSP90 to regulate the folding and stability of several cancer-driving proteins (**a**) and to facilitate the complex formation of molecules involved in aberrantly activated complexes (**b**, **c**). In these cases, HSP90 acts a scaffolding molecule that maintains protein complexes in an active configuration (**b**) or as a facilitator of protein complex assembly (**c**)

damage and repair, transcriptional and translational activity, and epigenetic regulation. These, while potentially partly dependent on HSP90 in normal cells, may become addicted to HSP90 for their function in the course of cell progression from a normal to a cancerous state. In such cases, the cancer cells require HSP90 to regulate the folding and stability of overexpressed (i.e., HER2), mutated (i.e., mutant B-Raf , mutant FLT3), activated (i.e., AKT), or chimeric proteins (i.e., BCR-ABL) [28] (Fig. 7.3a). Recent studies have started to unveil additional roles for HSP90, such as to facilitate complex formation of molecules involved in aberrantly activated complexes. In these cases, HSP90 acts as a scaffolding molecule that maintains protein complexes in an active configuration (i.e., STAT5) [27] (Fig. 7.3b) or as a facilitator of protein complex assembly (i.e., RNA polymerase II, RISC) [29] (Fig. 7.3c).

7.4.1 HSP90 as a Regulator of Onco-Protein Client Stability and Folding

Traditionally, HSP90 was known as a regulator of protein stability (Fig. 7.3a). It binds to near native proteins that have metastable characteristics and regulates their stability and helps to keep them in a folded, functional state [30, 31]. Among the proteins that require HSP90 for this activity are several kinases (Sect. 7.4.2), transcription factors (Sect. 7.4.3), epigenetic regulators (Sect. 7.4.4), and several proteins involved in DNA repair (Sect. 7.4.5). In general, inhibition of HSP90 leads to a reduction in the steady-state levels of these proteins. The major pathway by which the cell degrades these HSP90 client proteins is the proteasome. Specifically, upon HSP90 inhibition, an E3 ligase is recruited to the complex, and upon ubiquitination, the protein is directed to the proteasome for degradation, thus short-circuiting further cycles of HSP90-mediated refolding [22, 28, 32]. HSP90, therefore, sits at the crossroad of folding/stabilization and degradation pathways, with small molecule inhibitors pushing the equilibrium strongly toward the degradative fate.

7.4.2 HSP90 and Protein Kinases

Activation of kinases is necessary to maintain the increased signaling state characteristic of cancer cells, and thus, these proteins have emerged early on as important cancer targets. Concordantly, the nature of the HSP90-addicted kinases and the mechanism of such interrelationship were widely studied. In effect, kinases were the first known HSP90 onco-clients since HSP90 was identified in 1980s as part of pp60v-src-HSP90-p50 complexes immunoprecipitated from Rous sarcoma virus-transformed cells [33]. Since then, a number of mutated, activated, overexpressed, or chimeric kinases that are key mediators of disease have been shown to be regulated by HSP90 in a variety of cancers [32]. Among these are the HER-tyrosine kinases, AKT, Raf kinases, including c-Raf and B-Raf, JAK kinases, c-KIT, MET, FLT3, and a variety of chimeric kinases, including BCR-ABL and NPM-ALK, to list a few. Several key cell cycle regulator proteins such as cyclin-dependent kinases are also HSP90 regulated. In all of these cases, HSP90 is required to maintain the functional conformation of the kinase, and inhibition of HSP90 results in kinase inactivation and degradation. An updated list of potential HSP90-interacting kinases as well as other proteins can be found at the Picard laboratory website (www.picard.ch/downloads/Hsp90interactors.pdf). While HSP90 may help fold polymorphic variants of the same protein, the folding requirement for each variant may vary. This was observed with v-SRC, an unstable protein kinase, which has a greater requirement for HSP90 than its cellular counterpart c-SRC [34]. Similarly, mutant forms of p53 [35] and B-Raf protein kinase [36] and the chimeric BCR-ABL [27] have a greater requirement for chaperones than do their wild-type counterparts.

7.4.3 HSP90 and Transcription Factors

Transcription factors are the downstream effectors of many cancer-promoting pathways and thus play an important role in cancer therapy. Many of them, including the steroid receptors, the STAT family, and p53, appear to require HSP90 to facilitate increased signaling and transcriptional robustness.

7.4.3.1 Steroid Hormone Receptors

Most steroid hormone receptors such as glucocorticoid receptor (GR), mineralocorticoid receptor (MR), and androgen receptor (AR), validated anticancer targets in several cancers including breast and prostate cancer, appear to require HSP90 for activity [37, 38]. HSP90 has a complex role in integrating the subcellular distribution as well as the transcriptional activity of these receptors. One of the best studied is the GR. In the absence of a ligand, HSP90 binds to GR and maintains the receptor in a hormone-binding competent conformation. The dynein/dynactin motor complex also associates with the HSP90-FKBP52 complex bound to GR and MR, suggesting that this mega-complex may power the active retrograde movement of steroid receptors [39, 40]. As seen with HSP90 clients, treatment of cells with HSP90 inhibitors resulted in disruption of hormone-binding activity and increased proteasomal degradation of the receptors [41].

7.4.3.2 Tumor Suppressor p53

The tumor suppressor p53 is another transcription factor that requires HSP90. Somatic mutations in p53 are found in many types of human tumors and often result in the inactivation of its tumor suppressor function. These mutations result in conformation changes and prolonged protein half-life that require HSP90 for their stabilization [35]. Furthermore, binding of HSP90 to mutant p53 inhibits the ability of MDM2 to promote p53 ubiquitination and degradation, leading to the stabilization of both mutant p53 and MDM2 [42, 43]. Interestingly, while HSP90 associates with both the wild-type and the mutated p53, inhibition of HSP90 upregulates the wild-type protein but downregulates the mutant one, suggesting a distinct role for HSP90 in the regulation of the normal (wild-type) and aberrant (mutant) p53 [44].

7.4.3.3 HIF-1α

Hypoxia inducible factor-1α (HIF-1α) is a nuclear transcription factor involved in the transactivation of numerous target genes, many of which are implicated in the promotion of angiogenesis and adaptation to hypoxia. In normoxic conditions, HIF-1α is expressed at low levels being targeted for proteasome-dependent degradation by VHL, but hypoxia normally impairs VHL function, allowing HIF to accumulate [45]. In these conditions, HIF-1α interacts with HSP90. HIF protein from HSP90-inhibited cells is unable to bind DNA, suggesting that HSP90 is necessary for mediating the proper conformation of HIF and/or recruiting additional cofactors [46].

7.4.3.4 BCL6 Transcriptional Repressor

In diffuse large B-cell lymphoma (DLBCL), HSP90 was found to be an important regulator of the transcription repressor BCL6 [47]. The oncogenic effects of BCL6 are related to its ability to directly repress key genes such as *ATR* (encoding ataxia telangiectasia and Rad3-related) and *TP53* (encoding tumor suppressor protein p53). In these conditions, HSP90 stabilized BCL6 mRNA and protein and formed a complex with BCL6 at its target promoters. HSP90 inhibition derepressed BCL6 target genes, inducing reactivation of key BCL6 target genes and apoptosis.

7.4.4 HSP90 and the Epigenetic Machinery

Increasing evidence indicates that epigenetic alterations contribute to malignant states. These alterations result in gene silencing and activation as a consequence of chromatin remodeling driven by modifications including histone methylation, acetylation, and DNA methylation. Aberrant epigenetic changes can lead to tumor formation as a result of dysregulated gene expression. Thus, histone-modifying enzymes and DNA methyltransferases are important therapeutic targets. Of note, molecules involved in epigenetic regulation have been identified to be directly or indirectly affected by HSP90 inhibitory therapies.

7.4.4.1 DNA Methyl Transferase 1

Gene silencing by CpG island methylation is a common mechanism of tumor suppressor gene repression in cancers. The DNA methyltransferase DNMT1 plays a critical role in the maintenance of methylation. It has been shown that DNMT1 can cooperate with histone deacetylases (HDAC) to initiate and maintain epigenetic gene silencing [48]. Studies that blocked HSP90 activity revealed that it resulted in the ubiquitin–proteasome degradation of DNMT1 in breast cancer and in other tumor cells [49, 50]. By altering the stability of DNMT1, HSP90 inhibition resulted in decreased DNMT1 levels followed by the derepression of JunB, indicating DNMT1 as another cancer client of HSP90.

7.4.4.2 Histone Methyltransferases

HSP90 has also been shown to interact with proteins involved in chromatin regulation by modifying histone methylation, including EZH2, PRMT5, CARM1, SMYD3, and SMYD2. EZH2 is the catalytic core in the polycomb repressor complex 2 (PRC2) that is involved in the trimethylation of histone-3 lysine-27 (H3K27), resulting in chromatin condensation and gene silencing [51]. EZH2 is aberrantly regulated in a wide range of cancer types, and it is involved in stem cell maintenance and tumorigenesis [52, 53]. Thus, disruption of EZH2 is of interest as a therapeutic strategy for cancer. HSP90 was recently shown to regulate EZH2 in leukemia cells, with its inhibition leading to decreased EZH2 protein expression, suggesting that HSP90 is required to maintain the stability of EZH2 [54]. PRMT5 is a protein arginine methyltransferases (PRMTs) involved in arginine methylation of histones and other proteins that play a role in the regulation of major signaling pathways that control survival and malignant transformation [55]. PRMT5 is a client protein of HSP90 in ovarian cancer cells, and PRMT5 protein levels decreased upon treatment with an HSP90 inhibitor [56]. PRMT4/CARM1 (co-activator-associated arginine methyltransferase 1), an enzyme that catalyzes the transfer of a methyl groups to arginine residues resulting in the activation of signal transduction cascades and transcriptional activation [57], has been found to be deregulated in tumor cells [58]. CARM1 was recently reported to be a client protein of HSP90 in leukemia cells, and this interaction shown to be critical for their survival [27]. The SMYD (SET and MYND domain) family of lysine methyltransferases (KMTs) plays pivotal roles in various cellular processes, including gene expression regulation and DNA damage response. Initially identified as genuine histone methyltransferases, specific members of this family have recently been shown to methylate non-histone proteins such as p53 , VEGFR, and the retinoblastoma tumor suppressor (pRb). Both SMYD3 and SMYD2 (SET and MYND domain–containing protein 3) are histone methyltransferases that interact with HSP90 [59, 60]. HSP90 inhibition decreased the expression of SMYD3 and inhibited migration and proliferation of breast cancer cells. [60]

7.4.5 HSP90 and DNA Damage and Repair System

The DNA damage response refers to a series of tightly regulated, complex cellular mechanisms that repair DNA breaks occurring through normal cellular metabolism or environmental exposures. These pathways include base excision repair (BER), nucleotide excision repair (NER), mismatch repair (MMR), non-homologous end-joining (NHEJ), and homologous recombination (HR). The choice of repair pathway used is determined by the type of lesion and by the expression of repair proteins at specific cell cycle phases to ensure that the most accurate repair is performed [61]. DNA double-strand breaks (DSBs), which pose the most significant threat to genomic integrity, are repaired by two major DNA repair pathways, NHEJ and HR. Deficiency in the HR pathway results in repair of all DSBs by the error-prone NHEJ pathway leading to the accumulation of chromosomal aberrations. All steps of the DNA damage response require the recruitment and modification of multiple key proteins; several of which have been proposed as potential client proteins of the molecular chaperone HSP90 [62]. Several proteins crucial to the HR and NHEJ pathways for repair of DSBs including BRCA2, FANCA, ATR, CHK1, and DNA-PK are also HSP90-chaperoned [63–65]. Interestingly, HSP90 can be modified during the DNA damage response. HSP90 is phosphorylated by DNA-PK, and both phosphorylated HSP90 and DNA-PK have been found to co-localize to the H2AX apoptotic ring that contains activated ATM, CHK2, and H2AX [65, 66]. HSP90 inhibition may therefore promote the degradation of proteins involved in DNA repair and thereby potentiate DNA damage sustained by ionizing radiation (IR) or chemotherapy. These observations are supported by the enhancement of radiosensitivity of multiple human cancer cell lines including glioblastoma, breast, colon, sarcoma, and pancreatic cancer cell lines by treatment with inhibitors of HSP90 [67, 68].

7.4.6 HSP90 as a Scaffolding Molecule to Maintain Signaling Complexes in an Active Configuration

In addition to regulating protein stability and folding, HSP90 has been recently shown to act as a scaffolding molecule that increases the competency of protein complexes involved in signaling and in transcription (Fig. 7.3b). Specifically, in chronic myeloid leukemia (CML), HSP90 was found to facilitate increased STAT5 signaling by binding to and influencing the conformation of STAT5 and by maintaining STAT5 in an active conformation directly within STAT5-containing transcriptional complexes [27]. In the cytosol, HSP90 binding to STAT5 modulated the conformation of the protein to alter STAT5 phosphorylation and dephosphorylation kinetics. In addition, HSP90 maintained STAT5 in an active conformation in STAT5-containing transcriptional complexes. Unlike the classic HSP90 clients that require HSP90 for stability (Sect. 7.4.1), HSP90 inhibition led to inhibition but not degradation of STAT5.

7.4.7 HSP90 as a Protein Complex Assembly Facilitator

Recently, HSP90 has also been identified as a critical protein in promoting and maintaining the assembly of several cancer-related multi-protein complexes (Fig. 7.3c). The involvement of HSP90 in these complexes varies, and in some, it relates to stabilizing an unstable protein subunit and facilitating its incorporation into the mega-complex. In other cases, HSP90 promotes a change in the composition of a given complex. In any case, HSP90 is no longer present in the final assembled complex. Such effects of the chaperone on the assembly of the following seven complexes: snoRNP, RNA polymerase II, phosphatidylinositol-3 kinase–related protein kinase (PIKK), telomere complex, kinetochore, RNA-induced silencing complexes (RISC), and 26S proteasome was recently detailed in a review article [29].

7.4.8 Global HSP90 Proteome Investigations in Cancer

While the above studies led to the identification of important HSP90 client proteins, it is clear that due to the complexity of the altered proteome in any given cancer cell, large proteomic and genomic analyses combined with bioinformatic analyses are needed to understand the HSP90-addicted proteome. Because malignant evolution is characterized by alterations in a multitude of proteins and pathways, and perhaps no two tumors present an identical spectrum of defects, it is likely that addiction of cancer cells to HSP90 occurs in a tumor-specific manner, and thus, no two tumors will have an identical set of HSP90-sheltered proteins and protein networks. Indeed, several recent investigations have used such methods to demonstrate the complexity of HSP90-sheltered networks. A thorough review on this topic was recently published by the Matts group [69].

The yeast has been a useful system to study HSP90 functions, several of these findings being eventually validated in cancer cells. As such, several systematic proteomic and genomic methods were used to map HSP90 interactions in yeast [69]. In one such study, physical interactions were identified using genome-wide two-hybrid screens combined with large-scale affinity purification of HSP90-containing protein complexes [70]. Genetic interactions were uncovered using synthetic genetic array technology and by a microarray-based chemical-genetic screen of a set of about 4,700 viable yeast gene deletion mutants for hypersensitivity to the HSP90 inhibitor GM. These analyses proposed that HSP90 may interact with 10 % of the yeast proteome and that its interacting proteins were involved in a wide range of cellular functions, including cell cycle and DNA processing, cellular communication, fate/organization, and transport, as well as energy production, metabolism, transcription, translation, and transport facilitation. This study identified two novel HSP90 interactors, namely Pih1 (Protein interacting with HSP90) and Tah1 (TPR-containing protein associated with HSP90), which interact physically and functionally with the conserved AAA(+)-type DNA helicases Rvb1/Rvb2, forming the R2TP complex. The complex is required for chromatin remodeling, box C/D small nucleolar ribonucleoprotein (snoRNP) biogenesis, apoptosis, phosphatidylinositol-3 kinase–related protein kinase (PIKK) signaling, and RNA polymerase II assembly, implicating HSP90 in these processes.

In a yet another classic approach, changes in the proteome of HeLa cervical cancer cells were analyzed upon treatment with the HSP90 inhibitor 17-(dimethylaminoethylamino)-17-demethoxygeldanamycin (17-DMAG) using SILAC and high resolution, quantitative mass spectrometry [71]. This study found that the proteome most sensitive to HSP90 inhibition and thus most significantly downregulated by 17-DMAG was that fitting the functional categories of "protein kinase activity" and "DNA metabolic processes". The kinome regulated by HSP90 in HeLa was associated with a multitude of signaling and signaling-associated pathways including the PI3K -AKT-, IL-6-, HER-, FAK-, Ephrin receptor-, ERK -MAPK-, NFκB-, inositol phosphate metabolism, and G-protein-coupled receptor–mediated networks. In the "DNA metabolic processes" group, enriched were GO annotated processes such as "cellular response to DNA damage," "DNA modification," "response to DNA damage stimulus," and "regulation of transcription".

In a novel approach combining chemical proteomics with bioinformatic analyses, Moulick et al. analyzed the HSP90-sheltered proteome in the K562 CML cells [27]. In this approach, HSP90 in complex with its client proteins and co-chaperones was isolated by chemical purification using a solid-support-attached HSP90 inhibitor, namely the purine-scaffold PU-H71. PU-H71/HSP90 protein cargo was then analyzed by mass spectrometry. The method took advantage of the fact that PU-H71 bound preferentially to the fraction of tumor HSP90 associated with altered client proteins (i.e., "oncogenic HSP90"; Fig. 7.2a) and moreover, upon binding, locked HSP90 in an onco-client bound configuration. Together these features greatly facilitated the chemical affinity purification of tumor-associated protein clients by mass spectrometry, providing tumor-by-tumor global insights into the HSP90-sheltered proteome, including in primary patient specimens. Top scoring networks identified by this method to be HSP90-sheltered in CML were those used by BCR-ABL to propagate aberrant signaling: PI3K /AKT/mTOR -, Raf -MAPK-, NFκB-, STAT5-, and FAK-mediated signaling pathways. Other important transforming pathways in CML, driven by MYC and TGF-β as well as others involved in disease progression and aberrant cell cycle and proliferation of CML, were identified. In addition to signaling proteins, proteins that regulate carbohydrate and lipid metabolism, protein synthesis, epigenetics, gene expression, and cellular assembly and organization were identified, in accord with the postulated broad roles of HSP90 as an important mediator of cell transformation. The method also identified the histone-arginine methyltransferase CARM1 as a novel HSP90 interactor, implicating the HSP90-facilitated CARM1 activity in CML leukemogenesis.

7.5 Targeting HSP90 in Cancer

As delineated in Sect. 7.4, HSP90 shelters a complex proteome vital for cancer progression and maintenance. The HSP90-chaperoned networks and pathways differ from tumor-to-tumor, depending on the molecular wiring of the alterations that characterize a specific tumor. These are appealing features for a cancer target, because by inhibiting one protein, HSP90, one could have beneficial effects in a variety of tumors. Such effects were not appreciated early on when HSP90 was discovered, because as indicated above, HSP90 is an abundant protein expressed in most human cells. However, its persuasive roles in regulating the malignant proteome, the observation that it can be easily pharmacologically modulated by small molecule ligands, and findings demonstrating the differences between tumor and normal cell HSP90 have together given it credibility as a cancer target. Indeed, at the moment, more than 20 inhibitors have been or still are in clinical testing, with many more coming in the pipeline (Fig. 7.4) [20, 21].

As indicated, the first compounds identified as HSP90 inhibitors were those that modulate HSP90 chaperone activity by inhibiting the N-terminal domain ATP-binding site [25, 72]. Targeted approaches such as structure-based drug design, biochemical and cell-based screening, virtual screening, fragment-based drug design, and medicinal chemistry have since led to the identification and development of several novel inhibitors that act by similar or alternate mechanisms to inhibit HSP90 function [10]. The molecules currently advanced to clinical trials are those that bind to the regulatory pocket. These constitute a diverse array of structures, which at a close inspection classify according to their similarity to GM (Sect. 7.5.1), to the purine-scaffold (Sect. 7.5.2), or to radicicol (Sect. 7.5.3).

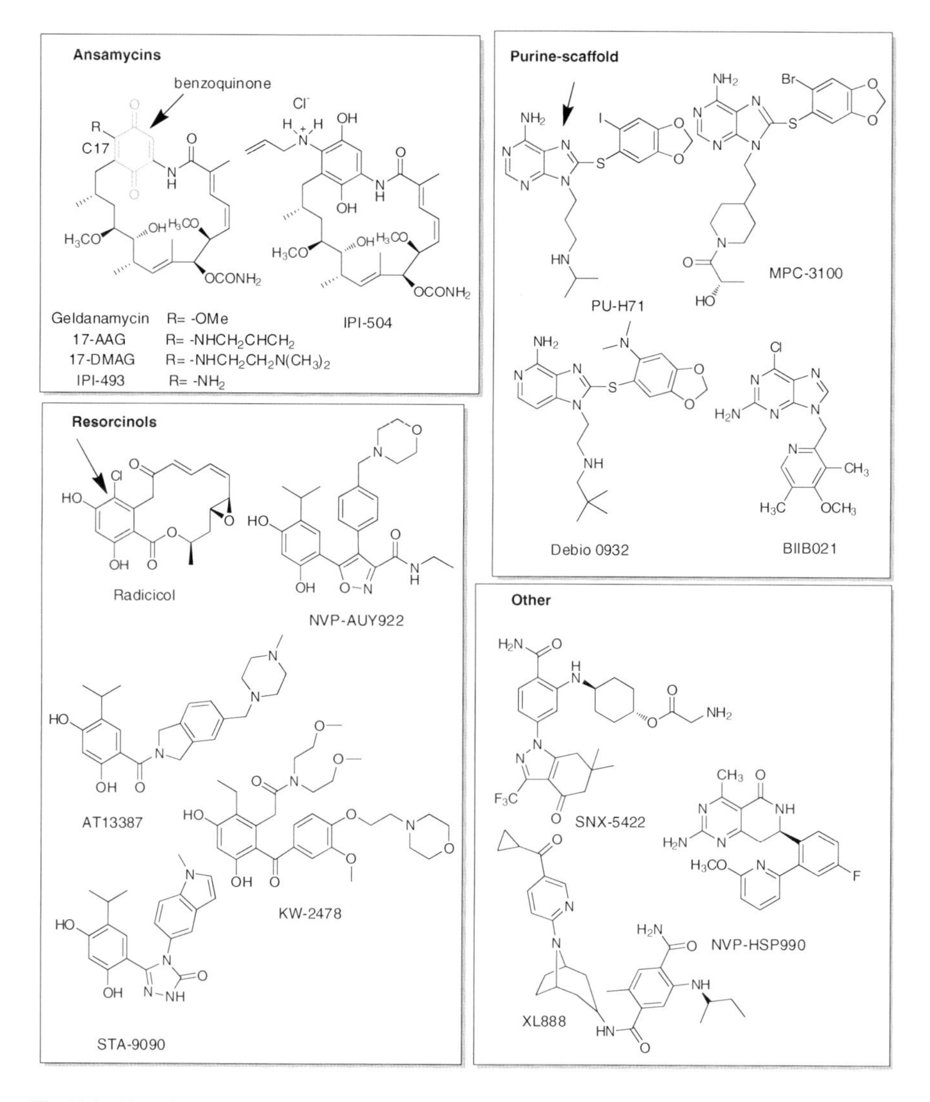

Fig. 7.4 Chemical structures of HSP90 inhibitors in clinical evaluation

7.5.1 Geldanamycin Derivatives

Initial efforts in the discovery of clinically translatable HSP90 inhibitors have focused on GM and several of its derivatives (Fig. 7.4, ansamycins). GM is a benzoquinone ansamycin first isolated from a fermentation broth of *Streptomyces hygroscopicus* in 1970 [25]. Inhibition of HSP90 was subsequently shown to be the mechanism of its anticancer activity [34] and more specifically was later shown to inhibit its ATPase activity by competing with

ATP for binding to the N-terminal nucleotide-binding pocket [25]. GM was never evaluated in clinical trials because of its poor pharmaceutical properties, including poor solubility, limited in vivo stability, and hepatotoxicity in animals [10, 25]. Despite the complexity of the molecule, attempts have been made to structurally modify geldanamycin so as to improve these properties and make it more “drug-like”. Early attempts focused on modifying the 17-position of the quinone ring. Synthetically, the C17 methoxy group can readily be substituted with amines, an approach that resulted in two clinically viable candidates, 17-AAG and 17-DMAG. 17-AAG, the first HSP90 inhibitor to enter clinical studies in cancers, has served to establish proof-of-principle for HSP90 inhibition in humans [19–21]. 17-DMAG is a related analog with an *N,N*-dimethylethylamino group in place of the C17-methoxy. This ionizable amine group led to improved solubility and better oral bioavailability for 17-DMAG when compared with 17-AAG. 17-DMAG showed activity in refractory metastatic breast cancer and in refractory acute myeloid leukemia patients, but its development was halted as a result of unacceptable side effects [19–21]. IPI-504, the reduced form of 17-AAG, was designed to resolve a number of issues concerning 17-AAG [10]. As a hydrochloride salt, IPI-504 was highly water soluble, and because it lacked the benzoquinone characteristic of 17-AAG, it showed an improved toxicity profile. In vivo, however, IPI-504 and 17-AAG interconvert through the action of oxidoreductases. IPI-504 showed clinical benefit in a study in patients with non-small-cell lung cancer (NSCLC) in tumors harboring ALK rearrangements [19–21].

7.5.2 The Purine and Purine-Like Scaffold HSP90 Inhibitors

The availability of crystal structures of ADP, GM, and radicicol bound to the N-terminal nucleotide-binding domain of HSP90 and the realization that HSP90 possesses a unique ATP-binding pocket made the rational design of synthetic inhibitors possible [10]. As mentioned, HSP90 is part of a small protein family called the GHKL (Gyrase, HSP90, histidine kinases, and MutL) family, which contains a unique fold in their regulatory pocket region, called the Bergerat fold [16], a feature that portends selectivity for its ligands over other ATP-binding pockets. Taking advantage of the distinct shape of the ATP pocket, Chiosis et al. [72] designed PU3 as the first reported synthetic inhibitor. PU3, a purine-scaffold compound, maintained the purine core of the endogenous ligands. This was linked at the 8-position to an aryl moiety via a methylene linker (Purine-CH_2-Aryl). This essential motif was maintained, with some variations, for essentially all the purine-scaffold HSP90 inhibitors (Fig. 7.4 purine-scaffold). From this interesting lead, a large effort from multiple groups resulted in four molecules to enter clinical evaluation in cancers: PU-H71, BIIB021, MPC-3100, and Debio 0932 (formerly CUDC-305) [20, 21].

7.5.3 The Resorcinol Class

The compounds within this class are distinguished by the presence of a resorcinol moiety, and for this, they are related to radicicol (Fig. 7.4 resorcinols), a natural product inhibitor of HSP90 isolated from *Monosporium bonorden*. Although a potent HSP90 inhibitor, radicicol is unstable in vivo, and no attempts to directly modify its structure have resulted in a clinical candidate. However, a number of agents currently investigated maintain the resorcinol moiety. One of these, NVP-AUY922, was developed from a lead molecule discovered during a high-throughput screen designed to measure the ability of a compound to inhibit yeast HSP90 ATPase activity [10, 21, 22]. AT13387, on the other hand, was developed using a fragment-based approach using NMR screening and X-ray crystallography. A number of other novel resorcinol compounds are currently being evaluated in clinical trials, including STA-9090 (ganetespib), KW-2478, and DS2248 [20, 21].

7.5.4 Other HSP90 Inhibitor Classes

Serenex used a chemo-proteomics approach in the discovery of a novel chemotype HSP90 inhibitor that ultimately resulted in the clinical candidate SNX-5422 [10, 20]. To accomplish this, purine-binding proteins from porcine lung or liver were loaded onto an ATP-affinity column and subsequently were challenged with a focused library of 8,000 compounds in a parallel search for both a suitable target protein and hit compound. The clinical development of this agent has been halted due to severe ocular toxicity [20–22]. Other HSP90 inhibitors advancing to clinical studies include XL888 and NVP-HSP990 (Fig. 7.4 others).

7.6 Clinical Experience with HSP90 Inhibitors

Given their potential to inactivate a number of different onco-client proteins and affect multiple cancer networks, HSP90 inhibitors have been hypothesized to be active, and consequently tested, in a wide variety of cancers [20, 21, 73]. To date, the greatest clinical activity as evidenced by objective tumor regressions has been observed with the GM analog tanespimycin (17-AAG) when given in combination with trastuzumab to patients with $HER2^+$ metastatic breast cancer in whom tumor progression was seen with trastuzumab alone (i.e., 59 % overall clinical benefit rate in trastuzumab-refractory disease). Preclinically, HER2 is among the most sensitive client proteins to HSP90 inhibition, and this may account for the activity observed in this setting. Recently, objective tumor responses were also reported with ganetespib (STA-9090) in NSCLC tumors that harbored the ALK rearrangement. ALK rearrangements, particularly the EML4-ALK fusion protein, found in a subset of NSCLC patients, are also a sensitive HSP90 client protein [20, 21, 73].

While these results validate HSP90 as a relevant anticancer target, the full potential of HSP90 inhibitors to be active in a broader spectrum of tumor types has yet to be achieved. This differential response could be because we are unable to select patients who might best benefit from therapy and perhaps have not successfully optimized the dosing and scheduling of this class of agents. As a class, HSP90 inhibitors demonstrate rapid clearance from normal tissues and the blood compartment with prolonged retention in tumors; hence, traditional serum pharmacokinetics is insufficient to guide dosing and scheduling [10]. Additionally, the current drug development paradigm based on identifying maximum tolerated dose may not be applicable in the case of these inhibitors where target modulation is critical to producing their anti-tumor effects. The limited value of plasma pharmacokinetics and the importance of tumor concentration in predicting therapeutic response suggest the need for the clinical development of an assay of tumor pharmacokinetics for HSP90 inhibitors. In addition, there is presently no clear consensus on how to identify those patients most likely to benefit from HSP90 therapy. Unlike for other targets, where either protein expression (i.e., HER2 levels for trastuzumab selection in treatment of breast cancer patients) or the presence of a mutation (i.e., EGFR mutation for Tarceva selection in treatment of NSCLC patients) drives patient selection, there is no such knowledge for HSP90. No mutation characterizes HSP90 in tumors nor is its expression considerably variable between responsive and un-responsive tumors. The lack of a biomarker for proper patient selection is especially problematic since successful development of targeted agents requires identification of the patient sub-population that should receive the drug (i.e., tumors with EGFR mutations for Tarceva). Such selection may reduce the number of patients receiving ineffective treatment, reduce costs to health care and facilitate the clinical development and route to approval for the HSP90 inhibitors [10, 20, 21, 73].

7.7 Conclusion

HSPs and HSP90, in particular, have emerged as important targets in cancer due to their ability to buffer wide proteome alterations such as are characteristic in malignant transformation. While not easily accepted initially as potential cancer targets, several studies now spanning almost three decades have revealed important differences in the biochemical and functional roles of HSP90 in cancer cells that are different from those in normal cells. These findings, together with the now almost explosive interest in the discovery and development of small molecule HSP90 inhibitors, and with the encouraging clinical studies with these agents, have cemented HSP90 as a cancer target. The journey is, however, yet to be complete.

Several issues, including the lack of a biomarker for patient selection and the inability to measure target inhibition in clinic, remain to be addressed.

Acknowledgments G Chiosis is funded in part by Mr. William H. and Mrs. Alice Goodwin and the Commonwealth Foundation for Cancer Research and "The Experimental Therapeutics Center of Memorial Sloan-Kettering Cancer Center", the Geoffrey Beene Cancer Research Center of MSKCC, Leukemia and Lymphoma Society, Breast Cancer Research Fund, the SPORE Pilot Award and Research and Therapeutics Program in Prostate Cancer, the Hirshberg Foundation for Pancreatic Cancer, the Byrne Fund, National Institutes of Health (1U01 AG032969, 1R01CA155226, 1R21AI090501, 1R21CA158609, 3P30CA008748, P50CA086438), MSKCC Society, Department of Defense (R03-BC085588), Susan G Komen for the Cure and the Institute for the Study of Aging and The Association for Frontotemporal Dementias (Grant #281207 AFTD). T Taldone discloses a grant support from the Department of Defense (PDF-BC093421). M Guzman is funded by the US National Institutes of Health (NIH) through the NIH Director's New Innovator Award Program, 1 DP2 OD007399-01, National Cancer Institute (R21 CA158728-01A1), Leukemia and Lymphoma Foundation (LLS 6330-11), and she is a V Foundation Scholar.

References

1. Hanahan D, Weinberg RA (2011) Hallmarks of cancer: the next generation. Cell 144(5):646–674 (Epub 2011/03/08)
2. Nahleh Z, Tfayli A, Najm A, El Sayed A, Nahle Z (2012) Heat shock proteins in cancer: targeting the chaperones. Future Med Chem 4(7):927–935 (Epub 2012/05/11)
3. Khalil AA, Kabapy NF, Deraz SF, Smith C (2011) Heat shock proteins in oncology: diagnostic biomarkers or therapeutic targets? Biochim Biophys 1816(2):89–104 (Epub 2011/05/25)
4. Calderwood SK (2010) Heat shock proteins in breast cancer progression–a suitable case for treatment? Int J Hyperth: official J Eur Soc Hyperth Oncol, North American Hyperthermia Group 26(7):681–685 (Epub 2010/07/27)
5. Soo ET, Yip GW, Lwin ZM, Kumar SD, Bay BH (2008) Heat shock proteins as novel therapeutic targets in cancer. In vivo 22(3):311–315 (Epub 2008/07/10)
6. Johnson JL (2012) Evolution and function of diverse Hsp90 homologs and cochaperone proteins. Biochim Biophys 1823(3):607–613 (Epub 2011/10/20)
7. Cotto JJ, Morimoto RI (1999) Stress-induced activation of the heat-shock response: cell and molecular biology of heat-shock factors. Biochem Soc Symp 64:105–118 (Epub 1999/04/20)
8. Solimini NL, Luo J, Elledge SJ (2007) Non-oncogene addiction and the stress phenotype of cancer cells. Cell 130(6):986–988 (Epub 2007/09/25)
9. Jego G, Hazoume A, Seigneuric R, Garrido C (2010) Targeting heat shock proteins in cancer. Cancer Lett Epub (2010/11/17)
10. Patel HJ, Modi S, Chiosis G, Taldone T (2011) Advances in the discovery and development of heat-shock protein 90 inhibitors for cancer treatment. Expert Opin Drug Discov 6(5):559–587 (Epub 2012/03/09)
11. Prodromou C, Roe SM, O'Brien R, Ladbury JE, Piper PW, Pearl LH (1997) Identification and structural characterization of the ATP/ADP-binding site in the Hsp90 molecular chaperone. Cell 90(1):65–75 (Epub 1997/07/11)
12. Sorger PK, Pelham HRB (1987) The glucose-regulated protein grp94 is related to heat shock protein hsp90. J Mol Biol 194(2):341–344

13. Altieri DC, Stein GS, Lian JB, Languino LR (2012) TRAP-1, the mitochondrial Hsp90. Biochim Biophys 1823(3):767–773 (Epub 2011/09/01)
14. Sreedhar AS, Kalmar E, Csermely P, Shen YF (2004) Hsp90 isoforms: functions, expression and clinical importance. FEBS Lett 562(1–3):11–15 (Epub 2004/04/09)
15. da Silva VC, Ramos CH (2012) The network interaction of the human cytosolic 90 kDa heat shock protein Hsp90: A target for cancer therapeutics. J Proteomics 75(10):2790–2802 (Epub 2012/01/13)
16. Chene P (2002) ATPases as drug targets: learning from their structure. Nat Rev Drug Discov 1(9):665–673 (Epub 2002/09/05)
17. Pearl LH, Prodromou C (2006) Structure and mechanism of the Hsp90 molecular chaperone machinery. Annu Rev Biochem 75:271–294 (Epub 2006/06/08)
18. Prodromou C (2012) The active life of Hsp90 complexes. Biochim Biophys 1823(3): 614–623 (Epub 2011/08/16)
19. Pearl LH, Prodromou C, Workman P (2008) The Hsp90 molecular chaperone: an open and shut case for treatment. Biochem J 410(3):439–453 (Epub 2008/02/23)
20. Jhaveri K, Taldone T, Modi S, Chiosis G (2012) Advances in the clinical development of heat shock protein 90 (Hsp90) inhibitors in cancers. Biochim Biophys 1823(3):742–755 (Epub 2011/11/09)
21. Travers J, Sharp S, Workman P (2012) HSP90 inhibition: two-pronged exploitation of cancer dependencies. Drug Discovery Today 17(5–6):242–252 (Epub 2012/01/17)
22. Neckers L (2006) Using natural product inhibitors to validate Hsp90 as a molecular target in cancer. Curr Top Med Chem 6(11):1163–1171 (Epub 2006/07/18)
23. Voss AK, Thomas T, Gruss P (2000) Mice lacking HSP90beta fail to develop a placental labyrinth. Development 127(1):1–11 (Epub 2000/02/02)
24. Ross JS, Schenkein DP, Pietrusko R, Rolfe M, Linette GP, Stec J et al (2004) Targeted therapies for cancer 2004. Am J Clin pathol 122(4):598–609 (Epub 2004/10/19)
25. Neckers L, Schulte TW, Mimnaugh E (1999) Geldanamycin as a potential anti-cancer agent: its molecular target and biochemical activity. Invest New Drugs 17(4):361–373 (Epub 2000/04/12)
26. Kamal A, Thao L, Sensintaffar J, Zhang L, Boehm MF, Fritz LC et al (2003) A high-affinity conformation of Hsp90 confers tumour selectivity on Hsp90 inhibitors. Nature 425(6956):407–410 (Epub 2003/09/26)
27. Moulick K, Ahn JH, Zong H, Rodina A, Cerchietti L, Gomes DaGama EM et al (2011) Affinity-based proteomics reveal cancer-specific networks coordinated by Hsp90. Nat Chem Biol 7(11):818–826 (Epub 2011/09/29)
28. Workman P, Burrows F, Neckers L, Rosen N (2007) Drugging the cancer chaperone HSP90: combinatorial therapeutic exploitation of oncogene addiction and tumor stress. Ann N Y Acad Sci 1113:202–216 (Epub 2007/05/22)
29. Makhnevych T, Houry WA (2012) The role of Hsp90 in protein complex assembly. Biochim Biophys 1823(3):674–682 (Epub 2011/09/29)
30. Walter S, Buchner J (2002) Molecular chaperones–cellular machines for protein folding. Angewandte Chemie 41(7):1098–1113 (Epub 2002/12/20)
31. Young JC, Moarefi I, Hartl FU (2001) Hsp90: a specialized but essential protein-folding tool. J Cell Biol 154(2):267–273 (Epub 2001/07/27)
32. Whitesell L, Lindquist SL (2005) HSP90 and the chaperoning of cancer. Nat Rev Cancer 5(10):761–772 (Epub 2005/09/22)
33. Perdew GH, Whitelaw ML (1991) Evidence that the 90-kDa heat shock protein (HSP90) exists in cytosol in heteromeric complexes containing HSP70 and three other proteins with Mr of 63,000, 56,000, and 50,000. J Biol Chem 266(11):6708–6713 (Epub 1991/04/15)
34. Whitesell L, Mimnaugh EG, De Costa B, Myers CE, Neckers LM (1994) Inhibition of heat shock protein HSP90-pp60v-src heteroprotein complex formation by benzoquinone ansamycins: essential role for stress proteins in oncogenic transformation. Proc Nat Acad Sci U. S. A. 91(18):8324–8328 (Epub 1994/08/30)

35. Blagosklonny MV, Toretsky J, Neckers L (1995) Geldanamycin selectively destabilizes and conformationally alters mutated p53. Oncogene 11(5):933–939 (Epub 1995/09/07)
36. Grbovic OM, Basso AD, Sawai A, Ye Q, Friedlander P, Solit D et al (2006) V600E B-Raf requires the Hsp90 chaperone for stability and is degraded in response to Hsp90 inhibitors. Proc Nat Acad Sci U. S. A. 103(1):57–62 (Epub 2005/12/24)
37. Pratt WB, Galigniana MD, Morishima Y, Murphy PJ (2004) Role of molecular chaperones in steroid receptor action. Essays Biochem 40:41–58 (Epub 2004/07/10)
38. Pratt WB, Toft DO (1997) Steroid receptor interactions with heat shock protein and immunophilin chaperones. Endocrine Rev 18(3):306–360 (Epub 1997/06/01)
39. Echeverria PC, Mazaira G, Erlejman A, Gomez-Sanchez C, Piwien Pilipuk G, Galigniana MD (2009) Nuclear import of the glucocorticoid receptor-hsp90 complex through the nuclear pore complex is mediated by its interaction with Nup62 and importin beta. Mol Cell Biol 29(17):4788–4797 (Epub 2009/07/08)
40. Galigniana MD, Erlejman AG, Monte M, Gomez-Sanchez C, Piwien-Pilipuk G (2010) The hsp90-FKBP52 complex links the mineralocorticoid receptor to motor proteins and persists bound to the receptor in early nuclear events. Mol Cell Biol 30(5):1285–1298 (Epub 2009/12/30)
41. Conway-Campbell BL, George CL, Pooley JR, Knight DM, Norman MR, Hager GL et al (2011) The HSP90 molecular chaperone cycle regulates cyclical transcriptional dynamics of the glucocorticoid receptor and its coregulatory molecules CBP/p300 during ultradian ligand treatment. Mol Endocrinol 25(6):944–954 (Epub 2011/04/23)
42. Nagata Y, Anan T, Yoshida T, Mizukami T, Taya Y, Fujiwara T et al (1999) The stabilization mechanism of mutant-type p53 by impaired ubiquitination: the loss of wild-type p53 function and the hsp90 association. Oncogene 18(44):6037–6049 (Epub 1999/11/11)
43. Peng Y, Chen L, Li C, Lu W, Chen J (2001) Inhibition of MDM2 by hsp90 contributes to mutant p53 stabilization. J Biol Chem 276(44):40583–40590 (Epub 2001/08/17)
44. Lin K, Rockliffe N, Johnson GG, Sherrington PD, Pettitt AR (2008) Hsp90 inhibition has opposing effects on wild-type and mutant p53 and induces p21 expression and cytotoxicity irrespective of p53/ATM status in chronic lymphocytic leukaemia cells. Oncogene 27(17):2445–2455 (Epub 2007 Nov 5)
45. Maxwell PH, Wiesener MS, Chang GW, Clifford SC, Vaux EC, Cockman ME et al (1999) The tumour suppressor protein VHL targets hypoxia-inducible factors for oxygen-dependent proteolysis. Nature 399(6733):271–275 (Epub 1999/06/03)
46. Isaacs JS, Jung YJ, Mimnaugh EG, Martinez A, Cuttitta F, Neckers LM (2002) Hsp90 regulates a von Hippel Lindau-independent hypoxia-inducible factor-1 alpha-degradative pathway. J Biol Chem 277(33):29936–29944 (Epub 2002/06/08)
47. Cerchietti LC, Lopes EC, Yang SN, Hatzi K, Bunting KL, Tsikitas LA et al (2009) A purine scaffold Hsp90 inhibitor destabilizes BCL-6 and has specific antitumor activity in BCL-6-dependent B cell lymphomas. Nat Med 15(12):1369–1376 (Epub 2009/12/08)
48. Bird A (2001) Molecular biology: Methylation talk between histones and DNA. Science 294(5549):2113–2115 (Epub 2001/12/12)
49. Zhou Q, Agoston AT, Atadja P, Nelson WG, Davidson NE (2008) Inhibition of histone deacetylases promotes ubiquitin-dependent proteasomal degradation of DNA methyltransferase 1 in human breast cancer cells. Mol Cancer Res: MCR 6(5):873–883 (Epub 2008/05/29)
50. Fiskus W, Buckley K, Rao R, Mandawat A, Yang Y, Joshi R et al (2009) Panobinostat treatment depletes EZH2 and DNMT1 levels and enhances decitabine mediated de-repression of JunB and loss of survival of human acute leukemia cells. Cancer Biol Ther 8(10):939–950 (Epub 2009/03/13)
51. Chang CJ, Hung MC (2012) The role of EZH2 in tumour progression. Br J Cancer 106(2):243–247 (Epub 2011/12/22)
52. He Y, Korboukh I, Jin J, Huang J (2012) Targeting protein lysine methylation and demethylation in cancers. Biochim Biophys Sinica 44(1):70–79 (Epub 2011/12/24)
53. Ho L, Crabtree GR (2008) An EZ mark to miss. Cell Stem Cell 3(6):577–578 (Epub 2008/12/02)

54. Fiskus W, Buckley K, Rao R, Mandawat A, Yang Y, Joshi R et al (2009) Panobinostat treatment depletes EZH2 and DNMT1 levels and enhances decitabine mediated de-repression of JunB and loss of survival of human acute leukemia cells. Cancer Biol Ther 8(10):939–950 (Epub 2009/03/13)
55. Karkhanis V, Hu YJ, Baiocchi RA, Imbalzano AN, Sif S (2011) Versatility of PRMT5-induced methylation in growth control and development. Trends Biochem Sci 36(12): 633–641 (Epub 2011/10/07)
56. Maloney A, Clarke PA, Naaby-Hansen S, Stein R, Koopman JO, Akpan A et al (2007) Gene and protein expression profiling of human ovarian cancer cells treated with the heat shock protein 90 inhibitor 17-allylamino-17-demethoxygeldanamycin. Cancer Res 67(7): 3239–3253 (Epub 2007/04/06)
57. Imhof A (2003) Histone modifications: an assembly line for active chromatin? Curr Biol: CB 13(1):R22–R24 (Epub 2003/01/16)
58. Kim YR, Lee BK, Park RY, Nguyen NT, Bae JA, Kwon DD et al (2010) Differential CARM1 expression in prostate and colorectal cancers. BMC Cancer 10:197 (Epub 2010/05/14)
59. Abu-Farha M, Lanouette S, Elisma F, Tremblay V, Butson J, Figeys D et al (2011) Proteomic analyses of the SMYD family interactomes identify HSP90 as a novel target for SMYD2. J Mol cell Biol 3(5):301–308 (Epub 2011/10/27)
60. Luo XG, Zou JN, Wang SZ, Zhang TC, Xi T (2010) Novobiocin decreases SMYD3 expression and inhibits the migration of MDA-MB-231 human breast cancer cells. IUBMB Life 62(3):194–199 (Epub 2009/12/30)
61. Lord CJ, Ashworth A (2012) The DNA damage response and cancer therapy. Nature 481(7381):287–294 (Epub 2012/01/20)
62. Samant RS, Clarke PA, Workman P (2012) The expanding proteome of the molecular chaperone HSP90. Cell Cycle 11(7):1301–1308 (Epub 2012/03/17)
63. Noguchi M, Yu D, Hirayama R, Ninomiya Y, Sekine E, Kubota N et al (2006) Inhibition of homologous recombination repair in irradiated tumor cells pretreated with Hsp90 inhibitor 17-allylamino-17-demethoxygeldanamycin. Biochem Biophys Res Commun 351(3):658–663 (Epub 2006/11/07)
64. Ha K, Lee GE, Palii SS, Brown KD, Takeda Y, Liu K et al (2011) Rapid and transient recruitment of DNMT1 to DNA double-strand breaks is mediated by its interaction with multiple components of the DNA damage response machinery. Human Mol Genet 20(1):126–40 (Epub 2010/10/14)
65. Solier S, Kohn KW, Scroggins B, Xu W, Trepel J, Neckers L et al (2012) Feature Article: Heat shock protein 90alpha (HSP90alpha), a substrate and chaperone of DNA-PK necessary for the apoptotic response. Proc Nat Acad Sci U S A (Epub 2012/07/04)
66. Solier S, Pommier Y (2009) The apoptotic ring: a novel entity with phosphorylated histones H2AX and H2B and activated DNA damage response kinases. Cell Cycle 8(12):1853–1859 (Epub 2009/05/19)
67. Camphausen K, Tofilon PJ (2007) Inhibition of Hsp90: a multitarget approach to radiosensitization. Clin Cancer Res Official J Am Assoc Cancer Res 13(15 Pt 1):4326–4330 (Epub 2007/08/03)
68. Stingl L, Niewidok N, Muller N, Selle M, Djuzenova CS, Flentje M (2012) Radiosensitizing effect of the novel Hsp90 inhibitor NVP-AUY922 in human tumour cell lines silenced for Hsp90alpha. Strahlentherapie und Onkologie : Organ Der Deutschen Rontgengesellschaft 188(6):507–515 (Epub 2012/03/24)
69. Hartson SD, Matts RL (2012) Approaches for defining the Hsp90-dependent proteome. Biochim Biophys 1823(3):656–667 (Epub 2011/09/13)
70. Zhao R, Davey M, Hsu YC, Kaplanek P, Tong A, Parsons AB et al (2005) Navigating the chaperone network: an integrative map of physical and genetic interactions mediated by the hsp90 chaperone. Cell 120(5):715–727 (Epub 2005/03/16)
71. Sharma K, Vabulas RM, Macek B, Pinkert S, Cox J, Mann M et al (2012) Quantitative proteomics reveals that Hsp90 inhibition preferentially targets kinases and the DNA damage response. Mol Cell Proteomics: MCP 11(3):M111 014654 (Epub 2011/12/15)

72. Chiosis G, Timaul MN, Lucas B, Munster PN, Zheng FF, Sepp-Lorenzino L et al (2001) A small molecule designed to bind to the adenine nucleotide pocket of Hsp90 causes Her2 degradation and the growth arrest and differentiation of breast cancer cells. Chem Biol 8(3):289–99 (Epub 2001/04/18)
73. Trepel J, Mollapour M, Giaccone G, Neckers L (2010) Targeting the dynamic HSP90 complex in cancer. Nat Rev Cancer 10(8):537–549 (Epub 2010/07/24)

Chapter 8
Sphingolipid Metabolism and Signaling as a Target for Cancer Treatment

Vinodh Rajagopalan and Yusuf A. Hannun

Abstract Sphingolipids play key roles in the regulation of several biological processes that are integral to cancer pathogenesis. Among the sphingolipid metabolites, ceramide and sphingosine-1-phosphate (S1P) have been shown to modulate cancer development and progression. The biological roles of other metabolites, such as sphingosine, and ceramide 1-phosphate, are also beginning to emerge. In general, ceramide plays a role as a tumor-suppressing lipid-inducing anti-proliferative response such as cell cycle arrest, induction of apoptosis, and senescence whereas S1P plays a role as a tumor-promoting lipid-inducing transformation, cellular proliferation, and inflammation in various cell models. Glycosphingolipids, another emerging class of bioactive sphingolipids, are believed to play anti-apoptotic roles and offer drug resistance to currently used chemotherapeutic drugs. These emerging biological roles of sphingolipids and its potential usefulness in treating cancer in the form of anticancer therapeutics are discussed in this chapter.

8.1 Sphingolipid Metabolism

Sphingolipids are a class of lipids with a sphingosine back bone that are formed from non-sphingolipid precursors in the ER and get metabolized further within different sub-cellular compartments thereby giving rise to a plethora of metabolites. Of all these metabolites, ceramide is one of the most widely studied bioactive molecules. It is formed through three distinct pathways (Fig 8.1) (1) de novo synthesis—synthesis from non-sphingolipid precursors; (2) turnover pathways—break down products from complex sphingolipids; and (3) recycling and salvage pathways—The de novo pathway starts with the condensation of serine

V. Rajagopalan
Department of Biochemistry and Molecular Biology, Medical University of South Carolina, Charleston, SC, USA

Y. A. Hannun (✉)
Stony Brook University, Stony Brook, NY, USA
e-mail: Yusuf.Hannun@stonybrookmedicine.edu

D. E. Johnson (ed.), *Cell Death Signaling in Cancer Biology and Treatment*,
Cell Death in Biology and Diseases, DOI: 10.1007/978-1-4614-5847-0_8,

X.-M. Yin and Z. Dong (Series eds.), *Cell Death in Biology and Diseases*

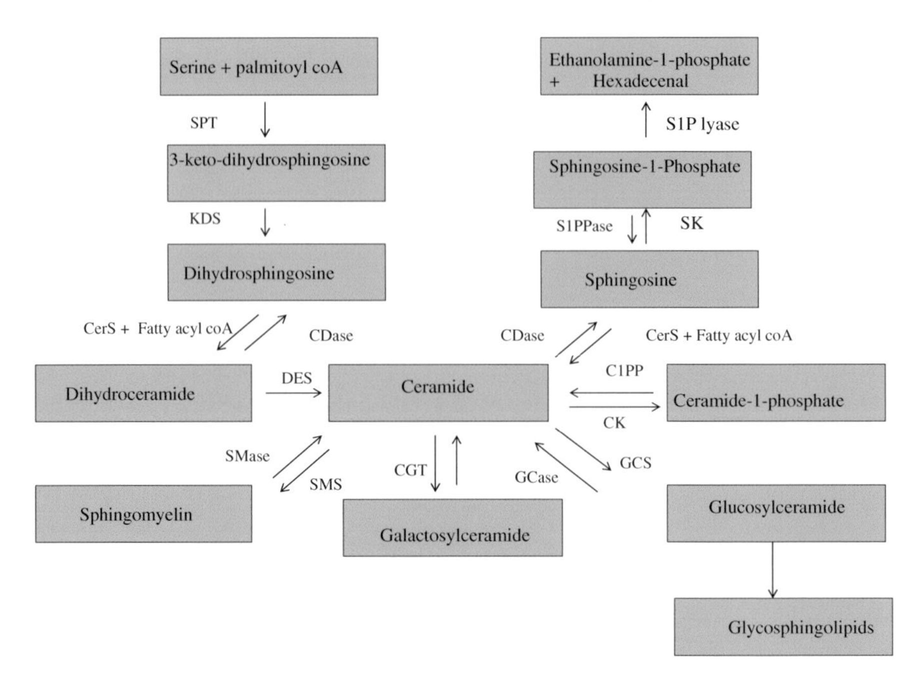

Fig. 8.1 Ceramide can be formed by de novo pathway from serine and palmitoyl-coA or from hydrolysis of sphingomyelin or cerebrosides (glucosyl or galactosyl ceramide). Ceramide, thus formed, can be phosphorylated by CK to yield ceramide-1-phosphate or serves as a substrate for the synthesis of sphingomyelin or glycosphingolipids. Ceramide can be deacylated by ceramidases to form sphingosine which can be phosphorylated by SKs to generate S1P which can be acted upon by phosphatases to generate sphingosine or by lyase to form ethanolamine-1-phosphate and hexadecanal, an aldehyde. Abbreviations: *SPT* serine palmitoyl transferase, *KDS* 3-keto-dihydrosphingosine reductase, *DES* dihydroceramide desaturase, *SPPase* Sph-1- phosphate phosphatase, *CK* cer kinase, *C1PP* C1P phosphatase, *SMS* SM synthase, *GCS* glucosylceramide synthase, *GCase* glucosyl CDase, *CGT* UDP-galactose ceramide-galactosyltransferase

and palmitoyl-CoA catalyzed by serine palmitoyl transferase (SPT) to generate 3-keto-dihydrosphingosine which is subsequently reduced to form dihydrosphingosine (sphinganine). Ceramide synthases (CerS) then act on dihydrosphingosine (or sphingosine) to form dihydroceramide (or ceramide) [1]. Dihydroceramide is subsequently desaturated by dihydroceramide desaturase (DES) which introduces a 4, 5-*trans*-double bond, thereby generating ceramide that occurs at the cytosolic face of the endoplasmic reticulum (ER) [2]. Ceramide, thus generated, can be used in biosynthetic reactions for the synthesis of sphingomyelin (SM), glucosylceramide (GluCer), galactosylceramide (GalCer) or ceramide 1-phosphate (C1P) by the attachment of head groups comprised of either phosphocholine, glucose, galactose or phosphate by sphingomyelin synthase (SMS), glucosyl ceramide synthase (GCS), UDP-galactose: ceramide-galactosyltransferase (CGT), or ceramide kinase, respectively (CK) [3, 4, 5]. In the turnover pathways of ceramide generation, sphingomyelinases act by cleaving sphingomyelin as a substrate [6] whereas

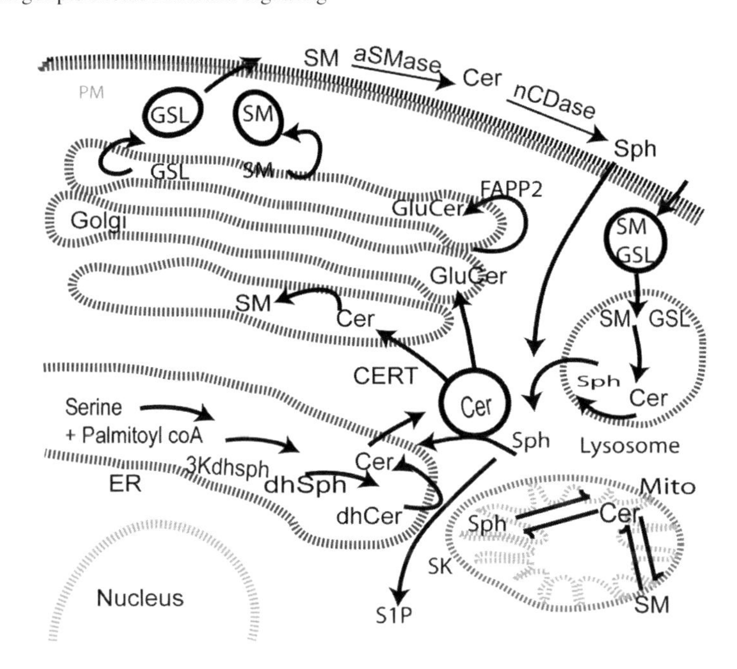

Fig. 8.2 Ceramide is generated in the ER by de novo pathway. It is transported to the Golgi membranes for SM synthesis in a CERT-dependent way or glucosylceramide (GlcCer) synthesis in FAPP2-dependent way. SM and complex GSLs are transported to the plasma membrane via vesicular trafficking where sphingomyelin gets metabolized by aSMase or neutral SMase to generate ceramide, or they are shuttled to lysosomes where aSMases and glucosidases metabolize SM and glucosylceramide, respectively, into ceramide. It is hydrolyzed by acid ceramidase to form sphingosine. Sphingosine may then traverse the lysosomal membrane to the cytosolic side where it might have two cellular fates. It can either be recycled into sphingolipid pathway in the ER or can be phosphorylated by SK1 or SK2. In mitochondria, ceramide is generated by the activation of n-SMase. Abbreviations: *3KdhSph* 3-keto-dihydrosphingosine, *dhSph* dihydrosphingosine, *CERT* ceramide transfer protein

hydrolases such as β-glucosidases and galactosidases act on glycosphingolipids such as glucosylceramide (GlcCer) and galactosylceramide as substrates to generate ceramide, respectively [7]. In the recycling and salvage pathways, ceramide generated in the lysosomes from the hydrolysis of complex sphingolipids is further broken down to sphingosine by the action of ceramidases [8] which is then re-acylated outside the lysosome by the action of ceramide synthases (CerS) to form ceramide. Since sphingosine derived from ceramide is salvaged to regenerate ceramide, it is referred to as the recycling and salvage pathway. Alternatively, sphingosine can be acted upon by sphingosine kinases (SK1 or SK2) [9] to form sphingosine-1 phosphate (S1P). S1P phosphatases can dephosphorylate S1P to regenerate sphingosine [10]. On the other hand, S1P lyase metabolizes S1P irreversibly to release ethanolamine phosphate and hexadecenal [11].

The sphingolipid enzymes discussed above are distributed in different intracellular locations. De novo ceramide synthesis takes place on the cytosolic surface of the ER and its associated membranes. Ceramide formed in the ER is transported through ceramide transfer protein (CERT) to the trans-Golgi wherein it serves as a substrate for sphingomyelin synthase for formation of SM [12], or through vesicular transport to the Golgi wherein it serves as substrate for GCS for the formation of glucosylceramide. The transport protein, four-phosphate-adaptor protein 2 (FAPP2) delivers glucosylceramide to appropriate sites in the Golgi for synthesis of more complex glycosphingolipids (GSL) [13]. Subsequently, SM and complex GSLs are transported to the plasma membrane via vesicular trafficking. In the plasma membrane, SM can be hydrolyzed to ceramide and metabolized further by acid sphingomyelinase (aSMase) possibly acting on the outer leaflet, or by neutral sphingomelinase (nSMase) residing on the inner leaflet of the plasma membrane [14, 15].

From the plasma membrane, sphingolipids are recycled through the endosomal pathway. In lysosomes, aSMases and glucosidases metabolize complex sphingolipids (SM and glucosylceramide, respectively) into ceramide which is hydrolyzed by acid ceramidase to form sphingosine. Sphingosine may then traverse the lysosomal membrane to the cytosolic side where it has two cellular fates. Cytosolic sphingosine is either recycled into the sphingolipid pathway in the ER or phosphorylated by SK1 or SK2 [9] (refer to Fig 8.2 [16]).

8.2 Biological Targets of Ceramide

Ceramide regulates many biological processes such as cancer cell growth, differentiation, apoptosis, and senescence [17, 18]. Many signals such as cytokines, anticancer drugs, and stress-inducers upregulate ceramide through the de novo or salvage pathways [19, 20]. Ceramide triggers signaling cascades by regulating phosphatases, cathepsin D, or kinase suppressor of RAS (KSR) as described below.

Phosphatases such as protein phosphatase 1 (PP1) and protein phosphatase 2A (PP2A) are activated by ceramide in vitro. Inhibition of these phosphatases inhibits the ability of ceramide to dephosphorylate (inactivate) several pro-proliferative proteins such as PKCα, Akt/PKB, c-Jun, Bcl-2, Rb, and SR [21, 22].

Many studies have documented the translocation of cathepsin D from lysosomes in response to oxidative stress followed by activation of caspase 3 and cell death [23]. Interestingly, cathepsin D was found to be a ceramide-binding and ceramide-activated protein [24]. Besides, acid sphingomyelinase-derived ceramide has been shown to favor autocatalytic proteolysis of inactive cathepsin D to enzymatically active cathepsin D isoform [25].

Similarly, KSR has been found to be ceramide responsive [26–30]. Mammalian KSR activates Raf, and activation of this pathway results in apoptosis [31, 32]. Also, ceramide has been shown to activate the zeta isoform of protein kinase C (PKCz) by phosphorylation [33]. Ceramide-activation of PKC-zeta has been

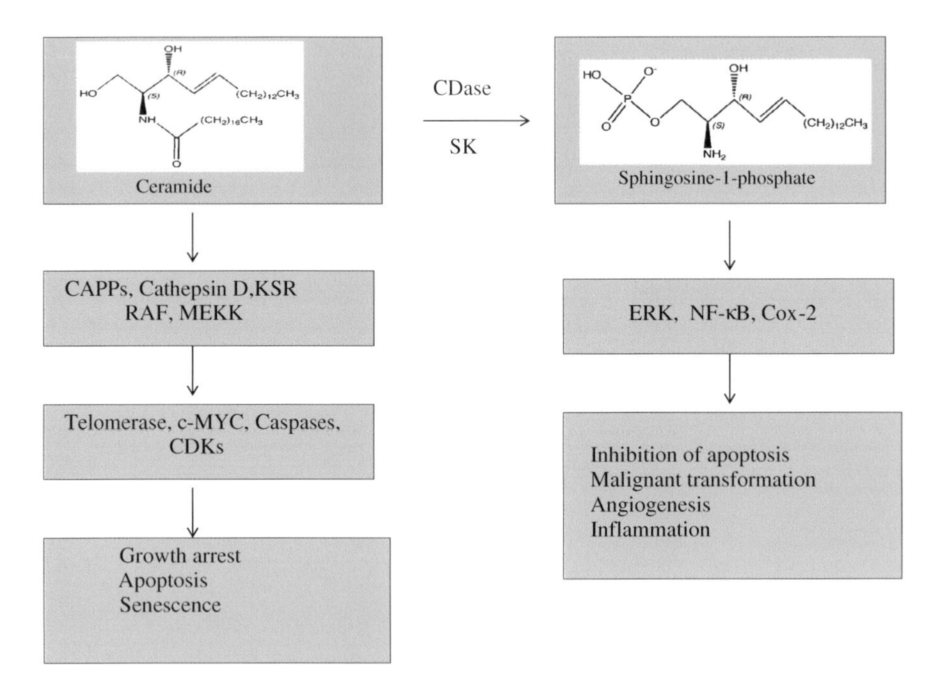

Fig. 8.3 Ceramide regulates many protein signaling molecules such as cathepsin D and ceramide-activated protein phosphatases (*CAPPs*), KSR, RAF, MEKK. Proteins that are modulated by these pathways include RB,SR, AKT, PKC-α, c-JUN, Bcl-2,telomerase, c-MYC, caspases, and cyclin-dependent kinases (*CDKs*) which in turn brings about cellular responses such as growth arrest, apoptosis, and/or senescence. Ceramide that gets metabolised to S1P by ceramidase and SK, regulates proteins such as ERK, NF-κB, Cox-2 which in turn brings about cellular responses such as inhibition of apoptosis, malignant transformation, angiogenesis, and inflammation

shown to be necessary for inactivation of Akt-dependent mitogenesis in vascular smooth muscle cells [34] (refer to Fig. 8.3).

In contrast to ceramide, S1P has emerged as a potential regulator of biological processes such as proliferation, inflammation, vasculogenesis, and tumor promotion [35]. S1P is a soluble molecule, secreted outside the cell which acts in autocrine and paracrine manner to bring about receptor-mediated cellular functions like cell motility and proliferation [35]. S1P acts directly on members of the S1P1-5 receptor family, which are G protein-coupled receptors. Non-receptor-mediated functions of S1P such as activation of TNF receptor–associated factor 2 (TRAF2) in the TNF pathway leading to NF-kB activation have been reported [36]. It has also been shown that interaction of TRAF2 with SK and activation of SK is critical for prevention of TNFα-mediated apoptosis [36] (refer to Fig. 8.3). Thus, these findings underscore the significance of uncovering bioactive sphingolipid-mediated cellular pathways that would help to conceive novel therapeutic strategies for many pathological conditions, importantly cancer.

8.3 Ceramide as a Tumor Suppressor

The involvement of sphingolipids in cancer pathogenesis is brought to light from the observation that the levels of sphingolipid metabolizing enzymes are altered in many tumors. For example, in a study, SK expression has been found to be upregulated in many human tumors originating from tissues such as breast, colon, lung, stomach, uterus, etc., [37]. In another study done by Riboni et al., an inverse correlation between the levels of ceramide and tumor malignancy in glial tumors (astrocytomas) was observed [38]. In general, exogenous ceramide-induced anti-proliferative responses like apoptosis, differentiation and growth inhibition, senescence, and autophagy. Therefore, it is considered as a tumor suppressor lipid. These anti-proliferative roles of ceramide are discussed in detail in the following section.

8.3.1 Ceramide and Apoptosis

Ceramide has been implicated in many cell death paradigms. Birbes et al. [39] showed that selective targeting of bacterial sphingomyelinase (bSMase) to mitochondria and not to any other compartments such as plasma membrane, ER, or Golgi resulted in apoptosis that was associated with generation of ceramide and release of cytochrome c in MCF-7 cells. Overexpression of Bcl-2 prevented the mitochondria-targeted bSMase effects on apoptosis. Dai et al. [40] showed that UV-induced apoptosis was marked by increase in SM in all sub-cellular locations, particularly mitochondria, in HeLa cells. Ceramide levels were found to be elevated in mitochondria at 2–6 h, consistent with the cell death time course. D609, an inhibitor of sphingomyelin synthase to a marked extent and fumonisin B1 (FB1), a ceramide synthase inhibitor to a lesser extent, rescued the cells from increases in SM and ceramide, and consequently cell death. On the other hand, the SPT inhibitor myriocin did not rescue the UV effects on cell death, suggesting the involvement of the turnover pathway-generated ceramide in bringing about UV-triggered cell death.

In another study in *Caenorhabditis elegans*, upon inactivation of ceramide synthase, somatic apoptosis was unaffected, but ionizing radiation-induced apoptosis of germ cells was obliterated, and this phenotype was reversed by microinjection of long-chain natural ceramide. Radiation-induced ceramide accumulation in mitochondria consequently activated CED-3 caspase and apoptosis [41].

In Ramos B cells, surface B-cell receptor (BcR)-triggered cell death was marked by an early increase in C16 ceramide. Pulse labeling with sphinoglipid precursor, palmitate, in the presence of ceramide synthase inhibitor, FB1, demonstrated that the de novo ceramide-generating pathway was activated following BcR activation. The apoptotic cell death induced by cross-linking of BcR was mediated through mitochondrial cell death pathways followed by caspase activation [42]. In LNCaP prostate cancer cells, androgen ablation, which is considered as one of the therapeutic modalities, was found to increase C16 ceramide level followed by G0/G1 cell cycle arrest and apoptosis. 5alpha-dihydrotestosterone (DHT) or fumonisin B1 treatment rescued LNCaP cells from apoptosis [43].

In another study, ceramide acting via PP1, dephosphorylated SR proteins that regulated the alternate splicing of Bcl-x(L) and caspase 9. In A549 lung adenocarcinoma cell lines, cell-permeable D-e-C(6) ceramide downregulated the mRNA levels of anti-apoptotic Bcl-x(L) and caspase 9b with concomitant increase in the mRNA of pro-apoptotic Bcl-x(s) and caspase 9. The chemotherapeutic agent, gemcitabine, induced de novo generation of ceramide and brought about aforementioned alternate splicing of Bcl-x(L) and caspase 9b and consequent loss of cell viability as measured by MTT assay [44].

In several studies, aSMase was shown to be necessary for radiation-induced apoptosis in endothelial cells and mice lacking aSMase were protected from gastrointestinal and CNS apoptosis [45–47]. In another study, the endolysosomal aspartate protease cathepsin D (CTSD) was identified as a target of ceramide generated by acid sphingomyelinase in response to TNFα. CTSD cleaved pro-apoptotic Bid and activated it in vitro. The lack of Bid activation in cathepsin-deficient fibroblasts suggested Bid is downstream of cathepsin D in bringing about apoptosis as a result of TNFα treatment [48].

In addition to aSMase, neutral sphingomyelinase has been implicated in stress response pathways initiated by TNFα in MCF-7 cells [49]; amyloid-β peptide in neuronal cells [50]; ethanol in HepG2 hepatoma cells [51]; and staurosporine in several neuronal cell lines [52].

In addition to the sphingomyelinases, ceramidases (CDases) were also found to regulate apoptosis. In one study, nitric oxide induced the degradation of nCDase, thereby, enabling ceramide accumulation and cell death [53]. In another study, the degeneration of photoreceptor cells was marked by an increase in ceramide which was rescued by overexpression of CDase that cleared the ceramide and prevented its apoptotic effect [54]. These studies clearly implicate ceramide in apoptosis and in mediating the cellular response to various stress causing stimuli.

8.3.2 Ceramide in Senescence

Ceramide has been implicated in cellular senescence. Their relationship stems from the observation that in WI-38 human diploid fibroblasts (HDF), there was increased neutral sphingomyelinase activity with generation of ceramide when cells entered senescence. These changes were not seen when cells entered quiescence achieved with serum withdrawal or contact inhibition [55]. Exogenous administration of ceramide (15uM) onto young WI-38 cells induced retinoblastoma protein dephosphorylation and inhibited serum-induced AP-1 activation, DNA synthesis, and mitosis, thereby inducing a senescence phenotype [55]. Involvement of ceramide in replicative senescence has been shown in human umbilical vein endothelial cells (HUVEC) as well [56]. In another study, gemcitabine induced senescence in pancreatic cancer cells, and sphingomyelin treatment enhanced chemosensitivity to the drug by reducing the induction of senescence and redirected the cells to enter apoptosis. The authors concluded that ceramide inhibited cell cycle progression at low levels, induced senescence at moderate levels, and apoptosis at high levels [57]. Besides these studies, the yeast aging genes lac 1 and lag1 were subsequently identified as ceramide synthases, thereby providing a genetic link between ceramide to senescence and aging [58].

Ceramide also regulates senescence by inhibiting telomerase which is the enzyme that prevents the shortening of the telomeres, the long tandem repeats of G-rich sequences (5′-TTAGGG-3′) found at the ends of chromosomes. Telomerase is found to be frequently activated in many immortal cells in culture representing different tissues and malignant tumors, suggesting its role in cellular immortalization and tumorigenesis [59, 60]. In the A549 lung carcinoma cell line, daunorubicin treatment or sphingomyelinase overexpression increased ceramide generation followed by inhibition of telomerase activity. Clearance of ceramide by overexpression of GCS prevented the telomerase inhibition [61]. These studies suggest ceramide is an upstream regulator of senescence and aging.

8.3.3 Ceramide in Cell Differentiation and Growth Inhibition

Historically, the role of ceramide in cellular differentiation was discovered with the observation that vitamin D3-induced monocytic, but not neutrophilic-type cell differentiation in HL-60, and U037 leukemia cells was accompanied by increase in nSMase activity and a concomitant spike in ceramide levels [62]. In turn, exogenous ceramide was found to induce monocytic differentiation of these cells. In neuronal cell lines, ceramide-induced differentiation in T9 glioma cells, purkinje and hippocampal neurons [63].

In another study, incubation of exponentially growing *Saccharomyces cerevisiae* with short-chain ceramide inhibited cell growth with the involvement of an okadaic acid-sensitive protein phosphatase [64]. Another study uncovered the mechanism of ceramide-induced growth suppression in that serum withdrawal in MOLT-4 cells resulted in significant dephosphorylation of Rb, correlating with the induction of G0/G1 cell cycle arrest [65]. Taken together, these studies implicate a role for ceramide in cell differentiation and cell cycle progression.

8.3.4 Ceramide and Autophagy

In mammalian cells, ceramide and/or dihydroceramide have been shown to induce autophagy. For example, in glioma cells, ceramide has been shown to activate the transcription of death-inducing mitochondrial protein, BNIP3, and subsequent autophagy [66]. In the human colon cancer HT-29 cells, C2 ceramide inhibited activation of protein kinase B, which is a negative regulator of interleukin 13-dependent macroautophic inhibition. In MCF-7 breast cancer cells, ceramide stimulated the expression of Beclin-1 which is an autophagy gene product. This study also showed that tamoxifen-induced autophagy was blocked using the SPT inhibitor myriocin (ISP1) [67]. In another study, ceramide was shown to induce autophagy by regulating calpain in MEFs [68]. In DU145 cells, fenretinide (4HPR) treatment favored autophagic induction possibly due to the increase in endogenous dihydroceramide [69].

In a study by Signorelli et al. [70], resveratrol induced autophagy in HGC-27 cells with an increase in dihydroceramides possibly by inhibition of dihydroceramide desaturase. Inhibitors of dihydroceramide desaturase mimicked the autophagic induction induced by resveratrol.

Mechanistically, Beclin-1 has been shown to be physiologically associated with the mammalian class III phosphatidylinositol 3-kinase (PI 3-kinase) Vps34, and the knockdown of Beclin-1 blunted the autophagic response of the cells to nutrient deprivation or C_2-ceramide treatment [71]. Class I PI3K and AKT pathway are known to suppress autophagy and ceramide has been shown to inhibit AKT by activation of PP2A, thereby establishing a possible mechanistic link between ceramide and autophagy induction [67, 72, 73].

Based on the above studies showing ceramide effects on mammalian autophagic regulation, combined with the observation that yeast subjected to heat stress exhibits growth suppression accompanied by upregulation of ceramide synthesis and downregulation of nutrient transporters on their cell surface [74, 75], Edinger and colleagues hypothesized that ceramide-induced mammalian autophagy might be mediated through a yeast-like response to heat stress by downregulation of nutrient transporters. They showed that C2 ceramide produced a profound downregulation of nutrient transporter proteins in mammalian cells. Inhibition of autophagy or acute limitation of extracellular nutrients increased the sensitivity of cells to ceramide. Supplementation of cells with the cell-permeable nutrient methyl pyruvate protected the cells from ceramide-induced cell death and delayed autophagic induction. So the authors concluded that ceramide killed cells (apoptosis) by provoking nutrient limitation via downregulation of nutrient transporters and subsequent autophagy [76]. Taken together, these studies implicate ceramide in the regulation of autophagic response.

8.4 S1P/S1PR as Tumor Promoters

S1P is considered a pro-survival lipid. For example, S1P stimulated the invasiveness of glioblastoma tumor cells [77], promoted estrogen-dependent tumorigenesis of breast cancer cells [78] and conferred resistance to the cytotoxic actions of TNF-α and daunorubicin [10]. A number of studies documented the role of S1P/S1PR in proliferation, inhibition of apoptosis, vasculogenesis/angiogenesis, and inflammation. These topics will be discussed in the following sections.

8.4.1 S1P in Proliferation and Inhibition of Apoptosis

Sphingosine kinase (SK) phosphorylates sphingosine to form S1P. Overexpression of the SK1 isoform induced oncogeneic transformation in NOD/SCID mice. Using inhibitors of SK, investigators implicated SK in the involvement of oncogenic H-Ras-mediated transformation [79]. In another study, addition of exogenous S1P reversed the cell death induced by ceramide [80]. Mechanistically, S1P counteracted ceramide-induced activation of stress-activated protein kinase (SAPK/JNK) and activated the extracellular signal-regulated kinase (ERK) pathway in governing the fate of the cell [80].

Neutralizing antibody to S1P substantially reduced tumor progression in murine xenograft and allograft models. The antibody arrested tumor-associated angiogenesis, neutralized S1P-induced proliferation, attenuated release of pro-angiogenic cytokines, and blocked the ability of S1P to protect tumor cells from apoptosis [81].

In yet another study, S1P-mediated inhibition of apoptosis in C3H10T 1/2 fibroblasts depended on ERK activation and MKP-1, which downregulated SAPK/JNK to bring about inhibition of apoptosis [82]. In male germ cells, S1P inhibited stress-induced cell death, possibly by inhibiting nuclear factor kappa B (NF-kappa B) and AKT phosphorylation [83].

8.4.2 S1P and Vasculogenesis/Angiogenesis

S1P promotes vasculogenesis and angiogenesis. S1P, the natural ligand for S1P3 receptor or KRX-725, a synthetic peptide that mimics S1P action on this receptor, favored angiogenesis, as demonstrated by assessment of vascular sprouting using aortic rings as an ex vivo model of angiogenesis. When S1P or KRX-725 were combined with other growth factors such as basic fibroblast growth factor (b-FGF), stem cell factor, or vascular endothelial growth factor (VEGF), the investigators observed synergistic induction of angiogenesis [84]. In a cultured mouse allantois explant model of blood vessel formation, Argraves et al. [85] showed that S1P, synthesized via the action of SK2, promoted vasculogenesis by promoting migratory activities of angioblasts and early endothelial cells to expand the vascular network.

VEGF has been shown to stimulate SK1 activity with an increase in the production of S1P and activation of H and N Ras oncogenes in T24 bladder tumor cell lines [86]. Endothelial cells undergo morphogenesis into capillary networks in response to S1P involving G protein receptors [87]. S1P has been shown to induce endothelial cell invasion and morphogenesis in physiologically relevant collagen and fibrin matrices [88]. Based on studies employing inhibitors and functional antagonists of S1P receptors, it has been hypothesized that the angiogenic function of S1P is mediated by S1P1 and S1P3 signaling [87, 89, 90].

Knockout of the S1P1 receptor resulted in vascular deficiencies in mice [91]. In a recent study, SphK1–SphK2 double-knockout mice manifested defective neural and vascular systems and exhibited embryonic lethality. The authors inferred that S1P was required for "functionally intertwined pathways of angiogenesis and neurogenesis" [92].

8.4.3 S1P in Inflammation

The SK1/S1P pathway has been implicated in inflammation. For instance, TNF-alpha resulted in activation of SK1/S1P pathway specifically leading to extracellular signal-regulated kinases and NF-kappa B activation [36] and consequently expression of vascular cell adhesion molecule (VCAM) and intercellular adhesion molecule (ICAM) [93]. S1P also induced cyclooxygenase 2 (COX2) and prostaglandin E2 (PGE_2) production in L929 fibrosarcoma and A549 lung adenocarcinoma cells and genetic knockdown using siRNA blocked their production. Additionally, preventing S1P clearance using siRNAs against S1P lyase/phosphatase resulted in increased production of COX2 and PGE_2, implicating a key role of S1P in this pathway [94]. Microglial activation has been implicated in neuroinflammation. LPS treatment increased SK1 mRNA and protein levels and consequently upregulated expression of proinflammatory cytokines such as TNF-alpha, IL-1beta, and iNOS in microglia. Chronic production of inflammatory cytokines by microglia has been implicated in neuroinflammation [95, 96]. Further, the SK/S1P pathway has been implicated in many other inflammatory disease paradigms such as asthma, rheumatoid arthritis, and inflammatory bowel diseases [97].

In summary, these data collectively demonstrate that S1P regulates cancer cell viability, angiogenesis, and inflammation which favor cancer pathogenesis.

8.5 Role of Other Metabolites of Sphingolipid Metabolism in Tumorigenesis

In addition to ceramide and S1P, ceramide -1 phosphate has been implicated in cell survival pathways. Chemotherapeutic agents changed the alternative splicing of caspase 9 and Bcl-x pre-mRNA into pro-apoptotic forms which was mediated by ceramide-dependent activation of PP1 [98]. C1P has been found to inhibit PP1, and therefore, it might antagonize ceramide-mediated apoptosis, functioning as a pro-survival lipid. In another study, C1P blocked apoptosis in part by activating PI3-K/PKB/NF-κB pathways and production of anti-apoptotic Bcl-X_L [99]. CERK might play an important role in regulating the balance between ceramide and C1P and therefore cell death and cell survival similar to S1P.

Sphingosine which is the breakdown product of ceramide is also implicated in apoptotic responses. In one study, gamma irradiation along with TNF-α induced sphingosine and S1P levels. The elevation of sphingosine by exogenous administration of sphingosine or by treatment with SK inhibitor induced apoptosis in LNCaP prostate cancer cell lines [100]. In another study, phorbol myristate acetate (PMA) and tumor necrosis factor (TNF) separately induced apoptosis marked by elevation of sphingosine in HL-60 cells and neutrophils, respectively. Exogenous administration of sphingosine or its methylated derivative N, N,-dimethylsphingosine (DMS) also induced apoptosis in cells of both hematopoietic and carcinoma origin [101]. In an attempt to find the mechanism of sphingosine-induced apoptosis, Domae and colleagues found that sphingosine induced c-Jun expression and apoptosis in HL-60 cells and inhibition of protein kinase A (PKA) potentiated this effect [102]. In another study by Houghton and co-workers, rhabdomyosarcoma cell lines were found to be more sensitive to the induction of apoptosis with an increase in the cellular levels of sphingosine. Mechanistically, sphingosine-mediated cell death involved mitochondrial events such as Bax activation and translocation to the mitochondria, release of cytochrome c and Smac/Diablo, but not apoptosis-inducing factor (AIF), endonuclease G, and HtrA2/Omi, from mitochondria and finally activation of caspase-3 and caspase-9 [103].

Gangliosphingolipids have been implicated in the epithelial-to-mesenchymal cell transition (EMT) which is believed to play a role in cancer progression. Pharmacological inhibition of GlcCer synthase has been shown to result in downregulation of E-cadherin, a major epithelial marker, and upregulation of vimentin and N-cadherin, major mesenchymal cell markers, with marked changes in gangliosphingolipids (Gg4 or GM2) and increased motility, implying these specific glycosphingoplipids (Gg4 or GM2) might play a role in inhibition of EMT [104]. Some glycosphingolipids have been recognized as tumor antigens and thus could participate in tumor cell regulation and as immune targets. In many studies, it was shown that GM3 modulated receptor tyrosine kinase activity in cells [105–107]. In a recent study, ganglioside GM3 inhibited autophosphorylation of the EGFR kinase domain, thereby inactivating it in response to ligand binding, and removal of neuraminic acid of the GM3 headgroup or expression of the K642G mutant released this inhibitory effect [108].

8.6 Sphingolipids in Cancer Therapy

Among the bioactive sphingolipids, ceramide and S1P act as pro-apoptotic and anti-apoptotic lipids, respectively, and therefore, modulation of these lipids may be effective as a treatment strategy for cancer (refer to Fig. 8.4 [109]). Such strategies to increase the accumulation of ceramide and attenuation of S1P are discussed in detail in the following section.

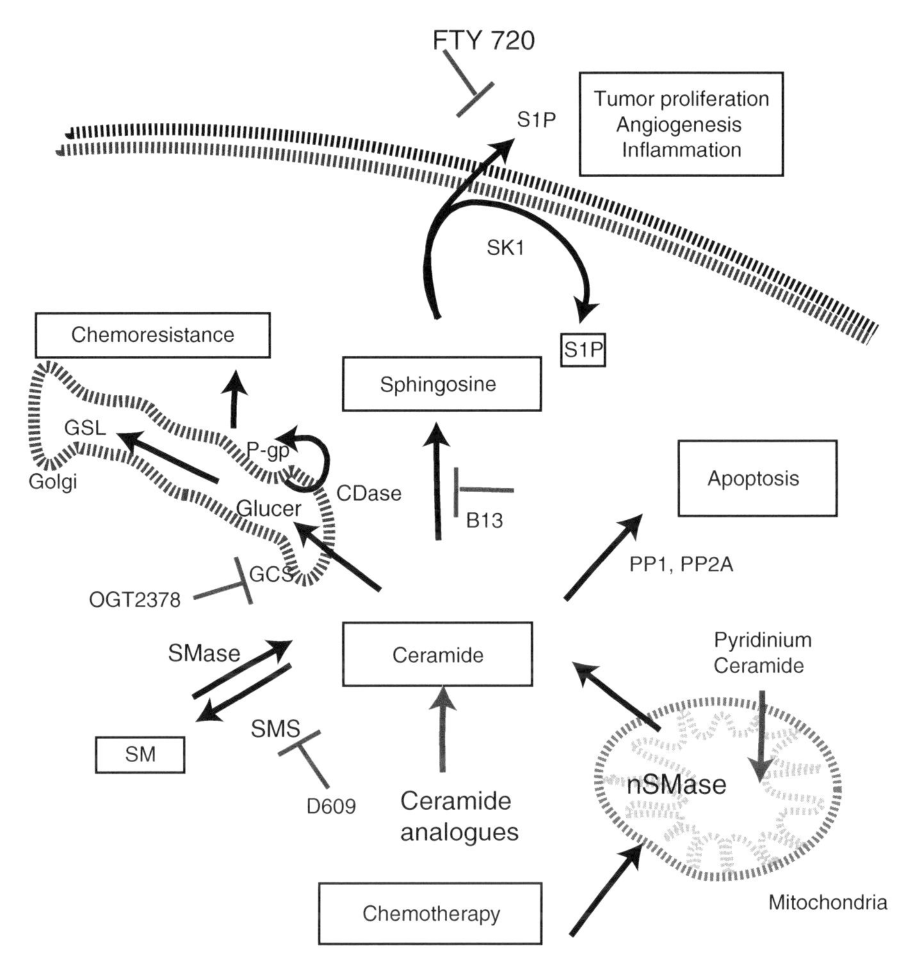

Fig. 8.4 Chemotherapeutic agents increases ceramide levels and induces apoptosis through the de novo pathway or through the neutral sphingomyelinase (*N-SMase*) pathway. Induction of SK1 in colon cancer leads to accumulation of S1P, possibly leading to tumor proliferation, angiogenesis, and inflammation. Clearance of ceramide to Glucosylceramide by GCS in breast cancer cells leads to the development of drug resistance. P-glycoprotein (*P-gp*) expression might potentiate the chemo resistant phenotype. Marked in red arrows are modulators of SPL metabolism. B13 is an inhibitor of acid CDase, OGT2378 is GCS inhibitor, D609 is an inhibitor of SMS, FTY720 is a sphingosine analogue, and ceramide analogues mimic endogenous ceramides, and Pyridinium ceramide targets mitochondria and promotes mitochondria mediated apoptosis. Abbreviations: *CDase* ceramidase, *ER* endoplasmic reticulum, *PP1* protein phosphatase 1, *PP2A* protein phosphatase 2A, *GSL* glycosphingolipid, *GCS* glucosylceramide synthase

8.6.1 Targeting Ceramide Generation

Cytotoxic chemotherapeutic agents such as daunorubicin, etoposide, camptothecin, fludarabine, and gemcitabine were shown to induce de novo ceramide generation, and inhibition of this pathway reduced the cytotoxic responses to these drugs meaning

Table 8.1 Sphingolipid analogues and inhibitors of ceramide metabolism

Compounds	Mode of action	Cancer types
B13	Acid ceramidase inhibitor	Prostate and colon
D609	Sphingomyelin synthase inhibitor	Monocytic leukemia
C16 Serinol, 4,6-diene-ceramide, 5R-OH-3E-C_8-ceramide, adamantyl-ceramide, and benzene-C_4-ceramide	Ceramide analogue	Neuroblastoma and breast
Pyridinium ceramide	Ceramide analogue	Head and neck squamous cell carcinoma, breast and colon
Pegylated liposomes with ceramide	Improved delivery	Breast
Vincristine in SM-liposomes	Improved delivery	Acute lymphoid leukemia
FTY720	Sphingosine analogue	Lymphoma, bladder, glioma, prostate
OGT2378	GCS inhibitor	Melanoma

that these drugs manifest their cytotoxic effects partly through ceramide production [110–112]. Besides the de novo pathway, certain drugs, including daunorubicin, also induce ceramide generation through the activation of nSMase which hydrolyzes sphingomyelin to generate ceramide. For instance, in leukemia cells, cytosine arabinoside (Ara-C) induced activation of nSMase [113]. Mechanistically, this induction of nSMase was brought about by generation of reactive oxygen species followed by Jun N-terminal kinase phosphorylation and apoptosis [114]. In another study, actinomycin D and etoposide induced nSMase activity in a p53- and ROS-dependent manner [115].

Alternatively, studies on reagents that inhibit the enzymes that favor the ceramide clearance pathway, leading to accumulation of ceramide and, thereby, potentiating the cytotoxic effects were tested. For instance, compounds such as B13 that inhibited acid CDase or tricyclodecan-9-yl-xanthogenate (D609) that inhibited SMS, induced apoptosis in colon cancer and in U937 human monocytic leukemia cells, respectively [116, 117] (refer to Fig. 8.4 and Table 8.1).

Interestingly, SM was found to potentiate the chemotherapeutic response of gemicitabine in prostate cancer cell lines [118]. In a study involving combination of SM with chemotherapeutic agents, doxorubicin, epirubicin, or topotecan, it was found that the combination therapy increased the cytotoxic effect of the drugs by increasing their bioavailability, possibly by modulation of plasma membrane lipophilicity, facilitating entry of these agents into the various cancer cell lines studied [119].

8.6.2 Mimicking Ceramide Action (Analogues of Ceramide)

Ceramide analogues (such as the soluble short-chain C_2- and C_6-ceramides) have been shown to bring about cell death in many types of cancer cell lines tested [120]. C_{16}-serinol, 4, 6-diene-ceramide, 5R-OH-3E-C_8-ceramide, adamantyl-ceramide, and benzene-C_4-ceramide (Table 8.1) are some of the ceramide analogues that induced cell death in cell lines such as neuroblastoma and breast cancer [121–124]. A novel, cationic, water soluble, pyridinium ceramide (Table 8.1) accumulated predominantly in cellular compartments that are negatively charged such as mitochondria and the nucleus and caused changes in mitochondrial structure and function and inhibited growth in various human head and neck cancer cell lines [125], while inducing apoptosis in squamous cell carcinoma (HNSCC) cell lines [126–128].

Experiments to uncover the most efficient means of delivery of these ceramide analogues have been tried extensively. Pegylated liposomes were very effective in bringing about ceramide-mediated cell death in breast cancer cell lines (Table 8.1). Liposomal delivery of ceramide decreased phosphorylated AKT and activation of caspase-3/7 more effectively than non-liposomal ceramide [129]. Vincristine incorporated in SM-liposomes called sphingosomes (Table 8.1) was found to be effective in animal models for treatment of acute lymphocytic leukemia (ALL) to the extent that it is currently in Phase II clinical trials [130]. These studies clearly demonstrate that targeting ceramide generation might be an effective method to bring about cancer cell death.

8.6.3 Attenuation of the S1P Pathway

Since S1P is found to be involved in angiogenesis and proliferation, it is intuitive to think that modulation of this pathway offers hope for the treatment of cancer. In fact, inhibition of SK1 resulted in increased cell death in many forms of cancer [131, 132] and increased the sensitivity to cell death stimuli such as TNFα and FAS ligand [133]. Dihydroxyaurone, an SK1 inhibitor, exhibits anti-tumor activity in mammary tumors [37]. Interestingly, in normal tissues of ovary and testis, exogenous S1P treatment seems to protect cells from chemotherapy-induced apoptosis [134, 135].

FTY720 (Table 8.1), an analogue of sphingosine, has been found to be phosphorylated in vivo, and the resulting FTY720 phosphate functioned as a ligand for sphingosine-1-phosphate receptors. This signaling mechanism enabled sequestration of lymphocytes in lymphoid tissues thereby causing immunosuppression [136–138]. This compound also induced apoptosis in various cancer cell lines such as lymphocyte and bladder cancer (T24, UMUC3 and HT1197), glioma (T98G), and prostate (DU145) cancer [37, 139–142]. Therefore, it is tempting to hypothesize that S1PR1 and S1PR3 antagonists might have similar anticancer effects. In another study, anti-S1P mAb greatly reduced tumor progression in murine xenograft and allograft models by inhibiting capillary formation and angiogenesis [81]. These data strongly support the candidacy of S1P as a potential therapeutic target for the treatment of cancer.

8.6.4 Dietary SM as a Cancer Therapeutic

Brasitus and co-workers demonstrated that there was a significant increase in SM levels and activity of SMS in rat colonic mucosa in response to 1, 2-dimethylhydrazine

(DMH), a chemical colonic carcinogen [143]. Merrill and co-workers found that milk sphingomyelin dietary supplementation reduced the incidence of DMH-induced pre-malignant lesions of colon tumors in CF1 mice [144]. In addition, mice fed with SM developed fewer adenocarcinomas. These findings suggest that milk SM might suppress advanced malignant tumors in colon [145]. Administration of synthetic SM and ceramide analogues also suppressed colonic crypt foci formation [146, 147]. In another study, azoxymethane (AOM)/dextran sulfate sodium (DSS)-induced colon carcinogenesis was modulated by dietary SM in the early stages by activation of peroxisome proliferator-activated receptor γ (PPAR-γ), but its anti-carcinogenic effect was independent of PPAR-γ [148]. Therefore, dietary SM might modulate the proteins expressed during early stages of colon carcinogenesis and therefore may be a potential therapeutic candidate in the context of colon carcinogenesis.

8.7 Sphingolipids in Drug Resistance

One of the reasons for the failure of chemotherapy in cancer is the development of tumor cell resistance. Part of the basis for chemoresistance might be attributed to a re-wiring of sphingolipid metabolism. For instance, in many cases of leukemia, breast cancer, and melanoma, chemotherapeutic agents increase the activity of GCS which thereby attenuates ceramide levels, resulting in a drug resistance phenotype [20, 149]. Overexpression of GCS offered increased resistance to doxorubicin whereas siRNA knockdown promoted increased sensitivity to doxorubicin, paclitaxel, and etoposide in breast cancer cells [150–152]. Mechanistically, GCS upregulated P-glycoprotein (P-gp) which is an ABC transporter implicated in drug resistance. Knockdown of GCS inhibited MDR1, a gene that encodes P-gp, reversing drug resistance [153, 154].

Based on the above studies, GCS inhibition has been predicted to improve the effectiveness of chemotherapeutic drugs. Some studies suggest that this hypothesis is in fact true. For instance, OGT2378 (Table 8.1), an inhibitor of GCS, inhibited melanoma growth in a syngeneic orthotopic murine model [155]. In separate studies, combination of fenretinide, a compound that induces accumulation of dihydroceramide [156] through direct inhibition of dihydroceramide desaturase [157], with GCS inhibitors resulted in synergistic suppression of the growth of various tumors. Additionally, fenretinide combined with SK inhibitors such as PPMP or safingol caused growth inhibition [158, 159].

Sphingosine kinase and S1P have also been implicated in drug resistance phenotypes. For instance, it has been brought to light that certain drug-resistant melanoma cell lines such as Mel-2a and M221 are resistant to Fas-induced cell death due to a decrease in ceramide and an increase in S1P compared with Fas-sensitive counterparts such as A-375 and M186. Downregulation of SK1 with siRNA decreased the resistance of Mel-2a cells to apoptosis [160]. Similar inference was made in camptothecin resistant prostate cancer cell lines [161]. In a recent study, SK1 was found to be upregulated in imatinib-resistant chronic myeloid leukemia cell line concomitant with increased BCR-ABL mRNA and protein levels. The PI3K/AKT/mTOR pathway was also found to be upregulated. Knocking down SK1 expression using siRNA reversed the imatinib resistance to apoptosis and returned BCR-ABL to normal levels [162], suggesting a role for SK1 in conferring drug resistance.

8.8 Conclusions

There is compelling evidence to suggest that sphingolipid metabolism plays an integral part of cancer pathogenesis and therapeutic response. Ceramide is a bioactive lipid that activates signaling pathways to induce apoptosis of various cancer cell lines. S1P, on the other hand, is emerging as a pro-proliferative lipid that is frequently upregulated in tumors. Therapeutic regimens targeting the balance of ceramide and S1P may prove useful in the treatment of many carcinomas.

In spite of our understanding of sphingolipid metabolism, and its relevance in cancer models, we still have a long way to go in understanding the intricacies of sphingolipid metabolism, how the metabolism proceeds in different cancer subtypes, how compartmentalization of metabolism may offer unique regulatory roles, and how different species of individual lipid molecules provide unique signaling functions. With the advent of sophisticated lipidomics and bioinformatic approaches, more and more sphingolipid functions/signaling mechanisms are being uncovered, and this area of research holds and will continue to hold promise as a potential avenue of therapy.

Acknowledgments We thank Benjamin Newcomb for his critical review of this chapter. We also thank the members of Yusuf Hannun and Lina Obeid laboratory for their helpful discussion. We apologize to those investigators whose important works were not included in this chapter because of the space limitations. The Yusuf Hannun laboratory is supported by research grants from the National Institutes of Health, USA.

References

1. Mandon EC, Ehses I, Rother J, van Echten G, Sandhoff K (1992) Subcellular localization and membrane topology of serine palmitoyltransferase, 3-dehydrosphinganine reductase, and sphinganine N-acyltransferase in mouse liver. J Biol Chem 267(16):11144–11148
2. Michel C, van Echten-Deckert G (1997) Conversion of dihydroceramide to ceramide occurs at the cytosolic face of the endoplasmic reticulum. FEBS Lett 416(2):153–155
3. Tafesse FG, Ternes P, Holthuis JC (2006) The multigenic sphingomyelin synthase family. J Biol Chem 281(40):29421–29425
4. Ichikawa S, Hirabayashi Y (1998) Glucosylceramide synthase and glycosphingolipid synthesis. Trends Cell Biol 8(5):198–202
5. Shinghal R, Scheller RH, Bajjalieh SM (1993) Ceramide 1-phosphate phosphatase activity in brain. J Neurochem 61(6):2279–2285

6. Marchesini N, Hannun YA (2004) Acid and neutral sphingomyelinases: roles and mechanisms of regulation. Biochem Cell Biol Biochimie et Biologie Cellulaire 82(1):27–44
7. Tettamanti G (2004) Ganglioside/glycosphingolipid turnover: new concepts. Glycoconj J 20(5):301–317
8. Mao C, Obeid LM (2008) Ceramidases: regulators of cellular responses mediated by ceramide, sphingosine, and sphingosine-1-phosphate. Biochim Biophys Acta 1781(9):424–434
9. Hait NC, Oskeritzian CA, Paugh SW, Milstien S, Spiegel S (2006) Sphingosine kinases, sphingosine 1-phosphate, apoptosis and diseases. Biochim Biophys Acta 1758(12):2016–2026
10. Johnson KR, Johnson KY, Becker KP, Bielawski J, Mao C, Obeid LM (2003) Role of human sphingosine-1-phosphate phosphatase 1 in the regulation of intra- and extracellular sphingosine-1-phosphate levels and cell viability. J Biol Chem 278(36):34541–34547
11. Bandhuvula P, Saba JD (2007) Sphingosine-1-phosphate lyase in immunity and cancer: silencing the siren. Trends Mol Med 13(5):210–217
12. Hanada K, Kumagai K, Tomishige N, Kawano M (2007) CERT and intracellular trafficking of ceramide. Biochim Biophys Acta 1771(6):644–653
13. Yamaji T, Kumagai K, Tomishige N, Hanada K (2008) Two sphingolipid transfer proteins, CERT and FAPP2: their roles in sphingolipid metabolism. IUBMB Life 60(8):511–518
14. Tani M, Hannun YA (2007) Analysis of membrane topology of neutral sphingomyelinase 2. FEBS Lett 581(7):1323–1328
15. Hannun YA, Obeid LM (2008) Principles of bioactive lipid signalling: lessons from sphingolipids. Nat Rev Mol Cell Biol 9(2):139–150
16. Bartke N, Hannun YA (2009) Bioactive sphingolipids: metabolism and function. J Lipid Res 50:S91–S96 Suppl
17. Hannun YA (1996) Functions of ceramide in coordinating cellular response to stress. Science 274:1855–1859
18. Hannun YA, Obeid LM (2002) The ceramide-centric universe of lipid-mediated cell regulation: stress encounters of the lipid kind. J Biol Chem 277:25847–25850
19. Andrieu-Abadie N (2001) Ceramide in apoptosis signaling: relationship with oxidative stress. Free Radic Biol Med 31:717–718
20. Ogretmen B, Hannun YA (2001) Updates on functions of ceramide in chemotherapy-induced cell death and in multidrug resistance. Drug Resist Updat 4:368–377
21. Hannun YA (1996) Functions of ceramide in coordinating cellular responses to stress. Science 274(5294):1855–1859
22. Chalfant CE, Ogretmen B, Galadari S, Kroesen BJ, Pettus BJ, Hannun YA (2001) FAS activation induces dephosphorylation of SR proteins; dependence on the de novo generation of ceramide and activation of protein phosphatase 1. J Biol Chem 276(48):44848–44855
23. Kagedal K, Johansson U, Ollinger K (2001) The lysosomal protease cathepsin D mediates apoptosis induced by oxidative stress. FASEB J: Official Publ Fed Am Soc Exp Biol 15(9):1592–1594
24. Heinrich M, Wickel M, Winoto-Morbach S, Schneider-Brachert W, Weber T, Brunner J, Saftig P, Peters C, Kronke M, Schutze S (2000) Ceramide as an activator lipid of cathepsin D. Adv Exp Med Biol 477:305–315
25. Heinrich M, Wickel M, Schneider-Brachert W, Sandberg C, Gahr J, Schwandner R, Weber T, Saftig P, Peters C, Brunner J et al (1999) Cathepsin D targeted by acid sphingomyelinase-derived ceramide. EMBO J 18(19):5252–5263
26. Basu S, Bayoumy S, Zhang Y, Lozano J, Kolesnick R (1998) BAD enables ceramide to signal apoptosis via Ras and Raf-1. J Biol Chem 273(46):30419–30426
27. Mathias S, Dressler KA, Kolesnick RN (1991) Characterization of a ceramide-activated protein kinase: stimulation by tumor necrosis factor alpha. Proc Nat Acad Sci USA 88(22):10009–10013
28. Yao B, Zhang Y, Delikat S, Mathias S, Basu S, Kolesnick R (1995) Phosphorylation of Raf by ceramide-activated protein kinase. Nature 378(6554):307–310

29. Liu J, Mathias S, Yang Z, Kolesnick RN (1994) Renaturation and tumor necrosis factor-alpha stimulation of a 97-kDa ceramide-activated protein kinase. J Biol Chem 269(4):3047–3052
30. Joseph CK, Byun HS, Bittman R, Kolesnick RN (1993) Substrate recognition by ceramide-activated protein kinase. Evidence that kinase activity is proline-directed. J Biol Chem 268(27):20002–20006
31. Xing HR, Kolesnick R (2001) Kinase suppressor of Ras signals through Thr269 of c-Raf-1. J Biol Chem 276(13):9733–9741
32. Basu S, Kolesnick R (1998) Stress signals for apoptosis: ceramide and c-Jun kinase. Oncogene 17(25):3277–3285
33. Fox TE, Houck KL, O'Neill SM, Nagarajan M, Stover TC, Pomianowski PT, Unal O, Yun JK, Naides SJ, Kester M (2007) Ceramide recruits and activates protein kinase C zeta (PKC zeta) within structured membrane microdomains. J Biol Chem 282(17):12450–12457
34. Bourbon NA, Sandirasegarane L, Kester M (2002) Ceramide-induced inhibition of Akt is mediated through protein kinase Czeta: implications for growth arrest. J Biol Chem 277(5):3286–3292
35. Payne SG, Milstien S, Spiegel S (2002) Sphingosine-1-phosphate: dual messenger functions. FEBS Lett 531:54–57
36. Xia P, Wang L, Moretti PA, Albanese N, Chai F, Pitson SM, D'Andrea RJ, Gamble JR, Vadas MA (2002) Sphingosine kinase interacts with TRAF2 and dissects tumor necrosis factor-alpha signaling. J Biol Chem 277(10):7996–8003
37. French KJ, Schrecengost RS, Lee BD, Zhuang Y, Smith SN, Eberly JL, Yun JK, Smith CD (2003) Discovery and evaluation of inhibitors of human sphingosine kinase. Cancer Res 63(18):5962–5969
38. Rylova SN, Somova OG, Dyatlovitskaya EV (1998) Comparative investigation of sphingoid bases and fatty acids in ceramides and sphingomyelins from human ovarian malignant tumors and normal ovary. Biochemistry 63:1057–1060
39. Birbes H, El Bawab S, Hannun YA, Obeid LM (2001) Selective hydrolysis of a mitochondrial pool of sphingomyelin induces apoptosis. FASEB J: Official Publ Fed Am Soc Exp Biol 15(14):2669–2679
40. Dai Q, Liu J, Chen J, Durrant D, McIntyre TM, Lee RM (2004) Mitochondrial ceramide increases in UV-irradiated HeLa cells and is mainly derived from hydrolysis of sphingomyelin. Oncogene 23(20):3650–3658
41. Deng X, Yin X, Allan R, Lu DD, Maurer CW, Haimovitz-Friedman A, Fuks Z, Shaham S, Kolesnick R (2008) Ceramide biogenesis is required for radiation-induced apoptosis in the germ line of C. elegans. Science 322(5898):110–115
42. Kroesen BJ, Pettus B, Luberto C, Busman M, Sietsma H, de Leij L, Hannun YA (2001) Induction of apoptosis through B-cell receptor cross-linking occurs via de novo generated C16-ceramide and involves mitochondria. J Biol Chem 276(17):13606–13614
43. Eto M, Bennouna J, Hunter OC, Hershberger PA, Kanto T, Johnson CS, Lotze MT, Amoscato AA (2003) C16 ceramide accumulates following androgen ablation in LNCaP prostate cancer cells. Prostate 57(1):66–79
44. Chalfant CE, Rathman K, Pinkerman RL, Wood RE, Obeid LM, Ogretmen B, Hannun YA (2002) De novo ceramide regulates the alternative splicing of caspase 9 and Bcl-x in A549 lung adenocarcinoma cells. Dependence on protein phosphatase-1. J Biol Chem 277(15):12587–12595
45. Santana P, Pena LA, Haimovitz-Friedman A, Martin S, Green D, McLoughlin M, Cordon-Cardo C, Schuchman EH, Fuks Z, Kolesnick R (1996) Acid sphingomyelinase-deficient human lymphoblasts and mice are defective in radiation-induced apoptosis. Cell 86(2):189–199
46. Paris F, Fuks Z, Kang A, Capodieci P, Juan G, Ehleiter D, Haimovitz-Friedman A, Cordon-Cardo C, Kolesnick R (2001) Endothelial apoptosis as the primary lesion initiating intestinal radiation damage in mice. Science 293(5528):293–297

47. Pena LA, Fuks Z, Kolesnick RN (2000) Radiation-induced apoptosis of endothelial cells in the murine central nervous system: protection by fibroblast growth factor and sphingomyelinase deficiency. Cancer Res 60(2):321–327
48. Heinrich M, Neumeyer J, Jakob M, Hallas C, Tchikov V, Winoto-Morbach S, Wickel M, Schneider-Brachert W, Trauzold A, Hethke A et al (2004) Cathepsin D links TNF-induced acid sphingomyelinase to Bid-mediated caspase-9 and -3 activation. Cell Death Differ 11(5):550–563
49. Luberto C, Hassler DF, Signorelli P, Okamoto Y, Sawai H, Boros E, Hazen-Martin DJ, Obeid LM, Hannun YA, Smith GK (2002) Inhibition of tumor necrosis factor-induced cell death in MCF7 by a novel inhibitor of neutral sphingomyelinase. J Biol Chem 277(43):41128–41139
50. Lee JT, Xu J, Lee JM, Ku G, Han X, Yang DI, Chen S, Hsu CY (2004) Amyloid-beta peptide induces oligodendrocyte death by activating the neutral sphingomyelinase-ceramide pathway. J Cell Biol 164(1):123–131
51. Liu JJ, Wang JY, Hertervig E, Cheng Y, Nilsson A, Duan RD (2000) Activation of neutral sphingomyelinase participates in ethanol-induced apoptosis in Hep G2 cells. Alcohol Alcohol 35(6):569–573
52. Testai FD, Landek MA, Dawson G (2004) Regulation of sphingomyelinases in cells of the oligodendrocyte lineage. J Neurosci Res 75(1):66–74
53. Franzen R (2002) Nitric oxide induces degradation of the neutral ceramidase in rat renal mesangial cells and is counterregulated by protein kinase C. J Biol Chem 277:46184–46190
54. Acharya U (2003) Modulating sphingolipid biosynthetic pathway rescues photoreceptor degeneration. Science 299:1740–1743
55. Venable ME, Lee JY, Smyth MJ, Bielawska A, Obeid LM (1995) Role of ceramide in cellular senescence. J Biol Chem 270(51):30701–30708
56. Venable ME, Yin X (2009) Ceramide induces endothelial cell senescence. Cell Biochem Funct 27(8):547–551
57. Modrak DE, Leon E, Goldenberg DM, Gold DV (2009) Ceramide regulates gemcitabine-induced senescence and apoptosis in human pancreatic cancer cell lines. Mol Cancer Res: MCR 7(6):890–896
58. Guillas I (2001) C26-CoA-dependent ceramide synthesis of Saccharomyces cerevisiae is operated by Lag1p and Lac1p. EMBO J 20:2655–2665
59. Kim NW, Piatyszek MA, Prowse KR, Harley CB, West MD, Ho PL, Coviello GM, Wright WE, Weinrich SL, Shay JW (1994) Specific association of human telomerase activity with immortal cells and cancer. Science 266(5193):2011–2015
60. Shay JW, Bacchetti S (1997) A survey of telomerase activity in human cancer. Eur J Cancer 33(5):787–791
61. Ogretmen B, Schady D, Usta J, Wood R, Kraveka JM, Luberto C, Birbes H, Hannun YA, Obeid LM (2001) Role of ceramide in mediating the inhibition of telomerase activity in A549 human lung adenocarcinoma cells. J Biol Chem 276(27):24901–24910
62. Okazaki T, Bell RM, Hannun YA (1989) Sphingomyelin turnover induced by vitamin D3 in HL-60 cells. Role in cell differentiation. J Biol Chem 264(32):19076–19080
63. Dobrowsky RT, Werner MH, Castellino AM, Chao MV, Hannun YA (1994) Activation of the sphingomyelin cycle through the low-affinity neurotrophin receptor. Science 265(5178):1596–1599
64. Fishbein JD, Dobrowsky RT, Bielawska A, Garrett S, Hannun YA (1993) Ceramide-mediated growth inhibition and CAPP are conserved in Saccharomyces cerevisiae. J Biol Chem 268(13):9255–9261
65. Dbaibo GS (1995) Retinoblastoma gene product as a downstream target for a ceramide-dependent pathway of growth arrest. Proc Natl Acad Sci USA 92:1347–1351
66. Daido S, Kanzawa T, Yamamoto A, Takeuchi H, Kondo Y, Kondo S (2004) Pivotal role of the cell death factor BNIP3 in ceramide-induced autophagic cell death in malignant glioma cells. Cancer Res 64(12):4286–4293

67. Scarlatti F, Bauvy C, Ventruti A, Sala G, Cluzeaud F, Vandewalle A, Ghidoni R, Codogno P (2004) Ceramide-mediated macroautophagy involves inhibition of protein kinase B and up-regulation of beclin 1. J Biol Chem 279(18):18384–18391
68. Demarchi F, Bertoli C, Copetti T, Tanida I, Brancolini C, Eskelinen EL, Schneider C (2006) Calpain is required for macroautophagy in mammalian cells. J Cell Biol 175(4):595–605
69. Bedia C, Triola G, Casas J, Llebaria A, Fabrias G (2005) Analogs of the dihydroceramide desaturase inhibitor GT11 modified at the amide function: synthesis and biological activities. Org Biomol Chem 3(20):3707–3712
70. Signorelli P, Munoz-Olaya JM, Gagliostro V, Casas J, Ghidoni R, Fabrias G (2009) Dihydroceramide intracellular increase in response to resveratrol treatment mediates autophagy in gastric cancer cells. Cancer Lett 282(2):238–243
71. Zeng X, Overmeyer JH, Maltese WA (2006) Functional specificity of the mammalian Beclin-Vps34 PI 3-kinase complex in macroautophagy versus endocytosis and lysosomal enzyme trafficking. J Cell Sci 119(Pt 2):259–270
72. Zhou H, Summers SA, Birnbaum MJ, Pittman RN (1998) Inhibition of Akt kinase by cell-permeable ceramide and its implications for ceramide-induced apoptosis. J Biol Chem 273(26):16568–16575
73. Schubert KM, Scheid MP, Duronio V (2000) Ceramide inhibits protein kinase B/Akt by promoting dephosphorylation of serine 473. J Biol Chem 275(18):13330–13335
74. Dickson RC (2008) Thematic review series: sphingolipids. New insights into sphingolipid metabolism and function in budding yeast. J Lipid Res 49(5):909–921
75. Cowart LA, Obeid LM (2007) Yeast sphingolipids: recent developments in understanding biosynthesis, regulation, and function. Biochim Biophys Acta 1771(3):421–431
76. Guenther GG, Peralta ER, Rosales KR, Wong SY, Siskind LJ, Edinger AL (2008) Ceramide starves cells to death by downregulating nutrient transporter proteins. Proc Nat Acad Sci USA 105(45):17402–17407
77. Van Brocklyn JR, Young N, Roof R (2003) Sphingosine-1-phosphate stimulates motility and invasiveness of human glioblastoma multiforme cells. Cancer Lett 199(1):53–60
78. Nava VE, Hobson JP, Murthy S, Milstien S, Spiegel S (2002) Sphingosine kinase type 1 promotes estrogen-dependent tumorigenesis of breast cancer MCF-7 cells. Exp Cell Res 281(1):115–127
79. Xia P, Gamble JR, Wang L, Pitson SM, Moretti PA, Wattenberg BW, D'Andrea RJ, Vadas MA (2000) An oncogenic role of sphingosine kinase. Curr Biol: CB 10(23):1527–1530
80. Cuvillier O, Pirianov G, Kleuser B, Vanek PG, Coso OA, Gutkind S, Spiegel S (1996) Suppression of ceramide-mediated programmed cell death by sphingosine-1-phosphate. Nature 381(6585):800–803
81. Visentin B, Vekich JA, Sibbald BJ, Cavalli AL, Moreno KM, Matteo RG, Garland WA, Lu Y, Yu S, Hall HS et al (2006) Validation of an anti-sphingosine-1-phosphate antibody as a potential therapeutic in reducing growth, invasion, and angiogenesis in multiple tumor lineages. Cancer Cell 9(3):225–238
82. Castillo SS, Teegarden D (2003) Sphingosine-1-phosphate inhibition of apoptosis requires mitogen-activated protein kinase phosphatase-1 in mouse fibroblast C3H10T 1/2 cells. J Nutr 133(11):3343–3349
83. Suomalainen L, Pentikainen V, Dunkel L (2005) Sphingosine-1-phosphate inhibits nuclear factor kappaB activation and germ cell apoptosis in the human testis independently of its receptors. Am J Pathol 166(3):773–781
84. Licht T, Tsirulnikov L, Reuveni H, Yarnitzky T, Ben-Sasson SA (2003) Induction of pro-angiogenic signaling by a synthetic peptide derived from the second intracellular loop of S1P3 (EDG3). Blood 102(6):2099–2107
85. Argraves KM, Wilkerson BA, Argraves WS, Fleming PA, Obeid LM, Drake CJ (2004) Sphingosine-1-phosphate signaling promotes critical migratory events in vasculogenesis. J Biol Chem 279(48):50580–50590
86. Wu W, Shu X, Hovsepyan H, Mosteller RD, Broek D (2003) VEGF receptor expression and signaling in human bladder tumors. Oncogene 22(22):3361–3370

87. Lee MJ, Thangada S, Claffey KP, Ancellin N, Liu CH, Kluk M, Volpi M, Sha'afi RI, Hla T (1999) Vascular endothelial cell adherens junction assembly and morphogenesis induced by sphingosine-1-phosphate. Cell 99(3):301–312
88. Bayless KJ, Davis GE (2003) Sphingosine-1-phosphate markedly induces matrix metalloproteinase and integrin-dependent human endothelial cell invasion and lumen formation in three-dimensional collagen and fibrin matrices. Biochem Biophys Res Commun 312(4):903–913
89. Yonesu K, Kawase Y, Inoue T, Takagi N, Tsuchida J, Takuwa Y, Kumakura S, Nara F (2009) Involvement of sphingosine-1-phosphate and S1P1 in angiogenesis: analyses using a new S1P1 antagonist of non-sphingosine-1-phosphate analog. Biochem Pharmacol 77(6):1011–1020
90. LaMontagne K, Littlewood-Evans A, Schnell C, O'Reilly T, Wyder L, Sanchez T, Probst B, Butler J, Wood A, Liau G et al (2006) Antagonism of sphingosine-1-phosphate receptors by FTY720 inhibits angiogenesis and tumor vascularization. Cancer Res 66(1):221–231
91. Liu Y, Wada R, Yamashita T, Mi Y, Deng CX, Hobson JP, Rosenfeldt HM, Nava VE, Chae SS, Lee MJ et al (2000) Edg-1, the G protein-coupled receptor for sphingosine-1-phosphate, is essential for vascular maturation. J Clin Investig 106(8):951–961
92. Mizugishi K, Yamashita T, Olivera A, Miller GF, Spiegel S, Proia RL (2005) Essential role for sphingosine kinases in neural and vascular development. Mol Cell Biol 25(24):11113–11121
93. Xia P, Gamble JR, Rye KA, Wang L, Hii CS, Cockerill P, Khew-Goodall Y, Bert AG, Barter PJ, Vadas MA (1998) Tumor necrosis factor-alpha induces adhesion molecule expression through the sphingosine kinase pathway. Proc Nat Acad Sci USA 95(24):14196–14201
94. Pettus BJ, Bielawski J, Porcelli AM, Reames DL, Johnson KR, Morrow J, Chalfant CE, Obeid LM, Hannun YA (2003) The sphingosine kinase 1/sphingosine-1-phosphate pathway mediates COX-2 induction and PGE2 production in response to TNF-alpha. FASEB J: Official Publ Fed Am Soc Exp Biol 17(11):1411–1421
95. Nayak D, Huo Y, Kwang WX, Pushparaj PN, Kumar SD, Ling EA, Dheen ST (2010) Sphingosine kinase 1 regulates the expression of proinflammatory cytokines and nitric oxide in activated microglia. Neuroscience 166(1):132–144
96. Hammad SM, Crellin HG, Wu BX, Melton J, Anelli V, Obeid LM (2008) Dual and distinct roles for sphingosine kinase 1 and sphingosine 1 phosphate in the response to inflammatory stimuli in RAW macrophages. Prostaglandins Other Lipid Mediat 85(3–4):107–114
97. Snider AJ (2010) Orr Gandy KA, Obeid LM: Sphingosine kinase: Role in regulation of bioactive sphingolipid mediators in inflammation. Biochimie 92(6):707–715
98. Massiello A, Salas A, Pinkerman RL, Roddy P, Roesser JR, Chalfant CE (2004) Identification of two RNA cis-elements that function to regulate the 5′ splice site selection of Bcl-x pre-mRNA in response to ceramide. J Biol Chem 279(16):15799–15804
99. Gomez-Munoz A, Kong JY, Parhar K, Wang SW, Gangoiti P, Gonzalez M, Eivemark S, Salh B, Duronio V, Steinbrecher UP (2005) Ceramide-1-phosphate promotes cell survival through activation of the phosphatidylinositol 3-kinase/protein kinase B pathway. FEBS Lett 579(17):3744–3750
100. Nava VE, Cuvillier O, Edsall LC, Kimura K, Milstien S, Gelmann EP, Spiegel S (2000) Sphingosine enhances apoptosis of radiation-resistant prostate cancer cells. Cancer Res 60(16):4468–4474
101. Sweeney EA, Sakakura C, Shirahama T, Masamune A, Ohta H, Hakomori S, Igarashi Y (1996) Sphingosine and its methylated derivative N, N-dimethylsphingosine (DMS) induce apoptosis in a variety of human cancer cell lines. Int J Cancer J Int du Cancer 66(3):358–366
102. Sawai H, Okazaki T, Domae N (2002) Sphingosine-induced c-jun expression: differences between sphingosine- and C2-ceramide-mediated signaling pathways. FEBS Lett 524(1–3):103–106
103. Phillips DC, Martin S, Doyle BT, Houghton JA (2007) Sphingosine-induced apoptosis in rhabdomyosarcoma cell lines is dependent on pre-mitochondrial Bax activation and post-mitochondrial caspases. Cancer Res 67(2):756–764
104. Guan F, Handa K, Hakomori SI (2009) Specific glycosphingolipids mediate epithelial-to-mesenchymal transition of human and mouse epithelial cell lines. Proc Nat Acad Sci USA 106(18):7461–7466

105. Bremer EG, Schlessinger J, Hakomori S (1986) Ganglioside-mediated modulation of cell growth. Specific effects of GM3 on tyrosine phosphorylation of the epidermal growth factor receptor. J Biol Chem 261(5):2434–2440
106. Zhou Q, Hakomori S, Kitamura K, Igarashi Y (1994) GM3 directly inhibits tyrosine phosphorylation and de-N-acetyl-GM3 directly enhances serine phosphorylation of epidermal growth factor receptor, independently of receptor–receptor interaction. J Biol Chem 269(3):1959–1965
107. Meuillet EJ, Kroes R, Yamamoto H, Warner TG, Ferrari J, Mania-Farnell B, George D, Rebbaa A, Moskal JR, Bremer EG (1999) Sialidase gene transfection enhances epidermal growth factor receptor activity in an epidermoid carcinoma cell line, A431. Cancer Res 59(1):234–240
108. Coskun U, Grzybek M, Drechsel D, Simons K (2011) Regulation of human EGF receptor by lipids. Proc Nat Acad Sci USA 108(22):9044–9048
109. Ogretmen B, Hannun YA (2004) Biologically active sphingolipids in cancer pathogenesis and treatment. Nat Rev Cancer 4(8):604–616
110. Bose R (1995) Ceramide synthase mediates daunorubicin-induced apoptosis: an alternative mechanism for generating death signals. Cell 82:405–414
111. Perry DK (2000) Serine palmitoyltransferase regulates de novo ceramide generation during etoposide-induced apoptosis. J Biol Chem 275:9078–9084
112. Chauvier D, Morjani H, Manfait M (2002) Ceramide involvement in homocamptothecin- and camptothecin-induced cytotoxicity and apoptosis in colon HT29 cells. Int J Oncol 20:855–863
113. Strum JC (1994) 1-[beta]-D-Arabinofuranosylcytosine stimulates ceramide and digylceride formation in HL-60 cells. J Biol Chem 269:15493–15497
114. Bezombes C (2001) Oxidative stress-induced activation of Lyn recruits sphingomyelinase and is requisite for its stimulation by Ara-C. FASEB J 15:1583–1585
115. Dbaibo GS (1998) p53-dependent ceramide response to genotoxic stress. J Clin Invest 102:329–339
116. Selzner M, Bielawska A, Morse MA, Rudiger HA, Sindram D, Hannun YA, Clavien PA (2001) Induction of apoptotic cell death and prevention of tumor growth by ceramide analogues in metastatic human colon cancer. Cancer Res 61(3):1233–1240
117. Meng A, Luberto C, Meier P, Bai A, Yang X, Hannun YA, Zhou D (2004) Sphingomyelin synthase as a potential target for D609-induced apoptosis in U937 human monocytic leukemia cells. Exp Cell Res 292(2):385–392
118. Modrak DE, Cardillo TM, Newsome GA, Goldenberg DM, Gold DV (2004) Synergistic interaction between sphingomyelin and gemcitabine potentiates ceramide-mediated apoptosis in pancreatic cancer. Cancer Res 64(22):8405–8410
119. Veldman RJ, Zerp S, van Blitterswijk WJ, Verheij M (2004) N-hexanoyl-sphingomyelin potentiates in vitro doxorubicin cytotoxicity by enhancing its cellular influx. Br J Cancer 90(4):917–925
120. Radin NS (2003) Killing tumours by ceramide-induced apoptosis: a critique of available drugs. Biochem J 371(Pt 2):243–256
121. Bieberich E, Kawaguchi T, Yu RK (2000) N-acylated serinol is a novel ceramide mimic inducing apoptosis in neuroblastoma cells. J Biol Chem 275(1):177–181
122. Bieberich E (2002) Synthesis and characterization of novel ceramide analogs for induction of apoptosis in human cancer cells. Cancer Lett 181:55–64
123. Struckhoff AP (2004) Novel ceramide analogs as potential chemotherapeutic agents in breast cancer. J Pharmacol Exp Ther 309:523–532
124. Crawford KW (2003) Novel ceramide analogs display selective cytotoxicity in drug-resistant breast tumor cell lines compared to normal breast epithelial cells. Cell Mol Biol 49:1017–1023
125. Dindo D, Dahm F, Szulc Z, Bielawska A, Obeid LM, Hannun YA, Graf R, Clavien PA (2006) Cationic long-chain ceramide LCL-30 induces cell death by mitochondrial targeting in SW403 cells. Mol Cancer Ther 5(6):1520–1529
126. Senkal CE, Ponnusamy S, Rossi MJ, Sundararaj K, Szulc Z, Bielawski J, Bielawska A, Meyer M, Cobanoglu B, Koybasi S et al (2006) Potent antitumor activity of a novel cationic pyridinium-ceramide alone or in combination with gemcitabine against human head and neck squamous cell carcinomas in vitro and in vivo. J Pharmacol Exp Ther 317(3):1188–1199

127. Rossi MJ, Sundararaj K, Koybasi S, Phillips MS, Szulc ZM, Bielawska A, Day TA, Obeid LM, Hannun YA, Ogretmen B (2005) Inhibition of growth and telomerase activity by novel cationic ceramide analogs with high solubility in human head and neck squamous cell carcinoma cells. Otolaryngol Head Neck Surg 132(1):55–62
128. Novgorodov SA, Szulc ZM, Luberto C, Jones JA, Bielawski J, Bielawska A, Hannun YA, Obeid LM (2005) Positively charged ceramide is a potent inducer of mitochondrial permeabilization. J Biol Chem 280(16):16096–16105
129. Stover T, Kester M (2003) Liposomal delivery enhances short-chain ceramide-induced apoptosis of breast cancer cells. J Pharmacol Exp Ther 307:468–475
130. Thomas DA, Sarris AH, Cortes J, Faderl S, O'Brien S, Giles FJ, Garcia-Manero G, Rodriguez MA, Cabanillas F, Kantarjian H (2006) Phase II study of sphingosomal vincristine in patients with recurrent or refractory adult acute lymphocytic leukemia. Cancer 106(1):120–127
131. Shirahama T, Sweeney EA, Sakakura C, Singhal AK, Nishiyama K, Akiyama S, Hakomori S, Igarashi Y (1997) In vitro and in vivo induction of apoptosis by sphingosine and N, N-dimethylsphingosine in human epidermoid carcinoma KB-3-1 and its multidrug-resistant cells. Clinical cancer research : an official journal of the American Association for Cancer Research 3(2):257–264
132. Sweeney EA (1996) Sphingosine and its methylated derivative N, N-dimethylsphingosine (DMS) induce apoptosis in a variety of human cancer cell lines. Int J Cancer 66:358–366
133. Cuvillier O, Levade T (2001) Sphingosine 1-phosphate antagonizes apoptosis of human leukemia cells by inhibiting release of cytochrome c and Smac/DIABLO from mitochondria. Blood 98(9):2828–2836
134. Tilly JL, Kolesnick RN (2002) Sphingolipids, apoptosis, cancer treatments and the ovary: investigating a crime against female fertility. Biochim Biophys Acta 1585(2–3):135–138
135. Suomalainen L, Hakala JK, Pentikainen V, Otala M, Erkkila K, Pentikainen MO, Dunkel L (2003) Sphingosine-1-phosphate in inhibition of male germ cell apoptosis in the human testis. J Clin Endocrinol Metab 88(11):5572–5579
136. Graler MH, Goetzl EJ (2004) The immunosuppressant FTY720 down-regulates sphingosine 1-phosphate G protein-coupled receptors. FASEB J 10
137. Billich A (2003) Phosphorylation of the immunomodulatory drug FTY720 by sphingosine kinases. J Biol Chem 278:47408–47415
138. Paugh SW (2003) The immunosuppressant FTY720 is phosphorylated by sphingosine kinase type 2. FEBS Lett 554:189–193
139. Azuma H (2002) Marked prevention of tumor growth and metastasis by a novel immunosuppressive agent, FTY720, in mouse breast cancer models. Cancer Res 62:1410–1419
140. Wang JD (1999) Early induction of apoptosis in androgen-independent prostate cancer cell line by FTY720 requires caspase-3 activation. Prostate 40:50–55
141. Sonoda Y (2001) FTY720, a novel immunosuppressive agent, induces apoptosis in human glioma cells. Biochem Biophys Res Commun 281:282–288
142. Azuma H (2003) Induction of apoptosis in human bladder cancer cells in vitro and in vivo caused by FTY720 treatment. J Urol 169:2372–2377
143. Dudeja PK, Dahiya R, Brasitus TA (1986) The role of sphingomyelin synthetase and sphingomyelinase in 1, 2-dimethylhydrazine-induced lipid alterations of rat colonic plasma membranes. Biochim Biophys Acta 863(2):309–312
144. Dillehay DL, Webb SK, Schmelz EM, Merrill AH Jr (1994) Dietary sphingomyelin inhibits 1, 2-dimethylhydrazine-induced colon cancer in CF1 mice. J Nutr 124(5):615–620
145. Schmelz EM, Dillehay DL, Webb SK, Reiter A, Adams J, Merrill AH Jr (1996) Sphingomyelin consumption suppresses aberrant colonic crypt foci and increases the proportion of adenomas versus adenocarcinomas in CF1 mice treated with 1, 2-dimethylhydrazine: implications for dietary sphingolipids and colon carcinogenesis. Cancer Res 56(21):4936–4941
146. Schmelz EM, Bushnev AS, Dillehay DL, Sullards MC, Liotta DC, Merrill AH Jr (1999) Ceramide-beta-D-glucuronide: synthesis, digestion, and suppression of early markers of colon carcinogenesis. Cancer Res 59(22):5768–5772

147. Schmelz EM, Bushnev AS, Dillehay DL, Liotta DC, Merrill AH Jr (1997) Suppression of aberrant colonic crypt foci by synthetic sphingomyelins with saturated or unsaturated sphingoid base backbones. Nutr Cancer 28(1):81–85
148. Mazzei JC, Zhou H, Brayfield BP, Hontecillas R, Bassaganya-Riera J, Schmelz EM (2011) Suppression of intestinal inflammation and inflammation-driven colon cancer in mice by dietary sphingomyelin: importance of peroxisome proliferator-activated receptor gamma expression. J Nutr Biochem 22(12):1160–1171
149. Bleicher RJ, Cabot MC (2002) Glucosylceramide synthase and apoptosis. Biochim Biophys Acta 1585:172–178
150. Liu X, Ryland L, Yang J, Liao A, Aliaga C, Watts R, Tan SF, Kaiser J, Shanmugavelandy SS, Rogers A et al (2010) Targeting of survivin by nanoliposomal ceramide induces complete remission in a rat model of NK-LGL leukemia. Blood 116(20):4192–4201
151. Liu YY (2001) Ceramide glycosylation potentiates cellular multidrug resistance. FASEB J 15:719–730
152. Liu YY (2004) Oligonucleotides blocking glucosylceramide synthase expression selectively reverse drug resistance in cancer cells. J Lipid Res 45:933–940
153. Gouaze V, Liu YY, Prickett CS, Yu JY, Giuliano AE, Cabot MC (2005) Glucosylceramide synthase blockade down-regulates P-glycoprotein and resensitizes multidrug-resistant breast cancer cells to anticancer drugs. Cancer Res 65(9):3861–3867
154. Gouaze-Andersson V, Yu JY, Kreitenberg AJ, Bielawska A, Giuliano AE, Cabot MC (2007) Ceramide and glucosylceramide upregulate expression of the multidrug resistance gene MDR1 in cancer cells. Biochim Biophys Acta 1771(12):1407–1417
155. Weiss M (2003) Inhibition of melanoma tumor growth by a novel inhibitor of glucosylceramide synthase. Cancer Res 63:3654–3658
156. Wang H, Maurer BJ, Liu YY, Wang E, Allegood JC, Kelly S, Symolon H, Liu Y, Merrill AH Jr, Gouaze-Andersson V et al (2008) N-(4-Hydroxyphenyl)retinamide increases dihydroceramide and synergizes with dimethylsphingosine to enhance cancer cell killing. Mol Cancer Ther 7(9):2967–2976
157. Rahmaniyan M, Curley RW Jr, Obeid LM, Hannun YA, Kraveka JM (2011) Identification of dihydroceramide desaturase as a direct in vitro target for fenretinide. J Biol Chem 286(28):24754–24764
158. Maurer BJ (1999) Increase of ceramide and induction of mixed apoptosis/necrosis by N-(4-hydroxyphenyl)-retinamide in neuroblastoma cell lines. J Natl Cancer Inst 91:1138–1146
159. Maurer BJ (2000) Synergistic cytotoxicity in solid tumor cell lines between N-(4-hydroxyphenyl)retinamide and modulators of ceramide metabolism. J Natl Cancer Inst 92:1897–1909
160. Bektas M, Jolly PS, Muller C, Eberle J, Spiegel S, Geilen CC (2005) Sphingosine kinase activity counteracts ceramide-mediated cell death in human melanoma cells: role of Bcl-2 expression. Oncogene 24(1):178–187
161. Akao Y, Banno Y, Nakagawa Y, Hasegawa N, Kim TJ, Murate T, Igarashi Y, Nozawa Y (2006) High expression of sphingosine kinase 1 and S1P receptors in chemotherapy-resistant prostate cancer PC3 cells and their camptothecin-induced up-regulation. Biochem Biophys Res Commun 342(4):1284–1290
162. Marfe G, Di Stefano C, Gambacurta A, Ottone T, Martini V, Abruzzese E, Mologni L, Sinibaldi-Salimei P, de Fabritis P, Gambacorti-Passerini C et al (2011) Sphingosine kinase 1 overexpression is regulated by signaling through PI3K, AKT2, and mTOR in imatinib-resistant chronic myeloid leukemia cells. Exp Hematol 39(6):653–665 e656

Chapter 9
Leading Small Molecule Inhibitors of Anti-Apoptotic Bcl-2 Family Members

Victor Y. Yazbeck and Daniel E. Johnson

Abstract Anti-apoptotic members of the Bcl-2 protein family are commonly overexpressed in human malignancies, where they contribute to tumorigenesis and the development of resistance to chemo-, radio-, and immunotherapies. Considerable effort is being invested in academic and pharmaceutical settings to identify and design effective small molecule inhibitors of the anti-apoptotic Bcl-2 family members. This chapter will focus on recent advances in the development and application of three small molecule inhibitors (ABT-737, ABT-263, GX15-070), with a particular emphasis on progress that has been made in the evaluation of these compounds in preclinical in vivo models and clinical trials.

9.1 Introduction

As detailed in Chap. 1, the anti-apoptotic Bcl-2 family members Bcl-2, Bcl-X_L, and Mcl-1 are frequently overexpressed in both hematopoietic and solid tumor malignancies. Importantly, overexpression of these proteins often correlates with chemotherapy and radiation resistance and poor clinical prognosis. Small molecule inhibitors of anti-apoptotic Bcl-2 family proteins have the potential to restore the sensitivity of tumors to apoptotic stimuli and may be particularly useful when used in combination with conventional chemotherapeutics or radiation therapy. This has spurred an intensive effort to identify or design small molecule inhibitors targeting Bcl-2, Bcl-X_L, and/or Mcl-1. A number of different approaches have been utilized to search for these inhibitors, including performance of high-throughput screening assays, virtual screening, structure-based

V. Y. Yazbeck (✉) · D. E. Johnson
Department of Medicine, University of Pittsburgh
and the University of Pittsburgh Cancer Institute, Pittsburgh, PA, USA
e-mail: yazbeckvy@upmc.edu

D. E. Johnson
Department of Pharmacology and Chemical Biology, University of Pittsburgh
and the University of Pittsburgh Cancer Institute, Pittsburgh, PA, USA

D. E. Johnson (ed.), *Cell Death Signaling in Cancer Biology and Treatment*,
Cell Death in Biology and Diseases, DOI: 10.1007/978-1-4614-5847-0_9,

X.-M. Yin and Z. Dong (Series eds.), *Cell Death in Biology and Diseases*

drug design, and fragment-based drug design. To date, several inhibitors have been isolated from natural sources, while others have been identified in chemical libraries or developed via medicinal chemistry efforts. Among the inhibitors that have been reported are peptides derived from BH3 domains of proapoptotic proteins [1–4], antimycin A_3 [5], HA14-1 [6], BH3I compounds [7], several tea polyphenol compounds [8, 9], chelerythrine [10], sanguinarine [11], (-)-gossypol [12–14], apogossypolone [15, 16], TW-37 [13], ABT-737 [17], ABT-263 [18], and GX15-070 [19]. This chapter will focus on three of these inhibitors, ABT-737, ABT-263, and GX15-070, that have shown particular promise in preclinical studies and have advanced rapidly to clinical testing. The identification and initial in vitro evaluation of these compounds will first be described. We will then summarize the current understanding regarding their mechanism of action and the mechanisms which lead to resistance to these inhibitors. Additionally, reported synergies between these inhibitors and conventional chemotherapeutic agents will be discussed. Lastly, results obtained from evaluation of the inhibitors in preclinical in vivo models and human clinical trials will be presented.

9.2 ABT-737: Discovery, Mechanism of Action, and Mechanisms of Resistance

ABT-737 (see Fig. 9.1a) was developed by Abbott Laboratories using a nuclear magnetic resonance (NMR)-based and fragment-based approach to rationally design inhibitors targeting the BH3-binding groove on anti-apoptotic Bcl-X_L. ABT-737 binds with high affinity to Bcl-X_L, Bcl-w, and Bcl-2 (Ki $\leq$ 1 nM), but with much lower affinity to Bcl-B, A1/Bfl-1, and Mcl-1 (Ki $\geq$ 0.46 μM) [17]. At 0.1 μM, ABT-737 disrupted nearly 50 % of a Gal4-Bcl-X_L/VP-16-Bcl-X_S complex in a mammalian two-hybrid system. Also, it competed with the binding of fluorescently labeled Bad BH3 peptide to Bcl-X_L and Bcl-2 with IC_{50} values of 35 and 103 nM, respectively [17]. In HL-60, an acute myeloid leukemia (AML) cell line, ABT-737 was capable of activating Bax through disruption of the Bcl-2/Bax heterodimeric complex [20]. In follicular lymphoma (FL) cells, primary B-cell lymphocytic leukemia (B-CLL) and multiple myeloma (MM) cells, ABT-737 released Bim from Bcl-2 sequestration, and partial Bim knockdown resulted in decreased sensitivity to the compound [21]. In an acute lymphoblastic leukemia (ALL) cell line, ABT-737 was found to induce oxidative stress characterized by decreased glutathione and increased hydrogen peroxide/superoxide levels, leading to the activation of caspase proteases [22]. Additionally, ABT-737 induces autophagy through disruption of physical interactions between the autophagy regulator Beclin-1 and Bcl-2 or Bcl-X_L [23]. As a single agent, ABT-737 demonstrates cytotoxic activity against a large number of hematological cell lines [2, 20, 24–26], but very weak activity against solid tumor cell lines [27], with the exception of small-cell lung cancer (SCLC) cell lines which are sensitive to ABT-737 alone (IC_{50} < 1 μM) [17, 28]. Resistance to ABT-737 is frequently associated with

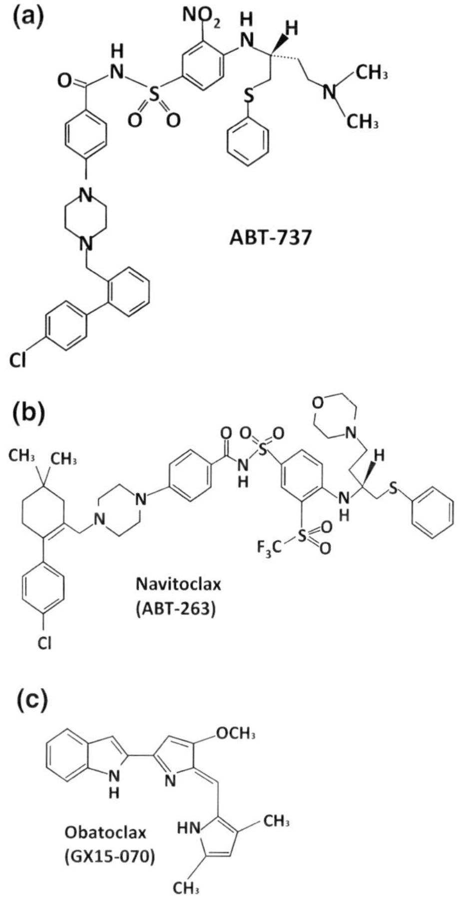

Fig. 9.1 Chemical structures of ABT-737, ABT-263, and GX15-070

overexpression of the anti-apoptotic Bcl-2 family member Mcl-1 [20, 28, 29]. For example, Mcl-1 is highly expressed in HeLa cells that are resistant to ABT-737, but reduction of Mcl-1 levels following CDK2 inhibition, DNA-damaging chemotherapy, or shRNA treatment, results in enhanced sensitivity to the compound [29]. In solid tumor cell lines, the induction of proapoptotic NOXA following the treatment with chemotherapy or radiation leads to the inhibition of Mcl-1 and resulting synergy with ABT-737 [27, 30]. Efficient cell killing by ABT-737 requires that binding of the compound to anti-apoptotic Bcl-2 family members results in the release of bound proapoptotic proteins (e.g., Bim, Bax, Bak). Moreover, the amount of released proapoptotic proteins must be sufficient to saturate empty Mcl-1 and A1/Bfl-1 that are present in the cell [21].

Deng et al. have studied the mechanism whereby lymphoma cell lines develop resistance to ABT-737 and have classified three forms of resistance. Class A resistance is observed in cells that express only low levels of BH3-only activator proteins, whereas Class B resistance is due to significant loss or mutation of Bax or Bak. On the other hand, Class C resistance is due to acute overexpression of anti-apoptotic proteins (Bcl-2, Bcl-X_L, Mcl-1, Bcl-w, A1/Bfl-1) [31]. As noted above, the importance of Mcl-1 overexpression in conferring resistance to ABT-737 has been reported by several different groups.

9.3 ABT-737 in Preclinical Models

As previously mentioned, ABT-737 induces apoptosis as a single agent in AML, diffuse large B-cell lymphoma (DLBCL), FL, MM, ALL, and SCLC cell lines, as well as primary tumor cells from multiple hematologic malignancies, including chronic lymphocytic leukemia (CLL) [17, 20, 21, 24, 25, 28, 32–34]. In these sensitive cell lines, ABT-737 typically exhibits IC_{50} values of less than 1 μM and often in the nanomolar range. By contrast, most solid tumor cell lines that have been studied are more resistant to the drug [27, 28]. However, ABT-737 has been found to synergize with chemotherapy in many solid tumor cell line models. For example, ABT-737 has been shown to synergize with paclitaxel in A549, a non-small-cell lung cancer (NSCLC) cell line [17]; carboplatin/etoposide in SCLC cells [28]; dexamethasone and melphalan in MM cells [25]; Ara-C or doxorubicin in OCI-AML3 leukemic cells that are resistant to single-agent ABT-737 [20]; the proteasome inhibitor MG-132 in melanoma [35]; MEK inhibitor in BRAF-mutant solid tumors [36]; transcriptional inhibitors (ARC [37], actinomycin [38]), gemcitabine [39], methylseleninic acid (second-generation selenium) [40] in multiple cancer cell lines [37–40]; and cisplatin/etoposide in head and neck cancer [27]; as well as tumor necrosis factor–related apoptosis-inducing ligand (TRAIL) in renal, prostate, and lung cancer cells [41].

9.3.1 Leukemia

In primary AML specimens, monotherapeutic ABT-737 has been shown to induce apoptosis at sub-micromolar levels. In AML specimens with activating FLT3 mutations, ABT-737 synergized with FLT3 inhibitors [24]. Synergy against AML has also been observed with ABT-737 in combination with sorafenib [42], the protein synthesis inhibitor silvestrol [43], or the MEK inhibitor PD0325901 [44]. Importantly, although AMLs exhibit sensitivity to ABT-737, normal peripheral blood mononuclear cells (PBMCs) appear to be resistant [20].

Testing of 7 ALL cell lines has revealed synergy between ABT-737 and L-asparaginase (7/7), vincristine (7/7), and dexamethasone (5/7), without an impact on normal PBMCs [26]. Synergy with the retinoid N-(4-hydroxyphenyl)retinamide (4-HPR) has also been observed [45]. In a mouse ALL xenograft model, the combination of ABT-737 and L-asparaginase significantly increased event-free survival when compared with each agent alone [26]. In pediatric ALL xenograft tumors,

ABT-737 showed single-drug activity and potentiated the anti-leukemic effects of several clinically relevant agents, including L-asparaginase, topotecan, vincristine, and etoposide [46]. Similarly, in adult T-cell leukemia, ABT-737 induced significant apoptosis as a single agent in cell lines, primary patient tumors, and xenograft tumors, and synergized with other cytotoxic agents [47].

Analysis of 30 primary CLL specimens demonstrated sensitivity to ABT-737 in 21 of the cases, while the remaining samples demonstrated heightened sensitivities following the addition of at least one cytotoxic agent [48]. However, cells with deletion of 17p, a poor prognostic indicator for CLL, were less sensitive [49]. In chronic myeloid leukemia (CML), ABT-737 induced apoptosis in cell lines and primary samples and prolonged the survival of mice harboring CML xenografts [50]. ABT-737 exhibits synergy with imatinib and INNO-406 in CML [51, 52].

9.3.2 Multiple Myeloma

In MM, ABT-737 demonstrates single-agent activity against a number of cell line models ($IC_{50} < 1\ \mu M$), including those characterized by glucocorticoid resistance [25, 33, 34], with the highest sensitivity observed in cells expressing $Bcl\text{-}2^{High}/Mcl\text{-}1^{Low}$ ratios [53]. An additive effect was seen when ABT-737 was combined with glucocorticoids, melphalan, and bortezomib [34], while synergy was observed in combination with Notch pathway inhibitor [54]. Importantly, ABT-737 did not affect the colony-forming potential of human PBMCs [54] and bone marrow-derived progenitors [33]. In a myeloma xenograft model, ABT-737 induced dose-dependent tumor regression [25], and combination with Notch inhibitor resulted in anti-tumor activity superior to either agent alone [54].

9.3.3 Lymphoma

In several lymphoma cell lines, including those representing MCL and DLBCL, ABT-737 showed dose-dependent cytotoxicity and synergized with proteasome inhibitors [55] and the histone deacetylase inhibitor vorinostat [56]. Combination with bortezomib also demonstrated synergy in primary samples of MCL, DLBCL, and CLL, but no significant cytotoxic effects were observed in PBMCs from healthy donors [55]. In Hodgkin's lymphoma (HL) cell lines, ABT-737 exhibits dose- and time-dependant cytotoxicity as a single agent and enhanced the activity of several conventional anti-lymphoma agents [57]. In cutaneous T-cell lymphoma (CTCL), ABT-737 was found to synergize with the pan-histone deacetylase inhibitor panobinostat [58]. Also, in a murine model of c-*myc*-driven B-cell lymphoma, ABT-737 improved overall survival when compared to vehicle alone, even in the face of Bcl-2 overexpression [29].

9.3.4 Lung Cancer

As mentioned above, ABT-737 exhibits single-agent activity against SCLC cell lines [28] and synergizes with actinomycin D in these models [59]. In mice harboring SCLC xenograft tumors, intraperitoneal administration of ABT-737 (75–100 mg/kg daily) for 3 weeks resulted in complete regression of the tumors [17]. Also, ABT-737 synergizes with actinomycin D [38] and enhances gefitinib-induced apoptosis [60] in NSCLC cell lines.

9.3.5 Gastrointestinal Malignancies

In hepatoblastoma cell lines, ABT-737 and GX15-070 (see Sects. 9.7, 9.8, 9.9) exhibit additive effects when combined with standard chemotherapy [61]. ABT-737 also synergizes with sorafenib against hepatocellular carcinoma (HCC) cell lines and HCC xenograft tumors [62, 63]. Inhibition of cholangiocarcinoma cell line proliferation with ABT-737 is observed with IC_{50} values in the 4–17 μM range, and synergy with zoledronic acid has also been reported [64]. In pancreatic cancer cell lines, ABT-737 synergizes with actinomycin D [38] and TRAIL [65]. Similarly, in human colorectal carcinoma (CRC) cell lines, ABT-737 induced dose-dependent apoptosis and synergized with irinotecan [66] and oxaliplatin [67]. ABT-737 also enhanced celecoxib-induced apoptosis and autophagy [68], and overcame resistance to immunotoxin-mediated apoptosis in CRC cell lines [69]. Single-agent ABT-737 activity against imatinib-sensitive and imatinib-resistant gastrointestinal stromal tumor (GIST) cell lines (IC_{50}s of 1–10 μM) has been reported. Furthermore, potent synergy with imatinib was observed in these models [70].

9.3.6 Breast and Ovarian Cancer

ABT-737 enhances cisplatin-induced apoptosis [71] and synergizes with GSIXII, a γ-secretase inhibitor [72], as well as GDC-0941, a phosphoinositide 3-kinase (PI3K) inhibitor, against breast cancer cell lines and xenograft tumors [73]. Additional studies have shown that ABT-737 alone is ineffective against breast cancer xenograft tumors exhibiting high Bcl-2 levels, but is effective in combination with docetaxel to significantly improve anti-tumor effects and overall survival [74]. In ovarian cancer cell lines and xenograft tumors, ABT-737 enhances sensitivity to carboplatin [75].

9.3.7 Central Nervous System Malignancies

In glioblastoma cells, the cytotoxic activities of vincristine, etoposide, bortezomib, and TRAIL have been reported to be enhanced by co-treatment with ABT-737 [76, 77]. Prolonged survival in glioblastoma xenograft tumor models was also observed [77].

ABT-737 synergizes with fenretinide in neuroblastoma cell lines, and the combination increased event-free survival in mice harboring neuroblastoma xenograft tumors [78].

9.3.8 Genitourinary Malignancies

Renal cell carcinoma (RCC) cell lines exhibit very little sensitivity to ABT-737 as a single agent. However, strong super-additive effects were observed when ABT-737 was combined with etoposide, vinblastine, or paclitaxel [79]. In prostate cancer, ABT-737 synergizes with docetaxel in vitro [80] and with Pim kinase inhibitors both in vitro and in vivo [81].

9.3.9 Other Malignancies

Single-agent ABT-737 is largely ineffective against melanoma cell lines, but demonstrates synergy when combined with temodar [82], dacarbazine, fotemustine, imiquimod [83], pseudomonas exotoxin A [84], and p38 MAPK inhibitor [85]. Similarly, ABT-737 alone is ineffective against head and neck squamous cell carcinoma (HNSCC) cell lines, but produces marked synergy characterized by NOXA upregulation when combined with cisplatin or etoposide [27]. In human erythroid cells expressing mutant JAK2, ABT-737 enhances apoptosis induced by JAK2 inhibitor [86]. ABT-737 also enhances the effect of interferon-α (IFN-α) against JAK2V617F-positive polycythemia vera hematopoietic progenitor cells [87]. Interestingly, ABT-737 induces apoptosis of mast cells in vitro and in vivo [88] and is effective in treating animal models of arthritis and lupus [89].

9.4 ABT-263 (Navitoclax): Discovery, Mechanism of Action, and Mechanisms of Resistance

Despite the potency of ABT-737 in targeting Bcl-2 and Bcl-X_L, the compound is not orally bioavailable and exhibits low solubility, hindering intravenous delivery. Hence, Abbott Laboratories sought to develop second-generation derivatives that would address these issues. Through medicinal chemistry efforts and further structure-based rational design, an effective ABT-737 derivative named ABT-263 (see Fig. 9.1b; ABT-263 is also called navitoclax) was developed [18]. ABT-263, like ABT-737, was shown to bind to purified anti-apoptotic Bcl-2 family proteins with high affinity (Ki's < 1 nM for Bcl-2, Bcl-X_L, and Bcl-w). The compound acts by disrupting the interactions of Bcl-2 and Bcl-X_L with proapoptotic Bcl-2 family members, leading to the Bax activation and induction of the intrinsic, mitochondrial-mediated apoptosis pathway. In preclinical models, the oral bioavailability of ABT-263 ranged between 20 and 50 %, depending on the formulation [18]. Also, like ABT-737, the new compound does not bind and inhibit Mcl-1 [18]. Therefore, overexpression of Mcl-1 represents a primary mechanism of resistance to single-agent ABT-263.

9.5 ABT-263 in Preclinical Models

Initial reports demonstrated that ABT-263 is cytotoxic for cell lines representing SCLC and various hematologic malignancies [18]. ABT-263 also induced complete tumor regression in xenograft models of SCLC [18, 90] and ALL [18]. However, in xenograft models of aggressive B-cell lymphoma and MM, single-agent ABT-263 exhibited only modest or no activity. On the other hand, ABT-263 significantly enhances the activity of other clinically relevant agents in xenograft tumor models [18], including enhancement of erlotinib activity and the activities of several conventional chemotherapy drugs [91, 92].

9.5.1 Solid Tumors

In NSCL adenocarcinoma cell lines, ABT-263 demonstrates synergy with Src inhibitors in promoting anoikis [93]. In SCLC, the combination of ABT-263 with an immunotoxin targeting the transferrin receptor was reported to promote synergistic killing of ABT-263-resistant cell lines and to have additive effects against ABT-263-sensitive lines [94]. ABT-263 also sensitizes HCC cell lines to TRAIL-induced apoptosis, while sparing normal liver cells [95]. In addition, ABT-263 enhances apoptosis induction when combined with a survivin inhibitor (YM-155) in HCC cell lines, but not in normal human hepatocytes [96]. When combined with the glycolysis inhibitor 2-deoxyglucose (2-DG), ABT-263 enhanced apoptotic death in multiple cell line models [97]. Moreover, the combination of 2-DG and ABT-263 resulted in improved survival of mice harboring tumors derived from PPC-1 cells, a highly chemoresistant prostate cancer cell line [97]. Lastly, in a panel of 27 ovarian cancer cell lines, ABT-263 demonstrated strong synergy in combination with paclitaxel in roughly half of the cell lines [98].

9.5.2 Hematologic Malignancies

ABT-263 demonstrates single-agent activity against DLBCL cell lines characterized by poor prognostic features (c-Myc and Bcl-2 overexpression) and additive/synergistic effects in combination with other conventional cytotoxic agents [99]. In MCL cell lines and primary patient specimens, ABT-263 synergizes with the histone deacetylase inhibitor, vorinostat [100]. The addition of ABT-263 to rapamycin enhances apoptosis in FL cell lines and promotes tumor regression in vivo [101]. ABT-263 also enhances the activity of bendamustine against DLBCL, MCL, and Burkitt's lymphoma in vivo and improves the response to a bendamustine/rituximab regimen in a subset of these xenograft tumors [102].

9.6 ABT-263 in Clinical Trials

9.6.1 Leukemia

A phase I dose-escalation study of single-agent ABT-263 has been conducted in 29 patients with relapsed or refractory CLL [103]. Patients received ABT-263 daily for 14 or 21 days of a 21-day cycle. At doses of the drug ≥110 mg/d, 35 % of patients manifested a partial response, while 27 % had stable disease for more than six months. ABT-263 activity was observed in poor risk patients, including those with bulky disease, deletion of 17p, and those refractory to fludarabine. Thrombocytopenia was dose-dependent and a major dose-limiting toxicity (DLT). The authors concluded that ABT-263 warrants further evaluation as a single agent or in combination with other agents in this population [103].

9.6.2 Lymphoma

In a phase I trial of relapsed or refractory lymphoid malignancies, 55 patients were enrolled in a dose-escalation study of single-agent ABT-263 [104]. ABT-263 was given in an intermittent schedule once daily (14 out of 21 days) or continuously once daily (21 out of 21 days). Grade 1/2 diarrhea and fatigue were the most common toxicities observed, while grade 3/4 thrombocytopenia and neutropenia were the most serious ones. Five DLTs were observed on the intermittent dosing schedule: one grade 3 cardiac arrhythmia, one grade 4 thrombocytopenia, one grade 3 transaminase elevation, and two hospital admissions for pleural effusion and bronchitis. Due to the rapid and dose-dependent occurrence of acute thrombocytopenia, the continuous schedule was preceded by a lead-in dose of 150 mg for 7–14 days on cycle one. There were three DLTs observed on the continuous schedule: one grade 4 thrombocytopenia, one grade 3 gastrointestinal bleed, and one grade 3 transaminase elevation. Partial responses were observed in 22 % of patients (10/46), with clinical responses seen across different histologies and doses. The authors selected the following regimen for a phase II study: 150 mg daily lead-in dose given for seven days before the first cycle, followed by 325 mg daily administered in a continuous schedule (21/21 days) on subsequent cycles [104].

9.6.3 Solid Tumors

A phase I dose-escalation trial of single-agent ABT-263 in solid tumors enrolled 47 patients, including 29 with SCLC or pulmonary carcinoid [105]. Thirty-five patients were enrolled in the intermittent cohort where ABT-263 was started on day minus 3, and 12 patients were enrolled in the continuous dosing cohort where ABT-263 was given as a 1-week lead-in dose of 150 mg daily followed by continuous daily administration. Common toxicities were grade 1/2 gastrointestinal side effects and fatigue. Dose- and schedule-dependent thrombocytopenia was observed in all patients. A confirmed partial response lasting more than two years was seen in one patient with SCLC, and stable disease was seen in eight patients with SCLC/carcinoid tumor. The MTD of 325 mg was reached. The authors concluded that ABT-263 was safe and well tolerated, with encouraging preliminary results in SCLC and dose-dependent thrombocytopenia as a side effect [105]. In the phase IIa trial, 36 patients received ABT-263 with an initial lead-in dose of 150 mg daily for seven days, followed by 325 mg daily for a 21-day cycle. Thrombocytopenia was the most common toxicity (41 % with grade 3/4). A partial response was observed in one (2.6 %) patient, while 9 (23 %) patients had stable disease. The authors concluded that ABT-263 exhibits limited single-agent activity in advanced and recurrent SCLC and future studies should focus on combination with other agents [106].

9.7 GX15-070 (Obatoclax): Discovery and Mechanism of Action

A screen of natural compounds for Bcl-2 family inhibitors by Gemin X Pharmaceuticals led to the identification of a chemotype from the polypyrrole class of molecules [107] that was further developed as a non-prodigiosin compound and given the name GX15-070 (Fig. 9.1c; also called obatoclax). The clinical formulation of GX15-070 is labeled obatoclax mesylate [108]. Unlike ABT-737, which binds to selected anti-apoptotic Bcl-2 family members (Bcl-2, Bcl-X_L, and Bcl-w), GX15-070 binds and inhibits all anti-apoptotic Bcl-2 family members (Bcl-2, Bcl-X_L, Bcl-w, Mcl-1, Bfl-1, Bcl-B) [109]. However, despite the broadened inhibitory activity of GX15-070, it exhibits somewhat reduced affinity for Bcl-2 and Bcl-X_L relative to ABT-737/ABT-263 (Ki = 220 nM for Bcl-2 and ~500 nM for Bcl-X_L, Mcl-1, and Bcl-w) [110]. None the less, the ability of GX15-070 to inhibit Mcl-1 represents a distinct advantage of this compound, since expression of Mcl-1 has been shown to mediate resistance to ABT-737/ABT-263. GX15-070 efficiently inhibits the interactions between Mcl-1 and proapoptotic proteins, including Bak, leading to the activation of the intrinsic apoptosis pathway [110]. Kidney epithelial cells derived from mice deficient in Bax and Bak expression fail to activate caspases when treated with GX15-070, underscoring the importance of these proteins and the intrinsic pathway in the mechanism of GX15-070 action [108].

In addition to inhibiting anti-apoptotic Bcl-2 family members, GX15-070 has also been reported to inhibit the activation of NFκB and to upregulate expression of the TRAIL receptor DR5 [111].

9.8 GX15-070 in Preclinical Models

9.8.1 Leukemia

Using AML cell lines, Konopleva et al. [20] determined that GX15-070 induces time- and dose-dependent cell death through activation of the intrinsic apoptotic pathway, with IC_{50} values in the low micromolar range (1.1–5.0 μM). At lower concentrations, GX15-070 induced cell cycle arrest at S-G2 phase. Treatment of AML progenitor cells with GX15-070 resulted in potent induction of apoptosis ($IC_{50} = 3.6 \pm 1.2$ μM), and clonogenicity was inhibited by concentrations in the 75–100 nM range. GX15-070 synergized with ABT-737 against AML cell lines and synergized with Ara-C against leukemia cell lines and primary AML specimens [20].

In other studies, GX15-070 demonstrated synergy with a histone deacetylase inhibitor [112], and the multi-kinase inhibitor sorafenib against AML cell lines and primary AML cells, but not against normal $CD34^+$ cells [113]. Combination of GX15-070 and sorafenib markedly reduced AML tumor growth in a xenograft model and significantly enhanced the survival of mice bearing these tumors when compared with either agent alone [113]. Single-agent GX15-070 was found to inhibit growth and activate both apoptosis and autophagy in ALL cell lines that were either sensitive or resistant to dexamethasone [114]. In primary CLL cells derived from patients who had not received treatment in the past 3 months, GX15-070 demonstrated killing activity in most specimens, including those characterized by poor prognostic features such as deletion of 11q or 17p [115]. However, ZAP-70-positive cases, associated with poor prognosis, were less sensitive to GX15-070 than were ZAP-70 negative cases. Additionally, inhibition of extracellular signal-regulated kinase (ERK)-1/2 was found to inhibit Bcl-2 phosphorylation in CLL cells, increasing the sensitivities of these cells to GX15-070 alone or in combination with proteasome inhibitors [115].

9.8.2 Lymphoma

In B-cell lymphoma cell lines, GX15-070 has been reported to enhance rituximab activity [116]. Moreover, GX15-070 synergized with chemotherapy to induce apoptosis in both rituximab/chemotherapy-sensitive (RSCL) and rituximab/chemotherapy-resistant cell lines and primary tumor cells from B-cell non-Hodgkin's lymphoma patients [116]. Via NFκB inhibition and inhibition of the DR5 repressor Yin Yang 1 (YY1), GX15-070 induced DR5 expression and sensitized B-cell lymphoma cells to TRAIL-induced apoptosis [111]. In Hodgkin's lymphomas, ABT-737 and GX15-070 demonstrated synergy with the histone deacetylase inhibitor SNDX-275 [117], while in MCL cell lines and primary cells, GX15-070 demonstrated single-agent activity and synergy with bortezomib [118].

9.8.3 Multiple Myeloma

As a single agent, GX15-070 reduced viability in 15 human MM cell lines, including those resistant to melphalan and dexamethasone, with a mean IC_{50} at 246 nM [19]. GX15-070 also induced cell death in primary patient specimens and demonstrated an additive effect with the myeloma drugs melphalan, dexamethasone, and bortezomib [19]. Recently, the pan-CDK inhibitor flavopiridol was found to synergize with GX15-070 in both drug-naïve and drug-resistant MM cells [119].

9.8.4 Solid Tumors

Single-agent GX15-070 is largely ineffective against melanoma cell lines and primary cells, although enhanced killing is observed when the drug is combined with ER stress inducers such as tunicamycin or thapsigargin [120]. By contrast, combination of ABT-737 with ER stress inducers resulted in only a very modest induction of apoptosis in melanoma cells [120]. In breast cancer cell lines, GX15-070 showed synergy in combination with the EGFR-HER-2/neu inhibitors lapatinib and GW2974 [121]. In CNS tumor cell lines, the combination of GX15-070 with lapatinib is also effective, except in cells lacking PTEN [122]. These studies further showed that pre-treatment with GX15-070 prior to addition of lapatinib resulted in improved killing compared with simultaneous treatment. The effectiveness of the GX15-070/lapatinib combination has also been demonstrated in vivo [122].

Treatment of esophageal cancer cell lines with GX15-070 alone results in dose-dependent growth inhibition, with IC_{50} values between 1.0 and 3.1 μM. Synergy was observed when the drug was combined with carboplatin or 5-fluorouracil [123]. In pancreatic cells, GX15-070 significantly reduced tumor viability and synergized with TRAIL [124]. Single-agent GX15-070 was reported in clonogenic survival assays of cholangiocarcinoma cell lines, with IC_{50}s ranging from 5 to 100 nM [125]. Experiments in a syngeneic rat orthotopic model confirmed the single-agent activity of GX15-070 in promoting overall survival when compared with vehicle control (44 versus 23 days) [125].

9.9 GX15-070 in Clinical Trials

9.9.1 Leukemia

The first phase I study of single-agent obatoclax mesylate (the clinical formulation of GX15-070; hereafter referred to as obatoclax) was undertaken in patients with refractory hematological malignancies [126]. A total of 44 patients with refractory AML, ALL, CLL, myelodysplasia (MDS), or CML in blast crisis were given 24-h infusions of the drug, with a median of five infusions per patient. The drug was found to be well tolerated, aside from grade 1/2 CNS symptoms, and the highest planned dose was reached without any DLT. One AML patient obtained complete remission for eight months, but subsequently relapsed, causing the authors to suggest that GX15-070 induced proliferation arrest, as opposed to differentiation. The final recommended dose for phase II was 28 mg/m^2, given as an infusion for 24 h up to four consecutive days. The authors concluded that obatoclax was well tolerated and suggested further investigation in patients with leukemia and myelodysplasia [126].

A subsequent phase I trial of single-agent obatoclax enrolled 26 CLL patients [127]. Patients were treated with single-agent obatoclax doses ranging from 3.5 to 14 mg/m^2 as a 1-h infusion and from 20 to 40 mg/m^2 as a 3-h infusion every 3 weeks. The observed neurological dose-limiting reactions (euphoria, ataxia, somnolence) were dose related, but quickly resolved after cessation of drug infusion. Only one patient achieved partial remission at an obatoclax dose of 3.5 mg/m^2. This patient had been pre-treated with fludarabine, Rituxan, and alemtuzumab but was the only patient in the trial who was naïve to alkylating agents. In other patients, there was reduction in transfusion dependence. The MTD was determined to be 28 mg/m^2 over 3 h every three weeks and was recommended for phase II studies. The authors concluded that the biologic activity of obatoclax is not only dependant on the dose, but also on tumor biology. The activity was modest in this population and suggested investigation of the drug in a less heavily pre-treated population and in combination with other agents [127].

9.9.2 Lymphoma

In another series of trials, 35 patients with previously treated lymphoma and solid tumors, who were not candidates for standard therapies, were enrolled in two phase I trials (GX001 and GX005) evaluating the safety and tolerability of single-agent obatoclax given as 1-h (GX001) or 3-h (GX005) weekly infusions [128]. With the 1-h infusion schedule, neurological DLT (somnolence) was observed at 5 and 7 mg/m^2. The MTD was 1.25 mg/m^2. Patients on the 3-h infusion schedule showed better tolerability, with an MTD of 20 mg/m^2. Stable disease by RECIST criteria was achieved in 25 % (2/8) of patients with the 1-h infusion and in 18 % (5/27) of patients with the 3-h infusion. A partial response was observed in one patient with large cell lymphoma stage IV assigned to the 28 mg/m^2 dose group, who received a total of 32 weeks of therapy. The authors concluded that the 3-h weekly infusion of obatoclax was better tolerated with an MTD of 20 mg/m^2 and evidence of clinical activity [128].

Recently, Oki et al. [129] reported a phase II trial in patients with relapsed or refractory classical Hodgkin's lymphoma (cHL) treated with single-agent obatoclax administered intravenously at 60 mg over 24 h and given every two weeks. Thirteen patients received at least one dose of obatoclax, with a median of four cycles per patient (range 1–24). The drug was found to be well tolerated, with grade 1 toxicities: dizziness ($n = 5$), euphoria ($n = 3$), and hypotension ($n = 1$). There were no objective responses; 38 % (5/13) of patients had stable disease. The authors concluded that obatoclax showed limited activity in this heavily pre-treated population of cHL. They recommended investigation of more potent Bcl-2 family inhibitors with pharmacodynamic studies to ensure target inhibition and biomarker analysis for plausible patient selection [129].

9.9.3 Myelofibrosis

Twenty-two patients who had previously been treated were enrolled in a multicenter, open-label, non-comparative phase II study of single-agent obatoclax administered as a 24-h infusion every 2 weeks at a fixed dose of 60 mg [130]. Patients received a median of 7 cycles. No objective responses were observed. Only one patient had clinical improvement (a decrease in transfusion requirement). The most common side effects were low-grade ataxia and fatigue, observed in 50 % of the patients. One patient had a dose reduction secondary to toxicity, and two patients were taken off the study due to grade 3 toxicity (ataxia and heart failure). The authors concluded that obatoclax showed no clinical activity in this patient population at the dose and schedule used in the study [130].

9.9.4 Solid Tumors

A phase I trial studied the safety, MTD, and early anti-tumor activity of the combination of topotecan and obatoclax in patients ($n = 14$) with solid tumors who would benefit from treatment with topotecan. Obatoclax was given at a starting dose of 14 mg/m^2 over a 3-h intravenous infusion on a 3-week cycle, while topotecan was given at 1.25 mg/m^2 on days 1–5 on a every 3-week cycle [131]. Of the observed toxicities, 88 % were grade 1 and 2, mainly neurologic, in the form of ataxia, somnolence, mood alterations, and cognitive dysfunction. Two-fifths of the patients developed grade 3 neurological DLTs at the 20 mg/m^2 dose. Two patients (SCLC) exhibited partial responses, and four (three with SCLC and one with pulmonary carcinoid) had stable disease. The recommended dose for phase

II trial was obatoclax at 14 mg/m^2 on days 1–3, in combination with topotecan at 1.25 mg/m^2 on days 1–5 of a 3-week cycle [131]. Given the preliminary efficacy data seen in SCLC, an open-label, single-arm, phase II extension trial was undertaken [132]. A total of 9 patients with recurrent SCLC were enrolled. They received a median of two cycles. The most common grade 3/4 toxicities were hematologic and ataxia. There were no objective responses, with 56 % (5/9) of patients exhibiting stable disease. The authors concluded that the combination did not achieve a better response rate than was previously seen with topotecan alone in relapsed SCLC after first-line platinum therapy [132].

In another phase I trial, the safety of obatoclax in combination with standard chemotherapy was evaluated in patients with extensive stage SCLC [133]. Twenty-five patients were treated with carboplatin (area under the curve of 5) on day 1 and etoposide (100 mg/m^2 on days 1–3) in combination with escalating doses of obatoclax given either as a 3- or 24-h infusion on days 1–3 of a 21-day cycle. Compared with the 24-h infusion cohort, there were more neurological DLTs (somnolence, euphoria, and disorientation) in the 3-h infusion cohort. However, these neurological toxicities were transient, resolving after the end of the infusion and without sequelae. The MTD for the 3-h cohort was 30 mg/day, while it was not reached for the 24-h cohort. The response rate was higher in the 3-h cohort (81 versus 44 %). The authors concluded that the 3-h infusion was associated with more neurological side effects, but better efficacy than the 24-h cohort. For further studies in SCLC, they recommended an obatoclax daily dose of 30 mg given intravenously over 3 h on three consecutive days [133].

Acknowledgments We apologize to those authors whose works have not been cited due to space limitations, or our oversight. This work was supported by National Institutes of Health grants R01 CA137260 and P50 CA097190.

References

1. Finnegan NM, Curtin JF, Prevost G, Morgan B, Cotter TG (2001) Induction of apoptosis in prostate carcinoma cells by BH3 peptides which inhibit Bak/Bcl-2 interactions. Br J Cancer 85(1):115–121 Epub 2001/07/05
2. Certo M, Del Gaizo Moore V, Nishino M, Wei G, Korsmeyer S, Armstrong SA et al (2006) Mitochondria primed by death signals determine cellular addiction to antiapoptotic BCL-2 family members. Cancer Cell 9(5):351–365 Epub 2006/05/16
3. Chen L, Willis SN, Wei A, Smith BJ, Fletcher JI, Hinds MG et al (2005) Differential targeting of prosurvival Bcl-2 proteins by their BH3-only ligands allows complementary apoptotic function. Mol Cell 17(3):393–403 Epub 2005/02/08
4. Willis SN, Fletcher JI, Kaufmann T, van Delft MF, Chen L, Czabotar PE et al (2007) Apoptosis initiated when BH3 ligands engage multiple Bcl-2 homologs, not Bax or Bak. Science 315(5813):856–859 Epub 2007/02/10
5. Tzung SP, Kim KM, Basanez G, Giedt CD, Simon J, Zimmerberg J et al (2001) Antimycin A mimics a cell-death-inducing Bcl-2 homology domain 3. Nat Cell Biol 3(2):183–191 Epub 2001/02/15

6. Wang JL, Liu D, Zhang ZJ, Shan S, Han X, Srinivasula SM et al (2000) Structure-based discovery of an organic compound that binds Bcl-2 protein and induces apoptosis of tumor cells. Proc Natl Acad Sci U S A 97(13):7124–7129 Epub 2000/06/22
7. Degterev A, Lugovskoy A, Cardone M, Mulley B, Wagner G, Mitchison T et al (2001) Identification of small-molecule inhibitors of interaction between the BH3 domain and Bcl-xL. Nat Cell Biol 3(2):173–182 Epub 2001/02/15
8. Nishikawa T, Nakajima T, Moriguchi M, Jo M, Sekoguchi S, Ishii M et al (2006) A green tea polyphenol, epigallocatechin-3-gallate, induces apoptosis of human hepatocellular carcinoma, possibly through inhibition of Bcl-2 family proteins. J Hepatol 44(6):1074–1082 Epub 2006/02/17
9. Leone M, Zhai D, Sareth S, Kitada S, Reed JC, Pellecchia M (2003) Cancer prevention by tea polyphenols is linked to their direct inhibition of antiapoptotic Bcl-2-family proteins. Cancer Res 63(23):8118–8121 Epub 2003/12/18
10. Chan SL, Lee MC, Tan KO, Yang LK, Lee AS, Flotow H et al (2003) Identification of chelerythrine as an inhibitor of BclXL function. J Biol Chem 278(23):20453–20456 Epub 2003/04/19
11. Zhang YH, Bhunia A, Wan KF, Lee MC, Chan SL, Yu VC et al (2006) Chelerythrine and sanguinarine dock at distinct sites on BclXL that are not the classic BH3 binding cleft. J Mol Biol 364(3):536–549 Epub 2006/10/03
12. Kitada S, Leone M, Sareth S, Zhai D, Reed JC, Pellecchia M (2003) Discovery, characterization, and structure-activity relationships studies of proapoptotic polyphenols targeting B-cell lymphocyte/leukemia-2 proteins. J Med Chem 46(20):4259–4264 Epub 2003/09/19
13. Wang G, Nikolovska-Coleska Z, Yang CY, Wang R, Tang G, Guo J et al (2006) Structure-based design of potent small-molecule inhibitors of anti-apoptotic Bcl-2 proteins. J Med Chem 49(21):6139–6142 Epub 2006/10/13
14. Qiu J, Levin LR, Buck J, Reidenberg MM (2002) Different pathways of cell killing by gossypol enantiomers. Exp Biol Med (Maywood) 227(6):398–401 Epub 2002/05/31
15. Sun Y, Wu J, Aboukameel A, Banerjee S, Arnold AA, Chen J et al (2008) Apogossypolone, a nonpeptidic small molecule inhibitor targeting Bcl-2 family proteins, effectively inhibits growth of diffuse large cell lymphoma cells in vitro and in vivo. Cancer Biol Ther 7(9):1418–1426 Epub 2008/09/05
16. Becattini B, Kitada S, Leone M, Monosov E, Chandler S, Zhai D et al (2004) Rational design and real time, in-cell detection of the proapoptotic activity of a novel compound targeting Bcl-X(L). Chem Biol 11(3):389–395 Epub 2004/05/05
17. Oltersdorf T, Elmore SW, Shoemaker AR, Armstrong RC, Augeri DJ, Belli BA et al (2005) An inhibitor of Bcl-2 family proteins induces regression of solid tumours. Nature 435(7042):677–681 Epub 2005/05/20
18. Tse C, Shoemaker AR, Adickes J, Anderson MG, Chen J, Jin S et al (2008) ABT-263: a potent and orally bioavailable Bcl-2 family inhibitor. Cancer Res 68(9):3421–3428 Epub 2008/05/03
19. Trudel S, Li ZH, Rauw J, Tiedemann RE, Wen XY, Stewart AK (2007) Preclinical studies of the pan-Bcl inhibitor obatoclax (GX015-070) in multiple myeloma. Blood 109(12):5430–5438 Epub 2007/03/03
20. Konopleva M, Contractor R, Tsao T, Samudio I, Ruvolo PP, Kitada S et al (2006) Mechanisms of apoptosis sensitivity and resistance to the BH3 mimetic ABT-737 in acute myeloid leukemia. Cancer Cell 10(5):375–388 Epub 2006/11/14
21. Del Gaizo Moore V, Brown JR, Certo M, Love TM, Novina CD, Letai A (2007) Chronic lymphocytic leukemia requires BCL2 to sequester prodeath BIM, explaining sensitivity to BCL2 antagonist ABT-737. J Clin Invest 117(1):112–121. Epub 2007/01/04
22. Howard AN, Bridges KA, Meyn RE, Chandra J (2009) ABT-737, a BH3 mimetic, induces glutathione depletion and oxidative stress. Cancer Chemother Pharmacol 65(1):41–54 Epub 2009/05/01
23. Maiuri MC, Criollo A, Tasdemir E, Vicencio JM, Tajeddine N, Hickman JA et al (2007) BH3-only proteins and BH3 mimetics induce autophagy by competitively disrupting the interaction between Beclin 1 and Bcl-2/Bcl-X(L). Autophagy 3(4):374–376 Epub 2007/04/18

24. Kohl TM, Hellinger C, Ahmed F, Buske C, Hiddemann W, Bohlander SK et al (2007) BH3 mimetic ABT-737 neutralizes resistance to FLT3 inhibitor treatment mediated by FLT3-independent expression of BCL2 in primary AML blasts. Leukemia 21(8):1763–1772 Epub 2007/06/08
25. Trudel S, Stewart AK, Li Z, Shu Y, Liang SB, Trieu Y et al (2007) The Bcl-2 family protein inhibitor, ABT-737, has substantial antimyeloma activity and shows synergistic effect with dexamethasone and melphalan. Clin Cancer Res 13(2 Pt 1):621–629 Epub 2007/01/27
26. Kang MH, Kang YH, Szymanska B, Wilczynska-Kalak U, Sheard MA, Harned TM et al (2007) Activity of vincristine, L-ASP, and dexamethasone against acute lymphoblastic leukemia is enhanced by the BH3-mimetic ABT-737 in vitro and in vivo. Blood 110(6):2057–2066 Epub 2007/05/31
27. Li R, Zang Y, Li C, Patel NS, Grandis JR, Johnson DE (2009) ABT-737 synergizes with chemotherapy to kill head and neck squamous cell carcinoma cells via a Noxa-mediated pathway. Mol Pharmacol 75(5):1231–1239 Epub 2009/02/28
28. Tahir SK, Yang X, Anderson MG, Morgan-Lappe SE, Sarthy AV, Chen J et al (2007) Influence of Bcl-2 family members on the cellular response of small-cell lung cancer cell lines to ABT-737. Cancer Res 67(3):1176–1183 Epub 2007/02/07
29. van Delft MF, Wei AH, Mason KD, Vandenberg CJ, Chen L, Czabotar PE et al (2006) The BH3 mimetic ABT-737 targets selective Bcl-2 proteins and efficiently induces apoptosis via Bak/Bax if Mcl-1 is neutralized. Cancer Cell 10(5):389–399 Epub 2006/11/14
30. Oda E, Ohki R, Murasawa H, Nemoto J, Shibue T, Yamashita T et al (2000) Noxa, a BH3-only member of the Bcl-2 family and candidate mediator of p53-induced apoptosis. Science 288(5468):1053–1058 Epub 2000/05/12
31. Deng J, Carlson N, Takeyama K, Dal Cin P, Shipp M, Letai A (2007) BH3 profiling identifies three distinct classes of apoptotic blocks to predict response to ABT-737 and conventional chemotherapeutic agents. Cancer Cell 12(2):171–185 Epub 2007/08/19
32. Chen S, Dai Y, Harada H, Dent P, Grant S (2007) Mcl-1 down-regulation potentiates ABT-737 lethality by cooperatively inducing Bak activation and Bax translocation. Cancer Res 67(2):782–791 Epub 2007/01/20
33. Kline MP, Rajkumar SV, Timm MM, Kimlinger TK, Haug JL, Lust JA et al (2007) ABT-737, an inhibitor of Bcl-2 family proteins, is a potent inducer of apoptosis in multiple myeloma cells. Leukemia 21(7):1549–1560 Epub 2007/04/27
34. Chauhan D, Velankar M, Brahmandam M, Hideshima T, Podar K, Richardson P et al (2007) A novel Bcl-2/Bcl-X(L)/Bcl-w inhibitor ABT-737 as therapy in multiple myeloma. Oncogene 26(16):2374–2380 Epub 2006/10/04
35. Miller LA, Goldstein NB, Johannes WU, Walton CH, Fujita M, Norris DA et al (2009) BH3 mimetic ABT-737 and a proteasome inhibitor synergistically kill melanomas through Noxa-dependent apoptosis. J Invest Dermatol 129(4):964–971 Epub 2008/11/07
36. Cragg MS, Jansen ES, Cook M, Harris C, Strasser A, Scott CL (2008) Treatment of B-RAF mutant human tumor cells with a MEK inhibitor requires Bim and is enhanced by a BH3 mimetic. J Clin Invest 118(11):3651–3659 Epub 2008/10/25
37. Bhat UG, Pandit B, Gartel AL (2010) ARC synergizes with ABT-737 to induce apoptosis in human cancer cells. Mol Cancer Ther 9(6):1688–1696 Epub 2010/06/03
38. Olberding KE, Wang X, Zhu Y, Pan J, Rai SN, Li C (2010) Actinomycin D synergistically enhances the efficacy of the BH3 mimetic ABT-737 by downregulating Mcl-1 expression. Cancer Biol Ther 10(9):918–929 Epub 2010/09/08
39. Zhang C, Cai TY, Zhu H, Yang LQ, Jiang H, Dong XW et al (2011) Synergistic antitumor activity of gemcitabine and ABT-737 in vitro and in vivo through disrupting the interaction of USP9X and Mcl-1. Mol Cancer Ther 10(7):1264–1275 Epub 2011/05/14
40. Yin S, Dong Y, Li J, Fan L, Wang L, Lu J et al (2012) Methylseleninic acid potentiates multiple types of cancer cells to ABT-737-induced apoptosis by targeting Mcl-1 and Bad. Apoptosis 17(4):388–399 Epub 2011/12/20
41. Song JH, Kandasamy K, Kraft AS (2008) ABT-737 induces expression of the death receptor 5 and sensitizes human cancer cells to TRAIL-induced apoptosis. J Biol Chem 283(36):25003–25013 Epub 2008/07/05

42. Zhang W, Konopleva M, Ruvolo VR, McQueen T, Evans RL, Bornmann WG et al (2008) Sorafenib induces apoptosis of AML cells via Bim-mediated activation of the intrinsic apoptotic pathway. Leukemia 22(4):808–818 Epub 2008/01/18
43. Cencic R, Carrier M, Trnkus A, Porco JA Jr, Minden M, Pelletier J (2010) Synergistic effect of inhibiting translation initiation in combination with cytotoxic agents in acute myelogenous leukemia cells. Leuk Res 34(4):535–541 Epub 2009/09/04
44. Konopleva M, Milella M, Ruvolo P, Watts JC, Ricciardi MR, Korchin B et al (2012) MEK inhibition enhances ABT-737-induced leukemia cell apoptosis via prevention of ERK-activated MCL-1 induction and modulation of MCL-1/BIM complex. Leukemia 26(4):778–787 Epub 2011/11/09
45. Kang MH, Wan Z, Kang YH, Sposto R, Reynolds CP (2008) Mechanism of synergy of N-(4-hydroxyphenyl)retinamide and ABT-737 in acute lymphoblastic leukemia cell lines: Mcl-1 inactivation. J Natl Cancer Inst 100(8):580–595 Epub 2008/04/10
46. High LM, Szymanska B, Wilczynska-Kalak U, Barber N, O'Brien R, Khaw SL et al (2010) The Bcl-2 homology domain 3 mimetic ABT-737 targets the apoptotic machinery in acute lymphoblastic leukemia resulting in synergistic in vitro and in vivo interactions with established drugs. Mol Pharmacol 77(3):483–494 Epub 2009/12/30
47. Ishitsuka K, Kunami N, Katsuya H, Nogami R, Ishikawa C, Yotsumoto F et al (2012) Targeting Bcl-2 family proteins in adult T-cell leukemia/lymphoma: in vitro and in vivo effects of the novel Bcl-2 family inhibitor ABT-737. Cancer Lett 317(2):218–225 Epub 2011/12/06
48. Mason KD, Khaw SL, Rayeroux KC, Chew E, Lee EF, Fairlie WD et al (2009) The BH3 mimetic compound, ABT-737, synergizes with a range of cytotoxic chemotherapy agents in chronic lymphocytic leukemia. Leukemia 23(11):2034–2041 Epub 2009/07/31
49. Kojima K, Duvvuri S, Ruvolo V, Samaniego F, Younes A, Andreeff M (2012) Decreased sensitivity of 17p-deleted chronic lymphocytic leukemia cells to a small molecule BCL-2 antagonist ABT-737. Cancer 118(4):1023–1031 Epub 2011/07/16
50. Kuroda J, Kimura S, Andreeff M, Ashihara E, Kamitsuji Y, Yokota A et al (2008) ABT-737 is a useful component of combinatory chemotherapies for chronic myeloid leukaemias with diverse drug-resistance mechanisms. Br J Haematol 140(2):181–190 Epub 2007/11/22
51. Kuroda J, Puthalakath H, Cragg MS, Kelly PN, Bouillet P, Huang DC et al (2006) Bim and Bad mediate imatinib-induced killing of Bcr/Abl+ leukemic cells, and resistance due to their loss is overcome by a BH3 mimetic. Proc Natl Acad Sci U S A 103(40):14907–14912 Epub 2006/09/26
52. Kuroda J, Kimura S, Strasser A, Andreeff M, O'Reilly LA, Ashihara E et al (2007) Apoptosis-based dual molecular targeting by INNO-406, a second-generation Bcr-Abl inhibitor, and ABT-737, an inhibitor of antiapoptotic Bcl-2 proteins, against Bcr-Abl-positive leukemia. Cell Death Differ 14(9):1667–1677 Epub 2007/05/19
53. Bodet L, Gomez-Bougie P, Touzeau C, Dousset C, Descamps G, Maiga S et al (2011) ABT-737 is highly effective against molecular subgroups of multiple myeloma. Blood 118(14):3901–3910 Epub 2011/08/13
54. Li M, Chen F, Clifton N, Sullivan DM, Dalton WS, Gabrilovich DI et al (2010) Combined inhibition of Notch signaling and Bcl-2/Bcl-xL results in synergistic antimyeloma effect. Mol Cancer Ther 9(12):3200–3209 Epub 2010/12/17
55. Paoluzzi L, Gonen M, Bhagat G, Furman RR, Gardner JR, Scotto L et al (2008) The BH3-only mimetic ABT-737 synergizes the antineoplastic activity of proteasome inhibitors in lymphoid malignancies. Blood 112(7):2906–2916 Epub 2008/07/02
56. Wiegmans AP, Alsop AE, Bots M, Cluse LA, Williams SP, Banks KM et al (2011) Deciphering the molecular events necessary for synergistic tumor cell apoptosis mediated by the histone deacetylase inhibitor vorinostat and the BH3 mimetic ABT-737. Cancer Res 71(10):3603–3615 Epub 2011/03/15
57. Jayanthan A, Howard SC, Trippett T, Horton T, Whitlock JA, Daisley L et al (2009) Targeting the Bcl-2 family of proteins in Hodgkin lymphoma: in vitro cytotoxicity, target modulation and drug combination studies of the Bcl-2 homology 3 mimetic ABT-737. Leuk Lymphoma 50(7):1174–1182 Epub 2009/06/27

58. Chen J, Fiskus W, Eaton K, Fernandez P, Wang Y, Rao R et al (2009) Cotreatment with BCL-2 antagonist sensitizes cutaneous T-cell lymphoma to lethal action of HDAC7-Nur77-based mechanism. Blood 113(17):4038–4048 Epub 2008/12/17
59. Xu H, Krystal GW (2010) Actinomycin D decreases Mcl-1 expression and acts synergistically with ABT-737 against small cell lung cancer cell lines. Clin Cancer Res 16(17):4392–4400 Epub 2010/08/25
60. Cragg MS, Kuroda J, Puthalakath H, Huang DC, Strasser A (2007) Gefitinib-induced killing of NSCLC cell lines expressing mutant EGFR requires BIM and can be enhanced by BH3 mimetics. PLoS Med 4(10):1681–1689. Discussion 90. Epub 2007/11/02
61. Lieber J, Ellerkamp V, Wenz J, Kirchner B, Warmann SW, Fuchs J et al (2012) Apoptosis sensitizers enhance cytotoxicity in hepatoblastoma cells. Pediatr Surg Int 28(2):149–159 Epub 2011/10/06
62. Galmiche A, Ezzoukhry Z, Francois C, Louandre C, Sabbagh C, Nguyen-Khac E et al (2010) BAD, a proapoptotic member of the BCL2 family, is a potential therapeutic target in hepatocellular carcinoma. Mol Cancer Res 8(8):1116–1125 Epub 2010/07/22
63. Hikita H, Takehara T, Shimizu S, Kodama T, Shigekawa M, Iwase K et al (2010) The Bcl-xL inhibitor, ABT-737, efficiently induces apoptosis and suppresses growth of hepatoma cells in combination with sorafenib. Hepatology 52(4):1310–1321 Epub 2010/08/28
64. Romani AA, Desenzani S, Morganti MM, Baroni MC, Borghetti AF, Soliani P (2011) The BH3-mimetic ABT-737 targets the apoptotic machinery in cholangiocarcinoma cell lines resulting in synergistic interactions with zoledronic acid. Cancer Chemother Pharmacol 67(3):557–567 Epub 2010/05/18
65. Huang S, Sinicrope FA (2008) BH3 mimetic ABT-737 potentiates TRAIL-mediated apoptotic signaling by unsequestering Bim and Bak in human pancreatic cancer cells. Cancer Res 68(8):2944–2951 Epub 2008/04/17
66. Okumura K, Huang S, Sinicrope FA (2008) Induction of Noxa sensitizes human colorectal cancer cells expressing Mcl-1 to the small-molecule Bcl-2/Bcl-xL inhibitor, ABT-737. Clin Cancer Res 14(24):8132–8142 Epub 2008/12/18
67. Raats DA, de Bruijn MT, Steller EJ, Emmink BL, Borel-Rinkes IH, Kranenburg O (2011) Synergistic killing of colorectal cancer cells by oxaliplatin and ABT-737. Cell Oncol (Dordr) 34(4):307–313 Epub 2011/04/07
68. Huang S, Sinicrope FA (2010) Celecoxib-induced apoptosis is enhanced by ABT-737 and by inhibition of autophagy in human colorectal cancer cells. Autophagy 6(2):256–269 Epub 2010/01/28
69. Traini R, Ben-Josef G, Pastrana DV, Moskatel E, Sharma AK, Antignani A et al (2010) ABT-737 overcomes resistance to immunotoxin-mediated apoptosis and enhances the delivery of pseudomonas exotoxin-based proteins to the cell cytosol. Mol Cancer Ther 9(7):2007–2015 Epub 2010/07/01
70. Reynoso D, Nolden LK, Yang D, Dumont SN, Conley AP, Dumont AG et al (2011) Synergistic induction of apoptosis by the Bcl-2 inhibitor ABT-737 and imatinib mesylate in gastrointestinal stromal tumor cells. Mol Oncol 5(1):93–104 Epub 2010/12/01
71. Chen ZJ, Zhang B, Pan SH, Zhao HM, Zhang Y, Feng WH et al (2011) Bcl-2 inhibitor ABT-737 enhances the cisplatin-induced apoptosis in breast cancer T47D cells. Zhonghua Zhong Liu Za Zhi 33(12):891–895 Epub 2012/02/22
72. Seveno C, Loussouarn D, Brechet S, Campone M, Juin P, Barille-Nion S (2012) Gamma-secretase inhibition promotes cell death, Noxa upregulation, and sensitization to BH3 mimetic ABT-737 in human breast cancer cells. Breast Cancer Res 14(3):R96 Epub 2012/06/19
73. Zheng L, Yang W, Zhang C, Ding WJ, Zhu H, Lin NM et al (2011) GDC-0941 sensitizes breast cancer to ABT-737 in vitro and in vivo through promoting the degradation of Mcl-1. Cancer Lett 309(1):27–36 Epub 2011/06/15

74. Oakes SR, Vaillant F, Lim E, Lee L, Breslin K, Feleppa F et al (2012) Sensitization of BCL-2-expressing breast tumors to chemotherapy by the BH3 mimetic ABT-737. Proc Natl Acad Sci U S A 109(8):2766–2771 Epub 2011/07/20
75. Witham J, Valenti MR, De-Haven-Brandon AK, Vidot S, Eccles SA, Kaye SB et al (2007) The Bcl-2/Bcl-XL family inhibitor ABT-737 sensitizes ovarian cancer cells to carboplatin. Clin Cancer Res 13(23):7191–7198 Epub 2007/12/07
76. Premkumar DR, Jane EP, DiDomenico JD, Vukmer NA, Agostino NR, Pollack IF (2012) ABT-737 synergizes with bortezomib to induce apoptosis, mediated by Bid cleavage, Bax activation, and mitochondrial dysfunction in an Akt-dependent context in malignant human glioma cell lines. J Pharmacol Exp Ther 341(3):859–872 Epub 2012/03/07
77. Tagscherer KE, Fassl A, Campos B, Farhadi M, Kraemer A, Bock BC et al (2008) Apoptosis-based treatment of glioblastomas with ABT-737, a novel small molecule inhibitor of Bcl-2 family proteins. Oncogene 27(52):6646–6656 Epub 2008/07/30
78. Fang H, Harned TM, Kalous O, Maldonado V, DeClerck YA, Reynolds CP (2011) Synergistic activity of fenretinide and the Bcl-2 family protein inhibitor ABT-737 against human neuroblastoma. Clin Cancer Res 17(22):7093–7104 Epub 2011/09/22
79. Zall H, Weber A, Besch R, Zantl N, Hacker G (2010) Chemotherapeutic drugs sensitize human renal cell carcinoma cells to ABT-737 by a mechanism involving the Noxa-dependent inactivation of Mcl-1 or A1. Mol Cancer 9:164 Epub 2010/06/26
80. Hao JW, Mao XP, Ding DG, Du GH, Liu ZH (2012) The effect of cell killing by ABT-737 synergized with docetaxel in human prostate cancer PC-3 cells. Zhonghua Wai Ke Za Zhi 50(2):161–165 Epub 2012/04/12
81. Song JH, Kraft AS (2012) Pim kinase inhibitors sensitize prostate cancer cells to apoptosis triggered by Bcl-2 family inhibitor ABT-737. Cancer Res 72(1):294–303 Epub 2011/11/15
82. Reuland SN, Goldstein NB, Partyka KA, Cooper DA, Fujita M, Norris DA et al (2011) The combination of BH3-mimetic ABT-737 with the alkylating agent temozolomide induces strong synergistic killing of melanoma cells independent of p53. PLoS ONE 6(8):e24294 Epub 2011/09/08
83. Weber A, Kirejczyk Z, Potthoff S, Ploner C, Hacker G (2009) Endogenous Noxa determines the strong proapoptotic synergism of the BH3-mimetic ABT-737 with chemotherapeutic agents in human melanoma cells. Transl Oncol 2(2):73–83 Epub 2009/05/05
84. Risberg K, Fodstad O, Andersson Y (2011) Synergistic anticancer effects of the 9.2.27PE immunotoxin and ABT-737 in melanoma. PLoS ONE 6(9):e24012 Epub 2011/09/15
85. Keuling AM, Andrew SE, Tron VA (2010) Inhibition of p38 MAPK enhances ABT-737-induced cell death in melanoma cell lines: novel regulation of PUMA. Pigment Cell Melanoma Res 23(3):430–440 Epub 2010/03/27
86. Will B, Siddiqi T, Jorda MA, Shimamura T, Luptakova K, Staber PB et al (2010) Apoptosis induced by JAK2 inhibition is mediated by Bim and enhanced by the BH3 mimetic ABT-737 in JAK2 mutant human erythroid cells. Blood 115(14):2901–2909 Epub 2010/02/18
87. Lu M, Wang J, Li Y, Berenzon D, Wang X, Mascarenhas J et al (2010) Treatment with the Bcl-xL inhibitor ABT-737 in combination with interferon alpha specifically targets JAK2V617F-positive polycythemia vera hematopoietic progenitor cells. Blood 116(20):4284–4287 Epub 2010/07/14
88. Karlberg M, Ekoff M, Huang DC, Mustonen P, Harvima IT, Nilsson G (2010) The BH3-mimetic ABT-737 induces mast cell apoptosis in vitro and in vivo: potential for therapeutics. J Immunol 185(4):2555–2562 Epub 2010/07/20
89. Bardwell PD, Gu J, McCarthy D, Wallace C, Bryant S, Goess C et al (2009) The Bcl-2 family antagonist ABT-737 significantly inhibits multiple animal models of autoimmunity. J Immunol 182(12):7482–7489 Epub 2009/06/06
90. Shoemaker AR, Mitten MJ, Adickes J, Ackler S, Refici M, Ferguson D et al (2008) Activity of the Bcl-2 family inhibitor ABT-263 in a panel of small cell lung cancer xenograft models. Clin Cancer Res 14(11):3268–3277 Epub 2008/06/04
91. Chen J, Jin S, Abraham V, Huang X, Liu B, Mitten MJ et al (2011) The Bcl-2/Bcl-X(L)/Bcl-w inhibitor, navitoclax, enhances the activity of chemotherapeutic agents in vitro and in vivo. Mol Cancer Ther 10(12):2340–2349 Epub 2011/09/15

92. Ackler S, Mitten MJ, Foster K, Oleksijew A, Refici M, Tahir SK et al (2010) The Bcl-2 inhibitor ABT-263 enhances the response of multiple chemotherapeutic regimens in hematologic tumors in vivo. Cancer Chemother Pharmacol 66(5):869–880 Epub 2010/01/26
93. Sakuma Y, Tsunezumi J, Nakamura Y, Yoshihara M, Matsukuma S, Koizume S et al (2011) ABT-263, a Bcl-2 inhibitor, enhances the susceptibility of lung adenocarcinoma cells treated with Src inhibitors to anoikis. Oncol Rep 25(3):661–667 Epub 2011/01/06
94. Fitzgerald AR (2012) Combination treatments with ABT-263 and an immunotoxin produce synergistic killing of ABT-263-resistant small cell lung cancer cell lines. Int J Cancer. Epub 2012/07/24
95. Wang G, Zhan Y, Wang H, Li W (2012) ABT-263 sensitizes TRAIL-resistant hepatocarcinoma cells by downregulating the Bcl-2 family of anti-apoptotic protein. Cancer Chemother Pharmacol 69(3):799–805 Epub 2011/11/01
96. Zhao X, Ogunwobi OO, Liu C (2011) Survivin inhibition is critical for Bcl-2 inhibitor-induced apoptosis in hepatocellular carcinoma cells. PLoS ONE 6(8):e21980 Epub 2011/08/11
97. Yamaguchi R, Janssen E, Perkins G, Ellisman M, Kitada S, Reed JC (2011) Efficient elimination of cancer cells by deoxyglucose-ABT-263/737 combination therapy. PLoS ONE 6(9):e24102 Epub 2011/09/29
98. Wong M, Tan N, Zha J, Peale FV, Yue P, Fairbrother WJ et al (2012) Navitoclax (ABT-263) reduces Bcl-x(L)-mediated chemoresistance in ovarian cancer models. Mol Cancer Ther 11(4):1026–1035 Epub 2012/02/04
99. Sasaki N, Kuroda J, Nagoshi H, Yamamoto M, Kobayashi S, Tsutsumi Y et al (2011) Bcl-2 is a better therapeutic target than c-Myc, but attacking both could be a more effective treatment strategy for B-cell lymphoma with concurrent Bcl-2 and c-Myc overexpression. Exp Hematol 39(8):817–28 e1. Epub 2011/06/07
100. Xargay-Torrent S, Lopez-Guerra M, Saborit-Villarroya I, Rosich L, Campo E, Roue G et al (2011) Vorinostat-induced apoptosis in mantle cell lymphoma is mediated by acetylation of proapoptotic BH3-only gene promoters. Clin Cancer Res 17(12):3956–3968 Epub 2011/06/10
101. Ackler S, Xiao Y, Mitten MJ, Foster K, Oleksijew A, Refici M et al (2008) ABT-263 and rapamycin act cooperatively to kill lymphoma cells in vitro and in vivo. Mol Cancer Ther 7(10):3265–3274 Epub 2008/10/15
102. Ackler S, Mitten MJ, Chen J, Clarin J, Foster K, Jin S et al (2012) Navitoclax (ABT 263) and bendamustine ± rituximab induce enhanced killing of non-Hodgkin's lymphoma tumors in vivo. Br J Pharmacol 167:881–891 Epub 2012/05/26
103. Roberts AW, Seymour JF, Brown JR, Wierda WG, Kipps TJ, Khaw SL et al (2012) Substantial susceptibility of chronic lymphocytic leukemia to BCL2 inhibition: results of a phase I study of navitoclax in patients with relapsed or refractory disease. J Clin Oncol 30(5):488–496 Epub 2011/12/21
104. Wilson WH, O'Connor OA, Czuczman MS, LaCasce AS, Gerecitano JF, Leonard JP et al (2010) Navitoclax, a targeted high-affinity inhibitor of BCL-2, in lymphoid malignancies: a phase 1 dose-escalation study of safety, pharmacokinetics, pharmacodynamics, and antitumour activity. Lancet Oncol 11(12):1149–1159 Epub 2010/11/26
105. Gandhi L, Camidge DR, Ribeiro de Oliveira M, Bonomi P, Gandara D, Khaira D et al (2011) Phase I study of Navitoclax (ABT-263), a novel Bcl-2 family inhibitor, in patients with small-cell lung cancer and other solid tumors. J Clin Oncol 29(7):909–916 Epub 2011/02/02
106. Rudin CM, Hann CL, Garon EB, Ribeiro de Oliveira M, Bonomi PD, Camidge DR et al (2012) Phase II study of single-agent navitoclax (ABT-263) and biomarker correlates in patients with relapsed small cell lung cancer. Clin Cancer Res 18(11):3163–3169 Epub 2012/04/13
107. Furstner A (2003) Chemistry and biology of roseophilin and the prodigiosin alkaloids: a survey of the last 2500 years. Angew Chem Int Ed Engl 42(31):3582–3603 Epub 2003/08/14

108. Shore GC, Viallet J (2005) Modulating the bcl-2 family of apoptosis suppressors for potential therapeutic benefit in cancer. Hematol Am Soc Hematol Educ Program 2005:226–230 Epub 2005/11/24
109. Zhai D, Jin C, Satterthwait AC, Reed JC (2006) Comparison of chemical inhibitors of antiapoptotic Bcl-2-family proteins. Cell Death Differ 13(8):1419–1421 Epub 2006/04/29
110. Nguyen M, Marcellus RC, Roulston A, Watson M, Serfass L, Murthy Madiraju SR et al (2007) Small molecule obatoclax (GX15-070) antagonizes MCL-1 and overcomes MCL-1-mediated resistance to apoptosis. Proc Natl Acad Sci U S A 104(49):19512–19517 Epub 2007/11/28
111. Martinez-Paniagua MA, Baritaki S, Huerta-Yepez S, Ortiz-Navarrete VF, Gonzalez-Bonilla C, Bonavida B et al (2011) Mcl-1 and YY1 inhibition and induction of DR5 by the BH3-mimetic Obatoclax (GX15-070) contribute in the sensitization of B-NHL cells to TRAIL apoptosis. Cell Cycle 10(16):2792–2805 Epub 2011/08/09
112. Wei Y, Kadia T, Tong W, Zhang M, Jia Y, Yang H et al (2010) The combination of a histone deacetylase inhibitor with the BH3-mimetic GX15-070 has synergistic antileukemia activity by activating both apoptosis and autophagy. Autophagy 6(7):976–978 Epub 2010/08/24
113. Rahmani M, Aust MM, Attkisson E, Williams DC Jr, Ferreira-Gonzalez A, Grant S (2012) Inhibition of Bcl-2 antiapoptotic members by obatoclax potently enhances sorafenib-induced apoptosis in human myeloid leukemia cells through a Bim-dependent process. Blood 119(25):6089–6098 Epub 2012/03/27
114. Heidari N, Hicks MA, Harada H (2010) GX15-070 (obatoclax) overcomes glucocorticoid resistance in acute lymphoblastic leukemia through induction of apoptosis and autophagy. Cell Death Dis 1:e76 Epub 2011/03/03
115. Perez-Galan P, Roue G, Lopez-Guerra M, Nguyen M, Villamor N, Montserrat E et al (2008) BCL-2 phosphorylation modulates sensitivity to the BH3 mimetic GX15-070 (Obatoclax) and reduces its synergistic interaction with bortezomib in chronic lymphocytic leukemia cells. Leukemia 22(9):1712–1720 Epub 2008/07/04
116. Brem EA, Thudium K, Khubchandani S, Tsai PC, Olejniczak SH, Bhat S et al (2011) Distinct cellular and therapeutic effects of obatoclax in rituximab-sensitive and -resistant lymphomas. Br J Haematol 153(5):599–611 Epub 2011/04/16
117. Jona A, Khaskhely N, Buglio D, Shafer JA, Derenzini E, Bollard CM et al (2011) The histone deacetylase inhibitor entinostat (SNDX-275) induces apoptosis in Hodgkin lymphoma cells and synergizes with Bcl-2 family inhibitors. Exp Hematol 39(10):1007–1017 e1. Epub 2011/07/20
118. Perez-Galan P, Roue G, Villamor N, Campo E, Colomer D (2007) The BH3-mimetic GX15-070 synergizes with bortezomib in mantle cell lymphoma by enhancing Noxa-mediated activation of Bak. Blood 109(10):4441–4449 Epub 2007/01/18
119. Chen S, Dai Y, Pei XY, Myers J, Wang L, Kramer LB et al (2012) CDK inhibitors up-regulate BH3-only proteins to sensitize human myeloma cells to BH3 mimetic therapies. Cancer Res 72(16):4225–4237 Epub 2012/06/14
120. Jiang CC, Wroblewski D, Yang F, Hersey P, Zhang XD (2009) Human melanoma cells under endoplasmic reticulum stress are more susceptible to apoptosis induced by the BH3 mimetic obatoclax. Neoplasia 11(9):945–955 Epub 2009/09/03
121. Witters LM, Witkoski A, Planas-Silva MD, Berger M, Viallet J, Lipton A (2007) Synergistic inhibition of breast cancer cell lines with a dual inhibitor of EGFR-HER-2/neu and a Bcl-2 inhibitor. Oncol Rep 17(2):465–469 Epub 2007/01/05
122. Cruickshanks N, Hamed HA, Bareford MD, Poklepovic A, Fisher PB, Grant S et al (2012) Lapatinib and obatoclax kill tumor cells through blockade of ERBB1/3/4 and through inhibition of BCL-XL and MCL-1. Mol Pharmacol 81(5):748–758 Epub 2012/02/24
123. Pan J, Cheng C, Verstovsek S, Chen Q, Jin Y, Cao Q (2010) The BH3-mimetic GX15-070 induces autophagy, potentiates the cytotoxicity of carboplatin and 5-fluorouracil in esophageal carcinoma cells. Cancer Lett 293(2):167–174 Epub 2010/02/16
124. Huang S, Okumura K, Sinicrope FA (2009) BH3 mimetic obatoclax enhances TRAIL-mediated apoptosis in human pancreatic cancer cells. Clin Cancer Res 15(1):150–159 Epub 2009/01/02

125. Smoot RL, Blechacz BR, Werneburg NW, Bronk SF, Sinicrope FA, Sirica AE et al (2010) A Bax-mediated mechanism for obatoclax-induced apoptosis of cholangiocarcinoma cells. Cancer Res 70(5):1960–1969 Epub 2010/02/18
126. Schimmer AD, O'Brien S, Kantarjian H, Brandwein J, Cheson BD, Minden MD et al (2008) A phase I study of the pan bcl-2 family inhibitor obatoclax mesylate in patients with advanced hematologic malignancies. Clin Cancer Res 14(24):8295–8301 Epub 2008/12/18
127. O'Brien SM, Claxton DF, Crump M, Faderl S, Kipps T, Keating MJ et al (2009) Phase I study of obatoclax mesylate (GX15-070), a small molecule pan-Bcl-2 family antagonist, in patients with advanced chronic lymphocytic leukemia. Blood 113(2):299–305 Epub 2008/10/22
128. Hwang JJ, Kuruvilla J, Mendelson D, Pishvaian MJ, Deeken JF, Siu LL et al (2010) Phase I dose finding studies of obatoclax (GX15-070), a small molecule pan-BCL-2 family antagonist, in patients with advanced solid tumors or lymphoma. Clin Cancer Res 16(15):4038–4045 Epub 2010/06/12
129. Oki Y, Copeland A, Hagemeister F, Fayad LE, Fanale M, Romaguera J et al (2012) Experience with obatoclax mesylate (GX15-070), a small molecule pan-Bcl-2 family antagonist in patients with relapsed or refractory classical Hodgkin lymphoma. Blood 119(9):2171–2172 Epub 2012/03/03
130. Parikh SA, Kantarjian H, Schimmer A, Walsh W, Asatiani E, El-Shami K et al (2010) Phase II study of obatoclax mesylate (GX15-070), a small-molecule BCL-2 family antagonist, for patients with myelofibrosis. Clin Lymphoma Myeloma Leuk 10(4):285–289 Epub 2010/08/17
131. Paik PK, Rudin CM, Brown A, Rizvi NA, Takebe N, Travis W et al (2010) A phase I study of obatoclax mesylate, a Bcl-2 antagonist, plus topotecan in solid tumor malignancies. Cancer Chemother Pharmacol 66(6):1079–1085 Epub 2010/02/19
132. Paik PK, Rudin CM, Pietanza MC, Brown A, Rizvi NA, Takebe N et al (2011) A phase II study of obatoclax mesylate, a Bcl-2 antagonist, plus topotecan in relapsed small cell lung cancer. Lung Cancer 74(3):481–485 Epub 2011/05/31
133. Chiappori AA, Schreeder MT, Moezi MM, Stephenson JJ, Blakely J, Salgia R et al (2012) A phase I trial of pan-Bcl-2 antagonist obatoclax administered as a 3-h or a 24-h infusion in combination with carboplatin and etoposide in patients with extensive-stage small cell lung cancer. Br J Cancer 106(5):839–845 Epub 2012/02/16

Chapter 10
SMAC IAP Addiction in Cancer

Matthew F. Brown, Kan He and Jian Yu

Abstract Deregulation of apoptosis is a hallmark of human cancer. Inhibitor of apoptosis proteins (IAPs) were first identified as inhibitors of caspases and apoptosis. They were later shown to play important roles in signal transduction to promote cell survival and proliferation beyond apoptosis, a function conserved in lower eukaryotes. Genetic alterations of cIAPs, as well as widespread altered expression in IAPs and their endogenous inhibitors such as SMAC, have been reported to be associated with disease progression and chemoresistance in cancer. Several strategies have been devised to target IAPs in cancer including rationally designed small-molecule IAP antagonists based on the conserved interaction between IAPs and SMAC. Known as SMAC mimetics, these agents are showing promise as novel cancer therapeutics and help reveal novel functions of IAPs.

10.1 Introduction

Apoptosis is an evolutionally conserved process essential for normal development, maintenance of tissue homeostasis and removal of unwanted or damaged cells following stress or injury. Deregulated cell death is a hallmark of cancer and contributes to chemoresistance [1, 2]. The inhibitor of apoptosis (IAP) family of proteins was originally identified as inhibitors of caspases in model organisms. Endogenous inhibitors of IAPs such as SMAC (second mitochondria-derived activator of caspase) and HrtA2 (high temperature requirement protein A2, also known as Omi) can directly bind to a conserved BIR (baculovirus interacting repeat) within IAPs. The binding of SMAC to IAPs is mediated through a

M. F. Brown · K. He · J. Yu (✉)
Department of Pathology, University of Pittsburgh School of Medicine,
University of Pittsburgh Cancer Institute, Pittsburgh, PA, USA
e-mail: yuj2@upmc.edu

M. F. Brown · J. Yu
Department of Pathology and Molecular and Cellular Pathology Graduate Training Program,
University of Pittsburgh School of Medicine, University of Pittsburgh Cancer Institute,
Pittsburgh, PA, USA

D. E. Johnson (ed.), *Cell Death Signaling in Cancer Biology and Treatment*,
Cell Death in Biology and Diseases, DOI: 10.1007/978-1-4614-5847-0_10,

X.-M. Yin and Z. Dong (Series eds.), *Cell Death in Biology and Diseases*

conserved tetrapeptide. Cumulative evidence in the last decade using biochemical, molecular and cellular, pharmacological, structural approaches, and genetically modified model systems has revealed a multitude of IAP functions in cell death, survival, proliferation, and immunity. Furthermore, several IAPs play a key role in regulating NF-wB (nuclear factor-κB) activation and non-apoptotic cell death. Overexpression and genetic alterations of IAPs, as well as reduced expression of their endogenous inhibitors, have been reported in solid tumors and hematological malignancies [3, 4]. In rare cases, loss of IAPs is associated with the development of certain types of hematological malignancies, suggesting a potential role as a tumor suppressor [5].

Agents targeting IAPs have been developed, with the most promising class being small molecules mimicking the IAP-binding domain in SMAC, also known as SMAC mimetics. These compounds have limited toxicity to cancer cells as a single agent, while synergizing with different classes of anticancer agents in a wide array of cancer cell types. The use of SMAC mimetics has helped elucidate signaling functions of IAPs, particularly cIAP1 and cIAP2, independent of their ability to suppress caspase activation or apoptosis.

This chapter provides an overview of the biology of IAP family members and the attempts to target them in cancer. Extensive studies have provided strong evidence that XIAP and cIAPs are key regulators of apoptosis, which are the focus of this book chapter. The roles of other IAPs such as survivin in apoptosis and cancer are less understood and will only be discussed briefly. More information can be found elsewhere [6, 7].

10.2 The IAP Family

Modern virology has produced some of the most valuable insights into the mechanisms of cell death. Many viruses encode gene products that suppress cell death by disabling tumor suppressor proteins and producing viral versions of anti-apoptotic Bcl-2 proteins and/or caspase inhibitors. This is critical for viral replication and carcinogenesis in certain patients [8–10]. The baculovirus protein, IAP, was the first inhibitor of apoptosis protein (IAP) discovered and shown to promote viral replication [11, 12]. Through sequence homology-based screens and other methods, eight IAP family members have been discovered in human, including NAIP, XIAP, cIAP1, cIAP2, Survivin, BRUCE (Apollon), ILP2, and ML-IAP. Seven of them have homologs in mice (Fig. 10.1 and Table 10.1). All IAP family members in viruses and metazoans contain at least one signature baculovirus IAP repeat (BIR) domain (Fig. 10.1 and Table 10.1), which is capable of interacting with caspases and various IAP antagonists. Several of these, including XIAP, cIAP1, cIAP2, survivin, and ILP2, directly bind to and suppress caspases through their BIR domains [13]. XIAP, cIAP1, and cIAP2 also contain a RING domain (Fig. 10.1), a signature of E3 ligases [13], and play critical regulatory roles in NF-κB signaling and necrosis. Certain IAP family members, including XIAP and

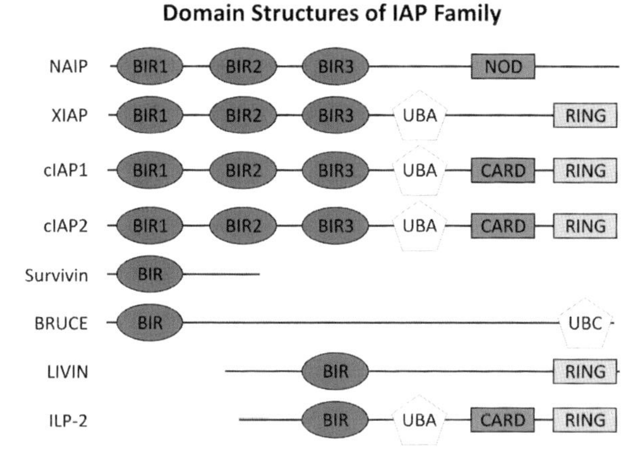

Fig. 10.1 Human IAP family members. Eight human IAP proteins have been identified and all contain at least one baculoviral IAP repeat (*BIR*) domain. Several IAPs contain a carboxy-terminal RING domain with E3 ubiquitin ligase activities, as well as ubiquitin-binding domains (UBA/C). Other domains such as leucine-rich repeats (*LRR*), nucleotide oligomerization domain (*NOD*) and caspase activation and recruitment domain (*CARD*) are present in some IAPs, which mediate protein–protein interactions

cIAP1/2, BRUCE and ILP-2 contain UBA/C (ubiquitin binding or conjugating) domains allowing them to bind or add mono- or poly-ubiquitin (Ub) to various substrates. Other domains such as LRR (leucine-rich repeats), NOD (nucleotide oligomerization domain), and CARD (caspase activation and recruitment domains) mediate protein–protein interactions and are found in IAPs and various proteins regulating apoptosis and inflammation. However, their roles in IAP regulation of apoptosis are not fully understood [7].

10.3 Apoptotic Pathways

Apoptosis is initiated and executed by a series of well-ordered biochemical events and regulated by complex signaling networks. There are two major apoptotic pathways, termed as the extrinsic and intrinsic pathways, which are responsible for processing stress signals and executing cell demise [14, 15]. Although with distinct regulatory processes, there is significant cross-talk in many situations among these two pathways.

The extrinsic apoptotic pathway was first found to be used by immune cells to kill infected or damaged cells and subsequently shown to operate in many cell types. This pathway is engaged upon binding of pro-apoptotic ligands to cell surface death receptors of the TNFR family. All death receptors contain a cysteine-rich extracellular domain, which is responsible for ligand binding, and

Table 10.1 IAPs and their endogenous inhibitors

Protein	Other names	Chr. location (Human)	AA (Human)	Chr. location (Mouse)	AA (Mouse)	Domains BIR	RING	Other
NAIP	BIRC1	5q13	1403	13 D1–D3	1403	3	0	NOD, LRR
cIAP1	BIRC2, API1	11q22	618	9 A1	612	3	1	CARD, UBA
cIAP2	BIRC3, API2	11q22	604	9 A2	602	3	1	CARD, UBA
XIAP	BIRC4, API3	Xq25	497	X A3–A5	496	3	1	UBA
Survivin	BIRC5, API4	17q25	142	11 E2	140	1	0	N/A
BRUCE	BIRC6, Apollon	2p22	4857	17 E3	4854	1	0	UBC
LIVIN	BIRC7	20q13	298	2 H4	285	1	1	N/A
ILP-2	BIRC8, Ts-IAP	19q13	236	N/A	N/A	1	1	CARD, UBA
						BIM	**RING**	**Other**
SMAC	DIABLO	12q24	239	5F	237	1	0	N/A
HtrA2	Omi	2p12	458	6 C3	458	1	0	Transmembrane
XAF1	BIRC4-binding protein	17p13	301	11 B4	273	0	0	Zinc finger

an intracellular death domain for transmitting signals through the recruitment of effectors. Pro-apoptotic ligands belong to the extended cytokine TNF superfamily and are either present on the cell surface or secreted into the extracellular space [14, 15]. Upon binding of pro-apoptotic ligands to their respective receptors, each receptor can independently form a death-inducing signaling complex (DISC) by recruiting the adapter protein FADD, along with pro-caspase-8 and pro-caspase-10 (Fig. 10.2) [15]. FADD recruitment and DISC formation lead to proximity-induced processing and activation of caspase-8 and caspase-10, and their release into the cytoplasm triggers subsequent activation of effector caspases to execute apoptosis (Fig. 10.2).

The intrinsic apoptotic pathway is triggered by stresses such as DNA damage, deregulated oncogenes or nutrient/growth factor deprivation and is largely regulated by the Bcl-2 family of proteins and mitochondria [16, 17] (Fig. 10.2). For example, following DNA damage and p53 activation, death signals are transmitted to BH3-only proteins to allow for the activation of Bax and Bak, which leads to mitochondrial outer membrane permeabilization (MOMP) [18–20] and

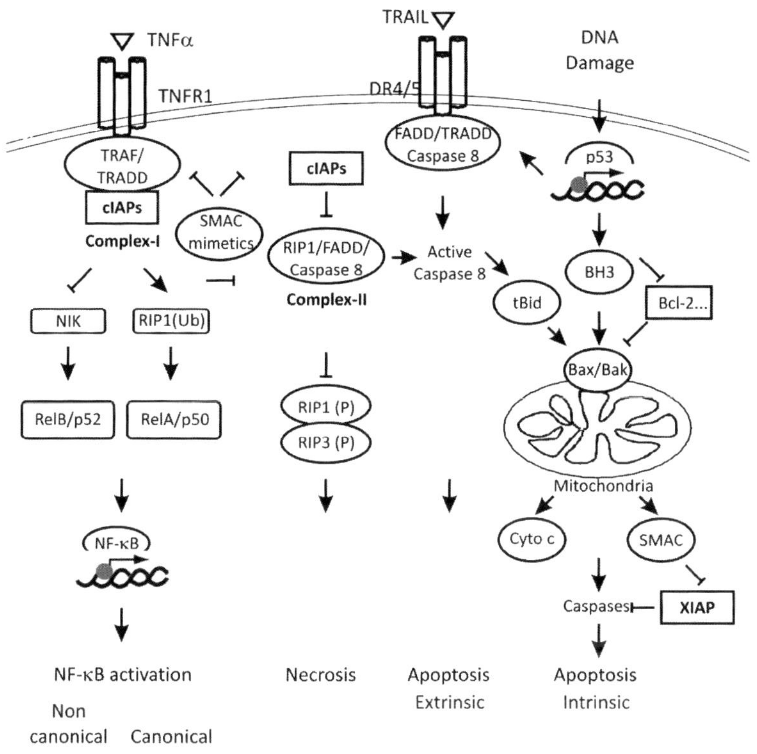

Fig. 10.2 IAP-mediated cell survival and signaling. IAP proteins serve as a signaling hub for cell survival. IAPs are negative regulators of programmed apoptosis and necrosis. XIAP and cIAPs can inhibit both the initiator and effector caspases to block apoptosis. In addition, cIAPs positively regulate canonical NF-κB signaling to allow TNFα production and inflammation, while negatively regulating non-canonical NF-κB signaling, necrosis and caspase-8 activation, all through complex assembly. SMAC or SMAC mimetics suppress XIAP and cIAP-dependent cell survival by relieving IAP caspase inhibition through competitive binding, as well as by depletion of cIAPs via activation of their E3 ligase activity (see text for details). (P)-phosphorylated and (Ub)-ubiquitinated

release of several mitochondrial apoptogenic proteins, including cytochrome c, SMAC/Diablo, Omi/HtrA2, AIF (apoptosis-inducing factor), and EndoG (endonuclease G) [21]. The release of cytochrome c promotes the assembly of the apoptosome and subsequent activation of caspase-9 [21]. Additionally, SMAC/Diablo and HtrA2/Omi facilitate apoptosis by binding to IAPs and relieving their inhibition of caspases [22, 23]. The release of AIF and Endo G promotes DNA degradation [24, 25]. Notably, in some cells where there are low levels of DISC, caspase-8-dependent cleavage of the BH3-only protein Bid generates truncated Bid (tBid) to amplify apoptotic signaling via the mitochondria (Fig. 10.2) [26, 27].

10.4 IAPs as Caspase Inhibitors

IAPs can inhibit caspases through two major mechanisms. Some IAPs bind directly to active sites of caspases to prevent processing or activity. Among all IAPs, mammalian XIAP is the only IAP that has high-affinity interactions with caspase-3, caspase-7, and caspase-9, and functions as a direct caspase inhibitor [28]. XIAP inhibits caspases-3, caspase-7, and caspase-9 via its BIR2 and BIR3 domains [29–33]. Specifically, insertion of the linker region between BIR1 and BIR2 into the catalytic pocket of active effector caspases (such as caspase-3 or caspase-7) prevents substrate entry. In addition, XIAP can prevent initiator caspases such as caspase-9 from dimerization and subsequent activation, by hindering a conformational change required for a functional catalytic pocket [34, 35].

IAPs can also induce ubiquitin-mediated degradation of caspases. The RING finger domain in DIAP1 regulates Ub conjugation of caspases, and therefore, apoptosis in *Drosophila*, though the precise role of mono- versus poly-ubiquitylation in this process, is not clear [36–39]. Consistent with these findings, cIAP1 and cIAP2, or their homologs in lower eukaryotes do not inhibit caspases efficiently in vitro [40]. Conversely, in living cells, XIAP promotes the degradation of active-form caspase-3 [30], but not pro-caspase-3 [38]. Interestingly, the RING domain in XIAP is not required for its inhibition of caspase-3, caspase-7, or caspase-9 in vitro or when overexpressed [41, 42]. These findings suggest multiple mechanisms are involved in IAP-mediated caspase inhibition.

10.5 cIAPs as Signaling Molecules in Cell Survival and Non-Apoptotic Cell Death

cIAP1 and cIAP2 were initially identified as interacting proteins of the scaffold protein TRAF2 (tumor necrosis factor receptor-associated factor 2). They are now known to play crucial roles in regulating NF-κB signaling, DNA damage response, and necrosis by regulating the assembly and stability of distinct signaling complexes through their E3 ligase function [4, 43]. This knowledge was largely obtained by using small molecules that deplete both cIAPs. In the canonical NF-κB pathway, cIAP1 and cIAP2 regulate Ub-dependent activation of NF-κB downstream of TNFR1, which in turn drives transcriptional programs important for cell survival and inflammation [4, 43]. In addition, cIAPs suppress non-canonical NF-κB signaling, necrosis, and the assembly of caspase-8 activation complexes in resting cells. Rapid degradation of cIAP1 and cIAP2 triggered by SMAC mimetics, or DNA damage can lead to activation of the non-canonical NF-κB pathway, apoptosis, or necrosis if caspase activation is blocked [44–49].

10.5.1 cIAPs in NF-κB Pathways

NF-κB regulates a wide variety of cellular functions, such as survival, proliferation, migration, and immune response. Constitutive activation of NF-κB and chronic inflammation plays a major role in tumor progression [50–52]. The NF-κB family consists of five transcription factors, RELA (also known as p65), RELB, CREL and the precursor proteins NF-κB1/p105 and NF-κB2/p100. NF-κB signaling is activated in response to receptor stimulation and various stresses [53, 54] (Fig. 10.2) and is regulated by Ub-dependent signal transduction cascades and assembly of protein complexes that lead to the processing of NF-κB1/p105 to p50 and NF-κB/p100 to p52, which is required for the transcription of target genes [53–55] (Fig. 10.2). NF-κB signaling can be divided into canonical or non-canonical pathways depending on the transcription factors used, either RELA/NF-κB1 (p50) or RELB/NF-κB2 (p52) (Fig. 10.2).

cIAPs are required for the activation of the canonical NF-κB pathway in response to TNFα. Binding of TNFα to TNFR1 triggers receptor trimerization and the recruitment of the adaptor protein TNFRSF1A-associated via death domain (TRADD), the Ub ligases TRAF2/5, cIAP1, and cIAP2, and the protein kinase RIP1 to a membrane-associated complex, referred to as complex-I (Fig. 10.2) [56]. Upon cIAP-mediated ubiquitination, RIP1 recruits additional components that promote poly-ubiquitination of IKKγ in complex-I. This leads to a phosphorylation cascade and degradation of the inhibitor of transcription factor NF-κB (IκB), allowing NF-κB to translocate to the nucleus. In *Drosophila*, DIAP2 is required for NF-κB activation in the immune deficiency (Imd) signaling cascade [57–60].

On the contrary, cIAPs negatively regulate non-canonical NF-κB signaling via the degradation of NF-κB-inducing kinase (NIK; also known as MAP3K14) [44] (Fig. 10.2). Non-canonical activation of NF-κB is activated upon receptor binding to the ligands of the TNF receptor superfamily, including CD40L, B-cell activating factor (BAFF) and TWEAK/TNFRSF12A [54, 61]. This pathway is suppressed in resting cells by the constitutive degradation of NIK through an Ub ligase complex consisting of TRAF2/3–cIAP1/2 and the proteasome (Fig. 10.2). NIK degradation requires the E3 ligase activity of cIAP1 and cIAP2 while TRAF2/3 serves as adaptors to recruit NIK to cIAP1/2. Depletion of IAPs or other components of the TRAF2/3–cIAP1/2 Ub ligase complex leads to stabilization of NIK and spontaneous activation of the non-canonical pathway and TNFα production [62, 63].

10.5.2 cIAPs and Necrosis

Necrosis is characterized by ATP depletion, swelling of organelles, and spilling of cellular contents. This form of cell death had long been thought to be a passive process until specific pathways came into light in recent years [10, 64]. In the absence of cIAPs, TNFα stimulates the formation of a secondary cytoplasmic complex, complex-II, which contains RIP1, Fas-associated via death domain (FADD) and caspase-8 [46, 65] (Fig. 10.2). Complex-II formation leads to a rapid activation of caspase-8 and subsequently apoptosis. In most cases, this form of cell death can be completely blocked by caspase inhibitors. In some cells, caspase inhibitors, or lack of caspase-8 activation can switch the apoptotic response to RIP1/RIP3-dependent necrosis [47, 48, 66–68]. Some evidence suggests that cIAPs can block necrosis triggered by CD95 stimulation perhaps in a fashion similar to TNFR1 stimulation [69].

10.5.3 cIAPs and DNA Damage Response

Radiation and common chemotherapeutics cause DNA damage and activation of NF-κB, ATM/p53 and other signaling pathways [70]. Extensive DNA damage depletes XIAP, cIAP1, and cIAP2 and promotes the assembly of a large ~2MDa cytoplasmic cell death-inducing platform referred to as the "ripoptosome". This complex contains RIP1, FADD, and caspase-8, and its formation can be stimulated by SMAC mimetics [49, 71] (Fig. 10.2). Ripoptosome assembly requires RIP1 kinase activity and can stimulate caspase-8-mediated apoptosis as well as caspase-independent necrosis without involvement of autocrine TNFα, a process similar to complex-II formation in response to TNFα (Fig. 10.2). In some cells, following excessive DNA damage, ATM employs NEMO (NF-κB essential modulator) and RIP1 kinase through autocrine TNFα signaling to switch on cytokine production and caspase activation [72]. These results illustrate novel p53-independent mechanisms of cell killing after DNA damage in cancer cells, which are negatively regulated by IAPs via both TNFα-dependent and TNFα-independent mechanisms.

10.6 Endogenous Inhibitors of IAPs

Several cellular proteins can antagonize the anti-apoptotic activities of IAPs. The best studied endogenous IAP inhibitor is second mitochondria-derived activator of caspase (SMAC), also known as direct inhibitor of apoptosis protein (IAP)-binding protein with low pI (Diablo) or SMAC/Diablo. Upon apoptotic induction, SMAC is released into the cytosol and binds to BIRs in IAPs through its amino (N)-terminus to facilitate caspase activation [22, 73]. The Drosophila death proteins Hid, Grim, and Reaper have limited homology with SMAC, in the BIR binding motif [22, 73, 74] (Fig. 10.3a), suggesting that regulation of caspase activation is an ancient function of these proteins [41, 75].

SMAC/Diablo functions as a general IAP inhibitor and binds to XIAP, cIAP1, cIAP2, survivin, livin, and BRUCE, but not NAIP [22, 73, 76–79]. The four amino-terminal residues of mature SMAC/DIABLO, Ala-Val-Pro-Ile (AVPI) are both necessary and sufficient for SMAC/DIABLO–IAP interaction [3, 80–83].

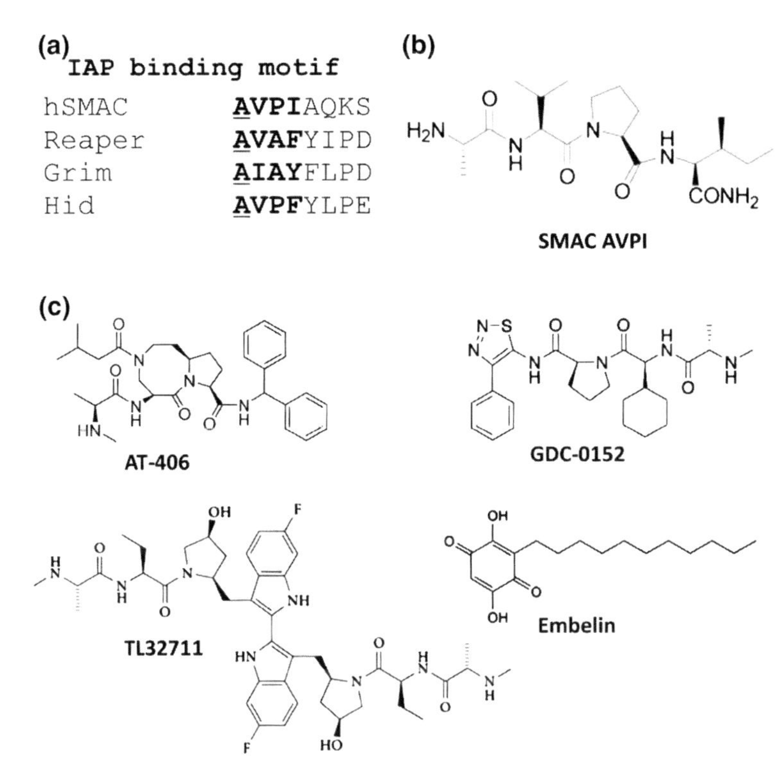

Fig. 10.3 IAP-binding motif (*IBM*) and SMAC mimetics. **a** The conserved BIR binding motif. This tetrapeptide motif has the consensus sequence of A-(V/T/I)-(P/A)-(F/Y/I/V) and is found in SMAC, the drosophila proteins Reaper, Grim, and Hid [41]. **b** The chemical structure of SMAC AVPI [151]. **c** Structures of several small-molecule (*IAP*) antagonists: monovalent SMAC mimetics GDC-0152 (Genentech) [200] and AT-406 (Aegera Therapeutics) [201], and the bivalent SMAC mimetic TL32711 [4] and natural product Embelin [151]

This tetrapeptide motif is exposed following SMAC release from the mitochondria after proteolytical removal of a 55-residue N terminal mitochondria-targeting sequence. Crystallography data revealed that SMAC/DIABLO homodimerizes through an extensive hydrophobic interface, which is essential for its high-affinity binding to IAPs and its pro-apoptotic functions [81]. Several SMAC isoforms have been reported to regulate IAPs through ubiquitination-dependent or caspase-dependent mechanisms to potentiate apoptosis in vitro [3].

Additional pro-apoptotic IAP-binding proteins have been identified and are capable of binding to several IAP members, including serine protease HtrA2 (high-temperature-requirement protein A2) also known as Omi [23, 84–86], and XIAP-associated factor 1 (XAF1) [87, 88]. However, a SMAC AVPI-like element is only found in HtrA2/Omi [86], and the role of these proteins as selective IAP antagonists in the regulation of apoptosis is much less understood compared with SMAC.

10.7 Phenotypes Associated with Gain- and Loss-of-Functions in IAPs and Their Inhibitors

The functions of IAPs, including XIAP, cIAPs, and their inhibitors have been examined extensively in human cells, mice, and other model organisms using both loss-of-function and overexpression systems (reviewed in [7, 60, 89]). In general, the findings support that IAP overexpression inhibits caspase activation and apoptosis mediated by both intrinsic and extrinsic apoptotic pathways. Overexpression of SMAC, HtrA2/Omi, or XAF1 increases apoptosis in a variety of cancer cell lines when combined with other anticancer agents, particularity with death receptor ligands such as TRAIL (TNF-related apoptosis-inducing ligand).

Knockout of IAP in mice or human cancer cell lines results in a range of phenotypes associated with altered apoptosis or cell survival, and also supports function of IAPs in regulating NF-κB activation and cell proliferation [7]. IAPs appear to regulate apoptosis with a great deal of redundancy. Total knockout of *XIAP* in mice resulted in highly elevated apoptosis in response to brain injury, increased NF-κB activation, and an overall reduced survival [90, 91]. *XIAP* knockout HCT116 colon cancer cells are sensitized to TRAIL-induced apoptosis [92]. Mice with total knockout of *cIAP1* have no discernable phenotype due to redundant pathway activation by cIAP2 [13]. Knockout of *cIAP2* in mice leads to increased macrophage death and resistance to LPS-induced sepsis [93]. Double knockout of *cIAP1/cIAP2* or *XIAP/cIAP1* is embryonic lethal due to apoptotic defects, while *XIAP/cIAP2* double-knockout mice showed little change in apoptosis due to compensation by cIAP1 [94]. Conditional *cIAP1/2* double-knockout mice show uncontrolled B-cell proliferation independent of growth factor [95].

Endogenous IAP inhibitors also exhibit redundant functions in apoptosis regulation [3, 7]. *SMAC* deficiency results in apoptosis resistance to a selective group of agents, including TRAIL and non-steroidal anti-inflammatory drugs (NSAID) in HCT116 colon cancer cells and mice, yet has little to no impact on apoptosis induced by most DNA-damaging agents [96–99]. *HtrA2/Omi* knockout mice or cells, or *SMAC/HtrA2* double-knockout mice have minimal changes in apoptosis [100].

10.8 IAPs and Their Endogenous Inhibitors in Human Cancer

Extensive studies have examined genetic and expression alterations in IAPs and their inhibitors in cancer, and their correlation with clinical outcomes (reviewed in [3, 4, 43, 101]). Genetic alterations in *cIAP1/2* have been reported in multiple cancers. Overexpression of IAP members, including XIAP, survivin, and cIAP1/2 and reduced expression of SMAC, HtrA2/Omi, or XFA1 are common in cancer and suggested to correlate with chemoresistance, disease progression, and poor prognosis. Although these results support targeting IAPs in cancer, they should be interpreted with caution for several reasons. For example, some studies reported nuclear but not cytosolic expression as having prognostic value. In certain cases, either a lack of correlation or a reverse correlation is reported in a different cohort of the same tumor type. Notably, the sample size was limited in several studies. Lastly, genetic evidence supports that loss of cIAPs can contribute to tumor initiation or progression in some cases.

10.8.1 IAPs and Cancer

While IAP overexpression can result from genetic changes such as gene amplification or chromosomal aberrations, the reason for their overexpression is not known in most cases. To date, genetic alterations have been reported for *cIAPs*, but not *survivin*, *XIAP* or *ML-IAP*. Amplification and translocation of cIAPs have been found in both solid tumors and hematological malignancies. The 11q21–q23 amplification, encompassing both *cIAP1* and *cIAP2* loci, has been reported in esophageal cancer [102], cervical cancer [103], liver cancer [104], lung cancer [105], pancreatic cancer [106], medulloblastoma [107], and glioblastoma [108]. Interestingly, recurrent amplification of *cIAP1* and *cIAP2* is found in Myc-driven liver cancer [104] as well as spontaneous osteosarcomas in mice [109]. Additionally, the cIAP2–MALT1 fusion protein resulting from t(11;18)(q21;q21) translocation is found in mucosa-associated lymphoid tissue lymphoma [110–114].

In rare cases, cIAPs can function as tumor suppressors. For example, 20 % of patients with multiple myeloma have activated non-canonical NF-κB signaling associated with frequent genetic alterations, including biallelic deletions of *cIAP1*, *cIAP2*, or *TRAF2*, *TRAF3*, *CYLD*, as well as enhanced expression of CD40, lymphotoxin-β receptor (LTβR), TNFRSF13B, NFKB2, and NIK [5, 115]. Mutational changes leading to increased NIK and non-canonical activation of NF-κB have also been reported in breast [116] and pancreatic cancers [117].

Elevated IAP expression has been reported to be a predictor of poor prognosis in many cancers. For example, elevated XIAP levels are correlated with disease progression, metastasis, and poor survival in colon cancer [118], liver cancer [119], gastric cancer [120], breast cancer [121], and melanoma [122]. High levels of cIAP1 are associated with poor overall survival and local recurrence-free survival in cervical squamous cell carcinomas [103], and nodal metastasis in squamous cell carcinoma of the tongue [123]. Furthermore, high levels of cIAP2 expression are correlated with reduced survival in colorectal cancer [124] and pancreatic ductal adenocarcinomas [125]. Increased ML-IAP expression in mRNA or protein is found in melanoma and renal cancer [126–129] and is associated with disease progression or poor prognosis in superficial bladder cancer [130], adult ALL [131], osteosarcoma [132], and a subset of *N-Myc*-amplified neuroblastoma [133]. Lastly, elevated survivin expression is prevalent in cancer and associated with poor prognosis [101, 134–136].

10.8.2 SMAC and Other Endogenous IAP Inhibitors in Cancer

No genetic alterations in *SMAC, Omi/HtrA1,* or *XFA1* have been reported in human cancer. A significant inverse correlation between SMAC expression with either stage or grade has been reported in a variety of cancers, including colon [137], esophageal [138], lung, renal cell, prostate, hepatocellular, and testicular germ cancers [139]. Moreover, reduced expression of HtrA2/Omi and XFA1 in cancer or cancer cell lines has been reported [7, 140]. While it is possible that reduced SMAC or XFA1 contributes to IAP upregulation, direct evidence is still lacking.

Reports on SMAC expression and prognosis in cancer are more divergent. Reduced or lack of SMAC expression was associated with poor prognosis, shorter survival, disease progression, and metastasis in colon [137], lung [141], breast [142], bladder [143], renal [144, 145], and endometrial cancers [146]. However, high SMAC expression was found to be associated with a worse prognosis in patients with cervical cancer [147] and acute myeloid leukemia [148].

10.9 IAP Antagonists as Anticancer Agents

Based on their widespread overexpression in human cancer and roles in promoting cell survival and chemoresistance, several strategies have been devised to target IAP-addicted cancer cells. These include siRNA or antisense targeting *XIAP* expression, small molecules targeting survivin expression, and peptidomimetics and small-molecule SMAC mimetics targeting several IAPs [149, 150]. Among these, small-molecule SMAC mimetics have garnered the most attention. As single agents, SMAC mimetics exert little or no toxicity in non-malignant human cells and can induce apoptosis in a limited number of cancer cell lines via TNFα production. However, most cancer cell lines are resistant to treatment with SMAC mimetics alone. On the other hand, the combination of enhanced expression of SMAC or SMAC mimetics sensitize human cancer cells to apoptosis induced by many classes of anticancer agents, including chemotherapeutics, radiation, death receptor agonists and kinase inhibitors, via both XIAP and cIAP-dependent mechanisms.

10.9.1 SMAC Mimetics

Detailed structural information on the binding between SMAC and IAPs prompted the development of peptidomimetics and small-molecule SMAC mimetics, providing an excellent example of rational drug design [151]. SMAC AVPI-derived peptides were first tested and found to effectively block IAP–caspase interactions and sensitize a variety of cancers to pro-apoptotic stimuli [76], as well as synergize with TRAIL in glioma xenografts [152]. However, SMAC peptidomimetics lacked favorable pharmacological properties for further development, as they have poor cell permeability, bind with low affinity to the XIAP BIR3 domain, and are sensitive to proteolytic degradation [153]. These early studies provided both critical rationale and tools for the subsequent development of small-molecule IAP antagonists [76, 152, 154, 155], which can overcome many of the mentioned limitations.

A number of small molecular SMAC mimetics have been developed using structure-based designs and synthetic and medicinal chemistry (Fig. 10.3) [151]. Most SMAC mimetics are dimers, containing two SMAC APVI-like structures connected by a chemical linker, resembling the higher-order structure of SMAC in solution [82]. The first bivalent SMAC mimetic was developed in 2004 [153], followed by development of many mono- and divalent SMAC mimetics [151]. Compared to the AVPI domain [156] or monovalent compounds, divalent SMAC mimetics have higher affinity to IAPs and are much more potent in inducing cytotoxicity in cancer cells or mouse xenografts [65, 157–159]. The increased potency of bivalent compounds may be attributable to better inhibition of XIAP via binding to BIR2 and BIR3 regions, allowing for prominent caspase-3 and caspase-7 activation [160, 161], and release of endogenous SMAC [162]. Furthermore, bivalent IAP mimetics effectively promote cIAP dimerization [44, 159] and facilitate conformational changes and activation of cIAP1 E3 ligase activity [163–165], leading to enhanced auto-ubiquitination of cIAPs and perhaps cross-ubiquitination of other signaling molecules such as RIP1 [157, 158].

10.9.2 Mechanisms of SMAC Mimetic-Induced Cell Killing

Induction of apoptosis is the major mechanism of IAP antagonists-mediated cell killing, and both XIAP- and cIAP-related mechanisms are involved [161, 166–170]. SMAC mimetics have been a particularly useful tool for probing cIAP functions, especially due to the difficulties in simultaneous and acute ablation of cIAP1 and cIAP2 in cells. The single agent sensitivity can be explained by production of TNFα following rapid loss of cIAP1/2 and activation of the non-canonical NF-κB pathway. SMAC mimetics can promote either apoptosis or necrosis of cancer cells in response to TNFα. In the absence of cIAPs and a suppressed canonical NF-κB pathway, TNFα/TNFR1 further stimulates the formation of a cytoplasmic RIP1/FADD/caspase-8 (complex-II), and apoptosis [45, 65, 157–159] (Fig. 10.2), or RIP1/RIP3-dependent necrosis when caspase activation is blocked [46–48] (Fig. 10.2). SMAC mimetics do not appear to sensitize normal primary cells to TNFα-induced killing [65, 159]. However, the underlying reasons why only certain cells produce TNFα and the selectivity against tumors cells are not well understood.

SMAC mimetics alone have limited toxicity to most cancer cell lines [65, 167]. The antitumor effects of SMAC mimetics have therefore been extensively examined in combination with other cytotoxic agents, including chemotherapeutic agents, radiation, death receptor ligands, kinase and proteasome inhibitors [4]. Synergistic or additive responses have been reported in numerous studies using in vitro and mouse xenograft models (reviewed in [4]). Perhaps, the most significant synergy observed is when SMAC mimetics are combined with death receptor ligands such as TRAIL [153, 167, 171–177], or with CD95 ligand or agonistic CD95-specific antibodies to trigger apoptotic cell death [69, 178]. This synergy might be explained by enhanced caspase-8 activation [96] as well as reduced c-FLIP (cellular FLICE-like inhibitory protein)[167], which can bypass the need for mitochondrial amplification to execute apoptosis in certain cells [167, 179]. The synergy with DNA-damaging agents is at least two-fold: the activation of the mitochondrial pathway [16] amplifying the death receptor signal, and DNA damage-induced cIAP depletion and subsequent caspase-8 action via TNFα-dependent [71] and TNFα-independent mechanisms [71, 180] (Fig. 10.2). Suppression of cancer cell invasion and metastasis might also contribute to the antitumor activities of SMAC mimetics without increased pro-apoptotic death receptor signaling in response to TRAIL [181].

10.9.3 IAP Antagonists in Clinical Development

Over 50 patents have been filed on IAP antagonists as potential anticancer agents, and some of these agents have entered clinical development [4, 43]. Currently, a large number of clinical trials are underway to assess their safety, pharmacological properties and efficacy in patients with advanced solid tumor or hematological malignancies (Table 10.2).

10.9.3.1 SMAC Mimetics in Clinical Development

Most of the IAP antagonists in clinical development are small-molecule SMAC mimetics, including both mono- and bivalent compounds that are currently in Phase I and Phase II trials (Table 10.2) [4, 43]. These include monovalent IAP antagonists GDC-0152 (Genentech) and AT-406 (Ascenta Therapeutics), as well as bivalent IAP antagonists LCL161 (Novartis Pharmaceuticals), TL32711 (TetraLogic Pharmaceuticals), and HGS1029 (Aegera Therapeutics/Human Genome Sciences). The structures of several of them have been published (examples in Fig. 10.3c). Preliminary data indicate that both monovalent compounds HGS1029 and bivalent compounds LCL161 and TL32711 are well tolerated and show predicted biomarker modulation, such as cIAP1 downregulation and increased levels of processed caspase-3 and caspase-7 in the serum. Manageable side effects such as transient lymphopaenia and neutrophilia have been observed with the use of bivalent IAP antagonists HGS1029 and TL32711 in some patients [4]. Despite more potent cytotoxicity in cancer cells and mouse xenograft models, divalent SMAC mimetics require intravenous administration due to their large size and present a challenge for administering in combination with agents taken orally

Table 10.2 IAP antagonists in clinical development

Target	Type	Name	Compound developer	Cancer	Phase
cIAP1/2, XIAP and others	Monovalent SMAC mimetic	GDC-0152	Genentech	Locally advanced or metastatic solid tumors, non-Hodgkin's lymphoma, breast cancer [200]	I
	Bivalent SMAC mimetic	LCL161	Novartis pharmaceuticals	Advanced solid tumors [202]	I
	Bivalent SMAC mimetic	TL32711	TetraLogic pharmaceuticals	Solid tumors, lymphoma [203]	II
	Monovalent SMAC Mimetic	AT-406	Ascenta therapeutics	Advanced solid tumors, lymphomas	I
	Bivalent SMAC mimetic	HGS1029	Human genome sciences	Advanced solid tumors [204]	I
XIAP	Antisense	AEG35156	Aegera therapeutics	Advanced solid tumors [169], AML [205],Lymphocytic, chronic, B-Cell, advanced hepatocellular, pancreatic [206], mammary carcinoma, non-small cell lung cancer	I/II

[4]. Therefore, additional studies are required to determine which classes of agents have more desirable pharmacodynamic or pharmacokinetic properties for clinical application.

10.9.3.2 XIAP Antagonists in Clinical Development

Two major strategies have been used to target XIAP by preventing it from binding to caspase-3 [182] or downregulation with antisense. A number of small-molecule XIAP antagonists have shown efficacy in vitro and are in preclinical development, including dTWX-024 [183], TPI-1396-34 [184], XAC 1396-11 [185], and the natural product embelin [186]. The XIAP antisense oligonucleotides AEG35156 (Aegera

Therapeutics) displays potent antitumor activities in vitro as a single agent and synergizes with various chemotherapeutic compounds, such as the death receptor ligand, TRAIL, and radiation to induce apoptosis in cell lines and mouse xenograft models [187–190]. Results from Phase I and II clinical trials indicated that AEG35156 is well tolerated and exhibits predictable pharmacokinetic properties, dose-dependent changes in circulating biomarkers of cell death, as well as some antitumor activities (Table 10.2) [191, 192], consistent with in vitro data on the suppression of XIAP mRNA and protein levels [187–190]. Future studies will be needed to determine whether modalities containing this agent improve efficacy over standard therapies.

10.10 Conclusions

During transformation, neoplastic cells become resistant to apoptosis due to acquired genetic and epigenetic alterations, which can in turn drive additional tumorigenic events and contribute to therapeutic resistance [1, 2, 17]. Mechanistic dissection of apoptotic pathways has stimulated intensive efforts to restore apoptosis in cancer cells for disease control [14–16]. Cancer cells appear to be addicted to overexpression of IAPs, which promotes carcinogenesis and drug resistance by inhibiting caspases and stimulating cell survival and inflammation. The interaction between IAP and SMAC guided the development of small-molecule SMAC mimetics and furthered the biology of IAPs.

IAP signaling has several prominent features; it is highly dependent on protein–protein, protein–ligand interactions such as BIR/IAP/and AVIP/SMAC, and ubiquitination mediated by E3 ligase complexes [193]. IAPs regulate distinct signaling complexes, including NF-κB, RIP1-associated death complex and Toll-like receptor (TLR) signaling [194] via competitive utilization of shared signaling or adaptor proteins, including cIAP1/2, FADD, TRADD, TRAFs, and RIP1 (Fig. 10.2). This mode of regulation is in contrast to p53-mediated apoptosis following DNA damage, which selectively activates the transcription of pro-apoptotic BH3 proteins, and death receptors, and ligands [18, 19].

A better understanding of IAP biology will certainly help the development and application of IAP antagonists in the clinic (Fig. 10.3), as a number of important questions still remain. For example, what is the relative contribution of TNFα/TNFR1-dependent and TNFα/TNFR1-independent mechanisms in SMAC mimetics and IAP depletion-induced cell killing? TNFα dependence is certainly worth exploration in inflammation-associated cancers, such as colon and liver cancer [52, 195, 196]. What influences the decision between distinct signaling complexes and whether there is a more predominant role of cIAP1 or cIAP2 [197]? What is the significance of necrosis induction in cancer treatment? Lastly, blockade in the mitochondrial apoptotic pathway is common in cancer due to frequent alterations in p53 and the Bcl-2 family of proteins [1, 16, 17]. Therefore, rational combination of IAP antagonists with BH3 mimetics [198, 199] might bring new hopes to patients whose cancers are addicted to IAPs.

Acknowledgments We are grateful to Dr. Lin Zhang for critical reading and comments, and Mrs. Laurice Vance-Carr for excellent secretarial assistance. The work in authors' laboratory is supported in part by NIH grants CA129829, UO1-DK085570, American Cancer Society grant RGS-10-124-01-CCE, and Flight Attendant Medical Research Institute (FAMRI).

References

1. Hanahan D, Weinberg RA (2011) Hallmarks of cancer: the next generation. Cell 144(5):646–674
2. Johnstone RW, Ruefli AA, Lowe SW (2002) Apoptosis: a link between cancer genetics and chemotherapy. Cell 108(2):153–164
3. LaCasse EC et al (2008) IAP-targeted therapies for cancer. Oncogene 27(48):6252–6275
4. Fulda S, Vucic D (2012) Targeting IAP proteins for therapeutic intervention in cancer. Nat Rev Drug Discov 11(2):109–124
5. Keats JJ et al (2007) Promiscuous mutations activate the noncanonical NF-kappaB pathway in multiple myeloma. Cancer Cell 12(2):131–144
6. Altieri DC (2006) Targeted therapy by disabling crossroad signaling networks: the survivin paradigm. Mol Cancer Ther 5(3):478–482
7. Hunter AM, LaCasse EC, Korneluk RG (2007) The inhibitors of apoptosis (IAPs) as cancer targets. Apoptosis 12(9):1543–1568
8. Moore PS, Chang Y (2010) Why do viruses cause cancer? Highlights of the first century of human tumour virology. Nat Rev Cancer 10(12):878–889
9. Vogelstein B, Lane D, Levine AJ (2000) Surfing the p53 network. Nature 408(6810):307–310
10. Mocarski ES, Upton JW, Kaiser WJ (2012) Viral infection and the evolution of caspase 8-regulated apoptotic and necrotic death pathways. Nat Rev Immunol 12(2):79–88
11. Crook NE, Clem RJ, Miller LK (1993) An apoptosis-inhibiting baculovirus gene with a zinc finger-like motif. J Virol 67(4):2168–2174
12. Birnbaum MJ, Clem RJ, Miller LK (1994) An apoptosis-inhibiting gene from a nuclear polyhedrosis virus encoding a polypeptide with Cys/His sequence motifs. J Virol 68(4):2521–2528
13. Vaux DL, Silke J (2005) IAPs, RINGs and ubiquitylation. Nat Rev Mol Cell Biol 6(4):287–297
14. Fulda S, Debatin KM (2006) Extrinsic versus intrinsic apoptosis pathways in anticancer chemotherapy. Oncogene 25(34):4798–4811
15. Ashkenazi A (2008) Targeting the extrinsic apoptosis pathway in cancer. Cytokine Growth Factor Rev 19(3–4):325–331
16. Adams JM, Cory S (2007) The Bcl-2 apoptotic switch in cancer development and therapy. Oncogene 26(9):1324–1337
17. Yu J, Zhang L (2004) Apoptosis in human cancer cells. Curr Opin Oncol 16(1):19–24
18. Vousden KH, Lu X (2002) Live or let die: the cell's response to p53. Nat Rev Cancer 2(8):594–604
19. Yu J, Zhang L (2005) The transcriptional targets of p53 in apoptosis control. Biochem Biophys Res Commun 331(3):851–858
20. Chipuk JE, Green DR (2008) How do BCL-2 proteins induce mitochondrial outer membrane permeabilization? Trends Cell Biol 18(4):157–164
21. Wang X (2001) The expanding role of mitochondria in apoptosis. Genes Dev 15(22):2922–2933
22. Du C et al (2000) Smac, a mitochondrial protein that promotes cytochrome c-dependent caspase activation by eliminating IAP inhibition. Cell 102(1):33–42
23. Suzuki Y et al (2001) A serine protease, HtrA2, is released from the mitochondria and interacts with XIAP, inducing cell death. Mol Cell 8(3):613–621

24. Joza N et al (2001) Essential role of the mitochondrial apoptosis-inducing factor in programmed cell death. Nature 410(6828):549–554
25. Wang X et al (2002) Mechanisms of AIF-mediated apoptotic DNA degradation in *Caenorhabditis elegans*. Science 298(5598):1587–1592
26. Luo X et al (1998) Bid, a Bcl2 interacting protein, mediates cytochrome c release from mitochondria in response to activation of cell surface death receptors. Cell 94(4):481–490
27. Li H et al (1998) Cleavage of BID by caspase 8 mediates the mitochondrial damage in the Fas pathway of apoptosis. Cell 94(4):491–501
28. Eckelman BP, Salvesen GS, Scott FL (2006) Human inhibitor of apoptosis proteins: why XIAP is the black sheep of the family. EMBO Rep 7(10):988–994
29. Sun C et al (1999) NMR structure and mutagenesis of the inhibitor-of-apoptosis protein XIAP. Nature 401(6755):818–822
30. Silke J et al (2001) Direct inhibition of caspase 3 is dispensable for the anti-apoptotic activity of XIAP. EMBO J 20(12):3114–3123
31. Huang Y et al (2001) Structural basis of caspase inhibition by XIAP: differential roles of the linker versus the BIR domain. Cell 104(5):781–790
32. Chai J et al (2001) Structural basis of caspase-7 inhibition by XIAP. Cell 104(5):769–780
33. Riedl SJ et al (2001) Structural basis for the inhibition of caspase-3 by XIAP. Cell 104(5):791–800
34. Srinivasula SM et al (2001) A conserved XIAP-interaction motif in caspase-9 and Smac/DIABLO regulates caspase activity and apoptosis. Nature 410(6824):112–116
35. Shiozaki EN et al (2003) Mechanism of XIAP-mediated inhibition of caspase-9. Mol Cell 11(2):519–527
36. Wang SL et al (1999) The *Drosophila* caspase inhibitor DIAP1 is essential for cell survival and is negatively regulated by HID. Cell 98(4):453–463
37. Ditzel M et al (2008) Inactivation of effector caspases through nondegradative polyubiquitylation. Mol Cell 32(4):540–553
38. Suzuki Y, Nakabayashi Y, Takahashi R (2001) Ubiquitin-protein ligase activity of X-linked inhibitor of apoptosis protein promotes proteasomal degradation of caspase-3 and enhances its anti-apoptotic effect in Fas-induced cell death. Proc Natl Acad Sci USA 98(15):8662–8667
39. Choi YE et al (2009) The E3 ubiquitin ligase cIAP1 binds and ubiquitinates caspase-3 and -7 via unique mechanisms at distinct steps in their processing. J Biol Chem 284(19):12772–12782
40. Riedl SJ, Shi Y (2004) Molecular mechanisms of caspase regulation during apoptosis. Nat Rev Mol Cell Biol 5(11):897–907
41. Shi Y (2002) Mechanisms of caspase activation and inhibition during apoptosis. Mol Cell 9(3):459–470
42. Deveraux QL et al (1999) Cleavage of human inhibitor of apoptosis protein XIAP results in fragments with distinct specificities for caspases. EMBO J 18(19):5242–5251
43. Gyrd-Hansen M, Meier P (2010) IAPs: from caspase inhibitors to modulators of NF-kappaB, inflammation and cancer. Nat Rev Cancer 10(8):561–574
44. Varfolomeev E et al (2007) IAP antagonists induce autoubiquitination of c-IAPs, NF-kappaB activation, and TNFalpha-dependent apoptosis. Cell 131(4):669–681
45. Gaither A et al (2007) A Smac mimetic rescue screen reveals roles for inhibitor of apoptosis proteins in tumor necrosis factor-alpha signaling. Cancer Res 67(24):11493–11498
46. Wang L, Du F, Wang X (2008) TNF-alpha induces two distinct caspase-8 activation pathways. Cell 133(4):693–703
47. Zhang DW et al (2009) RIP3, an energy metabolism regulator that switches TNF-induced cell death from apoptosis to necrosis. Science 325(5938):332–336
48. He S et al (2009) Receptor interacting protein kinase-3 determines cellular necrotic response to TNF-alpha. Cell 137(6):1100–1111
49. Feoktistova M et al (2012) Pick your poison: the ripoptosome, a cell death platform regulating apoptosis and necroptosis. Cell Cycle 11(3):460–467

50. Karin M, Greten FR (2005) NF-kappaB: linking inflammation and immunity to cancer development and progression. Nat Rev Immunol 5(10):749–759
51. Nathan C, Ding A (2010) Nonresolving inflammation. Cell 140(6):871–882
52. Grivennikov SI, Greten FR, Karin M (2010) Immunity, inflammation, and cancer. Cell 140(6):883–899
53. Perkins ND (2007) Integrating cell-signalling pathways with NF-kappaB and IKK function. Nat Rev Mol Cell Biol 8(1):49–62
54. Bonizzi G, Karin M (2004) The two NF-kappaB activation pathways and their role in innate and adaptive immunity. Trends Immunol 25(6):280–288
55. Bhoj VG, Chen ZJ (2009) Ubiquitylation in innate and adaptive immunity. Nature 458(7237):430–437
56. Micheau O, Tschopp J (2003) Induction of TNF receptor I-mediated apoptosis via two sequential signaling complexes. Cell 114(2):181–190
57. Leulier F et al (2006) The *Drosophila* inhibitor of apoptosis protein DIAP2 functions in innate immunity and is essential to resist gram-negative bacterial infection. Mol Cell Biol 26(21):7821–7831
58. Gesellchen V et al (2005) An RNA interference screen identifies Inhibitor of apoptosis protein 2 as a regulator of innate immune signalling in *Drosophila*. EMBO Rep 6(10):979–984
59. Kleino A et al (2005) Inhibitor of apoptosis 2 and TAK1-binding protein are components of the *Drosophila* Imd pathway. EMBO J 24(19):3423–3434
60. Orme M, Meier P (2009) Inhibitor of apoptosis proteins in *Drosophila*: gatekeepers of death. Apoptosis 14(8):950–960
61. Winkles JA (2008) The TWEAK-Fn14 cytokine-receptor axis: discovery, biology and therapeutic targeting. Nat Rev Drug Discov 7(5):411–425
62. Vallabhapurapu S et al (2008) Nonredundant and complementary functions of TRAF2 and TRAF3 in a ubiquitination cascade that activates NIK-dependent alternative NF-kappaB signaling. Nat Immunol 9(12):1364–1370
63. Zarnegar BJ et al (2008) Noncanonical NF-kappaB activation requires coordinated assembly of a regulatory complex of the adaptors cIAP1, cIAP2, TRAF2 and TRAF3 and the kinase NIK. Nat Immunol 9(12):1371–1378
64. Galluzzi L et al (2011) Programmed necrosis from molecules to health and disease. Int Rev Cell Mol Biol 289:1–35
65. Petersen SL et al (2007) Autocrine TNFalpha signaling renders human cancer cells susceptible to Smac-mimetic-induced apoptosis. Cancer Cell 12(5):445–456
66. Vercammen D et al (1998) Inhibition of caspases increases the sensitivity of L929 cells to necrosis mediated by tumor necrosis factor. J Exp Med 187(9):1477–1485
67. Vanden Berghe T et al (2003) Disruption of HSP90 function reverts tumor necrosis factor-induced necrosis to apoptosis. J Biol Chem 278(8):5622–5629
68. Zheng L et al (2006) Competitive control of independent programs of tumor necrosis factor receptor-induced cell death by TRADD and RIP1. Mol Cell Biol 26(9):3505–3513
69. Geserick P et al (2009) Cellular IAPs inhibit a cryptic CD95-induced cell death by limiting RIP1 kinase recruitment. J Cell Biol 187(7):1037–1054
70. Jackson SP, Bartek J (2009) The DNA-damage response in human biology and disease. Nature 461(7267):1071–1078
71. Tenev T et al (2011) The Ripoptosome, a signaling platform that assembles in response to genotoxic stress and loss of IAPs. Mol Cell 43(3):432–448
72. Biton S, Ashkenazi A (2011) NEMO and RIP1 control cell fate in response to extensive DNA damage via TNF-alpha feedforward signaling. Cell 145(1):92–103
73. Verhagen AM et al (2000) Identification of DIABLO, a mammalian protein that promotes apoptosis by binding to and antagonizing IAP proteins. Cell 102(1):43–53
74. Steller H (2008) Regulation of apoptosis in *Drosophila*. Cell Death Differ 15(7):1132–1138
75. Shi Y (2004) Caspase activation: revisiting the induced proximity model. Cell 117(7):855–858
76. Vucic D et al (2002) SMAC negatively regulates the anti-apoptotic activity of melanoma inhibitor of apoptosis (ML-IAP). J Biol Chem 277(14):12275–12279

77. Davoodi J et al (2004) Neuronal apoptosis-inhibitory protein does not interact with Smac and requires ATP to bind caspase-9. J Biol Chem 279(39):40622–40628
78. Hao Y et al (2004) Apollon ubiquitinates SMAC and caspase-9, and has an essential cytoprotection function. Nat Cell Biol 6(9):849–860
79. Qiu XB, Goldberg AL (2005) The membrane-associated inhibitor of apoptosis protein, BRUCE/Apollon, antagonizes both the precursor and mature forms of Smac and caspase-9. J Biol Chem 280(1):174–182
80. Shiozaki EN, Shi Y (2004) Caspases, IAPs and Smac/DIABLO: mechanisms from structural biology. Trends Biochem Sci 29(9):486–494
81. Chai J et al (2000) Structural and biochemical basis of apoptotic activation by Smac/DIABLO. Nature 406(6798):855–862
82. Liu Z et al (2000) Structural basis for binding of Smac/DIABLO to the XIAP BIR3 domain. Nature 408(6815):1004–1008
83. Wu G et al (2000) Structural basis of IAP recognition by Smac/DIABLO. Nature 408(6815):1008–1012
84. Hegde R et al (2002) Identification of Omi/HtrA2 as a mitochondrial apoptotic serine protease that disrupts inhibitor of apoptosis protein-caspase interaction. J Biol Chem 277(1):432–438
85. Verhagen AM et al (2002) HtrA2 promotes cell death through its serine protease activity and its ability to antagonize inhibitor of apoptosis proteins. J Biol Chem 277(1):445–454
86. Martins LM et al (2002) The serine protease Omi/HtrA2 regulates apoptosis by binding XIAP through a reaper-like motif. J Biol Chem 277(1):439–444
87. Liston P et al (2001) Identification of XAF1 as an antagonist of XIAP anti-Caspase activity. Nat Cell Biol 3(2):128–133
88. Arora V et al (2007) Degradation of survivin by the X-linked inhibitor of apoptosis (XIAP)-XAF1 complex. J Biol Chem 282(36):26202–26209
89. Kashkar H (2010) X-linked inhibitor of apoptosis: a chemoresistance factor or a hollow promise. Clin Cancer Res 16(18):4496–4502
90. West T et al (2009) Lack of X-linked inhibitor of apoptosis protein leads to increased apoptosis and tissue loss following neonatal brain injury. ASN Neuro 1(1):e00004
91. Bauler LD, Duckett CS, O'Riordan MX (2008) XIAP regulates cytosol-specific innate immunity to Listeria infection. PLoS Pathog 4(8):e1000142
92. Cummins JM et al (2004) X-linked inhibitor of apoptosis protein (XIAP) is a nonredundant modulator of tumor necrosis factor-related apoptosis-inducing ligand (TRAIL)-mediated apoptosis in human cancer cells. Cancer Res 64(9):3006–3008
93. Conte D et al (2006) Inhibitor of apoptosis protein cIAP2 is essential for lipopolysaccharide-induced macrophage survival. Mol Cell Biol 26(2):699–708
94. Moulin M et al (2012) IAPs limit activation of RIP kinases by TNF receptor 1 during development. EMBO J 31(7):1679–1691. doi: http://10.1038/ebomj.2012.18
95. Gardam S et al (2011) Deletion of cIAP1 and cIAP2 in murine B lymphocytes constitutively activates cell survival pathways and inactivates the germinal center response. Blood 117(15):4041–4051
96. Bank A et al (2008) SMAC mimetics sensitize nonsteroidal anti-inflammatory drug-induced apoptosis by promoting caspase-3-mediated cytochrome c release. Cancer Res 68(1):276–284
97. Kohli M et al (2004) SMAC/Diablo-dependent apoptosis induced by nonsteroidal antiinflammatory drugs (NSAIDs) in colon cancer cells. Proc Natl Acad Sci USA 101(48):16897–16902
98. Okada H et al (2002) Generation and characterization of Smac/DIABLO-deficient mice. Mol Cell Biol 22(10):3509–3517
99. Qiu W et al (2010) Chemoprevention by nonsteroidal anti-inflammatory drugs eliminates oncogenic intestinal stem cells via SMAC-dependent apoptosis. Proc Natl Acad Sci USA 107(46):20027–20032
100. Martins LM et al (2004) Neuroprotective role of the reaper-related serine protease HtrA2/Omi revealed by targeted deletion in mice. Mol Cell Biol 24(22):9848–9862

101. Altieri DC (2008) Survivin, cancer networks and pathway-directed drug discovery. Nat Rev Cancer 8(1):61–70
102. Imoto I et al (2001) Identification of cIAP1 as a candidate target gene within an amplicon at 11q22 in esophageal squamous cell carcinomas. Cancer Res 61(18):6629–6634
103. Imoto I et al (2002) Expression of cIAP1, a target for 11q22 amplification, correlates with resistance of cervical cancers to radiotherapy. Cancer Res 62(17):4860–4866
104. Zender L et al (2006) Identification and validation of oncogenes in liver cancer using an integrative oncogenomic approach. Cell 125(7):1253–1267
105. Dai Z et al (2003) A comprehensive search for DNA amplification in lung cancer identifies inhibitors of apoptosis cIAP1 and cIAP2 as candidate oncogenes. Hum Mol Genet 12(7):791–801
106. Bashyam MD et al (2005) Array-based comparative genomic hybridization identifies localized DNA amplifications and homozygous deletions in pancreatic cancer. Neoplasia 7(6):556–562
107. Reardon DA et al (1997) Extensive genomic abnormalities in childhood medulloblastoma by comparative genomic hybridization. Cancer Res 57(18):4042–4047
108. Weber RG et al (1996) Clinically distinct subgroups of glioblastoma multiforme studied by comparative genomic hybridization. Lab Invest 74(1):108–119
109. Ma O et al (2009) MMP13, Birc2 (cIAP1), and Birc3 (cIAP2), amplified on chromosome 9, collaborate with p53 deficiency in mouse osteosarcoma progression. Cancer Res 69(6):2559–2567
110. Dierlamm J et al (1999) The apoptosis inhibitor gene API2 and a novel 18q gene, MLT, are recurrently rearranged in the t(11;18)(q21;q21) associated with mucosa-associated lymphoid tissue lymphomas. Blood 93(11):3601–3609
111. Akagi T et al (1999) A novel gene, MALT1 at 18q21, is involved in t(11;18) (q21;q21) found in low-grade B-cell lymphoma of mucosa-associated lymphoid tissue. Oncogene 18(42):5785–5794
112. Zhou H, Du MQ, Dixit VM (2005) Constitutive NF-kappaB activation by the t(11;18) (q21;q21) product in MALT lymphoma is linked to deregulated ubiquitin ligase activity. Cancer Cell 7(5):425–431
113. Morgan JA et al (1999) Breakpoints of the t(11;18)(q21;q21) in mucosa-associated lymphoid tissue (MALT) lymphoma lie within or near the previously undescribed gene MALT1 in chromosome 18. Cancer Res 59(24):6205–6213
114. Varfolomeev E et al (2006) The inhibitor of apoptosis protein fusion c-IAP2.MALT1 stimulates NF-kappaB activation independently of TRAF1 AND TRAF2. J Biol Chem 281(39):29022–29029
115. Annunziata CM et al (2007) Frequent engagement of the classical and alternative NF-kappaB pathways by diverse genetic abnormalities in multiple myeloma. Cancer Cell 12(2):115–130
116. Yamaguchi N et al (2009) Constitutive activation of nuclear factor-kappaB is preferentially involved in the proliferation of basal-like subtype breast cancer cell lines. Cancer Sci 100(9):1668–1674
117. Wharry CE et al (2009) Constitutive non-canonical NF kappaB signaling in pancreatic cancer cells. Cancer Biol Ther 8(16):1567–1576
118. Xiang G et al (2009) Expression of X-linked inhibitor of apoptosis protein in human colorectal cancer and its correlation with prognosis. J Surg Oncol 100(8):708–712
119. Augello C et al (2009) Inhibitors of apoptosis proteins (IAPs) expression and their prognostic significance in hepatocellular carcinoma. BMC Cancer 9:125
120. Shibata T et al (2007) Disturbed expression of the apoptosis regulators XIAP, XAF1, and Smac/DIABLO in gastric adenocarcinomas. Diagn Mol Pathol 16(1):1–8
121. Zhang Y et al (2011) X-linked inhibitor of apoptosis positive nuclear labeling: a new independent prognostic biomarker of breast invasive ductal carcinoma. Diagn Pathol 6:49
122. Hiscutt EL et al (2010) Targeting X-linked inhibitor of apoptosis protein to increase the efficacy of endoplasmic reticulum stress-induced apoptosis for melanoma therapy. J Invest Dermatol 130(9):2250–2258

123. Qi S et al (2008) Expression of cIAP-1 correlates with nodal metastasis in squamous cell carcinoma of the tongue. Int J Oral Maxillofac Surg 37(11):1047–1053
124. Krajewska M et al (2005) Analysis of apoptosis protein expression in early-stage colorectal cancer suggests opportunities for new prognostic biomarkers. Clin Cancer Res 11(15):5451–5461
125. Esposito I et al (2007) Overexpression of cellular inhibitor of apoptosis protein 2 is an early event in the progression of pancreatic cancer. J Clin Pathol 60(8):885–895
126. Vucic D et al (2000) ML-IAP, a novel inhibitor of apoptosis that is preferentially expressed in human melanomas. Curr Biol 10(21):1359–1366
127. Wagener N et al (2007) Expression of inhibitor of apoptosis protein Livin in renal cell carcinoma and non-tumorous adult kidney. Br J Cancer 97(9):1271–1276
128. Gong J et al (2005) Melanoma inhibitor of apoptosis protein is expressed differentially in melanoma and melanocytic naevus, but similarly in primary and metastatic melanomas. J Clin Pathol 58(10):1081–1085
129. Kempkensteffen C et al (2007) Expression of the apoptosis inhibitor livin in renal cell carcinomas: correlations with pathology and outcome. Tumour Biol 28(3):132–138
130. Gazzaniga P et al (2003) Expression and prognostic significance of LIVIN, SURVIVIN and other apoptosis-related genes in the progression of superficial bladder cancer. Ann Oncol 14(1):85–90
131. El-Mesallamy HO, Hegab HM, Kamal AM (2011) Expression of inhibitor of apoptosis protein (IAP) livin/BIRC7 in acute leukemia in adults: correlation with prognostic factors and outcome. Leuk Res 35(12):1616–1622
132. Nedelcu T et al (2008) Livin and Bcl-2 expression in high-grade osteosarcoma. J Cancer Res Clin Oncol 134(2):237–244
133. Kim DK et al (2005) Expression of inhibitor-of-apoptosis protein (IAP) livin by neuroblastoma cells: correlation with prognostic factors and outcome. Pediatr Dev Pathol 8(6):621–629
134. Duffy MJ et al (2007) Survivin: a promising tumor biomarker. Cancer Lett 249(1):49–60
135. Ambrosini G, Adida C, Altieri DC (1997) A novel anti-apoptosis gene, survivin, expressed in cancer and lymphoma. Nat Med 3(8):917–921
136. Ambrosini G et al (1998) Induction of apoptosis and inhibition of cell proliferation by survivin gene targeting. J Biol Chem 273(18):11177–11182
137. Endo K et al (2009) Clinical significance of Smac/DIABLO expression in colorectal cancer. Oncol Rep 21(2):351–355
138. Xu Y et al (2011) Role of smac in determining the chemotherapeutic response of esophageal squamous cell carcinoma. Clin Cancer Res 17(16):5412–5422
139. Martinez-Ruiz G et al (2008) Role of Smac/DIABLO in cancer progression. J Exp Clin Cancer Res 27:48
140. Fong WG et al (2000) Expression and genetic analysis of XIAP-associated factor 1 (XAF1) in cancer cell lines. Genomics 70(1):113–122
141. Sekimura A et al (2004) Expression of Smac/DIABLO is a novel prognostic marker in lung cancer. Oncol Rep 11(4):797–802
142. Pluta P et al (2011) Correlation of Smac/DIABLO protein expression with the clinico-pathological features of breast cancer patients. Neoplasma 58(5):430–435
143. Mizutani Y, Katsuoka Y, Bonavida B (2010) Prognostic significance of second mitochondria-derived activator of caspase (Smac/DIABLO) expression in bladder cancer and target for therapy. Int J Oncol 37(2):503–508
144. Mizutani Y et al (2005) Downregulation of Smac/DIABLO expression in renal cell carcinoma and its prognostic significance. J Clin Oncol 23(3):448–454
145. Kempkensteffen C et al (2008) Expression levels of the mitochondrial IAP antagonists Smac/DIABLO and Omi/HtrA2 in clear-cell renal cell carcinomas and their prognostic value. J Cancer Res Clin Oncol 134(5):543–550
146. Dobrzycka B et al (2010) Prognostic significance of smac/DIABLO in endometrioid endometrial cancer. Folia Histochem Cytobiol 48(4):678–681
147. Arellano-Llamas A et al (2006) High Smac/DIABLO expression is associated with early local recurrence of cervical cancer. BMC Cancer 6:256

148. Pluta A et al (2010) Influence of high expression of Smac/DIABLO protein on the clinical outcome in acute myeloid leukemia patients. Leuk Res 34(10):1308–1313
149. Ndubaku C et al (2009) Targeting inhibitor of apoptosis proteins for therapeutic intervention. Future Med Chem 1(8):1509–1525
150. Vucic D, Fairbrother WJ (2007) The inhibitor of apoptosis proteins as therapeutic targets in cancer. Clin Cancer Res 13(20):5995–6000
151. Wang S (2011) Design of small-molecule Smac mimetics as IAP antagonists. Curr Top Microbiol Immunol 348:89–113
152. Fulda S, Meyer E, Debatin KM (2002) Inhibition of TRAIL-induced apoptosis by Bcl-2 overexpression. Oncogene 21(15):2283–2294
153. Li L et al (2004) A small molecule Smac mimic potentiates TRAIL- and TNFalpha-mediated cell death. Science 305(5689):1471–1474
154. Arnt CR et al (2002) Synthetic Smac/DIABLO peptides enhance the effects of chemotherapeutic agents by binding XIAP and cIAP1 in situ. J Biol Chem 277(46):44236–44243
155. Yang L et al (2003) Predominant suppression of apoptosome by inhibitor of apoptosis protein in non-small cell lung cancer H460 cells: therapeutic effect of a novel polyarginine-conjugated Smac peptide. Cancer Res 63(4):831–837
156. Sun H et al (2007) Design, synthesis, and characterization of a potent, nonpeptide, cell-permeable, bivalent Smac mimetic that concurrently targets both the BIR2 and BIR3 domains in XIAP. J Am Chem Soc 129(49):15279–15294
157. Bertrand MJ et al (2008) cIAP1 and cIAP2 facilitate cancer cell survival by functioning as E3 ligases that promote RIP1 ubiquitination. Mol Cell 30(6):689–700
158. Varfolomeev E et al (2008) c-IAP1 and c-IAP2 are critical mediators of tumor necrosis factor alpha (TNFalpha)-induced NF-kappaB activation. J Biol Chem 283(36):24295–24299
159. Vince JE et al (2007) IAP antagonists target cIAP1 to induce TNFalpha-dependent apoptosis. Cell 131(4):682–693
160. Gao Z et al (2007) A dimeric Smac/diablo peptide directly relieves caspase-3 inhibition by XIAP. dynamic and cooperative regulation of XIAP by Smac/Diablo. J Biol Chem 282(42):30718–30727
161. Varfolomeev E et al (2009) X chromosome-linked inhibitor of apoptosis regulates cell death induction by proapoptotic receptor agonists. J Biol Chem 284(50):34553–34560
162. Sun Q et al (2011) Smac Modulates chemosensitivity in head and neck cancer cells through the mitochondrial apoptotic pathway. Clin Cancer Res 17(8):2361–2372 [Epub ahead of print]
163. Feltham R et al (2011) Smac mimetics activate the E3 ligase activity of cIAP1 protein by promoting RING domain dimerization. J Biol Chem 286(19):17015–17028
164. Mace PD et al (2008) Structures of the cIAP2 RING domain reveal conformational changes associated with ubiquitin-conjugating enzyme (E2) recruitment. J Biol Chem 283(46):31633–31640
165. Dueber EC et al (2011) Antagonists induce a conformational change in cIAP1 that promotes autoubiquitination. Science 334(6054):376–380
166. Dineen SP et al (2010) Smac mimetic increases chemotherapy response and improves survival in mice with pancreatic cancer. Cancer Res 70(7):2852–2861
167. Cheung HH et al (2009) Down-regulation of c-FLIP enhances death of cancer cells by smac mimetic compound. Cancer Res 69(19):7729–7738
168. Lu J et al (2011) Therapeutic potential and molecular mechanism of a novel, potent, nonpeptide, Smac mimetic SM-164 in combination with TRAIL for cancer treatment. Mol Cancer Ther 10(5):902–914
169. Foster FM et al (2009) Targeting inhibitor of apoptosis proteins in combination with ErbB antagonists in breast cancer. Breast Cancer Res 11(3):R41
170. Aird KM et al (2010) X-linked inhibitor of apoptosis protein inhibits apoptosis in inflammatory breast cancer cells with acquired resistance to an ErbB1/2 tyrosine kinase inhibitor. Mol Cancer Ther 9(5):1432–1442
171. Fulda S et al (2002) Smac agonists sensitize for Apo2L/TRAIL- or anticancer drug-induced apoptosis and induce regression of malignant glioma in vivo. Nat Med 8(8):808–815

172. Guo F et al (2002) Ectopic overexpression of second mitochondria-derived activator of caspases (Smac/DIABLO) or cotreatment with N-terminus of Smac/DIABLO peptide potentiates epothilone B derivative-(BMS 247550) and Apo-2L/TRAIL-induced apoptosis. Blood 99(9):3419–3426
173. Ren X et al (2007) Bypass NFkappaB-mediated survival pathways by TRAIL and Smac. Cancer Biol Ther 6(7):1031–1035
174. Stadel D et al (2010) TRAIL-induced apoptosis is preferentially mediated via TRAIL receptor 1 in pancreatic carcinoma cells and profoundly enhanced by XIAP inhibitors. Clin Cancer Res 16(23):5734–5749
175. Siegelin MD, Gaiser T, Siegelin Y (2009) The XIAP inhibitor embelin enhances TRAIL-mediated apoptosis in malignant glioma cells by down-regulation of the short isoform of FLIP. Neurochem Int 55(6):423–430
176. Mori T et al (2007) Effect of the XIAP inhibitor embelin on TRAIL-induced apoptosis of pancreatic cancer cells. J Surg Res 142(2):281–286
177. Loeder S et al (2009) A novel paradigm to trigger apoptosis in chronic lymphocytic leukemia. Cancer Res 69(23):8977–8986
178. Kater AP et al (2005) Inhibitors of XIAP sensitize CD40-activated chronic lymphocytic leukemia cells to CD95-mediated apoptosis. Blood 106(5):1742–1748
179. Jost PJ et al (2009) XIAP discriminates between type I and type II FAS-induced apoptosis. Nature 460(7258):1035–1039
180. Feoktistova M et al (2011) cIAPs block ripoptosome formation, a RIP1/caspase-8 containing intracellular cell death complex differentially regulated by cFLIP isoforms. Mol Cell 43(3):449–463
181. Fingas CD et al (2010) A smac mimetic reduces TNF related apoptosis inducing ligand (TRAIL)-induced invasion and metastasis of cholangiocarcinoma cells. Hepatology 52(2):550–561
182. Schimmer AD et al (2006) Targeting XIAP for the treatment of malignancy. Cell Death Differ 13(2):179–188
183. Wu TY et al (2003) Development and characterization of nonpeptidic small molecule inhibitors of the XIAP/caspase-3 interaction. Chem Biol 10(8):759–767
184. Schimmer AD et al (2004) Small-molecule antagonists of apoptosis suppressor XIAP exhibit broad antitumor activity. Cancer Cell 5(1):25–35
185. Dean EJ et al (2010) A small molecule inhibitor of XIAP induces apoptosis and synergises with vinorelbine and cisplatin in NSCLC. Br J Cancer 102(1):97–103
186. Cheng YJ et al (2010) XIAP-mediated protection of H460 lung cancer cells against cisplatin. Eur J Pharmacol 627(1–3):75–84
187. LaCasse EC et al (2006) Preclinical characterization of AEG35156/GEM 640, a second-generation antisense oligonucleotide targeting X-linked inhibitor of apoptosis. Clin Cancer Res 12(17):5231–5241
188. Holt SV et al (2011) Down-regulation of XIAP by AEG35156 in paediatric tumour cells induces apoptosis and sensitises cells to cytotoxic agents. Oncol Rep 25(4):1177–1181
189. Hu Y et al (2003) Antisense oligonucleotides targeting XIAP induce apoptosis and enhance chemotherapeutic activity against human lung cancer cells in vitro and in vivo. Clin Cancer Res 9(7):2826–2836
190. Amantana A et al (2004) X-linked inhibitor of apoptosis protein inhibition induces apoptosis and enhances chemotherapy sensitivity in human prostate cancer cells. Mol Cancer Ther 3(6):699–707
191. Dean E et al (2009) Phase I trial of AEG35156 administered as a 7 and 3 day continuous intravenous infusion in patients with advanced refractory cancer. J Clin Oncol 27(10):1660–1666
192. Schimmer AD et al (2009) Phase I/II trial of AEG35156 X-linked inhibitor of apoptosis protein antisense oligonucleotide combined with idarubicin and cytarabine in patients with relapsed or primary refractory acute myeloid leukemia. J Clin Oncol 27(28):4741–4746

193. Crnkovic-Mertens I et al (2006) Isoform-specific silencing of the livin gene by RNA interference defines Livin beta as key mediator of apoptosis inhibition in HeLa cells. J Mol Med (Berl) 84(3):232–240
194. Tseng PH et al (2010) Different modes of ubiquitination of the adaptor TRAF3 selectively activate the expression of type I interferons and proinflammatory cytokines. Nat Immunol 11(1):70–75
195. Kim S et al (2009) Carcinoma-produced factors activate myeloid cells through TLR2 to stimulate metastasis. Nature 457(7225):102–106
196. Wu Y, Zhou BP (2010) TNF-alpha/NF-kappaB/Snail pathway in cancer cell migration and invasion. Br J Cancer 102(4):639–644
197. Petersen SL et al (2010) Overcoming cancer cell resistance to Smac mimetic induced apoptosis by modulating cIAP-2 expression. Proc Natl Acad Sci USA 107(26):11936–11941
198. Fesik SW (2005) Promoting apoptosis as a strategy for cancer drug discovery. Nat Rev Cancer 5(11):876–885
199. Zhang L, Ming L, Yu J (2007) BH3 mimetics to improve cancer therapy; mechanisms and examples. Drug Resist Updat 10(6):207–217
200. Flygare J et al (2012) The discovery of a potent small-molecule antagonist of inhibitor of apoptosis (IAP) proteins and clinical candidate for the treatment of cancer (GDC-0152). J Med Chem 55(9):4101–4113
201. Cai Q et al (2011) A potent and orally active antagonist (SM-406/AT-406) of multiple inhibitor of apoptosis proteins (IAPs) in clinical development for cancer treatment. J Med Chem 54(8):2714–2726
202. Infante JR (2010) A phase I study of LCL-161, an oral IAP inhibitor, in patients with advanced cancer. In: Proceedings of the 101st annual meeting of the american association for cancer research. Washington
203. Sikic B (2011) Safety, pharmacokinetics (PK), and pharmacodynamics (PD) of HGS1029, an inhibitor of apoptosis protein (IAP) inhibitor, in patients (Pts) with advanced solid tumors: Results of a phase I study. J Clin Oncol (Meeting abstract)
204. Amaravadi RK (2011) Phase 1 study of the Smac mimetic TL32711 in adult subjects with advanced solid tumors and lymphoma to evaluate safety, pharmacokinetics, pharmacodynamics and antitumor activity. In: Proceedings of the 102nd annual meeting of the american association for cancer research. Orlando, Florida
205. Weisberg E et al (2010) Smac mimetics: implications for enhancement of targeted therapies in leukemia. Leukemia 24(12):2100–2109
206. Mahadevan D et al (2012) Phase I trial of AEG35156 an antisense oligonucleotide to XIAP plus gemcitabine in patients with metastatic pancreatic ductal adenocarcinoma. Am J Clin Oncol

Chapter 11
Harnessing Death Receptor Signaling for Cancer Treatment

Simone Fulda

Abstract Apoptosis, the cell's intrinsic cell death program, is a key regulator of tissue homeostasis. Accordingly, tilting the balance between cell death on one side and cell proliferation on the other side toward survival promotes tumor formation. The death receptor (extrinsic) pathway represents one of the major apoptosis signaling cascades, which links exogenous stimuli via transmembrane surface receptors to the intracellular signaling machinery that mediates and executes the death signal. Since defects in death receptor signaling can confer resistance to apoptosis, a better understanding of the regulation of the signaling events and their perturbation in human cancers may lead to the identification of new molecular targets that can be exploited for therapeutic purposes. This strategy is expected to open new perspectives to target the death receptor pathway for cancer therapy.

11.1 Introduction

Programmed cell death (apoptosis) represents an evolutionary highly conserved intrinsic cell death program that is critically involved in the regulation of various physiological and pathological processes [1]. For example, tissue homeostasis is maintained by a delicate balance of cell growth on one side and cell death on the other side [2]. Tipping this balance toward one or the other side results in too little or too much apoptosis and can foster either tumor formation or tissue loss [3]. In addition, the efficacy of most current cancer treatments including chemo-, radio- or immunotherapy, largely depends on intact cell death programs in cancer cells [4–6]. Therefore, defective apoptosis signaling pathways can lead to treatment resistance, one of the major challenges nowadays in clinical oncology. The identification of the molecular mechanisms that are responsible for cancer cell's evasion of apoptosis is expected to open new perspectives to specifically exploit

S. Fulda (✉)
Institute for Experimental Cancer Research in Pediatrics, Goethe-University Frankfurt, Frankfurt, Germany
e-mail: simone.fulda@kgu.de

D. E. Johnson (ed.), *Cell Death Signaling in Cancer Biology and Treatment*, Cell Death in Biology and Diseases, DOI: 10.1007/978-1-4614-5847-0_11,

X.-M. Yin and Z. Dong (Series eds.), *Cell Death in Biology and Diseases*

cell death pathways for therapeutic purposes. The current review focuses on the opportunities to target death receptor signaling in order to develop new therapeutic strategies for the treatment for cancer.

11.2 Death Receptors

Death receptors belong to the superfamily of tumor necrosis factor (TNF) receptors, which consist of more than 20 members with a wide spectrum of biological functions including regulation of cell death, survival, differentiation, and immune regulation [7]. All TNF receptor family members share a similar, cytoplasmic motif of about 80 amino acids, the so-called death domain, which is critical for transmitting the death signal from the cell's surface to intracellular signaling pathways. In addition, death receptors harbor cysteine-rich extracellular domains for ligand binding. Among the death receptors CD95 (APO-1/Fas), TNF receptor 1 (TNFR1) and TNF-related apoptosis-inducing ligand (TRAIL) receptors have been extensively studied in the last two decades [7]. The corresponding death receptor ligands of the TNF superfamily comprise among others CD95 ligand, TNFα, lymphotoxin-α (the latter two bind to TNFR1), TRAIL and TWEAK, a ligand for DR3 [7]. The CD95 receptor/CD95 ligand system represents a major signaling pathway that mediates apoptosis in several different cell types, for example in the immune system [8]. TRAIL was identified in 1995 based on its sequence homology to other members of the TNF superfamily and is constitutively expressed in a wide range of tissues [7]. There are two agonistic TRAIL receptors, that is, TRAIL-R1 and TRAIL-R2, that contain a death domain, and therefore a signal to cell death upon ligand binding, whereas TRAIL-R3 to R-5 are antagonistic decoy receptors, which bind TRAIL, but are not able to transmit a death signal [7].

11.3 Apoptosis Pathways

Two major apoptosis signaling pathways have been identified, that is the death receptor (extrinsic) and the mitochondrial (intrinsic) pathway [4]. Stimulation of death receptors such as CD95, TRAIL-R1 or TRAIL-R2 by CD95 ligand, TRAIL or agonistic antibodies results in receptor oligomerization and recruitment of FADD and caspase-8 to activated death receptors to form the death-inducing signaling complex (DISC) [7]. This multimeric complex drives the activation of caspase-8, which in turn transmits the apoptosis signal. To this end, caspase-8 can directly activate effector pathways of apoptosis by cleaving caspase-3. Alternatively, caspase-8 can initiate a crosstalk to the mitochondrial pathway of apoptosis by processing Bid into its active form tBid [9]. Bid belongs to the proapoptotic proteins of the Bcl-2 family that contains a BH3-only domain [10].

tBid translocates from the cytosol to the mitochondria and triggers mitochondrial outer membrane permeabilization. This results in the release of mitochondrial proteins from the intermembrane space of the mitochondria into the cytosol, for example of cytochrome c or second mitochondrial activator of caspases (Smac) [10]. Cytochrome c forms a complex in the cytosol together with Apaf-1 and caspase-9 to trigger caspase-9 and subsequently caspase-3 activation [10]. Smac antagonizes Inhibitor of Apoptosis (IAP) proteins, which function as endogenous inhibitors of caspases [11]. The release of caspases from the inhibition of IAP proteins by Smac promotes caspase activation and apoptosis. The engagement of either the extrinsic or the intrinsic apoptosis pathway results in activation of caspases, a family of proteases that function as executioners in multiple modes of cell death [12].

Cell death pathways are tightly regulated by pro- and antiapoptotic factors. This should ensure that they are rapidly activated upon stimulation, for example upon death receptor ligation. Vice versa, this tight control should prevent their accidental engagement, which could have detrimental effects on cellular survival. Importantly, cancer cells have adopted many of these antiapoptotic mechanisms to escape programmed cell death. Accordingly, evasion of apoptosis represents one of the hallmarks of cancer cells. This also implies that targeting defective cell death pathways bears the potential to tackle one of the key properties of malignant cells.

11.4 Apoptosis and Cancer Biology

One of the hallmarks of human cancers is their ability to evade apoptosis in order to survive environmental or oncogenic stress signals. This favors the progressive growth of a tumor and, as such, cooperates with proliferative signals to foster tumor development as well as its progression. On a theoretical ground, apoptosis programs can be disrupted either via a reduction in the apoptosis promoting factors or via the dominance of processes that block apoptosis. Both genetic as well as epigenetic events can cause inactivation of apoptosis pathways, implying that at least a proportion of these events is in principle reversible and amenable for therapeutic interventions.

11.4.1 Aberrant Death Receptor Signaling in Cancers

11.4.1.1 Alterations in the CD95 Pathway

Signaling via the death receptor pathway of apoptosis may be disturbed at multiple levels in human cancers. Along the signaling cascade, the surface levels of death receptors have been described to be downregulated in human cancers. CD95 was reduced in CD95-resistant tumor cells [13, 14] as well as in cells that were refractory to various anticancer drugs [13, 14], indicating that CD95 expression also regulates drug responsiveness. Genetic alterations of CD95 have also been implicated in tumorigenesis. Hematological malignancies, as well as solid tumors, were reported to harbor CD95 gene mutations [15]. In addition to genetic lesions, epigenetic alterations including hypermethylation of the CD95 promoter have also been implicated as an underlying cause for reduced CD95 expression in cancers [16, 17]. This mechanism may contribute to tumor immune escape, as cancer with epigenetically inactivated CD95 displayed low sensitivity to immune cell-mediated killing. Consequently, restoration of CD95 expression by treatment with epigenetic drugs such as histone deacetylase inhibitors concomitantly enhanced NK cell-dependent tumor cell killing as well as the response to chemotherapy [18].

11.4.1.2 Alterations in the TRAIL Pathway

Along the same lines, resistance toward TRAIL has been linked to low or absent surface expression of one of the two agonistic TRAIL receptors TRAIL-R1 and TRAIL-R2. Interestingly, the chromosomal localization of these TRAIL receptors on chromosome 8p falls within a region that is often genetically altered in human cancers, for example by the loss of heterozygosity (LOH) [19]. Furthermore, the loss of both copies of TRAIL-R1 or TRAIL-R2 due to deletions or mutations has been detected in a small percentage of various cancers, for example several carcinomas (colorectal, breast, head and neck, lung), non-Hodgkin's lymphoma and osteosarcoma [20, 21]. In addition to these genetic events, aberrations in the subcellular distribution of TRAIL receptors may account for the evasion of TRAIL-induced apoptosis. In this respect, it has been reported in colon carcinoma cells that the apoptosis-inducing TRAIL receptors, TRAIL-R1 and TRAIL-R2, are retained in intracellular stores, for example the endoplasmic reticulum, and are not properly transported to the cell surface, the location where they usually engage with their ligand TRAIL to initiate proapoptotic signaling [22].

Another mechanism to block death receptor signaling resides in the relative abundance of decoy receptors. Both the CD95 and TRAIL cascade can be impaired by such decoy receptors. Decoy receptor 3 (DcR3) has been shown to competitively bind to CD95 ligand, thereby blocking CD95-induced apoptosis [23, 24]. Of note, high expression levels of DcR3 or genetic amplification of CcR3 were detected in lung or colon cancer and in glioblastoma [23, 24], indicating that DcR3 may contribute to CD95 resistance in these cancers. TRAIL-R3 and TRAIL-R4 represent the decoy receptors that bind TRAIL, but are not able to transmit a death signal, since they are devoid of a functional death domain. The death domain is the intracellular region of the receptor that is required for the recruitment of signaling proteins to activated death receptors and the subsequent activation of initiator caspases. Interestingly, it was described that TRAIL-R3 and TRAIL-R4 inhibit TRAIL-R1 and TRAIL-R2-mediated apoptosis upon treatment with the soluble ligand TRAIL via distinct mechanisms [25]. While TRAIL-R3 inhibits the assembly of the DISC complex by sequestrating TRAIL within lipid rafts, TRAIL-R4 is co-recruited together with TRAIL-R2 into the

DISC and interferes with the activation of initiator caspases [25]. Overexpression of TRAIL-R3 was found in several cancers, for example in gastrointestinal cancers or leukemia [26–28]. In colorectal cancer, concomitant high TRAIL-R3 and low/medium TRAIL-R1 expression was shown to correlate with a poor response to 5-FU-based first-line chemotherapy and with shorter progression-free survival [27]. AML blasts were recently reported to express TRAIL-R3 in a substantial proportion of patients in addition to the proapoptotic TRAIL receptors [26]. Of note, co-expression of this decoy receptor correlated with a significant shortened overall survival. [26]. Knockdown of TRAIL-R3 resulted in TRAIL-induced cell death confirming the decoy function of TRAIL-R3 on AML blasts [26]. Also, treatment with TRAIL-R2-specific antibodies resulted in higher cell death rates [26]. This underlines that specific targeting of agonistic TRAIL receptors is required in cancers that express TRAIL decoy receptors in order to exploit the apoptosis-inducing activities of TRAIL [26].

Furthermore, DNA-damaging events such as anticancer drugs or ionizing radiation were found to transcriptionally activate expression levels of TRAIL-R3 [28, 29]. This p53-stimulated transactivation occurred via a p53 consensus element located within the first intron of the human TRAIL-R3 gene [29]. Similarly, TRAIL-R4 was found to be induced by p53 upon ectopic expression of p53 [30]. This indicates that genotoxic drugs not only induce proapoptotic TRAIL receptors [31], but may also stimulate antiapoptotic genes such as TRAIL-R3 and TRAIL-R4.

TRAIL-R4 has recently been demonstrated to stimulate activation of signaling pathways in an Akt-dependent manner in addition to its ability to inhibit TRAIL-mediated signaling and cell death at the membrane by forming a heteromeric complex with the agonistic receptor TRAIL-R2 [32]. Overexpression of TRAIL-R4 triggered morphological changes such as cell rounding, loss of adherence, increased cell proliferation in vitro, and promoted tumor growth in vivo, indicating that it contributes to carcinogenesis [32].

11.4.1.3 Overexpression of Death Domain-Containing Proteins

In addition to blocking death receptor signaling at the level of surface expression of the receptors, the cascade can also be blocked by factors that interfere with the formation of the DISC complex upon ligand binding by preventing the recruitment of signaling molecules to activated death receptors. c-FLIP or phosphoprotein enriched in diabetes/phosphoprotein enriched in astrocytes-15 kDa (PED/PEA-15) are two death domain-containing proteins that bind to the cytoplasmatic domains of CD95 or TRAIL receptors and subsequently impair the recruitment of procaspase-8 or procaspase-10 to the DISC [33, 34]. High expression levels of c-FLIP occur in a variety of cancers and have been shown to mediate resistance to death receptor- and also to chemotherapy-induced apoptosis [35, 36]. For example, in pancreatic cancer, c-FLIP was shown to be expressed in pancreatic intraepithelial neoplasm (PanIN) lesions and in pancreatic ductal adenocarcinomas, whereas c-FLIP was not detected in normal pancreatic ducts [37]. Concomitant knockdown of both c-$FLIP_L$ and c-$FLIP_S$ isoforms and individual silencing of either c-$FLIP_L$ or c-$FLIP_S$ by RNA interference significantly increased TRAIL- and CD95-induced apoptosis [37]. In addition, downregulation of c-FLIP by pretreatment with chemotherapeutic drugs, for example 5-fluorouracil (5-FU), sensitized pancreatic carcinoma cells to death receptor-mediated apoptosis [37]. Of note, primary cultured pancreatic cancer cells were similarly primed for TRAIL-triggered apoptosis by pre-exposure to anticancer drugs [37].

11.4.1.4 Silencing of Caspase-8

The notion that caspase-8, a key factor of death receptor-triggered apoptosis, may restrict tumor development is underlined by a study, showing that a deficiency in caspase-8 can facilitate cellular transformation [38]. Furthermore, expression of caspase-8 can be downregulated by epigenetic inactivation. Caspase-8 expression was shown to be silenced by hypermethylation of a regulatory sequence of the caspase-8 gene mapped to the boundary between exon 3 and intron 3 in several malignancies, for example neuroectodermal tumors, sarcoma, and lung carcinoma [39–43]. While this regulatory region of caspase-8 is not a classical CpG island and is devoid of promoter activity, its methylation status was associated with caspase-8 expression levels in cell lines and primary tumor specimens [11, 13, 14, 44–46]. Interestingly, co-methylation for caspase-8 and FLIP was identified in neuroblastoma in one study, suggesting that caspase-8 is epigenetically silenced in a non-random fashion [47]. Hypermethylation of caspase-8 was recently shown to be associated with relapse susceptibility in neuroblastoma [48]. In medulloblastoma, the loss of caspase-8 expression correlated with unfavorable outcome in childhood medulloblastoma [42]. The loss of caspase-8 protein expression was identified in the majority of neuroblastoma tumor samples and was not restricted to advanced disease stages [46]. No correlation was observed between caspase-8 expression and MYCN amplification or other variables of high-risk disease (e.g., 1p36 aberrations, disease stage, age at diagnosis, or tumor histology) [46]. Also, the loss of caspase-8 protein had no effect on event-free or overall survival in the overall study population or in distinct subgroups of patients [46], indicating that inactivation of caspase-8 is not a characteristic feature of aggressive neuroblastoma.

Moreover, caspase-8L is a dominant-negative variant of caspase-8 that is produced by alternative splicing, for example in $CD34^+$ progenitor cells as well as in leukemia and neuroblastoma [49–52]. Caspase-8L was found to block the binding of caspase-8 to FADD [49]. In addition, caspase-8L was recruited to the DISC after CD95 stimulation instead of caspase-8, thereby interfering with CD95 signaling at the receptor level by preventing caspase-8 activation [52].

Further, the tyrosine kinase Src was identified as a kinase that phosphorylates caspase-8 on tyrosine 308 in the linker loop of caspase-8, resulting in the suppression of the proapoptotic activity of caspase-8 [53]. Phosphorylation of caspase-8 occurred constitutively in cells with aberrant Src activity or alternatively in response to receptor tyrosine kinase stimulation, following the growth factor ligand binding [53]. In addition, integrin-mediated adhesion was described to increase the phosphorylation of caspase-8 on tyrosine 380 [54]. Interestingly, this phosphorylation step resulted in increased cell migration independent of its protease activity [54]. The linker region of caspase-8 encompassing T380 was shown to function as a Src homology 2 binding site that is required for the interaction of caspase-8 with Src homology 2 domains and the recruitment of caspase-8 to lamella of migrating cells [54]. In addition, phosphorylation of caspase-8 on T380 was shown to promote its interaction with the p85 alpha subunit of phosphatidylinositol 3-kinase to regulate cell adhesion and motility [55]. Also, caspase-8 was reported to be necessary for efficient adhesion-induced stimulation of the extracellular signal-regulated kinase (Erk)-1/2 pathway independently of its proteolytic activity [44]. This function of caspase-8 was demonstrated to require specific residues within the caspase-8 "RXDLL motif" that mediates complex formation with Src [44]. In addition, caspase-8 was found to interact with the focal adhesion complex, thereby promoting cleavage of focal adhesion substrates and subsequently cell migration [56]. Rab5 has been implicated as an important integrator of caspase-8-mediated signal transduction downstream of integrins during cell migration [57]. Together, these data demonstrate that procaspase-8 exerts important functions in cell adhesion and motility independently of its catalytic activity. Phosphorylation of caspase-8 that prevents the conversion of procaspase-8 into its mature catalytically active cleavage fragments constitutes an important switch between these non-apoptotic and apoptotic functions of caspase-8.

11.5 Therapeutic Strategies to Target the Death Receptor Pathway in Cancers

Since the death receptor pathway of apoptosis represents a signaling cascade that directly connects to an intrinsic cell death machinery and that is amenable to therapeutic targeting from the outside, it has attracted much attention in the last decade for the development of molecular cancer therapeutics. Most strategies focused on targeting the two agonistic TRAIL receptors as discussed in more detail in the following paragraphs.

11.5.1 TRAIL Receptor Agonists

Death receptors represent promising targets to directly engage the apoptotic machinery. Most approaches for cancer therapy have so far focused to develop agents directed against the proapoptotic TRAIL receptors. Tumor selectivity is one of the characteristics of TRAIL receptor agonists, as they have been shown to predominantly trigger apoptosis in malignant cells with little effect on normal cells [7].

There is now a large body of evidence from preclinical studies that recombinant soluble TRAIL or antibodies against the proapoptotic TRAIL receptors TRAIL-R1 or TRAIL-R2 trigger apoptosis in a wide range of cancer cell lines and human cancer xenograft models [19, 58, 59]. Of note, TRAIL-R2 antibodies were found to trigger tumor-specific T-cell memory besides their cytotoxic effects against cancer cells, thereby protecting from tumor recurrence [60]. In addition to soluble recombinant TRAIL ligand or TRAIL receptor specific antibodies, gene therapy approaches have been launched to deliver TRAIL to the tumor site. An adenoviral vector-based system based on the hTERT promoter yielded high levels of TRAIL, leading to tumor-specific induction of apoptosis, suppression of tumor growth in a xenograft model of breast cancer, and increased tumor-free survival of mice [61]. Furthermore, the apoptosis-inducing activity of TRAIL has been combined with the ability of mesenchymal stem cells (MSCs) to infiltrate tumors as well as lymphatic tissues in order to deliver TRAIL to the primary tumor site as well as to disseminated cancer cells [62]. In a lung cancer model, MSCs expressing TRAIL were shown to cause tumor growth inhibition by triggering apoptosis [63].

11.5.2 TRAIL-Based Combination Therapies

Since a substantial proportion of tumors displays low sensitivity or even resistance toward TRAIL, although at least one of the agonistic TRAIL receptors was found to be expressed on the surface, a range of different TRAIL-based combination therapies have been designed. They comprise chemo-, radio- or immunotherapy as well as various targeted therapeutics. Both hypothesis-driven evaluations and exploratory screening approaches have led to the identification of synergistic interaction between TRAIL receptor agonists and other cytotoxic principles in various human cancers [64–71]. The cooperative interaction of TRAIL and anticancer drugs has been attributed to various mechanistic events, for example transactivation and increased surface expression [72–74] or enhanced aggregation of proapoptotic death receptors [75]. Recently, chemotherapeutic drugs have been reported to restore the sensitivity to TRAIL-induced apoptosis in TRAIL-R4-expressing cells by enhancing the recruitment of caspase-8 to the TRAIL DISC, thereby promoting caspase-8 activation [76].

Inhibition of the proteasome presents another strategy to increase the sensitivity of malignant cells toward TRAIL. For example, Bortezomib (PS-341, VELCADE), a dipeptidyl boronic acid, is a reversible inhibitor of the proteolytic activity of the proteasome that has been shown to synergistically induce apoptosis together with TRAIL in a large variety of human cancers. This cooperative interaction has been attributed to a number of molecular events, including increased TRAIL-R1 or TRAIL-R2 expression; enhanced formation of the TRAIL DISC, reduced expression of c-FLIP or XIAP; reduced degradation of caspase-8, Bid, Bax or Bim; upregulation of Bim, Bik, Puma, Bax or p53; enhanced release of Smac from the mitochondria; accumulation of p21; or inhibition of NF-κB [77–107].

Moreover, HDAC inhibitors have been shown in a set of different human malignancies to prime cancer cells toward TRAIL-induced apoptosis [108]. This HDAC inhibitor-mediated sensitization toward TRAIL may involve modulation of death receptor (extrinsic) or mitochondrial (intrinsic) signal transduction pathways. For example, HDAC inhibitors have been reported to cause upregulation of TRAIL-R1 or TRAIL-R2 surface expression [109–117], redistribution of TRAIL receptors into lipid rafts [118], increased expression of caspase-8 [119], downregulation of c-FLIP [120–124], enhancement of caspase-9 activation [125], downregulation of antiapoptotic Bcl-2 family proteins [112, 126–130] accompanied by the upregulation of proapoptotic Bcl-2 family members [116, 127, 130]. There is also growing evidence in a variety of cancers that the inhibition of IAP proteins, for example by small-molecule inhibitors, may present a particularly promising approach to sensitize cancer cells to TRAIL-induced apoptosis [47, 62, 131–140].

11.5.3 Upregulation of Caspase-8

Furthermore, several therapeutic strategies have been developed to restore the function of caspase-8 in cancer cells by upregulating its expression levels. Since caspase-8 is frequently silenced by epigenetic mechanisms, several approaches have been developed to reverse these events. For example, the demethylating agent 5-aza-2-deoxycytidine (5-AZA) caused demethylation of the regulatory sequence of caspase-8 and enhanced caspase-8 promoter activity and caspase-8 expression in several cancers with epigenetically inactivated caspase-8 [39, 141–143]. Since the caspase-8 promoter was found to harbor several interferon-sensitive response elements that can regulate caspase-8 gene expression, interferon-γ was used to stimulate re-expression of caspase-8 [144–146]. Pretreatment with interferon-γ resulted in enhanced TRAIL-, chemotherapy- or radiotherapy-induced apoptosis [42, 145–158]. Interferon-γ-mediated upregulation of caspase-8 was mediated, at least in part, via activation of signal transducer and activator of transcription 1 (STAT-1) [144, 146, 159]. This transcription factor transactivates interferon-γ-inducible genes via interferon-γ activation sites (GAS) that also occur in the caspase-8 promoter [144, 146, 159]. Overexpression of dominant-negative STAT-1 prevented interferon-γ-stimulated upregulation of caspase-8 [150], underscoring that STAT-1 acts as a mediator of interferon-γ-stimulated re-expression of caspase-8. Administration of interferon-γ also caused upregulation of caspase-8 expression in an in vivo model of Ewing's sarcoma [156]. Of note, results from a phase I clinical trial in neuroblastoma showed that treatment with interferon-γ within an immunotherapeutic regimen resulted in upregulation of caspase-8 protein levels in tumor cells in vivo [160]. Combining interferon-γ together with demethylating agents (i.e., 5-Aza) proved to be even more effective compared with single agents to stimulate re-expression of caspase-8, resulting in enhanced apoptosis [45]. Further, interferon-γ was recently shown to cooperate with histone deacetylase inhibitors to restore caspase-8 expression resistance in cancers with silencing of caspase-8, thereby overcoming resistance to TRAIL [119]. Of note, interferon-γ acted in concert with histone deacetylase inhibitors to re-express caspase-8 expression and to enhance sensitivity to TRAIL-induced apoptosis also in primary medulloblastoma samples and in a medulloblastoma model in vivo [119]. In addition to interferon-γ, interferon-α has been shown to stimulate caspase-8 expression [161]. Moreover, several cytotoxic drugs including methotrexate and 5-fluorouracil were described to sensitize resistant tumor cells for apoptosis by p53-mediated upregulation of caspase-8 [162]. Experiments using p53 decoy oligonucleotides as well as p53 or caspase-8 RNA interference vectors underlined the requirement of p53 and caspase-8 for this chemotherapy-mediated sensitization toward TRAIL [162]. Upregulation of caspase-8 and sensitization to TRAIL-induced apoptosis was observed in a panel of tumor cell lines with caspase-8 silencing as well as in TRAIL-resistant primary acute leukemia cells [162], underscoring both the general implications and the clinical relevance of these findings.

Retinoic acid represents another agent that has been shown to trigger caspase-8 expression in neuroblastoma cells [163]. This involves transcription via an intronic region of the caspase-8 gene through a CREB binding site, leading to sustained increase in caspase-8 levels for several days [163]. As a consequence, retinoic acid in combination with doxorubicin or TNFα resulted in enhanced apoptosis [163]. Together, these reports demonstrate that demethylating agents, interferons, retinoic acid, or anticancer drugs can be used to stimulate caspase-8 re-expression, resulting in the sensitization to apoptotic cell death.

11.5.4 Clinical Studies with TRAIL Receptor Agonists

TRAIL receptor agonists have been evaluated in a number of early clinical trials over the last years. This includes protocols with recombinant human TRAIL (e.g., dulanermin) [164–169] that is directed against both agonistic TRAIL receptors TRAIL-R1 and TRAIL-R2 as well as regimens with agonistic monoclonal antibodies targeting selectively either TRAIL-R1, for example mapatumumab [170–178], or TRAIL-R2, for example lexatumumab [179–182], conatumumab [166, 169, 183–189], drozitumab [190–194], tigatuzumab [195], and LBY135 [196]. Initially, these trials were conducted with TRAIL or TRAIL receptor antibodies as single agents. Subsequently, combination studies were launched to exploit additive or even synergistic interactions by simultaneously targeting both death receptor and additional signaling pathways. The design of these combination protocols was based on a large body of data from preclinical studies showing that the incorporation of additional agents that concomitantly trigger, for example, mitochondrial apoptosis such as chemotherapeutic drugs significantly enhances the antitumor activity of TRAIL receptor agonists. In addition, targeted therapies, for example kinase inhibitors or antiangiogenic drugs, have been combined with TRAIL receptor agonists. Ongoing trials are testing TRAIL receptor antibodies in different combination protocols, for example, together with the proteasome inhibitor Bortezomib, the histone deacetylase inhibitor vorinostat or interferon-γ (see clinicaltrials.gov).

One of the key challenges for the successful application of TRAIL receptor agonists for the treatment for cancer is the identification of suitable biomarkers that can be used to rationally select patients for the inclusion in clinical trials that most likely benefit from TRAIL-based protocols. In addition, biomarkers are required to guide the treatment course with TRAIL receptor agonists, for example to monitor treatment response and to adjust the dosing. While there has been a discussion whether or not expression levels of TRAIL receptors may serve as a suitable biomarker, several preclinical studies could not detect a correlation of TRAIL receptor surface levels with TRAIL sensitivity. Similarly, immunohistochemical detection of TRAIL receptors in tumor tissue samples did not correlate with outcome in clinical studies [170, 172, 173, 177, 178]. Thus, additional studies are required to identify and validate suitable biomarkers for TRAIL-based regimens.

11.6 Conclusion

Targeting death receptor pathways represents a promising strategy to exploit endogenous cell death programs for the treatment for cancer. This approach has been extensively studied in preclinical cancer models and has more recently also been transferred to the phase of clinical evaluation. Harnessing the possibilities to engage the cell's intrinsic death program will hopefully pave the avenue for innovative treatment options for patients suffering from cancer.

Acknowledgments Work in the author's laboratory is supported by grants from the Deutsche Forschungsgemeinschaft, the Deutsche Krebshilfe, the Bundesministerium für Forschung und Technologie (01GM0871, 01GM1104C), Wilhelm-Sander-Stiftung, Else Kröner-Fresenius-Stiftung, Novartis Stiftung für therapeutische Forschung, the European Community (ApopTrain, APO-SYS), and IAP6/18.

References

1. Taylor RC, Cullen SP, Martin SJ (2008) Apoptosis: controlled demolition at the cellular level. Nat Rev Mol Cell Biol 9:231–241
2. Evan GI, Vousden KH (2001) Proliferation, cell cycle and apoptosis in cancer. Nature 411:342–348
3. Fulda S (2009) Tumor resistance to apoptosis. Int J Cancer 124:511–515
4. Fulda S, Debatin KM (2006) Extrinsic versus intrinsic apoptosis pathways in anticancer chemotherapy. Oncogene 25:4798–4811
5. Johnstone RW, Ruefli AA, Lowe SW (2002) Apoptosis: a link between cancer genetics and chemotherapy. Cell 108:153–164

6. Makin G, Dive C (2001) Apoptosis and cancer chemotherapy. Trends Cell Biol 11:22–26
7. Ashkenazi A (2008) Targeting the extrinsic apoptosis pathway in cancer. Cytokine Growth Factor Rev 19:325–331
8. Lavrik IN, Krammer PH (2012) Regulation of CD95/Fas signaling at the DISC. Cell Death Differ 19:36–41
9. Kroemer G, Galluzzi L, Brenner C (2007) Mitochondrial membrane permeabilization in cell death. Physiol Rev 87:99–163
10. Fulda S, Galluzzi L, Kroemer G (2010) Targeting mitochondria for cancer therapy. Nat Rev Drug Discov 9:447–464
11. Fulda S, Vucic D (2012) Targeting IAP proteins for therapeutic intervention in cancer. Nat Rev Drug Discov 11:109–124
12. Logue SE, Martin SJ (2008) Caspase activation cascades in apoptosis. Biochem Soc Trans 36:1–9
13. Friesen C, Fulda S, Debatin KM (1997) Deficient activation of the CD95 (APO-1/Fas) system in drug-resistant cells. Leukemia 11:1833–1841
14. Fulda S, Scaffidi C, Susin SA et al (1998) Activation of mitochondria and release of mitochondrial apoptogenic factors by betulinic acid. J Biol Chem 273:33942–33948
15. Fulda S (2009) Inhibitor of apoptosis proteins in hematological malignancies. Leukemia 23:467–476
16. Petak I, Danam RP, Tillman DM et al (2003) Hypermethylation of the gene promoter and enhancer region can regulate Fas expression and sensitivity in colon carcinoma. Cell Death Differ 10:211–217
17. Van Noesel MM, Van Bezouw S, Salomons GS et al (2002) Tumor-specific down-regulation of the tumor necrosis factor-related apoptosis-inducing ligand decoy receptors DcR1 and DcR2 is associated with dense promoter hypermethylation. Cancer Res 62:2157–2161
18. Maecker HL, Yun Z, Maecker HT et al (2002) Epigenetic changes in tumor Fas levels determine immune escape and response to therapy. Cancer Cell 2:139–148
19. Ashkenazi A (2008) Directing cancer cells to self-destruct with pro-apoptotic receptor agonists. Nat Rev Drug Discov 7:1001–1012
20. Dechant MJ, Fellenberg J, Scheuerpflug CG et al (2004) Mutation analysis of the apoptotic "death-receptors" and the adaptors TRADD and FADD/MORT-1 in osteosarcoma tumor samples and osteosarcoma cell lines. Int J Cancer 109:661–667
21. Pai SI, Wu GS, Ozoren N et al (1998) Rare loss-of-function mutation of a death receptor gene in head and neck cancer. Cancer Res 58:3513–3518
22. Jin Z, Mcdonald ER 3rd, Dicker DT et al (2004) Deficient tumor necrosis factor-related apoptosis-inducing ligand (TRAIL) death receptor transport to the cell surface in human colon cancer cells selected for resistance to TRAIL-induced apoptosis. J Biol Chem 279:35829–35839
23. Pitti RM, Marsters SA, Lawrence DA et al (1998) Genomic amplification of a decoy receptor for Fas ligand in lung and colon cancer. Nature 396:699–703
24. Roth W, Isenmann S, Nakamura M et al (2001) Soluble decoy receptor 3 is expressed by malignant gliomas and suppresses CD95 ligand-induced apoptosis and chemotaxis. Cancer Res 61:2759–2765
25. Merino D, Lalaoui N, Morizot A et al (2006) Differential inhibition of TRAIL-mediated DR5-DISC formation by decoy receptors 1 and 2. Mol Cell Biol 26:7046–7055
26. Chamuleau ME, Ossenkoppele GJ, Van Rhenen A et al (2011) High TRAIL-R3TRAIL-R3 expression on leukemic blasts is associated with poor outcome and induces apoptosis-resistance which can be overcome by targeting TRAIL-R2TRAIL-R2. Leuk Res 35:741–749
27. Granci V, Bibeau F, Kramar A et al (2008) Prognostic significance of TRAIL-R1 and TRAIL-R3 expression in metastatic colorectal carcinomas. Eur J Cancer 44:2312–2318
28. Sheikh MS, Huang Y, Fernandez-Salas EA et al (1999) The antiapoptotic decoy receptor TRID/TRAIL-R3 is a p53-regulated DNA damage-inducible gene that is overexpressed in primary tumors of the gastrointestinal tract. Oncogene 18:4153–4159

29. Ruiz De Almodovar C, Ruiz-Ruiz C, Rodriguez A et al (2004) Tumor necrosis factor-related apoptosis-inducing ligand (TRAIL) decoy receptor TRAIL-R3 is up-regulated by p53 in breast tumor cells through a mechanism involving an intronic p53-binding site. J Biol Chem 279:4093–4101
30. Meng RD, Mcdonald ER 3rd, Sheikh MS et al (2000) The TRAIL decoy receptor TRUNDD (DcR2, TRAIL-R4) is induced by adenovirus-p53 overexpression and can delay TRAIL-, p53-, and KILLER/DR5-dependent colon cancer apoptosis. Mol Ther 1:130–144
31. Sheikh MS, Burns TF, Huang Y et al (1998) P53-dependent and -independent regulation of the death receptor KILLER/DR5 gene expression in response to genotoxic stress and tumor necrosis factor alpha. Cancer Res 58:1593–1598
32. Lalaoui N, Morle A, Merino D et al (2011) TRAIL-R4 promotes tumor growth and resistance to apoptosis in cervical carcinoma HeLa cells through AKT. PLoS One 6:19679
33. Hao C, Beguinot F, Condorelli G et al (2001) Induction and intracellular regulation of tumor necrosis factor-related apoptosis-inducing ligand (TRAIL) mediated apoptosis in human malignant glioma cells. Cancer Res 61:1162–1170
34. Krueger A, Baumann S, Krammer PH et al (2001) FLICE-inhibitory proteins: regulators of death receptor-mediated apoptosis. Mol Cell Biol 21:8247–8254
35. Fulda S, Meyer E, Debatin KM (2000) Metabolic inhibitors sensitize for CD95 (APO-1/Fas)-induced apoptosis by down-regulating Fas-associated death domain-like interleukin 1-converting enzyme inhibitory protein expression. Cancer Res 60:3947–3956
36. Longley DB, Wilson TR, Mcewan M et al (2006) c-FLIP inhibits chemotherapy-induced colorectal cancer cell death. Oncogene 25:838–848
37. Haag C, Stadel D, Zhou S et al (2011) Identification of c-FLIP(L) and c-FLIP(S) as critical regulators of death receptor-induced apoptosis in pancreatic cancer cells. Gut 60:225–237
38. Krelin Y, Zhang L, Kang TB et al (2008) Caspase-8 deficiency facilitates cellular transformation in vitro. Cell Death Differ 15:1350–1355
39. Fulda S, Kufer MU, Meyer E et al (2001) Sensitization for death receptor- or drug-induced apoptosis by re-expression of caspase-8 through demethylation or gene transfer. Oncogene 20:5865–5877
40. Harada K, Toyooka S, Shivapurkar N et al (2002) Deregulation of caspase 8 and 10 expression in pediatric tumors and cell lines. Cancer Res 62:5897–5901
41. Hopkins-Donaldson S, Ziegler A, Kurtz S et al (2003) Silencing of death receptor and caspase-8 expression in small cell lung carcinoma cell lines and tumors by DNA methylation. Cell Death Differ 10:356–364
42. Pingoud-Meier C, Lang D, Janss AJ et al (2003) Loss of caspase-8 protein expression correlates with unfavorable survival outcome in childhood medulloblastoma. Clin Cancer Res 9:6401–6409
43. Teitz T, Wei T, Valentine MB et al (2000) Caspase 8 is deleted or silenced preferentially in childhood neuroblastomas with amplification of MYCN. Nat Med 6:529–535
44. Finlay D, Howes A, Vuori K (2009) Critical role for caspase-8 in epidermal growth factor signaling. Cancer Res 69:5023–5029
45. Fulda S, Debatin KM (2006) 5-Aza-2′-deoxycytidine and IFN-gamma cooperate to sensitize for TRAIL-induced apoptosis by upregulating caspase-8caspase-8. Oncogene 25:5125–5133
46. Fulda S, Poremba C, Berwanger B et al (2006) Loss of caspase-8 expression does not correlate with MYCN amplification, aggressive disease, or prognosis in neuroblastoma. Cancer Res 66:10016–10023
47. Fulda S, Wick W, Weller M et al (2002) Smac agonists sensitize for Apo2L/TRAIL- or anticancer drug-induced apoptosis and induce regression of malignant glioma in vivo. Nat Med 8:808–815
48. Grau E, Martinez F, Orellana C et al (2011) Hypermethylation of apoptotic genes as independent prognostic factor in neuroblastoma disease. Mol Carcinog 50:153–162
49. Himeji D, Horiuchi T, Tsukamoto H et al (2002) Characterization of caspase-8L: a novel isoform of caspase-8 that behaves as an inhibitor of the caspase cascade. Blood 99:4070–4078
50. Horiuchi T, Himeji D, Tsukamoto H et al (2000) Dominant expression of a novel splice variant of caspase-8 in human peripheral blood lymphocytes. Biochem Biophys Res Commun 272:877–881

51. Miller MA, Karacay B, Zhu X et al (2006) Caspase 8L, a novel inhibitory isoform of caspase 8, is associated with undifferentiated neuroblastoma. Apoptosis 11:15–24
52. Mohr A, Zwacka RM, Jarmy G et al (2005) Caspase-8L expression protects CD34+ hematopoietic progenitor cells and leukemic cells from CD95-mediated apoptosis. Oncogene 24:2421–2429
53. Cursi S, Rufini A, Stagni V et al (2006) Src kinase phosphorylates Caspase-8 on Tyr380: a novel mechanism of apoptosis suppression. EMBO J 25:1895–1905
54. Barbero S, Barila D, Mielgo A et al (2008) Identification of a critical tyrosine residue in caspase 8 that promotes cell migration. J Biol Chem 283:13031–13034
55. Senft J, Helfer B, Frisch SM (2007) Caspase-8 interacts with the p85 subunit of phosphatidylinositol 3-kinase to regulate cell adhesion and motility. Cancer Res 67:11505–11509
56. Barbero S, Mielgo A, Torres V et al (2009) Caspase-8 association with the focal adhesion complex promotes tumor cell migration and metastasis. Cancer Res 69:3755–3763
57. Torres VA, Mielgo A, Barbero S et al (2010) Rab5 mediates caspase-8-promoted cell motility and metastasis. Mol Biol Cell 21:369–376
58. Chuntharapai A, Dodge K, Grimmer K et al (2001) Isotype-dependent inhibition of tumor growth in vivo by monoclonal antibodies to death receptor 4. J Immunol 166:4891–4898
59. Ichikawa K, Liu W, Zhao L et al (2001) Tumoricidal activity of a novel anti-human DR5 monoclonal antibody without hepatocyte cytotoxicity. Nat Med 7:954–960
60. Takeda K, Yamaguchi N, Akiba H et al (2004) Induction of tumor-specific T cell immunity by anti-DR5 antibody therapy. J Exp Med 199:437–448
61. Lin T, Huang X, Gu J et al (2002) Long-term tumor-free survival from treatment with the GFP-TRAIL fusion gene expressed from the hTERT promoter in breast cancer cells. Oncogene 21:8020–8028
62. Mohr A, Albarenque SM, Deedigan L et al (2010) Targeting of XIAP combined with systemic mesenchymal stem cell-mediated delivery of sTRAIL ligand inhibits metastatic growth of pancreatic carcinoma cells. Stem Cells 28:2109–2120
63. Mohr A, Lyons M, Deedigan L et al (2008) Mesenchymal stem cells expressing TRAIL lead to tumour growth inhibition in an experimental lung cancer model. J Cell Mol Med 12:2628–2643
64. Belka C, Schmid B, Marini P et al (2001) Sensitization of resistant lymphoma cells to irradiation-induced apoptosis by the death ligand TRAIL. Oncogene 20:2190–2196
65. Chinnaiyan AM, Prasad U, Shankar S et al (2000) Combined effect of tumor necrosis factor-related apoptosis-inducing ligand and ionizing radiation in breast cancer therapy. Proc Natl Acad Sci USA 97:1754–1759
66. Gliniak B, Le T (1999) Tumor necrosis factor-related apoptosis-inducing ligand's antitumor activity in vivo is enhanced by the chemotherapeutic agent CPT-11. Cancer Res 59:6153–6158
67. Keane MM, Rubinstein Y, Cuello M et al (2000) Inhibition of NF-kappaB activity enhances TRAIL mediated apoptosis in breast cancer cell lines. Breast Cancer Res Treat 64:211–219
68. Nagane M, Pan G, Weddle JJ et al (2000) Increased death receptor 5 expression by chemotherapeutic agents in human gliomas causes synergistic cytotoxicity with tumor necrosis factor-related apoptosis-inducing ligand in vitro and in vivo. Cancer Res 60:847–853
69. Ray S, Almasan A (2003) Apoptosis induction in prostate cancer cells and xenografts by combined treatment with Apo2 ligand/tumor necrosis factor-related apoptosis-inducing ligand and CPT-11. Cancer Res 63:4713–4723
70. Rohn TA, Wagenknecht B, Roth W et al (2001) CCNU-dependent potentiation of TRAIL/Apo2L-induced apoptosis in human glioma cells is p53-independent but may involve enhanced cytochrome c release. Oncogene 20:4128–4137
71. Singh TR, Shankar S, Chen X et al (2003) Synergistic interactions of chemotherapeutic drugs and tumor necrosis factor-related apoptosis-inducing ligand/Apo-2 ligand on apoptosis and on regression of breast carcinoma in vivo. Cancer Res 63:5390–5400
72. Meng RD, El-Deiry WS (2001) p53-independent upregulation of KILLER/DR5 TRAIL receptor expression by glucocorticoids and interferon-gamma. Exp Cell Res 262:154–169

73. Takimoto R, El-Deiry WS (2000) Wild-type p53 transactivates the KILLER/DR5 gene through an intronic sequence-specific DNA-binding site. Oncogene 19:1735–1743
74. Wang S, El-Deiry WS (2003) Requirement of p53 targets in chemosensitization of colonic carcinoma to death ligand therapy. Proc Natl Acad Sci USA 100:15095–15100
75. Lacour S, Micheau O, Hammann A et al (2003) Chemotherapy enhances TNF-related apoptosis-inducing ligand DISC assembly in HT29 human colon cancer cells. Oncogene 22:1807–1816
76. Morizot A, Merino D, Lalaoui N et al (2011) Chemotherapy overcomes TRAIL-R4-mediated TRAIL resistance at the DISC level. Cell Death Differ 18:700–711
77. Baritaki S, Suzuki E, Umezawa K et al (2008) Inhibition of Yin Yang 1-dependent repressor activity of DR5 transcription and expression by the novel proteasome inhibitor NPI-0052 contributes to its TRAIL-enhanced apoptosis in cancer cells. J Immunol 180:6199–6210
78. Brooks AD, Jacobsen KM, Li W et al (2010) Bortezomib sensitizes human renal cell carcinomas to TRAIL apoptosis through increased activation of caspase-8caspase-8 in the death-inducing signaling complex. Mol Cancer Res 8:729–738
79. Concannon CG, Koehler BF, Reimertz C et al (2007) Apoptosis induced by proteasome inhibition in cancer cells: predominant role of the p53/PUMA pathway. Oncogene 26:1681–1692
80. Conticello C, Adamo L, Giuffrida R et al (2007) Proteasome inhibitors synergize with tumor necrosis factor-related apoptosis-induced ligand to induce anaplastic thyroid carcinoma cell death. J Clin Endocrinol Metab 92:1938–1942
81. Ding WX, Ni HM, Chen X et al (2007) A coordinated action of Bax, PUMA, and p53 promotes MG132-induced mitochondria activation and apoptosis in colon cancer cells. Mol Cancer Ther 6:1062–1069
82. Ganten TM, Koschny R, Haas TL et al (2005) Proteasome inhibition sensitizes hepatocellular carcinoma cells, but not human hepatocytes, to TRAIL. Hepatology 42:588–597
83. He Q, Huang Y, Sheikh MS (2004) Proteasome inhibitor MG132 upregulates death receptor 5 and cooperates with Apo2L/TRAIL to induce apoptosis in Bax-proficient and -deficient cells. Oncogene 23:2554–2558
84. Hetschko H, Voss V, Seifert V et al (2008) Upregulation of DR5 by proteasome inhibitors potently sensitizes glioma cells to TRAIL-induced apoptosis. FEBS J 275:1925–1936
85. Inoue T, Shiraki K, Fuke H et al (2006) Proteasome inhibition sensitizes hepatocellular carcinoma cells to TRAIL by suppressing caspase inhibitors and AKT pathway. Anticancer Drugs 17:261–268
86. Johnson TR, Stone K, Nikrad M et al (2003) The proteasome inhibitor PS-341 overcomes TRAIL resistance in Bax and caspase 9-negative or Bcl-xL overexpressing cells. Oncogene 22:4953–4963
87. Kandasamy K, Kraft AS (2008) Proteasome inhibitor PS-341 (VELCADE) induces stabilization of the TRAIL receptor DR5 mRNA through the 3′-untranslated region. Mol Cancer Ther 7:1091–1100
88. Kashkar H, Deggerich A, Seeger JM et al (2007) NF-kappaB-independent down-regulation of XIAP by bortezomib sensitizes HL B cells against cytotoxic drugs. Blood 109:3982–3988
89. Khanbolooki S, Nawrocki ST, Arumugam T et al (2006) Nuclear factor-kappaB maintains TRAIL resistance in human pancreatic cancer cells. Mol Cancer Ther 5:2251–2260
90. Koschny R, Ganten TM, Sykora J et al (2007) TRAIL/bortezomib cotreatment is potentially hepatotoxic but induces cancer-specific apoptosis within a therapeutic window. Hepatology 45:649–658
91. Koschny R, Holland H, Sykora J et al (2007) Bortezomib sensitizes primary human astrocytoma cells of WHO grades I to IV for tumor necrosis factor-related apoptosis-inducing ligand-induced apoptosis. Clin Cancer Res 13:3403–3412
92. Lashinger LM, Zhu K, Williams SA et al (2005) Bortezomib abolishes tumor necrosis factor-related apoptosis-inducing ligand resistance via a p21-dependent mechanism in human bladder and prostate cancer cells. Cancer Res 65:4902–4908

93. Leverkus M, Sprick MR, Wachter T et al (2003) Proteasome inhibition results in TRAIL sensitization of primary keratinocytes by removing the resistance-mediating block of effector caspase maturation. Mol Cell Biol 23:777–790
94. Liu FT, Agrawal SG, Gribben JG et al (2008) Bortezomib blocks Bax degradation in malignant B cells during treatment with TRAIL. Blood 111:2797–2805
95. Liu X, Yue P, Chen S et al (2007) The proteasome inhibitor PS-341 (bortezomib) up-regulates DR5 expression leading to induction of apoptosis and enhancement of TRAIL-induced apoptosis despite up-regulation of c-FLIP and surviving expression in human NSCLC cells. Cancer Res 67:4981–4988
96. Nagy K, Szekely-Szuts K, Izeradjene K et al (2006) Proteasome inhibitors sensitize colon carcinoma cells to TRAIL-induced apoptosis via enhanced release of Smac/DIABLO from the mitochondria. Pathol Oncol Res 12:133–142
97. Naumann I, Kappler R, Von Schweinitz D et al (2011) Bortezomib primes neuroblastoma cells for TRAIL-induced apoptosis by linking the death receptor to the mitochondrial pathway. Clin Cancer Res 17:3204–3218
98. Nikrad M, Johnson T, Puthalalath H et al (2005) The proteasome inhibitor bortezomib sensitizes cells to killing by death receptor ligand TRAIL via BH3-only proteins Bik and Bim. Mol Cancer Ther 4:443–449
99. Sayers TJ, Brooks AD, Koh CY et al (2003) The proteasome inhibitor PS-341 sensitizes neoplastic cells to TRAIL-mediated apoptosis by reducing levels of c-FLIP. Blood 102:303–310
100. Shanker A, Brooks AD, Tristan CA et al (2008) Treating metastatic solid tumors with bortezomib and a tumor necrosis factor-related apoptosis-inducing ligand receptor agonist antibody. J Natl Cancer Inst 100:649–662
101. Tan TT, Degenhardt K, Nelson DA et al (2005) Key roles of BIM-driven apoptosis in epithelial tumors and rational chemotherapy. Cancer Cell 7:227–238
102. Thorpe JA, Christian PA, Schwarze SR (2008) Proteasome inhibition blocks caspase-8 degradation and sensitizes prostate cancer cells to death receptor-mediated apoptosis. Prostate 68:200–209
103. Unterkircher T, Cristofanon S, Vellanki SH et al (2011) Bortezomib primes glioblastoma, including glioblastoma stem cells, for TRAIL by increasing tBid stability and mitochondrial apoptosis. Clin Cancer Res 17:4019–4030
104. Voortman J, Resende TP, Abou El Hassan MA et al (2007) TRAIL therapy in non-small cell lung cancer cells: sensitization to death receptor-mediated apoptosis by proteasome inhibitor bortezomib. Mol Cancer Ther 6:2103–2112
105. Yoshida T, Shiraishi T, Nakata S et al (2005) Proteasome inhibitor MG132 induces death receptor 5 through CCAAT/enhancer-binding protein homologous protein. Cancer Res 65:5662–5667
106. Zhao X, Qiu W, Kung J et al (2008) Bortezomib induces caspase-dependent apoptosis in Hodgkin lymphoma cell lines and is associated with reduced c-FLIP expression: a gene expression profiling study with implications for potential combination therapies. Leuk Res 32:275–285
107. Zhu H, Guo W, Zhang L et al (2005) Proteasome inhibitors-mediated TRAIL resensitization and Bik accumulation. Cancer Biol Ther 4:781–786
108. Fulda S (2012) Histone deacetylase (HDAC) inhibitors and regulation of TRAIL-induced apoptosis. Exp Cell Res 318:1208–1212
109. Butler LM, Liapis V, Bouralexis S et al (2006) The histone deacetylase inhibitor, suberoylanilide hydroxamic acid, overcomes resistance of human breast cancer cells to Apo2L/TRAIL. Int J Cancer 119:944–954
110. Chopin V, Slomianny C, Hondermarck H et al (2004) Synergistic induction of apoptosis in breast cancer cells by cotreatment with butyrate and TNF-alpha, TRAIL, or anti-FasFas agonist antibody involves enhancement of death receptor's signaling and requires P21(waf1). Exp Cell Res 298:560–573
111. Earel JK Jr, Vanoosten RL, Griffith TS (2006) Histone deacetylase inhibitors modulate the sensitivity of tumor necrosis factor-related apoptosis-inducing ligand-resistant bladder tumor cells. Cancer Res 66:499–507

112. Guo F, Sigua C, Tao J et al (2004) Cotreatment with histone deacetylase inhibitor LAQ824 enhances Apo-2L/tumor necrosis factor-related apoptosis inducing ligand-induced death inducing signaling complex activity and apoptosis of human acute leukemia cells. Cancer Res 64:2580–2589
113. Inoue S, Macfarlane M, Harper N et al (2004) Histone deacetylase inhibitors potentiate TNF-related apoptosis-inducing ligand (TRAIL)-induced apoptosis in lymphoid malignancies. Cell Death Differ 11(Suppl 2):193–206
114. Kim YH, Park JW, Lee JY et al (2004) Sodium butyrate sensitizes TRAIL-mediated apoptosis by induction of transcription from the DR5 gene promoter through Sp1 sites in colon cancer cells. Carcinogenesis 25:1813–1820
115. Nakata S, Yoshida T, Horinaka M et al (2004) Histone deacetylase inhibitors upregulate death receptor 5/TRAIL-R2 and sensitize apoptosis induced by TRAIL/APO2-L in human malignant tumor cells. Oncogene 23:6261–6271
116. Singh TR, Shankar S, Srivastava RK (2005) HDAC inhibitors enhance the apoptosis-inducing potential of TRAIL in breast carcinoma. Oncogene 24:4609–4623
117. Vanoosten RL, Moore JM, Karacay B et al (2005) Histone deacetylase inhibitors modulate renal cell carcinoma sensitivity to TRAIL/Apo-2L-induced apoptosis by enhancing TRAIL-R2 expression. Cancer Biol Ther 4:1104–1112
118. Vanoosten RL, Moore JM, Ludwig AT et al (2005) Depsipeptide (FR901228) enhances the cytotoxic activity of TRAIL by redistributing TRAIL receptor to membrane lipid rafts. Mol Ther 11:542–552
119. Hacker S, Dittrich A, Mohr A et al (2009) Histone deacetylase inhibitors cooperate with IFN-gamma to restore caspase-8 expression and overcome TRAIL resistance in cancers with silencing of caspase-8. Oncogene 28:3097–3110
120. Aron JL, Parthun MR, Marcucci G et al (2003) Depsipeptide (FR901228) induces histone acetylation and inhibition of histone deacetylase in chronic lymphocytic leukemia cells concurrent with activation of caspase 8-mediated apoptosis and down-regulation of c-FLIPc-FLIP protein. Blood 102:652–658
121. Hernandez A, Thomas R, Smith F et al (2001) Butyrate sensitizes human colon cancer cells to TRAIL-mediated apoptosis. Surgery 130:265–272
122. Pathil A, Armeanu S, Venturelli S et al (2006) HDAC inhibitor treatment of hepatoma cells induces both TRAIL-independent apoptosis and restoration of sensitivity to TRAIL. Hepatology 43:425–434
123. Schuchmann M, Schulze-Bergkamen H, Fleischer B et al (2006) Histone deacetylase inhibition by valproic acid down-regulates c-FLIP/CASH and sensitizes hepatoma cells towards CD95- and TRAIL receptor-mediated apoptosis and chemotherapy. Oncol Rep 15:227–230
124. Watanabe K, Okamoto K, Yonehara S (2005) Sensitization of osteosarcoma cells to death receptor-mediated apoptosis by HDAC inhibitors through downregulation of cellular FLIP. Cell Death Differ 12:10–18
125. Reddy RM, Yeow WS, Chua A et al (2007) Rapid and profound potentiation of Apo2L/TRAIL-mediated cytotoxicity and apoptosis in thoracic cancer cells by the histone deacetylase inhibitor Trichostatin A: the essential role of the mitochondria-mediated caspase activation cascade. Apoptosis 12:55–71
126. El-Zawahry A, Lu P, White SJ et al (2006) In vitro efficacy of AdTRAIL gene therapy of bladder cancer is enhanced by trichostatin A-mediated restoration of CAR expression and downregulation of cFLIP and Bcl-XLBcl-XL. Cancer Gene Ther 13:281–289
127. Gillespie S, Borrow J, Zhang XD et al (2006) Bim plays a crucial role in synergistic induction of apoptosis by the histone deacetylase inhibitor SBHA and TRAIL in melanoma cells. Apoptosis 11:2251–2265
128. Muhlethaler-Mottet A, Flahaut M, Bourloud KB et al (2006) Histone deacetylase inhibitors strongly sensitise neuroblastoma cells to TRAIL-induced apoptosis by a caspases-dependent increase of the pro- to anti-apoptotic proteins ratio. BMC Cancer 6:214

129. Neuzil J, Swettenham E, Gellert N (2004) Sensitization of mesothelioma to TRAIL apoptosis by inhibition of histone deacetylase: role of Bcl-xL down-regulation. Biochem Biophys Res Commun 314:186–191
130. Zhang XD, Gillespie SK, Borrow JM et al (2004) The histone deacetylase inhibitor suberic bishydroxamate regulates the expression of multiple apoptotic mediators and induces mitochondria-dependent apoptosis of melanoma cells. Mol Cancer Ther 3:425–435
131.Abhari BA, Cristofanon S, Kappler R et al (2012) RIP1 is required for IAP inhibitor-mediated sensitization for TRAIL-induced apoptosis via a RIP/FADD/caspase-8 cell death complex. Oncogene: Aug 13 [E-pub ahead of print]
132. Fakler M, Loeder S, Vogler M et al (2009) Small molecule XIAP inhibitors cooperate with TRAIL to induce apoptosis in childhood acute leukemia cells and overcome Bcl-2-mediated resistance. Blood 113:1710–1722
133. Hiscutt EL, Hill DS, Martin S et al (2010) Targeting X-linked inhibitor of apoptosis protein to increase the efficacy of endoplasmic reticulum stress-induced apoptosis for melanoma therapy. J Invest Dermatol 130:2250–2258
134. Loeder S, Drensek A, Jeremias I et al (2010) Small molecule XIAP inhibitors sensitize childhood acute leukemia cells for CD95-induced apoptosis. Int J Cancer 126:2216–2228
135. Loeder S, Zenz T, Schnaiter A et al (2009) A novel paradigm to trigger apoptosis in chronic lymphocytic leukemia. Cancer Res 69:8977–8986
136. Stadel D, Mohr A, Ref C et al (2010) TRAIL-induced apoptosis is preferentially mediated via TRAIL receptor 1 in pancreatic carcinoma cells and profoundly enhanced by XIAP inhibitors. Clin Cancer Res 16:5734–5749
137. Varfolomeev E, Alicke B, Elliott JM et al (2009) X chromosome-linked inhibitor of apoptosis regulates cell death induction by proapoptotic receptor agonists. J Biol Chem 284:34553–34560
138. Vogler M, Durr K, Jovanovic M et al (2007) Regulation of TRAIL-induced apoptosis by XIAP in pancreatic carcinoma cells. Oncogene 26:248–257
139. Vogler M, Walczak H, Stadel D et al (2009) Small molecule XIAP inhibitors enhance TRAIL-induced apoptosis and antitumor activity in preclinical models of pancreatic carcinoma. Cancer Res 69:2425–2434
140. Vogler M, Walczak H, Stadel D et al (2008) Targeting XIAP bypasses Bcl-2-mediated resistance to TRAIL and cooperates with TRAIL to suppress pancreatic cancer growth in vitro and in vivo. Cancer Res 68:7956–7965
141. Eggert A, Grotzer MA, Zuzak TJ et al (2001) Resistance to tumor necrosis factor-related apoptosis-inducing ligand (TRAIL)-induced apoptosis in neuroblastoma cells correlates with a loss of caspase-8 expression. Cancer Res 61:1314–1319
142. Grotzer MA, Eggert A, Zuzak TJ et al (2000) Resistance to TRAIL-induced apoptosis in primitive neuroectodermal brain tumor cells correlates with a loss of caspase-8 expression. Oncogene 19:4604–4610
143. Hopkins-Donaldson S, Bodmer JL, Bourloud KB et al (2000) Loss of caspase-8 expression in highly malignant human neuroblastoma cells correlates with resistance to tumor necrosis factor-related apoptosis-inducing ligand-induced apoptosis. Cancer Res 60:4315–4319
144. Casciano I, De Ambrosis A, Croce M et al (2004) Expression of the caspase-8caspase-8 gene in neuroblastoma cells is regulated through an essential interferon-sensitive response element (ISRE). Cell Death Differ 11:131–134
145. Ruiz-Ruiz C, Ruiz De Almodovar C, Rodriguez A et al (2004) The up-regulation of human caspase-8 by interferon-gamma in breast tumor cells requires the induction and action of the transcription factor interferon regulatory factor-1. J Biol Chem 279:19712–19720
146. Yang X, Merchant MS, Romero ME et al (2003) Induction of caspase 8 by interferon gamma renders some neuroblastoma (NB) cells sensitive to tumor necrosis factor-related apoptosis-inducing ligand (TRAIL) but reveals that a lack of membrane TR1/TR2 also contributes to TRAIL resistance in NB. Cancer Res 63:1122–1129
147. Casciano I, Banelli B, Croce M et al (2004) Caspase-8 gene expression in neuroblastoma. Ann N Y Acad Sci 1028:157–167

148. Das A, Banik NL, Ray SK (2009) Molecular mechanisms of the combination of retinoid and interferon-gamma for inducing differentiation and increasing apoptosis in human glioblastoma T98G and U87MG cells. Neurochem Res 34:87–101
149. De Ambrosis A, Casciano I, Croce M et al (2007) An interferon-sensitive response element is involved in constitutive caspase-8caspase-8 gene expression in neuroblastoma cells. Int J Cancer 120:39–47
150. Fulda S, Debatin KM (2002) IFN gamma sensitizes for apoptosis by upregulating caspase-8 expression through the Stat1 pathway. Oncogene 21:2295–2308
151. Johnsen JI, Pettersen I, Ponthan F et al (2004) Synergistic induction of apoptosis in neuroblastoma cells using a combination of cytostatic drugs with interferon-gamma and TRAIL. Int J Oncol 25:1849–1857
152. Kim KB, Choi YH, Kim IK et al (2002) Potentiation of Fas- and TRAIL-mediated apoptosis by IFN-gamma in A549 lung epithelial cells: enhancement of caspase-8 expression through IFN-response element. Cytokine 20:283–288
153. Kontny HU, Hammerle K, Klein R et al (2001) Sensitivity of Ewing's sarcoma to TRAIL-induced apoptosis. Cell Death Differ 8:506–514
154. Lissat A, Vraetz T, Tsokos M et al (2007) Interferon-gamma sensitizes resistant Ewing's sarcoma cells to tumor necrosis factor apoptosis-inducing ligand-induced apoptosis by up-regulation of caspase-8 without altering chemosensitivity. Am J Pathol 170:1917–1930
155. Meister N, Shalaby T, Von Bueren AO et al (2007) Interferon-gamma mediated up-regulation of caspase-8 sensitizes medulloblastoma cells to radio- and chemotherapy. Eur J Cancer 43:1833–1841
156. Merchant MS, Yang X, Melchionda F et al (2004) Interferon gamma enhances the effectiveness of tumor necrosis factor-related apoptosis-inducing ligand receptor agonists in a xenograft model of Ewing's sarcoma. Cancer Res 64:8349–8356
157. Ruiz-Ruiz C, Munoz-Pinedo C, Lopez-Rivas A (2000) Interferon-gamma treatment elevates caspase-8 expression and sensitizes human breast tumor cells to a death receptor-induced mitochondria-operated apoptotic program. Cancer Res 60:5673–5680
158. Tekautz TM, Zhu K, Grenet J et al (2006) Evaluation of IFN-gamma effects on apoptosis and gene expression in neuroblastoma–preclinical studies. Biochim Biophys Acta 1763:1000–1010
159. Levy DE, Darnell JE Jr (2002) Stats: transcriptional control and biological impact. Nat Rev Mol Cell Biol 3:651–662
160. Wexler L, Thiele CJ, Mcclure L et al (1992) Adoptive immunotherapy of refractory neuroblastoma with tumor-infiltrating lymphocytes, interferon-{gamma}, and interleukin-2. Proc Ann Meet Am Soc Clin Oncol 11:368
161. Liedtke C, Groger N, Manns MP et al (2006) Interferon-alpha enhances TRAIL-mediated apoptosis by up-regulating caspase-8 transcription in human hepatoma cells. J Hepatol 44:342–349
162. Ehrhardt H, Hacker S, Wittmann S et al (2008) Cytotoxic drug-induced, p53-mediated upregulation of caspase-8 in tumor cells. Oncogene 27:783–793
163. Jiang M, Zhu K, Grenet J et al (2008) Retinoic acid induces caspase-8 transcription via phospho-CREB and increases apoptotic responses to death stimuli in neuroblastoma cells. Biochim Biophys Acta 1783:1055–1067
164. Belada D, Smolej L, Stepankova P et al (2010) Diffuse large B-cell lymphoma in a patient with hyper-IgE syndrome: Successful treatment with risk-adapted rituximab-based immunochemotherapy. Leuk Res 34:232–234
165. Fanale MA, Younes A (2008) Nodular lymphocyte predominant Hodgkin's lymphoma. Cancer Treat Res 142:367–381
166. Herbst RS, Eckhardt SG, Kurzrock R et al (2010) Phase I Dose-Escalation study of recombinant human Apo2L/TRAIL, a dual proapoptotic receptor agonist, in patients with advanced cancer. J Clin Oncol 28:2839–2846
167. Soria JC, Mark Z, Zatloukal P et al (2011) Randomized phase II study of dulanermin in combination with paclitaxel, carboplatin, and bevacizumab in advanced non-small-cell lung cancer. J Clin Oncol 29:4442–4451

168. Soria JC, Smit E, Khayat D et al (2010) Phase 1b study of dulanermin (recombinant human Apo2L/TRAIL) in combination with paclitaxel, carboplatin, and bevacizumab in patients with advanced non-squamous non-small-cell lung cancer. J Clin Oncol 28:1527–1533
169. Yee YK, Tan VP, Chan P et al (2009) Epidemiology of colorectal cancer in Asia. J Gastroenterol Hepatol 24:1810–1816
170. Greco FA, Bonomi P, Crawford J et al (2008) Phase 2 study of mapatumumab, a fully human agonistic monoclonal antibody which targets and activates the TRAIL receptor-1, in patients with advanced non-small cell lung cancer. Lung Cancer 61:82–90
171. Hotte SJ, Hirte HW, Chen EX et al (2008) A Phase 1 Study of Mapatumumab (Fully Human Monoclonal Antibody to TRAIL-R1) in Patients with Advanced Solid Malignancies. Clin Cancer Res 14:3450–3455
172. Leong S, Cohen RB, Gustafson DL et al (2009) Mapatumumab, an antibody targeting TRAIL-R1, in combination with paclitaxel and carboplatin in patients with advanced solid malignancies: results of a phase I and pharmacokinetic study. J Clin Oncol 27:4413–4421
173. Mom CH, Verweij J, Oldenhuis CN et al (2009) Mapatumumab, a fully human agonistic monoclonal antibody that targets TRAIL-R1, in combination with gemcitabine and cisplatin: a phase I study. Clin Cancer Res 15:5584–5590
174. Pawel JV, Harvey JH, Spigel DR et al (2010) A randomized phase II trial of mapatumumab, a TRAIL-R1 agonist monoclonal antibody, in combination with carboplatin and paclitaxel in patients with advanced NSCLC. J Clin Oncol 28 suppl
175. Sun W, Sohal D, Haller DG et al (2011) Phase 2 trial of bevacizumab, capecitabine, and oxaliplatin in treatment of advanced hepatocellular carcinoma. Cancer 117:3187–3192
176. Tolcher AW, Mita M, Meropol NJ et al (2007) Phase I pharmacokinetic and biologic correlative study of mapatumumab, a fully human monoclonal antibody with agonist activity to tumor necrosis factor-related apoptosis-inducing ligand receptor-1. J Clin Oncol 25:1390–1395
177. Trarbach T, Moehler M, Heinemann V et al (2010) Phase II trial of mapatumumab, a fully human agonistic monoclonal antibody that targets and activates the tumour necrosis factor apoptosis-inducing ligand receptor-1 (TRAIL-R1), in patients with refractory colorectal cancer. Br J Cancer 102:506–512
178. Younes A, Vose JM, Zelenetz AD et al (2010) A Phase 1b/2 trial of mapatumumab in patients with relapsed/refractory non-Hodgkin's lymphoma. Br J Cancer 103:1783–1787
179. Merchant MS, Chou AJ, Price A et al (2010) Lexatumumab: Results of a phase I trial in pediatric patients with advanced solid tumors. J Clin Oncol 28:9500 (Meeting Abstracts)
180. Plummer R, Attard G, Pacey S et al (2007) Phase 1 and pharmacokinetic study of lexatumumab in patients with advanced cancers. Clin Cancer Res 13:6187–6194
181. Sikic BI, Wakelee HA, Mehren MV et al (2007) A phase 1b study to assess the safety of lexatumumab, a human monoclonal antibody that activates TRAIL-R2, in 32 combination with gemcitabine, pemetrexed, doxorubicin or FOLFIRI. J Clin Oncol 25:14006 (Meeting Abstracts)
182. Wakelee HA, Patnaik A, Sikic BI et al (2010) Phase I and pharmacokinetic study of lexatumumab (HGS-ETR2) given every 2 weeks in patients with advanced solid tumors. Ann Oncol 21:376–381
183. Chawla SP, Tabernero J, Kindler HL et al (2010) Phase I evaluation of the safety of conatumumab (AMG 655) in combination with AMG 479 in patients (pts) with advanced, refractory solid tumors. J Clin Oncol 28:3102 (Meeting Abstracts)
184. Doi T, Murakami H, Ohtsu A et al (2011) Phase 1 study of conatumumab, a pro-apoptotic death receptor 5 agonist antibody, in Japanese patients with advanced solid tumors. Cancer Chemother Pharmacol 68:733–741
185. Kindler HL, Garbo L, Stephenson J et al (2009) A phase 1b study to evaluate the safety and efficacy of AMG 655 in combination with gemcitabine (G) in patients (pts) with metastatic pancreatic cancer (PC). J Clin Oncol 27:4501 (Meeting Abstracts)
186. Kindler HL, Richards DA, Stephenson J et al (2010) A placebo-controlled, randomized phase 34 II study of conatumumab (C) or AMG 479 (A) or placebo (P) plus gemcitabine (G) in patients (pts) with metastatic pancreatic cancer (mPC). Clin Oncol 28:4035 (Meeting Abstracts)

187. Paz-Ares L, Torres JMS, Diaz-Padilla I et al (2009) Safety and efficacy of AMG 655 in 33 combination with paclitaxel and carboplatin (PC) in patients with advanced non-small cell lung cancer (NSCLC). J Clin Oncol 27:19048 (Meeting Abstracts)
188. Peeters M, Infante P, PLR J et al (2010) Phase Ib/II trial of conatumumab and panitumumab (pmab) for the treatment (tx) of metastatic colorectal cancer (mCRC): Safety and efficacy, ASCO Gastrointestinal Cancers Symposium, abstract 443
189. Saltz L, Infante J, Schwartzberg L et al (2009) Safety and efficacy of AMG 655 plus modified FOLFOX6 (mFOLFOX6) and bevacizumab (B) for the first-line treatment of patients (pts) with metastatic colorectal cancer (mCRC). J Clin Oncol 27:4079 (Meeting Abstracts)
190. Baron AD, O'bryant CL, Choi Y et al (2011) Phase 1b study of drozitumab combined with cetuximab (CET) plus irinotecan (IRI) or with FOLFIRI±bevacizumab (BV) in previously treated patients (pts) with metastatic colorectal cancer (mCRC). Clin Oncol 29:3581 (Meeting Abstracts)
191. Camidge DR, Herbst RS, Gordon MS et al (2010) A phase I safety and pharmacokinetic study of the death receptor 5 agonistic antibody PRO95780 in patients with advanced malignancies. Clin Cancer Res 16:1256–1263
192. Karapetis CS, Clingan PR, Leighl NB et al (2010) Phase II study of PRO95780 plus paclitaxel, carboplatin, and bevacizumab (PCB) in non-small cell lung cancer (NSCLC). J Clin Oncol 28:7535 (Meeting Abstracts)
193. Rocha CSL, Baranda JC, Wallmark J et al (2011) Phase 1b study of drozitumab combined with first-line FOLFOX plus bevacizumab (BV) in patients (pts) with metastatic colorectal cancer (mCRC). J Clin Oncol 29:546 (Meeting Abstracts)
194. Wittebol S, Ferrant A, Wickham NW et al (2010) Phase II study of PRO95780 plus rituximab in patients with relapsed follicular non-Hodgkin's lymphoma (NHL). J Clin Oncol 28:18511 (Meeting Abstracts)
195. Forero-Torres A, Shah J, Wood T et al (2010) Phase I trial of weekly tigatuzumab, an agonistic humanized monoclonal antibody targeting death receptor 5 (DR5). Cancer biother radiopharm 25:13–19
196. Sharma S, Vries EGD, Infante JR et al (2008) Phase I trial of LBY135, a monoclonal antibody agonist to DR5, alone and in combination with capecitabine in advanced solid tumors. J Clin Oncol 26:3538 (Meeting Abstracts)

Chapter 12
Proteasome Inhibition as a Novel Strategy for Cancer Treatment

Min Shen and Q. Ping Dou

Abstract The proteasome is a multi-subunit protease complex, responsible for the degradation of misfolded, damaged, or short-lived proteins. Indeed, more than 90 % of intracellular proteins are degraded through the ubiquitin–proteasome system (UPS). The UPS is extensively involved in various cellular events, including cell cycle, cell signaling, stress response, and apoptosis. Inhibition of proteasome function could result in growth arrest and/or cell death. The involved molecular mechanisms include dysregulation of the cell cycle, inactivation of NF-κB pathway, disturbance of the pro- and anti-apoptotic balance, induction of unfolded protein response, and induction of oxidative stress. The observation that suppression of proteasome function by small chemical inhibitors was able to induce apoptosis in cancer cells but not in normal cells supports the hypothesis that the proteasome could be a valuable target for cancer treatment. This idea has been validated from benchtop to bedside. In 2003, the first proteasome inhibitor anticancer drug, bortezomib, was approved in the United States for the treatment of multiple myeloma and mantle cell lymphoma. While bortezomib achieved great success in clinical applications, problems such as resistance, dose-limiting toxicities, unsatisfied efficacy in solid tumors, and interaction with some natural compounds have been observed. Therefore, it is necessary to further investigate the proteasome inhibition-mediated mechanism of cell death as well as the development of novel, new-generation proteasome inhibitors with lower toxicity and wider applications.

12.1 Introduction

Cellular protein homeostasis is precisely regulated by a series of cellular processes including protein synthesis, folding, trafficking, and degradation. Protein degradation by the UPS was initially underestimated as a "garbage disposal" system, but its role has since been appreciated, as it plays an important regulatory

M. Shen · Q. P. Dou (✉)
Karmanos Cancer Institute and Department of Pharmacology,
Wayne State University School of Medicine, Detroit, MI, USA
e-mail: doup@karmanos.org

D. E. Johnson (ed.), *Cell Death Signaling in Cancer Biology and Treatment*,
Cell Death in Biology and Diseases, DOI: 10.1007/978-1-4614-5847-0_12,

X.-M. Yin and Z. Dong (Series eds.), *Cell Death in Biology and Diseases*

role in various cellular events, such as protein quality control, cell cycle progression, cell differentiation, embryonic development, apoptosis, signal transduction, gene expression, circadian clocks, as well as immune and inflammatory responses [1, 2]. In all eukaryotic cells, protein degradation is executed by two major pathways, namely the lysosomal pathway, which mainly degrades extracellular and transmembrane proteins, and the ubiquitin–proteasome pathway (UPP), which mainly degrades intracellular proteins [2]. Dysregulation of proteasome function is implicated in several pathological processes, including neoplastic disease and autoimmune disease which exhibit increased proteasome function. Furthermore, neurodegenerative disease has been found to be associated with decreased proteasome function. These properties grant the proteasome potential to be a therapeutic target. Indeed, the first proteasome inhibitor bortezomib was launched in the United States in 2003 and subsequently in Europe in 2004 for the treatment of multiple myeloma (MM) and mantle cell lymphoma (MCL) [3].

12.2 Ubiquitin–Proteasome System

The ubiquitin–proteasome system (UPS) was discovered in the late 1970s and early 1980s. The importance of this discovery was acknowledged through the award of the 2004 Nobel Prize in Chemistry to its discoverers, Aaron Ciechanover, Avram Hershko, and Irwin Rose.

12.2.1 Ubiquitin–Proteasome Pathway

Protein degradation by the ubiquitin–proteasome pathway (UPP) usually includes two steps, ubiquitin conjugation and proteasomal degradation. Ubiquitin (8.5 kDa) is a highly conserved small regulatory protein. Only proteins conjugated with a Lys48-linked poly-ubiquitin chain can be recognized and processed by the proteasome for degradation. Ubiquitin conjugation occurs through an enzymatic cascade that involves three distinct enzymes, Ub-activating (E1), Ub-conjugating (E2), and Ub-ligating (E3) enzymes. The poly-ubiquitinated proteins with four or more Lys48-linked ubiquitin are then directed to the 26S proteasome complex where the poly-ubiquitin chain will be removed and recycled, while the protein substrates will be progressively degraded into oligopeptides that are 3–25 amino acids long [4, 5] (Fig. 12.1).

12.2.2 Proteasome

The 26S proteasome (2.4 MDa) is a multi-subunit protease complex composed of the 20S catalytic core (700 kDa) and two 19S regulatory particles (700 kDa).

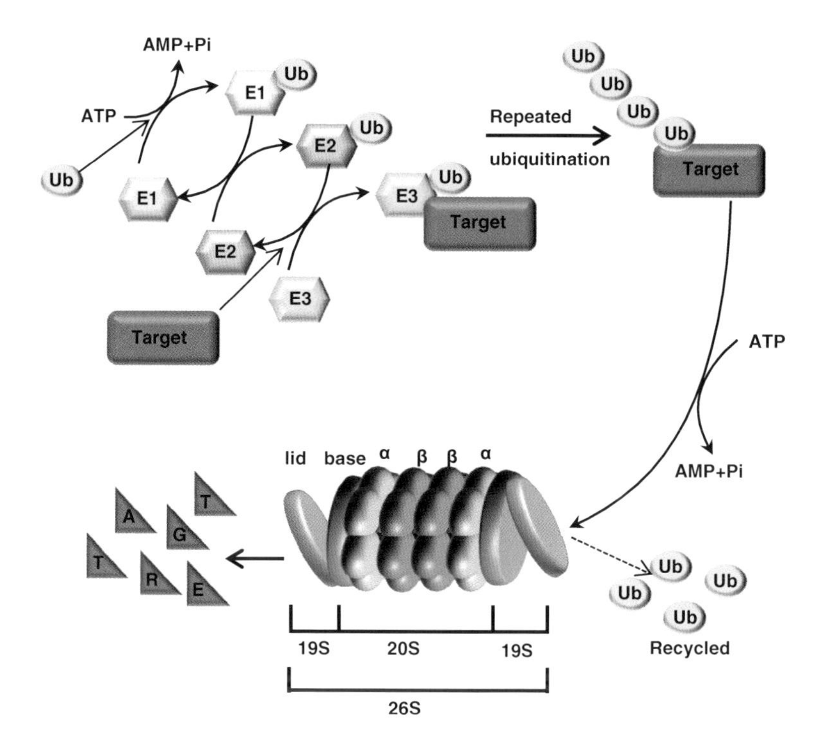

Fig. 12.1 A schematic diagram of the UPP and the structure of the 26S proteasome

The 20S core is formed by two identical α rings and two identical β rings stacked in a symmetric manner with the outside α rings surrounding the inner β rings (Fig. 12.1). Each α or β ring contains seven different subunits, named α1–α7 or β1–β7, among which only β1, β2, and β5 possess proteolytic activity. The β1 subunit possesses caspase-like or peptidyl-glutamyl peptide-hydrolyzing-like (PGPH-like) activity which preferentially cleaves after acidic residues such as aspartate and glutamate; the β2 subunit possesses trypsin-like activity which preferentially cleaves after basic residues such as arginine and lysine; and the β5 subunit possesses chymotrypsin-like activity which preferentially cleaves after hydrophobic residues such as tyrosine and phenylalanine [3, 6]. The proteasome also exhibits two other enzymatic activities, the "branched chain amino acid–preferring" (BrAAP) activity and the "small neutral amino acid–preferring" (SNAAP) activity [7]. The 19S regulatory particle binds to both ends of the 20S core proteasome and is composed of a "lid" and a "base" (Fig. 12.1). The "lid" contains at least nine non-ATPase subunits, which remove the poly-ubiquitin chain from the substrate by a process called deubiquitylation. The "base" contains six ATPase and four non-ATPase subunits and opens the gate of the 20S proteasome, unfolds the substrate proteins, and promotes their entry into the 20S proteasome [3].

In addition to the 19S regulatory particle, the 20S core can alternatively interact with other regulatory particles at one or both ends, resulting in the formation of asymmetric or symmetric isoforms of the proteasome. These regulatory particles include PA28 (also referred to as the 11S regulatory particle) and PA200 [5, 8].

12.2.3 Immunoproteasome

In addition to the constitutive proteasome described thus far, which contains catalytic subunits β1, β2, and β5, there also exists the immunoproteasome, which harbors different sets of catalytic subunits designated as β1i, β2i, and β5i, correspondingly. Upon cytokine stimulation, especially interferon-γ and tumor necrosis factor-α, the expression of β1i, β2i, and β5i dramatically increases and cooperatively assembles into nascent 20S core particles to form immunoproteasomes. Interferon-γ stimulation also induces the expression of regulatory subunits PA28α and PA28β, both of which are non-ATPase subunits. Three PA28α and four PA28β comprise the heteroheptameric PA28αβ regulatory particle, which facilitates the assembly of the immunoproteasome and enhances its activity. The immunoproteasome is a key component of antigen processing. Compared with the constitutive proteasome, the immunoproteasome possesses enhanced chymotrypsin- and trypsin-like activities and reduced caspase-like activity. This specialized enzymatic property endows the immunoproteasome the ability to generate peptides that are suitable for MHC class I-mediated antigen presentation [9, 10].

12.3 Proteasome Inhibitors

12.3.1 Synthetic Proteasome Inhibitors

Based on their chemical structures, the known synthetic proteasome inhibitors fall into three classes: peptide aldehydes, peptide boronates, and peptide vinyl sulfones.

Peptide aldehydes are widely used proteasome inhibitors. The representatives of this class, MG132 (A-LLL-al, Fig. 12.2a) and PSI, were among the first proteasome inhibitors developed. This class of compounds inhibits the proteasome by forming a covalent hemiacetal bond with the hydroxyl group of the N-terminal threonine (Thr1) in the catalytic β subunits. This inhibition is reversible, and the dissociation rate is relatively fast. MG132 is an analog of Calpain Inhibitor I, which shows a 10-fold higher preference toward the proteasome over calpains and cathepsins. Although it is a well-known research tool in protesome studies in vitro and *in cellulo*, the fact that MG132 is rapidly oxidized into inactive carbonic acid in vivo largely prevents its development as a therapeutic drug [3, 11, 12].

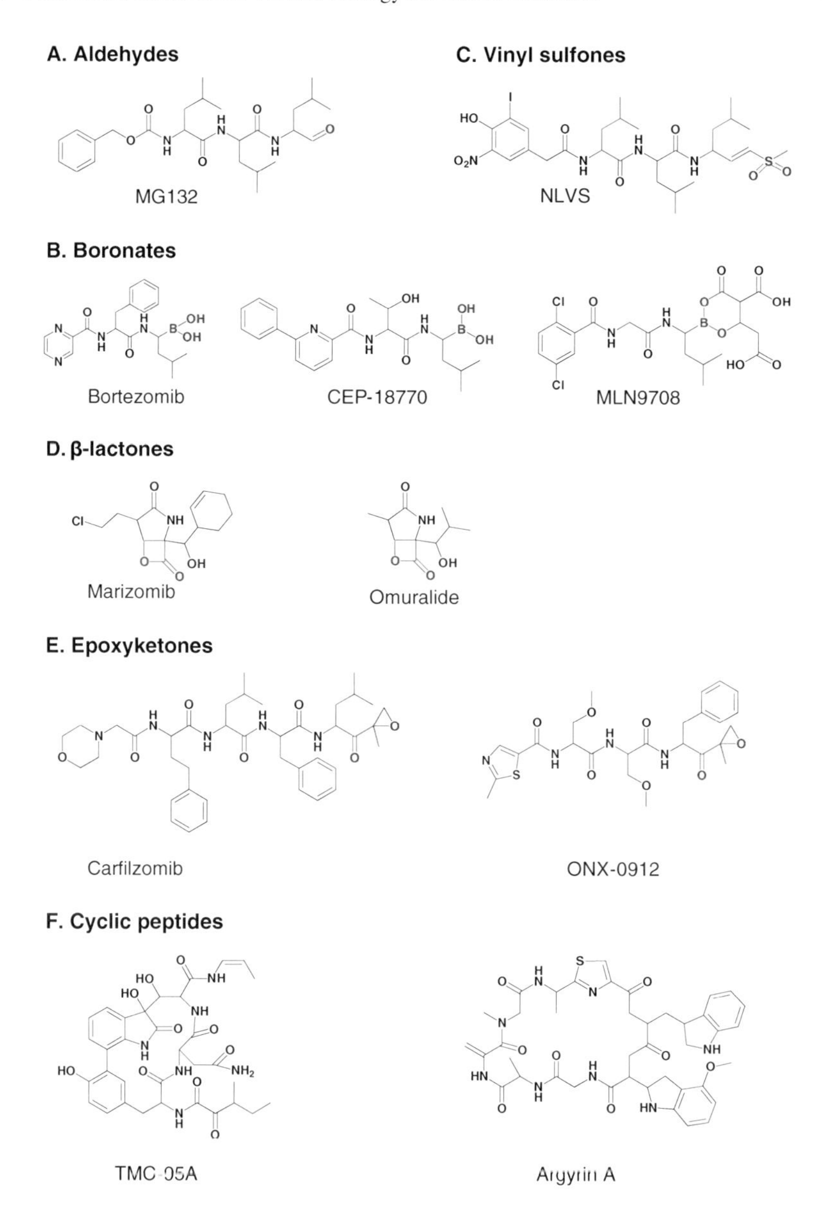

Fig. 12.2 Representatives of the major classes of proteasome inhibitors. Electrophilic functional groups that directly interact with the proteasome are depicted in *red*. **a** Aldehydes, **b** Boronates, **c** Vinyl Sulfones, **d** β-lactones, **e** Epoxy Ketones, **f** Cyclic Peptides

Peptide boronates are aldehyde surrogates with improved potency and specificity for the proteasome. For example, MG262 (Z-LLL-Boronate), the boronate surrogate of MG132, was shown to be 100-fold more potent than MG132. Besides MG262,

Table 12.1 A summary of proteasome inhibitors in clinical use or clinical trials [11, 135]

Inhibitor	Developer	Chemical nature	Binding mode and kinetics	Target selectivity	Route of administration	Development stage
Bortezomib (PS341)	Millennium	Boronate	Covalent, slowly reversible	$\beta5 > \beta1$	IV/SC	FDA approved
Carfilzomib (PR-171)	ONXY	Epoxyketone	Covalent, irreversible	$\beta5$, $\beta5i$	IV	Phase III
Marizomib (NPI-0052)	Nereus	β-lactone	Covalent, irreversible	$\beta5 > \beta2 > \beta1$	IV	Phase I
MLN-9708	Millennim	Boronate	Covalent, rapidly reversible	$\beta5 > \beta1$	IV/oral	Phase I-II
Delanzomib (CEP-18770)	Cephalon	Boronate	Covalent, slowly reversible	$\beta5 > \beta1$	IV/oral	Phase I-II
Oprozomib (ONX-0912)	ONXY	Epoxyketone	Covalent, irreversible	$\beta5$, $\beta5i$	Oral	Phase I-II

IV intravenous; *SC* subcutaneous; *FDA* Food and Drug Administration

bortezomib (PS341), delanzomib (CEP-18770), and MLN9708 are the most successful representatives in this class (Fig. 12.2b). Peptide boronates form tetrahedral adducts with Thr1 in catalytic β subunits, which are further stabilized by two extra hydrogen bonds. Therefore, although the inhibition is reversible, the dissociation rate is very slow. Furthermore, unlike peptide aldehydes, peptide boronates have increased bioavailability and are metabolically stable under physiological conditions. These characteristics make peptide boronates suitable for therapeutic development. Indeed, bortezomib (Velcade®) was the first proteasome inhibitor launched in 2003 (Table 12.1). Both delanzomib and MLN9708 are currently under clinical trials. MLN9708 is sophisticatedly designed as a prodrug in the form of a boronic ester so that it is orally bioavailable [11, 12].

Peptide vinyl sulfones are irreversible proteasome inhibitors. They were first described as inhibitors of cysteine proteases. Not surprisingly, they exhibit less potency and specificity toward the proteasome. Nevertheless, this feature of irreversible inhibition offers them certain advantages of being activity-based proteasome probes in basic research. The representatives of this class are ZLVS (Z-LLL-vs), NLVS (NIP-LLL-vs) (Fig. 12.2c), and AdaAhx3-LLL-vs [3, 11].

12.3.2 Natural Proteasome Inhibitors

A large portion of proteasome inhibitors were initially derived from natural products, and the number of such is still growing. This large family can be further divided into four groups based on their mechanisms of action: covalent inhibitors, non-covalent inhibitors, non-proteasome-specific inhibitors, and allosteric inhibitors.

12.3.2.1 Covalent Inhibitors

β-lactones are non-peptide natural products showing appreciable proteasome-inhibitory activity. The representatives of this class are lactacystin, marizomib (salinosporamide A, NPI-0052) (Fig. 12.2d), and belactosins. The carbonyl group of the β-lactones can interact with the hydroxyl group of Thr1 in catalytic β subunits, leading to the formation of an acyl-ester adduct. Although this is a reversible process, the dissociation rate of the covalent adduct is extremely slow. Lactacystin, is a metabolite produced by the bacteria *Streptomyces*, it was the first identified natural proteasome inhibitor. Lactacystin itself is an inactive precursor, which will be spontaneously hydrolyzed into *clasto*-lactacystin-β-lactone (omuralide) (Fig. 12.2d), the active form, in aqueous solutions at pH 8. Marizomib, a secondary metabolite produced by the bacteria *Salinispora tropica*, is a very promising compound, which is currently under investigation into a phase I clinical trial. In addition to the acyl-ester bond, marizomib also forms a cyclic tetrahydrofuran ring with the proteasome, making the reaction irreversible [11, 13].

Peptide epoxyketones are the most specific and potent proteasome inhibitors thus far. The representatives of this class are epoxomicin, carfilzomib (PR-171), and oprozomib (ONX-0912 or PR-047) (Fig. 12.2e). The α,β-epoxyketone moiety interacts with both the hydroxyl group and the free α–amino group of Thr1 in the catalytic β subunits, leading to the formation of the morpholino adduct in an irreversible way. Since the free α–amino group required for adduct formation is not present in serine and cysteine proteases, these proteases are not affected by epoxyketones. Epoxomicin was originally isolated from an unidentified strain of Actinobacteria. Since its identification as a proteasome inhibitor, little, if any off-target effects have been reported. Carfilzomib and oprozomib are derivatives of epoxomicin, both currently under clinical development. As irreversible proteasome inhibitors, peptide epoxyketones are, like peptide vinyl sulfones, also widely used as activity-based proteasome probes in basic research [11, 13].

12.3.2.2 Non-covalent Inhibitors

Some natural cyclic peptides have been found to inhibit proteasome function in a non-covalent manner. The representatives of this class are TMC-95A (Fig. 12.2f), argyrin A (Fig. 12.2f), and more recently scytonemides [14]. The TMC-95 family is a group of cyclic tripeptides produced by the fungus *Apiospora montagnei*. They interact with the active sites of catalytic β subunits through five hydrogen bonds, thereby reversibly inhibiting proteasome function [15]. TMC-95A is able to selectively and competitively inhibit all three catalytic β subunits in the low nanomolar range [15]. The shortcoming for the development of these large inhibitors is that their de novo synthesis is very difficult and expensive. Argyrin A is a cyclic octapeptide produced by the bacteria *Archangium gephyra*. It also reversibly inhibits proteasomal activity in the low nanomolar range, making itself a promising candidate [16]. However, given its large size, the exact mechanism by which argyrin A interacts with the proteasome remains unclear.

12.3.2.3 Non-specific Proteasome Inhibitors

Besides the natural proteasome inhibitors discussed above, most of which are metabolites of bacteria or fungus, a rapidly growing group of natural proteasome inhibitors has also been isolated from the plant kingdom. These natural products include some flavonoids, triterpenoids, and others such as curcumin [17] and shikonin [18]. Unlike the proteasome inhibitors discussed hitherto that mainly target the proteasome, plant-derived natural products usually have multiple cellular targets including the proteasome.

Flavonoids are a group of natural products with a common structure of two aromatic rings linked by three carbons (C_6–C_3–C_6) [19]. They are abundant in various vegetables and fruits. Flavonoids that have been found to possess proteasome-inhibitory activity include (-)-epigallocatechin-3-gallate [(-)-EGCG] (Fig. 12.3a), apigenin, quercetin, and genistein. Among them, (-)-EGCG, the major catechin in green tea, is most potent in terms of proteasome inhibition. It inhibits the proteasomal β5 subunit irreversibly through covalent binding. Notably, (-)-EGCG is able to interact with not only the β5 subunit of the constitutive proteasome but also the β5i subunit of the immunoproteasome with an even greater affinity [20]. However, (-)-EGCG has been found to have the potential to decrease the efficacy of bortezomib and other boronate proteasome inhibitors. This neutralizing effect is mainly due to the direct interaction between the pyrocatechol moieties on (-)-EGCG and the boronic acid moiety on bortezomib [21]. Whether patients receiving bortezomib should consume green tea needs a clarification by further clinical studies.

Triterpenoids are a group of compounds comprised of six isoprene units with 30 carbons. They occur naturally as complex cyclic structures, with the pentacyclic structure being most common. Some have been reported to have proteasome-inhibitory activity, including celastrol [22] (Fig. 12.3b), withaferin A [23], and pristimerin [24]. These compounds are present in many herbal medicines.

12.3.2.4 Allosteric Inhibitors

Allosteric inhibitors refer to inhibitors that bind to the allosteric site (a site other than the enzyme's active site) of an enzyme and cause its inactivation. The advantage of

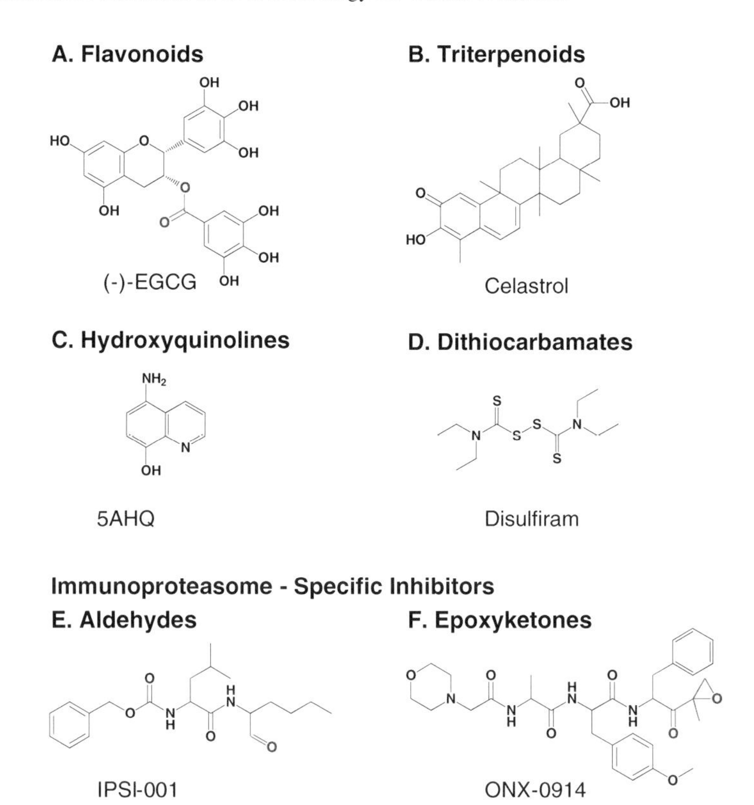

Fig. 12.3 Chemical structures of some natural compounds with proteasome-inhibitory activity and immunoproteasome-specific inhibitors. Electrophilic functional groups that directly interact with the immunoproteasome are depicted in *red*. **a** Flavonoids, **b** Triterpenoids, **c** Hydroxyquinolines, **d** Dithiocarbamates, **e** Aldehydes, **f** Epoxyketones

allosteric inhibitors lies in their ability to overcome some particular types of resistance, for example, resistance caused by mutations at the active site. A number of different metal-binding organic compounds have been shown to be allosteric proteasome inhibitors in vitro and in cellulo. However, like proteasome inhibitors derived from plants, most of the organometallic compounds have multiple cellular targets with the proteasome being one of them.

Two biggest groups of metal-binding compounds that have proteasome-inhibitory activity are hydroxyquinolines [such as 5-amino-8-hydroxyquinoline (5AHQ) (Fig. 12.3c) and clioquinol] and dithiocarbamates [such as disulfiram (Fig. 12.3d), diethyldithiocarbamate, and pyrrolidine dithiocarbamate] [25–27]. NMR data indicated that 5AHQ might bind to proteasomal α subunits [27]. Copper is the metal most commonly used to form complex with these ligands. Other metals that have been studied include zinc, gold, and gallium.

12.3.3 Immunoproteasome-Specific Inhibitors

In recent years, the immunoproteasome has been attracting increasing attention, as are immunoproteasome-specific inhibitors. To date, the immunoproteasome-specific inhibitors IPSI-001 (Fig. 12.3e), ONX-0914 (PR-957) (Fig. 12.3f), and PR-924 are the most advanced developed.

IPSI-001 belongs to the peptide aldehydes class of compounds. It selectively inhibits β5i and β1i subunits of the immunoproteasome. It has a 100-fold preference for the immunoproteasome (Kis of 1.03 and 1.45 μM against β5i and β1i, respectively) *over* the constitutive proteasome (Kis of 105 and 239 μM against β5 and β1, respectively) in vitro. Moreover, IPSI-001 potently inhibits proliferation and induces apoptosis in MM cell lines as well as in patient samples. More importantly, it was able to overcome conventional and novel drug resistance, including resistance to bortezomib [28].

ONX-0914 and PR-924 belong to the class of epoxyketones. Like other epoxyketones, the inhibition caused by ONX-0914 and PR-924 is irreversible. Both of the inhibitors selectively inhibit the β5i subunit. ONX-0914 is 20- to 40-fold more selective toward the β5i subunit than both the β5 and β1i subunits [29].

12.4 Mechanisms of Proteasome Inhibition-Induced Growth Arrest and/or Cell Death

Proteasome function can be suppressed or inhibited by either endogenous protein inhibitors or exogenous chemical inhibitors. Inhibition of proteasomal activity will result in the accumulation of substrate proteins and disruption of normal cellular function.

12.4.1 Proteasome Inhibition and Cell Cycle Arrest: p27 and p53

Cell cycle progression is tightly regulated by cyclin-dependent kinases (CDKs), cyclin-dependent kinase inhibitors (CDKIs), and other proteins involved in cell cycle checkpoints. The UPS is critical in the regulation of the cell cycle. Basically, the cell cycle is driven by CDKs whose activity is determined by the level of cyclins present in the cell. Oscillations in the level of cyclins are achieved by periodic protein synthesis and proteasome-dependent degradation. Inhibition of proteasome function can result in the accumulation of two cyclins at the same time, which results in two contradictory signals. These conflicting signals are only resolved through cellular self-destruction. Proteasome inhibition stabilizes two important negative regulators of cell cycle, $p27^{KIP1}$ and p53, both of which are known proteasome substrates. $P27^{KIP1}$ is an inhibitor of CDK2 and CDK4, and during cell cycle progression, it is degraded by the proteasome to release CDK2 and CDK4 for G_1/S phase transition, whereas accumulation of $p27^{KIP1}$ by proteasome inhibition results in G_1 phase arrest [30]. Tambyrajah et al. [31] reported that $p27^{KIP1}$ is degraded by the proteasomal β1-mediated activity, which is upregulated as cells enter the cell cycle without concomitant changes in the protein levels of β1, β2, or β5 subunits. Furthermore, one of the novel proteasome inhibitors discussed above, argyrin A, was actually discovered in a screen for compounds that prevent degradation of $p27^{KIP1}$ [16]. P53 is a bona fide tumor suppressor. As a transcription factor, accumulation of p53 by proteasome inhibition could lead to increased expression of its target genes such as $p21^{WAF1}$, another inhibitor of CDK2 [32, 33], and PUMA, a proapoptotic Bcl-2 family member [34]. However, p53 is not indispensable in proteasome inhibitor-induced apoptosis, at least in certain types of cancer cells [35–37].

12.4.2 Proteasome Inhibition and Suppression of the Cell Survival NF-κB Pathway

The transcription factor NF-κB is a critical player against apoptosis in cells and proteasome function is essential to two proteolytic processes required for the activation of NF-κB. First, the NF-κB dimer is trapped in the cytoplasm by inhibitor of NF-κB (IκB) protein. Degradation of IκB by the proteasome is required to release the NF-κB complex for its nuclear translocation and activation. Inhibition of proteasome function stabilizes IκB and thereby prevents NF-κB activation [38, 39]. Secondly, p105 and p100, two precursors of NF-κB subunits p50 and p52, respectively, need to be processed by the proteasome; otherwise they function like IκB and confine their dimeric partners to the cytoplasm [39–41]. In addition, it is well established that conventional cytotoxic drugs as well as inflammatory factors such as TNFα and IL-1β often activate NF-κB, which is at least partially responsible for the development of drug resistance. Moreover, constitutive activation of NF-κB was observed in a large fraction of advanced cancers that are very resistant to most therapies. Pioneering work done by Delic et al. [42] demonstrated that a low dose of proteasome inhibitor lactacystin was sufficient to prevent NF-κB activation and sensitize chemo- and radio-resistant cancer cells to TNFα-induced apoptosis. Synergistic apoptosis-inducing effects were also observed when combining a proteasome inhibitor with TNF-related apoptosis-inducing ligand (TRAIL) [43, 44], paclitaxel [45], or etoposide [46]. Therefore, combining a proteasome inhibitor with other chemotherapeutic agents may represent an effective way to overcome NF-κB-mediated drug resistance through resensitizing cancer cells to apoptosis [47]. However, a study by Hideshima et al. [48] pointed out that under some experimental conditions, bortezomib actually induced canonical NF-κB activation in MM cells through phosphorylation of IκB kinase (IKKβ) and its upstream receptor-interacting protein 2. More intriguingly, Amschler et al. [49] reported that NF-κB inhibition by either proteasome inhibition or IKKβ blockade sensitized melanoma cells to camptothecin through distinct pathways. Therefore, more mechanistic studies are needed to unveil the regulation of NF-κB pathway by the proteasome.

12.4.3 Proteasome Inhibition and Mitochondrial Apoptotic Pathway: The Bcl-2 Family Members

The Bcl-2 family plays a pivotal role in the mitochondrial apoptotic pathway (the intrinsic pathway). It is comprised of both pro-apoptotic (e.g., Bax, Bak, Bid, Bim, Bik, NOXA, and PUMA) and anti-apoptotic (e.g., Bcl-2, Bcl-X_L, A1, and Mcl-1) members, the ratio of which is tightly regulated by different factors. The proteasomal activity influences the level of Bcl-2 family members both directly and indirectly. Inhibition of proteasome function by different types of inhibitors was reported to directly induce the accumulation of its substrates Bax [50, 51], Bik [52], and Bim [53–55]. However, in other studies, conformational change and mitochondrial translocation, instead of accumulation, of Bax was observed under proteasome-inhibitory conditions [56, 57]. Proteasome inhibition was also reported to indirectly induce the transcriptional upregulation of NOXA, PUMA, and Mcl-1. Although both NOXA and PUMA are target genes of p53, only upregulation of PUMA, but not NOXA, upon proteasome inhibition was found to be p53 dependent [34, 36, 51, 58]. A study by Nikiforov et al. [59] suggested that the oncogene c-Myc, another transcription factor, was responsible for tumor cell-selective upregulation of NOXA in response to proteasome inhibition. Dysregulation of c-Myc was listed among the most pivotal events by a genome-wide siRNA screen looking for modulators of cell death induced by bortezomib [60]. Intriguingly, the anti-apoptotic protein Mcl-1 was also upregulated upon proteasome inhibition and was found to be mediated by activating transcription factor-4 (ATF4), an important effector of the unfolded protein response (UPR) [61]. However, the effect of Mcl-1 was mostly counteracted by the upregulation of NOXA, which binds to and functionally represses Mcl-1 [62, 63]. Moreover, regardless of its accumulation, proteasome inhibition caused Mcl-1 cleavage in a caspase-3-dependent manner [63–65].

12.4.4 Proteasome Inhibition and Unfolded Protein Response

Accumulation of unfolded or misfolded proteins in the endoplasmic reticulum (ER) lumen causes ER stress and subsequently the activation of unfolded protein response (UPR). ER stress is sensed, and the activation of UPR is controlled by three ER transmembrane proteins, PERK, IRE1 and ATF6. Initially, the UPR can support cell survival through ATF6-mediated or IRE1- and XBP1-mediated induction of ER chaperone proteins and proteins involved in ER-associated protein degradation

(ERAD), both of which help to relieve ER stress. However, prolonged UPR can eventually lead to apoptosis through the PERK-, ATF4-, and CHOP/GADD153-mediated pathway or the IRE1- and JNK-mediated pathway [66]. Although the proteasome does not directly reside in the ER lumen, it participates in the UPR in multiple levels. First, increased expression of the ER chaperones GRP78/Bip and GRP94/gp96 was observed upon proteasome inhibition, suggesting activation of UPR [67]. Secondly, proteasome inhibitor treatment was found to suppress the activation of IRE1- and XBP1-mediated survival pathways during UPR [68]. Thirdly, proteasome inhibitors were reported to induce the proapoptotic/terminal UPR mediated by PERK, ATF4, and CHOP/GADD153 [69]. However, the exact mechanism by which proteasome inhibition induces ER stress remains unclear and debatable.

12.4.5 Proteasome Inhibition and Oxidative Stress

Generation of reactive oxygen species (ROS) upon proteasome inhibition has been observed in different cell types including glioma, thyroid cancer, head and neck squamous cell carcinoma, and non-small cell lung cancer [70–73]. Pre-treating cells with an antioxidant was able to relieve oxidative stress and suppress apoptosis induction by proteasome inhibition [71, 74]. Consistently, Du et al. [72] reported that high-level intracellular glutathione protected cells from proteasome inhibition-induced oxidative stress, suggesting that a glutathione-dependent redox system might play an important role in the sensitivity of certain cells to proteasome inhibition-induced apoptosis. More intriguingly, Lee et al. [75] reported that low-dose lactacystin led to increased oxidative/nitrosative protein damage and lipid peroxidation with little cell death, while high-dose lactacystin further caused oxidative DNA damage and induced apoptotic cell death. Similar effects were observed by using another proteasome inhibitor epoxomicin [75]. A comprehensive proteomic and transcriptomic analysis by Bieler et al. [76] further revealed that low-dose proteasome inhibition induced a transcriptional profile reminiscent of a physiological stress response that preconditions and protects cells from oxidative stress, while high-dose proteasome inhibition induced massive transcriptional dysregulation and pronounced oxidative stress, triggering apoptosis.

12.5 Clinical Development of Bortezomib

12.5.1 Clinical Development of Bortezomib for Treatment of Multiple Myeloma and Mantle Cell Lymphoma

Bortezomib (Velcade®) is the first and currently the only proteasome inhibitor approved by the U.S. Food and Drug Administration (FDA). Based on two phase

II trials, the CREST trial [77, 78] and the SUMMIT trial [79], bortezomib received a fast-track approval from the FDA in 2003 for the treatment of patients with relapsed or refractory MM who had received at least two prior lines of therapy and progressed on their last therapy. In 2005, bortezomib was promoted by the FDA to treat patients with MM who had received at least one prior therapy based on the phase III APEX trial [80, 81]. In 2008, bortezomib successfully became the front-line therapy for MM patients who are newly diagnosed or previously untreated based on the phase III VISTA trial [82, 83]. Bortezomib was initially approved for intravenous injection, but very recently, in 2012, it received FDA approval for subcutaneous administration based on the results from a phase III, non-inferiority trial conducted in bortezomib-naive patients with relapsed MM [84].

The rational combinations of bortezomib with other chemotherapeutic agents have been explored. High-dose dexamethasone is a mainstay of therapy for MM. The combination of bortezomib and dexamethasone was studied in a phase II trial, where it was used as the first-line treatment in untreated MM patients. The results suggest that bortezomib with or without dexamethasone is an effective and well-tolerated induction regimen for the frontline treatment of MM [85, 86]. In a global phase IIIb trial, the similar regimen was extended to heavily pre-treated patients with relapsed or refractory MM and proved to be safe and effective in these patients as well [87]. Adding a third drug to the combination of bortezomib and dexamethasone has also been studied. For example, a combination of bortezomib, doxorubicin, and dexamethasone used as induction therapy before stem-cell transplantation induced >90 % of responses in newly diagnosed MM patients with well-tolerated and manageable toxicities [88, 89]. Likewise, a combination of bortezomib, thalidomide, and dexamethasone also achieved favorable outcome when used as induction regimen in newly diagnosed MM patients [90].

For patients ineligible for stem-cell transplantation, melphalan plus prednisone has long been the standard treatment with response rates around 50 % and median survival around 3 years [91, 92]. The previously mentioned pivotal VISTA trial compared the use of melphalan and prednisone with or without bortezomib in untreated MM patients who were ineligible for high-dose therapy and eventually supported the approval of bortezomib as frontline therapy in these patients. In this trial, adding bortezomib to melphalan and prednisone (VMP) achieved much higher complete response (CR) rate and partial response (PR) rate as well as significantly prolonged overall survival [82, 83]. Replacing melphalan with thalidomide in this triple-drug regimen resulted in comparable efficacy but more serious adverse events and discontinuations, suggesting that the VMP regimen should remain the upfront treatment for untreated MM patients who are ineligible for stem-cell transplantation [93]. Rather than replacing melphalan with thalidomide, adding thalidomide to the VMP regimen as a fourth drug was found to be superior to VMP alone [94, 95].

In 2006, the FDA extended the application of bortezomib to the treatment of mantle cell lymphoma (MCL) based on the results from the pivotal phase II PINNACLE trial where 155 patients with relapsed or refractory MCL received bortezomib monotherapy [96, 97]. The overall response rate (ORR) was 32 %.

The median time to progression and overall survival time in responders were much longer than those in the entire patient population, suggesting that single-agent bortezomib therapy is associated with lengthy responses and notable survival in relapsed or refractory MCL patients. In another phase II trial, the combination of bortezomib and gemcitabine was investigated and found to be active in patients with relapsed or refractory MCL with the ORR being 60 %. This regimen also offers a chemotherapy backbone to which other less myelosuppressive agents may be added [98].

12.5.2 Clinical Trials of Bortezomib in Other Hematological Malignancies and Solid Tumors

Dozens of clinical trials are currently undergoing in order to further expand the application of bortezomib to other hematological malignancies such as different types of B-cell lymphoma as well as some advanced-stage solid tumors [99].

The application of bortezomib in different B-cell lymphomas has been explored. Besides MCL, which accounts for 5 % of all B-cell lymphomas, two major types of B-cell lymphomas are diffuse large B-cell lymphoma and follicular lymphoma, accounting for 30–40 and 20 %, respectively [100]. In a phase II trial, a regimen of weekly or twice-weekly bortezomib plus rituximab was evaluated in patients with relapsed or refractory follicular or marginal-zone B-cell lymphoma [101]. Based on the encouraging results, the regimen of weekly bortezomib plus rituximab was further evaluated in patients with relapsed follicular lymphoma in a phase III trial. However, comparing with rituximab alone, the improvement elicited by adding bortezomib was not as great as expected, suggesting that this regimen might be useful for only some subgroups of patients [102]. In the phase II VERTICAL trial, a combination of bortezomib, bendamustine, and rituximab was evaluated and proved to be highly active (ORR of 88 % and CR of 53 %) in patients with relapsed or refractory follicular lymphoma [103]. In a phase I/II study, bortezomib was introduced to the standard CHOP (cyclophosphamide, doxorubicin, vincristine, and prednisone) plus rituximab regimen (R-CHOP). In diffuse large B-cell lymphoma patients, the evaluable ORR was 100 % with 86 % CR/unconfirmed CR. The numbers were 91 and 72 % in MCL patients, respectively. These results strongly support the bortezomib plus R-CHOP regimen to enter phase III trial [104].

The application of bortezomib in solid tumors has also been extensively investigated. Unfortunately, in contrast to its potent efficacy in hematological malignancies, the efficacy of bortezomib in solid tumors is disappointing. In patients with castration resistant metastatic prostate cancer, neither bortezomib alone nor bortezomib plus prednisone, exhibited significant antitumor effects [105]. Similarly, docetaxel plus bortezomib used as first-line treatment showed no improved efficacy versus docetaxel alone [106]. In heavily pre-treated metastatic breast cancer patients, bortezomib plus pegylated liposomal doxorubicin was well tolerated but had minimal activity [107]. A trial testing the bortezomib monotherapy in chemotherapy-naïve patients with advanced-stage non-small cell lung cancer was terminated in the first stage due to lack of response in all patients [108]. The bortezomib monotherapy was inactive in patients with unresectable or metastatic gastric and gastroesophageal junction adenocarcinoma, either [109]. A first-line regimen of bortezomib, paclitaxel, and carboplatin also failed in the treatment of patients with metastatic esophageal, gastric, and gastroesophageal cancer [110].

Therefore, continuous efforts are required both on the benchtop and at the bedside. A mechanistic explanation regarding the failure of bortezomib in solid tumors will help in designing a rational regimen and eventually lead to clinical success.

12.6 Advantages and Disadvantages of Bortezomib and the Development of Second Generation Proteasome Inhibitors

12.6.1 Advantages of Bortezomib

According to the preclinical and clinical studies comparing proteasome inhibitors with other anticancer drugs, proteasome inhibitors, represented by bortezomib, possess several advantages with respect to cancer treatment.

First, as a proteasome inhibitor, bortezomib exhibits favorable selectivity toward tumor cells over normal cells, which is an important criterion for being a good anticancer drug. Numerous studies have proved the selectivity of bortezomib in tumor cells versus normal cells. Multiple factors have been implicated in contributing to this tumor cell selectivity. (1) Cancer cells are rapidly dividing cells compared with non-cancerous cells. Therefore, they more heavily rely on proteasomal turnover of cell cycle regulatory proteins to promote cell cycle progression than normal cells. (2) Cancer cells appear to generate misfolded or damaged proteins much faster than non-cancerous cells due to their uncontrolled cell proliferation. (3) The amount of proteasomes and/or the proteasomal activity have been implicated to be upregulated in many types of cancers and this upregulation is important for the maintenance of the malignant phenotype [111, 112].

Secondly, the outstanding efficacy of bortezomib in the treatment of MM and MCL has been observed in clinical trials and applications. More importantly, clinical trials suggest that bortezomib holds activity in about one third of patients who were heavily pre-treated. This could be attributed to the fact that bortezomib is targeting a unique, previously unaffected target, the proteasome; therefore, the mechanism of action of bortezomib differs from any existing chemotherapeutic agents [113, 114].

Thirdly, bortezomib is capable of enhancing the sensitivity of cancer cells to conventional chemotherapeutic agents or radiation. In several previously discussed clinical trials, bortezomib was introduced into the standard therapy and largely improved the outcome in terms of ORR, time to progression, etc. The chemo- and radio-sensitizing effect of bortezomib could also be attributed to its unique target and mechanism of action. It is conceivable that two drugs with different targets or mechanisms of action may have better efficacy and less toxicity than two drugs with the same target or similar mechanism of action [115, 116].

12.6.2 Disadvantages of Bortezomib

On the other hand, some disadvantages or limitations of bortezomib have been observed during its preclinical and clinical development.

First, although it is generally well tolerated, bortezomib still generates some toxicity and in some cases, even causes discontinuation of the regimen. The most frequently occurring side effects are nausea, diarrhea, and fatigue. More serious adverse drug reactions include peripheral neuropathy, thrombocytopenia, neutropenia, and lymphopenia. It was estimated that more than 40 % of patients were afflicted with peripheral neuropathy [117]. The newly approved subcutaneous administration was found to significantly alleviate peripheral neuropathy compared with intravenous injection [84].

Another shortcoming of bortezomib is its narrow therapeutic window. According to a phase I trial, the therapeutic dose of bortezomib is 1.3 mg/m^2, while dose-limiting toxic effects appear at 1.5 mg/m^2. However, the latest results from the phase II CREST trial, which compared two doses of bortezomib (1.3 vs. 1.0 mg/m^2), indicate that if bortezomib dose reduction is required, the 1.0 mg/m^2 dose still offers patients a substantial survival benefit [78].

In addition, despite the appreciable therapeutic outcome of bortezomib, like almost all anticancer drugs, drug resistance occurs after a certain period. Even bortezomib resistance was observed in patients who were newly diagnosed and received bortezomib treatment as monotherapy for the first time. These clinical observations indicate that bortezomib resistance could be either acquired or inherent. The results from cell-based studies suggest that bortezomib resistance could occur either at the level of the proteasome itself or its downstream. Increased mRNA and protein expression of proteasomal β5 subunit was observed in bortezomib-resistant leukemic cell lines as well as patient samples [118–120]. Mutations in β5 subunits which impair bortezomib binding were also reported [120, 121]. In addition, constitutively active NF-κB pathway, located downstream of the proteasome pathway, was found in some bortezomib-resistant cell lines [122].

Moreover, during the clinical application of bortezomib, it was noticed that some natural products, including green tea polyphenols, are able to reduce the efficacy of bortezomib. This was due to the direct interaction between the boronic acid structure of bortezomib and the catechol structure of green tea polyphenols, resulting in the formation of a borate ester and inactivation of bortezomib [21].

Further clinical studies are needed for evidence-based recommendations of green tea consumption in patients receiving bortezomib treatment. Fortunately, this issue only affects boronic acid-based proteasome inhibitors.

Finally, as discussed earlier, the efficacy of bortezomib, either alone or in a multi-drug regimen, is unsatisfactory in the treatment of solid tumors. Addressing this issue will largely extend the application of proteasome inhibitors in cancer treatment.

12.6.3 Second Generation Proteasome Inhibitors

At least five second generation proteasome inhibitors are currently under clinical investigation (Table 12.1).

Carfilzomib is an irreversible proteasome inhibitor, with new protein synthesis being required for recovery of proteasome activity, which gives it greater potency. Whether it has decreased side effects compared with bortezomib needs to be determined. In addition to targeting the β5 subunit in the constitutive proteasome, carfilmozib also targets the correlated β5i subunit in the immunoproteasome, which appears to be preferentially expressed in MM. Moreover, carfilzomib is shown to be more specific than borzetomib, with little or no off-target activity outside of the proteasome [123]. However, like bortezomib, the administration of carfilzomib is intravenous twice weekly, which is inconvenient for patients.

Carfilzomib has been evaluated in two phase II trials, PX-171-003-A1 [124] and PX-171-004 [125, 126], as monotherapy for the treatment of relapsed and refractory MM patients. Both trials observed durable response to carfilzomib with well-tolerated and manageable side effects, regardless of prior exposure to bortezomib. Analysis of 136 patients in the above two trials indicated that peripheral neuropathy occurred in 15 % of patients, among which 9 % was attributed to carfilzomib. None of the patients required discontinuation or dose adjustments due to neurotoxicity, allowing long-term treatment and prolonged disease control by carfilzomib [127]. These data also indicate that peripheral neuropathy is not a class effect of proteasome inhibitors [127]. A recent study comparing bortezomib and carfilzomib further pointed out that the neurotoxicity of bortezomib was related to the off-target inhibition of HtrA2/Omi by bortezomib, a factor known to be involved in neuronal survival [128].

Carfilzomib has also been evaluated in combination with lenalidomide, and low-dose dexamethasone (CRd) in both relapsed and/or refractory MM patients [129] and newly diagnosed MM patients [130] in two separate phase II trials. CRd regimen is highly active and well tolerated in both trials. In newly diagnosed patients, the responses are rapid and improve over time reaching 100 % very good PR or better, which compare favorably with the best frontline regimens in MM therapy [130]. In relapsed and/or refractory patients, the ORR was 78 % [129]. Based on these encouraging results, the phase III ASPIRE trial of this regimen is underway for previously bortezomib-treated MM patients.

Marizomib is another irreversible proteasome inhibitor administrated intravenously twice weekly. Compared with other proteasome inhibitors, marizomib produces rapid, broad, and prolonged inhibition of all three 20S proteasome catalytic activities. Data from a phase I trial suggest that responses to marizomib were found in patients with bortezomib-refractory MM. The safety profile of marizomib clearly differs from bortezomib, with no significant treatment-emergent peripheral neuropathy, myelosuppression or thrombocytopenia reported and dose-limiting toxicities being transient hallucinations, cognitive changes, and loss of balance. In addition, marizomib exhibited interesting pharmacokinetic and pharmacodynamic properties and tissue distribution, supporting a possible role for marizomib in patients with different disease characteristics such as extramedullary spread [131].

MLN-9708, a boronate-based reversible proteasome inhibitor, has the advantage of being the first oral proteasome inhibitor to enter clinical investigation in MM patients. Oral administration is not only convenient but also seems to produce milder side effects. Data from a phase I trial suggest that single-agent MLN-9708 may result in clinical activity in heavily pre-treated relapsed and/or refractory MM patients including durable disease control and is generally well tolerated with infrequent peripheral neuropathy [132].

Delanzomib (CEP-18770) is another boronate-based reversible proteasome inhibitor with oral bioavailability. However, current phase I/II clinical investigations are still using intravenous administration. Delanzomib has shown proteasome-inhibitory activity similar to that of bortezomib in hematological and solid tumor cell lines, as well as in primary cells from MM patients [133].

Oprozomib (ONX-0912) is an epoxyketone-based irreversible proteasome inhibitor with oral bioavailability. It showed similar potency to carfilzomib in cytotoxicity assays. More excitingly, orally administered oprozomib had equivalent antitumor activity to intravenously administered carfilzomib in human tumor xenograft and mouse syngeneic models [134]. It is currently being investigated in early clinical trials via oral administration twice daily.

12.7 Conclusion

The proteasome has been shown to be a validated, novel and valuable target for cancer treatment. Suppression of proteasome function by synthetic or natural compounds is proven to be an effective way to induce cancer cell death with minimal effects on normal healthy cells. The first FDA-approved proteasome inhibitor bortezomib has achieved great success in the treatment of MM and MCL. However, some limitations and side effects have been noted during clinical application, such as the occurrence of bortezomib resistance, dose-limiting toxicities, and interaction with some natural compounds, all of which compromise the clinical benefits of bortezomib. Furthermore, the efficaciousness of bortezomib in solid tumors is disappointing. Therefore, the development of novel, new-generation proteasome inhibitors is necessary to improve the efficacy, reduce toxicity, and expand the application of proteasome inhibition-based strategies in cancer therapy. Meanwhile, continuous exploration of the effective combination of a proteasome inhibitor with other chemotherapeutic agents is essential to optimize the clinical outcome. A better understanding of the mechanism of proteasome inhibition-induced cell death will greatly promote the development and clinical application of novel anticancer drugs which will help illuminate the bright future of cancer treatment.

Acknowledgments The authors thank Sara Schmitt and Daniela Buac for critical reading of the manuscript. This work was partially supported by the National Cancer Institute (1R01CA120009, 3R01CA120009-04S1, and 5R01CA127258-05, to QPD).

References

1. Hershko A (2005) The ubiquitin system for protein degradation and some of its roles in the control of the cell division cycle. Cell Death Differ 12(9):1191–1197
2. Spataro V, Norbury C, Harris AL (1998) The ubiquitin-proteasome pathway in cancer. Br J Cancer 77(3):448–455
3. de Bettignies G, Coux O (2010) Proteasome inhibitors: Dozens of molecules and still counting. Biochimie 92(11):1530–1545
4. Adams J (2004) The development of proteasome inhibitors as anticancer drugs. Cancer Cell 5(5):417–421
5. Sorokin AV, Kim ER, Ovchinnikov LP (2009) Proteasome system of protein degradation and processing. Biochemistry (Mosc) 74(13):1411–1442
6. Adams J (2004) The proteasome: a suitable antineoplastic target. Nat Rev Cancer 4(5):349–360
7. Orlowski M, Cardozo C, Michaud C (1993) Evidence for the presence of five distinct proteolytic components in the pituitary multicatalytic proteinase complex. Properties of two components cleaving bonds on the carboxyl side of branched chain and small neutral amino acids. Biochemistry 32(6):1563–1572
8. Nandi D, Tahiliani P, Kumar A, Chandu D (2006) The ubiquitin-proteasome system. J Biosci 31(1):137–155
9. Strehl B, Seifert U, Kruger E, Heink S, Kuckelkorn U, Kloetzel PM (2005) Interferon-gamma, the functional plasticity of the ubiquitin-proteasome system, and MHC class I antigen processing. Immunol Rev 207:19–30
10. Angeles A, Fung G, Luo H (2012) Immune and non-immune functions of the immunoproteasome. Front Biosci 17:1904–1916
11. Kisselev AF, van der Linden WA, Overkleeft HS (2012) Proteasome inhibitors: an expanding army attacking a unique target. Chem Biol 19(1):99–115
12. Borissenko L, Groll M (2007) 20S proteasome and its inhibitors: crystallographic knowledge for drug development. Chem Rev 107(3):687–717
13. Tsukamoto S, Yokosawa H (2010) Inhibition of the ubiquitin-proteasome system by natural products for cancer therapy. Planta Med 76(11):1064–1074
14. Krunic A, Vallat A, Mo S, Lantvit DD, Swanson SM, Orjala J (2010) Scytonemides A and B, cyclic peptides with 20S proteasome inhibitory activity from the cultured cyanobacterium Scytonema hofmanii. J Nat Prod 73(11):1927–1932
15. Koguchi Y, Kohno J, Nishio M, Takahashi K, Okuda T, Ohnuki T et al (2000) TMC-95A, B, C, and D, novel proteasome inhibitors produced by Apiospora montagnei Sacc. TC 1093. Taxonomy, production, isolation, and biological activities. J Antibiot 53(2):105–109

16. Nickeleit I, Zender S, Sasse F, Geffers R, Brandes G, Sorensen I et al (2008) Argyrin a reveals a critical role for the tumor suppressor protein p27(kip1) in mediating antitumor activities in response to proteasome inhibition. Cancer Cell 14(1):23–35
17. Milacic V, Banerjee S, Landis-Piwowar KR, Sarkar FH, Majumdar AP, Dou QP (2008) Curcumin inhibits the proteasome activity in human colon cancer cells in vitro and in vivo. Cancer Res 68(18):7283–7292
18. Yang H, Zhou P, Huang H, Chen D, Ma N, Cui QC et al (2009) Shikonin exerts antitumor activity via proteasome inhibition and cell death induction in vitro and in vivo. Int J Cancer 124(10):2450–2459
19. Ross JA, Kasum CM (2002) Dietary flavonoids: bioavailability, metabolic effects, and safety. Annu Rev Nutr 22:19–34
20. Mozzicafreddo M, Cuccioloni M, Cecarini V, Eleuteri AM, Angeletti M (2009) Homology modeling and docking analysis of the interaction between polyphenols and mammalian 20S proteasomes. J Chem Inf Model 49(2):401–409
21. Golden EB, Lam PY, Kardosh A, Gaffney KJ, Cadenas E, Louie SG et al (2009) Green tea polyphenols block the anticancer effects of bortezomib and other boronic acid-based proteasome inhibitors. Blood 113(23):5927–5937
22. Yang H, Chen D, Cui QC, Yuan X, Dou QP (2006) Celastrol, a triterpene extracted from the Chinese "Thunder of God Vine," is a potent proteasome inhibitor and suppresses human prostate cancer growth in nude mice. Cancer Res 66(9):4758–4765
23. Yang H, Shi G, Dou QP (2007) The tumor proteasome is a primary target for the natural anticancer compound Withaferin A isolated from "Indian winter cherry". Mol Pharmacol 71(2):426–437
24. Yang H, Landis-Piwowar KR, Lu D, Yuan P, Li L, Reddy GP et al (2008) Pristimerin induces apoptosis by targeting the proteasome in prostate cancer cells. J Cell Biochem 103(1):234–244
25. Chen D, Cui QC, Yang H, Dou QP (2006) Disulfiram, a clinically used anti-alcoholism drug and copper-binding agent, induces apoptotic cell death in breast cancer cultures and xenografts via inhibition of the proteasome activity. Cancer Res 66(21):10425–10433
26. Daniel KG, Chen D, Orlu S, Cui QC, Miller FR, Dou QP (2005) Clioquinol and pyrrolidine dithiocarbamate complex with copper to form proteasome inhibitors and apoptosis inducers in human breast cancer cells. Breast Cancer Res 7(6):R897–R908
27. Li X, Wood TE, Sprangers R, Jansen G, Franke NE, Mao X et al (2010) Effect of noncompetitive proteasome inhibition on bortezomib resistance. J Natl Cancer Inst 102(14):1069–1082
28. Kuhn DJ, Hunsucker SA, Chen Q, Voorhees PM, Orlowski M, Orlowski RZ (2009) Targeted inhibition of the immunoproteasome is a potent strategy against models of multiple myeloma that overcomes resistance to conventional drugs and nonspecific proteasome inhibitors. Blood 113(19):4667–4676
29. Muchamuel T, Basler M, Aujay MA, Suzuki E, Kalim KW, Lauer C et al (2009) A selective inhibitor of the immunoproteasome subunit LMP7 blocks cytokine production and attenuates progression of experimental arthritis. Nat Med 15(7):781–787
30. Pagano M, Tam SW, Theodoras AM, Beer-Romero P, Del Sal G, Chau V et al (1995) Role of the ubiquitin-proteasome pathway in regulating abundance of the cyclin-dependent kinase inhibitor p27. Science 269(5224):682–685
31. Tambyrajah WS, Bowler LD, Medina-Palazon C, Sinclair AJ (2007) Cell cycle-dependent caspase-like activity that cleaves p27(KIP1) is the beta(1) subunit of the 20S proteasome. Arch Biochem Biophys 466(2):186–193
32. Zhu Q, Wani G, Yao J, Patnaik S, Wang QE, El-Mahdy MA et al (2007) The ubiquitin-proteasome system regulates p53-mediated transcription at p21waf1 promoter. Oncogene 26(29):4199–4208
33. Chen F, Chang D, Goh M, Klibanov SA, Ljungman M (2000) Role of p53 in cell cycle regulation and apoptosis following exposure to proteasome inhibitors. Cell Growth Differ 11(5):239–246

34. Concannon CG, Koehler BF, Reimertz C, Murphy BM, Bonner C, Thurow N et al (2007) Apoptosis induced by proteasome inhibition in cancer cells: predominant role of the p53/PUMA pathway. Oncogene 26(12):1681–1692
35. Strauss SJ, Higginbottom K, Juliger S, Maharaj L, Allen P, Schenkein D et al (2007) The proteasome inhibitor bortezomib acts independently of p53 and induces cell death via apoptosis and mitotic catastrophe in B-cell lymphoma cell lines. Cancer Res 67(6):2783–2790
36. Pandit B, Gartel AL (2011) Proteasome inhibitors induce p53-independent apoptosis in human cancer cells. Am J Pathol 178(1):355–360
37. Wagenknecht B, Hermisson M, Eitel K, Weller M (1999) Proteasome inhibitors induce p53/p21-independent apoptosis in human glioma cells. Cell Physiol Biochem 9(3):117–125
38. Traenckner EB, Wilk S, Baeuerle PA (1994) A proteasome inhibitor prevents activation of NF-kappa B and stabilizes a newly phosphorylated form of I kappa B-alpha that is still bound to NF-kappa B. EMBO J 13(22):5433–5441
39. Palombella VJ, Rando OJ, Goldberg AL, Maniatis T (1994) The ubiquitin-proteasome pathway is required for processing the NF-kappa B1 precursor protein and the activation of NF-kappa B. Cell 78(5):773–785
40. Sears C, Olesen J, Rubin D, Finley D, Maniatis T (1998) NF-kappa B p105 processing via the ubiquitin-proteasome pathway. J Biol Chem 273(3):1409–1419
41. Heusch M, Lin L, Geleziunas R, Greene WC (1999) The generation of nfkb2 p52: mechanism and efficiency. Oncogene 18(46):6201–6208
42. Delic J, Masdehors P, Omura S, Cosset JM, Dumont J, Binet JL et al (1998) The proteasome inhibitor lactacystin induces apoptosis and sensitizes chemo- and radioresistant human chronic lymphocytic leukaemia lymphocytes to TNF-alpha-initiated apoptosis. Br J Cancer 77(7):1103–1107
43. Voortman J, Resende TP (2007) Abou El Hassan MA, Giaccone G, Kruyt FA. TRAIL therapy in non-small cell lung cancer cells: sensitization to death receptor-mediated apoptosis by proteasome inhibitor bortezomib. Mol Cancer Ther 6(7):2103–2112
44. Nencioni A, Wille L, Dal Bello G, Boy D, Cirmena G, Wesselborg S et al (2005) Cooperative cytotoxicity of proteasome inhibitors and tumor necrosis factor-related apoptosis-inducing ligand in chemoresistant Bcl-2-overexpressing cells. Clin Cancer Res 11(11):4259–4265
45. Oyaizu H, Adachi Y, Okumura T, Okigaki M, Oyaizu N, Taketani S et al (2001) Proteasome inhibitor 1 enhances paclitaxel-induced apoptosis in human lung adenocarcinoma cell line. Oncol Rep 8(4):825–829
46. von Metzler I, Heider U, Mieth M, Lamottke B, Kaiser M, Jakob C et al (2009) Synergistic interaction of proteasome and topoisomerase II inhibition in multiple myeloma. Exp Cell Res 315(14):2471–2478
47. Mitsiades N, Mitsiades CS, Richardson PG, Poulaki V, Tai YT, Chauhan D et al (2003) The proteasome inhibitor PS-341 potentiates sensitivity of multiple myeloma cells to conventional chemotherapeutic agents: therapeutic applications. Blood 101(6):2377–2380
48. Hideshima T, Ikeda H, Chauhan D, Okawa Y, Raje N, Podar K et al (2009) Bortezomib induces canonical nuclear factor-kappaB activation in multiple myeloma cells. Blood 114(5):1046–1052
49. Amschler K, Schon MP, Pletz N, Wallbrecht K, Erpenbeck L, Schon M (2010) NF-kappaB inhibition through proteasome inhibition or IKKbeta blockade increases the susceptibility of melanoma cells to cytostatic treatment through distinct pathways. J Invest Dermatol. 130(4):1073–1086
50. Li B, Dou QP (2000) Bax degradation by the ubiquitin/proteasome-dependent pathway: involvement in tumor survival and progression. Proc Natl Acad Sci U S A 97(8):3850–3855
51. Ding WX, Ni HM, Chen X, Yu J, Zhang L, Yin XM (2007) A coordinated action of Bax, PUMA, and p53 promotes MG132-induced mitochondria activation and apoptosis in colon cancer cells. Mol Cancer Ther 6(3):1062–1069

52. Zhu H, Zhang L, Dong F, Guo W, Wu S, Teraishi F et al (2005) Bik/NBK accumulation correlates with apoptosis-induction by bortezomib (PS-341, Velcade) and other proteasome inhibitors. Oncogene 24(31):4993–4999
53. Nikrad M, Johnson T, Puthalalath H, Coultas L, Adams J, Kraft AS (2005) The proteasome inhibitor bortezomib sensitizes cells to killing by death receptor ligand TRAIL via BH3-only proteins Bik and Bim. Mol Cancer Ther 4(3):443–449
54. Li C, Li R, Grandis JR, Johnson DE (2008) Bortezomib induces apoptosis via Bim and Bik up-regulation and synergizes with cisplatin in the killing of head and neck squamous cell carcinoma cells. Mol Cancer Ther 7(6):1647–1655
55. Pigneux A, Mahon FX, Moreau-Gaudry F, Uhalde M, de Verneuil H, Lacombe F et al (2007) Proteasome inhibition specifically sensitizes leukemic cells to anthracyclin-induced apoptosis through the accumulation of Bim and Bax pro-apoptotic proteins. Cancer Biol Ther 6(4):603–611
56. Dewson G, Snowden RT, Almond JB, Dyer MJ, Cohen GM (2003) Conformational change and mitochondrial translocation of Bax accompany proteasome inhibitor-induced apoptosis of chronic lymphocytic leukemic cells. Oncogene 22(17):2643–2654
57. Lang-Rollin I, Maniati M, Jabado O, Vekrellis K, Papantonis S, Rideout HJ et al (2005) Apoptosis and the conformational change of Bax induced by proteasomal inhibition of PC12 cells are inhibited by bcl-xL and bcl-2. Apoptosis 10(4):809–820
58. Perez-Galan P, Roue G, Villamor N, Montserrat E, Campo E, Colomer D (2006) The proteasome inhibitor bortezomib induces apoptosis in mantle-cell lymphoma through generation of ROS and Noxa activation independent of p53 status. Blood 107(1):257–264
59. Nikiforov MA, Riblett M, Tang WH, Gratchouck V, Zhuang D, Fernandez Y et al (2007) Tumor cell-selective regulation of NOXA by c-MYC in response to proteasome inhibition. Proc Natl Acad Sci U S A 104(49):19488–19493
60. Chen S, Blank JL, Peters T, Liu XJ, Rappoli DM, Pickard MD et al (2010) Genome-wide siRNA screen for modulators of cell death induced by proteasome inhibitor bortezomib. Cancer Res 70(11):4318–4326
61. Hu J, Dang N, Menu E, De Bryune E, Xu D, Van Camp B et al (2012) Activation of ATF4 mediates unwanted Mcl-1 accumulation by proteasome inhibition. Blood 119(3):826–837
62. Ri M, Iida S, Ishida T, Ito A, Yano H, Inagaki A et al (2009) Bortezomib-induced apoptosis in mature T-cell lymphoma cells partially depends on upregulation of Noxa and functional repression of Mcl-1. Cancer Sci 100(2):341–348
63. Gomez-Bougie P, Wuilleme-Toumi S, Menoret E, Trichet V, Robillard N, Philippe M et al (2007) Noxa up-regulation and Mcl-1 cleavage are associated to apoptosis induction by bortezomib in multiple myeloma. Cancer Res 67(11):5418–5424
64. Podar K, Gouill SL, Zhang J, Opferman JT, Zorn E, Tai YT et al (2008) A pivotal role for Mcl-1 in Bortezomib-induced apoptosis. Oncogene 27(6):721–731
65. Yuan BZ, Chapman J, Reynolds SH (2009) Proteasome inhibitors induce apoptosis in human lung cancer cells through a positive feedback mechanism and the subsequent Mcl-1 protein cleavage. Oncogene 28(43):3775–3786
66. Liu Y, Ye Y (2011) Proteostasis regulation at the endoplasmic reticulum: a new perturbation site for targeted cancer therapy. Cell Res 21(6):867–883
67. Bush KT, Goldberg AL, Nigam SK (1997) Proteasome inhibition leads to a heat-shock response, induction of endoplasmic reticulum chaperones, and thermotolerance. J Biol Chem 272(14):9086–9092
68. Lee AH, Iwakoshi NN, Anderson KC, Glimcher LH (2003) Proteasome inhibitors disrupt the unfolded protein response in myeloma cells. Proc Natl Acad Sci U S A 100(17):9946–9951
69. Obeng EA, Carlson LM, Gutman DM, Harrington WJ Jr, Lee KP, Boise LH (2006) Proteasome inhibitors induce a terminal unfolded protein response in multiple myeloma cells. Blood 107(12):4907–4916

70. Fan WH, Hou Y, Meng FK, Wang XF, Luo YN, Ge PF (2011) Proteasome inhibitor MG-132 induces C6 glioma cell apoptosis via oxidative stress. Acta Pharmacol Sin 32(5):619–625
71. Ling YH, Liebes L, Zou Y, Perez-Soler R (2003) Reactive oxygen species generation and mitochondrial dysfunction in the apoptotic response to Bortezomib, a novel proteasome inhibitor, in human H460 non-small cell lung cancer cells. J Biol Chem 278(36):33714–33723
72. Du ZX, Zhang HY, Meng X, Guan Y, Wang HQ (2009) Role of oxidative stress and intracellular glutathione in the sensitivity to apoptosis induced by proteasome inhibitor in thyroid cancer cells. BMC Cancer 9:56
73. Fribley A, Zeng Q, Wang CY (2004) Proteasome inhibitor PS-341 induces apoptosis through induction of endoplasmic reticulum stress-reactive oxygen species in head and neck squamous cell carcinoma cells. Mol Cell Biol 24(22):9695–9704
74. Papa L, Gomes E, Rockwell P (2007) Reactive oxygen species induced by proteasome inhibition in neuronal cells mediate mitochondrial dysfunction and a caspase-independent cell death. Apoptosis 12(8):1389–1405
75. Lee MH, Hyun DH, Jenner P, Halliwell B (2001) Effect of proteasome inhibition on cellular oxidative damage, antioxidant defences and nitric oxide production. J Neurochem 78(1):32–41
76. Bieler S, Meiners S, Stangl V, Pohl T, Stangl K (2009) Comprehensive proteomic and transcriptomic analysis reveals early induction of a protective anti-oxidative stress response by low-dose proteasome inhibition. Proteomics 9(12):3257–3267
77. Jagannath S, Barlogie B, Berenson J, Siegel D, Irwin D, Richardson PG et al (2004) A phase 2 study of two doses of bortezomib in relapsed or refractory myeloma. Br J Haematol 127(2):165–172
78. Jagannath S, Barlogie B, Berenson JR, Siegel DS, Irwin D, Richardson PG et al (2008) Updated survival analyses after prolonged follow-up of the phase 2, multicenter CREST study of bortezomib in relapsed or refractory multiple myeloma. Br J Haematol 143(4):537–540
79. Richardson PG, Barlogie B, Berenson J, Singhal S, Jagannath S, Irwin D et al (2003) A phase 2 study of bortezomib in relapsed, refractory myeloma. N Engl J Med 348(26):2609–2617
80. Richardson PG, Sonneveld P, Schuster MW, Irwin D, Stadtmauer EA, Facon T et al (2005) Bortezomib or high-dose dexamethasone for relapsed multiple myeloma. N Engl J Med 352(24):2487–2498
81. Richardson PG, Sonneveld P, Schuster M, Irwin D, Stadtmauer E, Facon T et al (2007) Extended follow-up of a phase 3 trial in relapsed multiple myeloma: final time-to-event results of the APEX trial. Blood 110(10):3557–3560
82. San Miguel JF, Schlag R, Khuageva NK, Dimopoulos MA, Shpilberg O, Kropff M et al (2008) Bortezomib plus melphalan and prednisone for initial treatment of multiple myeloma. N Engl J Med 359(9):906–917
83. Mateos MV, Richardson PG, Schlag R, Khuageva NK, Dimopoulos MA, Shpilberg O et al (2010) Bortezomib plus melphalan and prednisone compared with melphalan and prednisone in previously untreated multiple myeloma: updated follow-up and impact of subsequent therapy in the phase III VISTA trial. J Clin Oncol 28(13):2259–2266
84. Moreau P, Pylypenko H, Grosicki S, Karamanesht I, Leleu X, Grishunina M et al (2011) Subcutaneous versus intravenous administration of bortezomib in patients with relapsed multiple myeloma: a randomised, phase 3, non-inferiority study. Lancet Oncol 12(5):431–440
85. Jagannath S, Durie BG, Wolf J, Camacho E, Irwin D, Lutzky J et al (2005) Bortezomib therapy alone and in combination with dexamethasone for previously untreated symptomatic multiple myeloma. Br J Haematol 129(6):776–783
86. Jagannath S, Durie BG, Wolf JL, Camacho ES, Irwin D, Lutzky J et al (2009) Extended follow-up of a phase 2 trial of bortezomib alone and in combination with dexamethasone for the frontline treatment of multiple myeloma. Br J Haematol 146(6):619–626

87. Mikhael JR, Belch AR, Prince HM, Lucio MN, Maiolino A, Corso A et al (2009) High response rate to bortezomib with or without dexamethasone in patients with relapsed or refractory multiple myeloma: results of a global phase 3b expanded access program. Br J Haematol 144(2):169–175
88. Oakervee HE, Popat R, Curry N, Smith P, Morris C, Drake M et al (2005) PAD combination therapy (PS-341/bortezomib, doxorubicin and dexamethasone) for previously untreated patients with multiple myeloma. Br J Haematol 129(6):755–762
89. Popat R, Oakervee HE, Hallam S, Curry N, Odeh L, Foot N et al (2008) Bortezomib, doxorubicin and dexamethasone (PAD) front-line treatment of multiple myeloma: updated results after long-term follow-up. Br J Haematol 141(4):512–516
90. Cavo M, Tacchetti P, Patriarca F, Petrucci MT, Pantani L, Galli M et al (2010) Bortezomib with thalidomide plus dexamethasone compared with thalidomide plus dexamethasone as induction therapy before, and consolidation therapy after, double autologous stem-cell transplantation in newly diagnosed multiple myeloma: a randomised phase 3 study. Lancet 376(9758):2075–2085
91. Driscoll JJ, Burris J, Annunziata CM (2012) Targeting the proteasome with bortezomib in multiple myeloma: update on therapeutic benefit as an upfront single agent, induction regimen for stem-cell transplantation and as maintenance therapy. Am J Ther 19(2):133–144
92. Raab MS, Podar K, Breitkreutz I, Richardson PG, Anderson KC (2009) Multiple myeloma. Lancet 374(9686):324–339
93. Mateos MV, Oriol A, Martinez-Lopez J, Gutierrez N, Teruel AI, de Paz R et al (2010) Bortezomib, melphalan, and prednisone versus bortezomib, thalidomide, and prednisone as induction therapy followed by maintenance treatment with bortezomib and thalidomide versus bortezomib and prednisone in elderly patients with untreated multiple myeloma: a randomised trial. Lancet Oncol 11(10):934–941
94. Palumbo A, Ambrosini MT, Benevolo G, Pregno P, Pescosta N, Callea V et al (2007) Bortezomib, melphalan, prednisone, and thalidomide for relapsed multiple myeloma. Blood 109(7):2767–2772
95. Palumbo A, Bringhen S, Rossi D, Cavalli M, Larocca A, Ria R et al (2010) Bortezomib-melphalan-prednisone-thalidomide followed by maintenance with bortezomib-thalidomide compared with bortezomib-melphalan-prednisone for initial treatment of multiple myeloma: a randomized controlled trial. J Clin Oncol 28(34):5101–5109
96. Fisher RI, Bernstein SH, Kahl BS, Djulbegovic B, Robertson MJ, de Vos S et al (2006) Multicenter phase II study of bortezomib in patients with relapsed or refractory mantle cell lymphoma. J Clin Oncol 24(30):4867–4874
97. Goy A, Bernstein SH, Kahl BS, Djulbegovic B, Robertson MJ, de Vos S et al (2009) Bortezomib in patients with relapsed or refractory mantle cell lymphoma: updated time-to-event analyses of the multicenter phase 2 PINNACLE study. Ann Oncol 20(3):520–525
98. Kouroukis CT, Fernandez LA, Crump M, Gascoyne RD, Chua NS, Buckstein R et al (2011) A phase II study of bortezomib and gemcitabine in relapsed mantle cell lymphoma from the National Cancer Institute of Canada Clinical Trials Group (IND 172). Leukemia lymphoma 52(3):394–399
99. Yang H, Zonder JA, Dou QP (2009) Clinical development of novel proteasome inhibitors for cancer treatment. Expert Opin Investig Drugs 18(7):957–971
100. Kuppers R (2005) Mechanisms of B-cell lymphoma pathogenesis. Nat Rev Cancer 5(4):251–262
101. de Vos S, Goy A, Dakhil SR, Saleh MN, McLaughlin P, Belt R et al (2009) Multicenter randomized phase II study of weekly or twice-weekly bortezomib plus rituximab in patients with relapsed or refractory follicular or marginal-zone B-cell lymphoma. J Clin Oncol 27(30):5023–5030
102. Coiffier B, Osmanov EA, Hong X, Scheliga A, Mayer J, Offner F et al (2011) Bortezomib plus rituximab versus rituximab alone in patients with relapsed, rituximab-naive or rituximab-sensitive, follicular lymphoma: a randomised phase 3 trial. Lancet Oncol. 12(8):773–784

103. Fowler N, Kahl BS, Lee P, Matous JV, Cashen AF, Jacobs SA et al (2011) Bortezomib, bendamustine, and rituximab in patients with relapsed or refractory follicular lymphoma: the phase II VERTICAL study. J Clin Oncol 29(25):3389–3395
104. Ruan J, Martin P, Furman RR, Lee SM, Cheung K, Vose JM et al (2011) Bortezomib plus CHOP-rituximab for previously untreated diffuse large B-cell lymphoma and mantle cell lymphoma. J Clin Oncol 29(6):690–697
105. Morris MJ, Kelly WK, Slovin S, Ryan C, Eicher C, Heller G et al (2007) A phase II trial of bortezomib and prednisone for castration resistant metastatic prostate cancer. J Urology 178(6):2378–2383, discussion 83-4
106. Hainsworth JD, Meluch AA, Spigel DR, Barton J Jr, Simons L, Meng C et al (2007) Weekly docetaxel and bortezomib as first-line treatment for patients with hormone-refractory prostate cancer: a Minnie Pearl Cancer Research Network phase II trial. Clin Genitourinary Cancer 5(4):278–283
107. Irvin WJ Jr, Orlowski RZ, Chiu WK, Carey LA, Collichio FA, Bernard PS et al (2010) Phase II study of bortezomib and pegylated liposomal doxorubicin in the treatment of metastatic breast cancer. Clin Breast Cancer 10(6):465–470
108. Li T, Ho L, Piperdi B, Elrafei T, Camacho FJ, Rigas JR et al (2010) Phase II study of the proteasome inhibitor bortezomib (PS-341, Velcade) in chemotherapy-naive patients with advanced stage non-small cell lung cancer (NSCLC). Lung Cancer 68(1):89–93
109. Shah MA, Power DG, Kindler HL, Holen KD, Kemeny MM, Ilson DH et al (2011) A multicenter, phase II study of bortezomib (PS-341) in patients with unresectable or metastatic gastric and gastroesophageal junction adenocarcinoma. Invest New Drugs 29(6):1475–1481
110. Jatoi A, Dakhil SR, Foster NR, Ma C, Rowland KM Jr, Moore DF Jr et al (2008) Bortezomib, paclitaxel, and carboplatin as a first-line regimen for patients with metastatic esophageal, gastric, and gastroesophageal cancer: phase II results from the North Central Cancer Treatment Group (N044B). J Thoracic Oncol 3(5):516–520
111. Pleban E, Bury M, Mlynarczuk I, Wojcik C (2001) Effects of proteasome inhibitor PSI on neoplastic and non-transformed cell lines. Folia Histochem Cytobiol 39(2):133–134
112. Adams J (2003) Potential for proteasome inhibition in the treatment of cancer. Drug Discov Today 8(7):307–315
113. Chen D, Frezza M, Schmitt S, Kanwar J, Dou QP (2011) Bortezomib as the first proteasome inhibitor anticancer drug: current status and future perspectives. Curr Cancer Drug Targets 11(3):239–253
114. Shah JJ, Orlowski RZ (2009) Proteasome inhibitors in the treatment of multiple myeloma. Leukemia 23(11):1964–1979
115. Orlowski RZ (2004) Bortezomib in combination with other therapies for the treatment of multiple myeloma. J Natl Compr Canc Netw (Suppl 4):S16–S20
116. Reddy N, Czuczman MS (2010) Enhancing activity and overcoming chemoresistance in hematologic malignancies with bortezomib: preclinical mechanistic studies. Ann Oncol 21(9):1756–1764
117. Appel A (2011) Drugs: More shots on target. Nature 480(7377):S40–S42
118. Lu S, Chen Z, Yang J, Chen L, Gong S, Zhou H et al (2008) Overexpression of the PSMB5 gene contributes to bortezomib resistance in T-lymphoblastic lymphoma/leukemia cells derived from Jurkat line. Exp Hematol 36(10):1278–1284
119. Shuqing L, Jianmin Y, Chongmei H, Hui C, Wang J (2011) Upregulated expression of the PSMB5 gene may contribute to drug resistance in patient with multiple myeloma when treated with bortezomib-based regimen. Exp Hematol 39(12):1117–1118
120. Oerlemans R, Franke NE, Assaraf YG, Cloos J, van Zantwijk I, Berkers CR et al (2008) Molecular basis of bortezomib resistance: proteasome subunit beta5 (PSMB5) gene mutation and overexpression of PSMB5 protein. Blood 112(6):2489–2499
121. Franke NE, Niewerth D, Assaraf YG, van Meerloo J, Vojtekova K, van Zantwijk CH et al (2011) Impaired bortezomib binding to mutant beta5 subunit of the proteasome is the underlying basis for bortezomib resistance in leukemia cells. Leukemia 26(4):757–768

122. Markovina S, Callander NS, O'Connor SL, Kim J, Werndli JE, Raschko M et al (2008) Bortezomib-resistant nuclear factor-kappaB activity in multiple myeloma cells. Mol Cancer Res 6(8):1356–1364
123. Ruschak AM, Slassi M, Kay LE, Schimmer AD (2011) Novel proteasome inhibitors to overcome bortezomib resistance. J Natl Cancer Inst 103(13):1007–1017
124. Samuel D, Martin T, Wang M, Vij R, Jakubowiak AJ, Jagannath S et al (2010) Results of PX-171-003-A1, An Open-Label, Single-Arm, Phase 2 (Ph 2) Study of Carfilzomib (CFZ) In Patients (pts) with Relapsed and Refractory Multiple Myeloma (MM). Blood 116(21):433
125. Vii R, Kaufman JL, Jakubowiak AJ, Wang M, Jagannath S, Kukreti V et al (2011) Final results from the bortezomib-naive group of PX-171-004, a phase 2 study of single-agent carfilzomib in patients with relapsed and/or refractory MM. Blood 118(21):369–370
126. Stewart K, Siegel D, Wang M, Kaufman J, Jakubowiak A, Jagannath S et al (2010) Results of Px-171-004, an ongoing open-label, phase ii study of carfilzomib in patients with relapsed and/or refractory multiple myeloma (R/R Mm) with or without prior bortezomib exposure. Haematologica 95:452
127. Vij R, Wang LH, Orlowski RZ, Stewart AK, Jagannath S, Lonial S et al (2009) Carfilzomib (CFZ), a novel proteasome inhibitor for relapsed or refractory multiple myeloma, is associated with minimal peripheral neuropathic effects. Blood 114(22):178–179
128. Arastu-Kapur S, Anderl JL, Kraus M, Parlati F, Shenk KD, Lee SJ et al (2011) Nonproteasomal targets of the proteasome inhibitors bortezomib and carfilzomib: a link to clinical adverse events. Clin Cancer Res 17(9):2734–2743
129. Wang M, Bensinger W, Martin T, Alsina M (2011) Interim results from PX-171-006, a phase (Ph) II multicenter dose-expansion study of carfilzomib (CFZ), lenalidomide (LEN), and low-dose dexamethasone (loDex) in relapsed and/or refractory multiple myeloma (R/R MM). J Clin Oncol 2011(suppl):abstr 8025
130. Jakubowiak AJ, Dytfeld D, Jagannath S, Vesole DH, Anderson TB, Nordgren BK et al (2011) Final results of a frontline phase 1/2 Study of carfilzomib, lenalidomide, and low-dose dexamethasone (CRd) in multiple myeloma (MM). Blood 118(21):288–289
131. Richardson PG, Spencer A, Cannel P, Harrison SJ, Catley L, Underhill C et al (2011) Phase 1 clinical evaluation of twice-weekly marizomib (NPI-0052), a novel proteasome inhibitor, in patients with relapsed/refractory multiple myeloma (MM). Blood 118(21):140–141
132. Richardson PG, Baz R, Wang LH, Jakubowiak AJ, Berg D, Liu GH et al (2011) Investigational agent MLN9708, an oral proteasome inhibitor, in patients (Pts) with relapsed and/or refractory multiple myeloma (MM): Results from the expansion cohorts of a phase 1 dose-escalation study. Blood 118(21):140
133. Molineaux SM (2012) Molecular pathways: targeting proteasomal protein degradation in cancer. Clin Cancer Res 18(1):15–20
134. Chauhan D, Singh AV, Aujay M, Kirk CJ, Bandi M, Ciccarelli B et al (2010) A novel orally active proteasome inhibitor ONX 0912 triggers in vitro and in vivo cytotoxicity in multiple myeloma. Blood 116(23):4906–4915
135. Bedford L, Lowe J, Dick LR, Mayer RJ, Brownell JE (2011) Ubiquitin-like protein conjugation and the ubiquitin-proteasome system as drug targets. Nat Rev Drug Discov 10(1):29–46

Chapter 13
New Agents and Approaches for Targeting the RAS/RAF/MEK/ERK and PI3K/AKT/mTOR Cell Survival Pathways

James A. McCubrey, Linda S. Steelman, William H. Chappell, Stephen L. Abrams, Richard A. Franklin, Giuseppe Montalto, Melchiorre Cervello, Ferdinando Nicoletti, Graziella Malaponte, Clorinda Massarino, Massimo Libra, Jörg Bäsecke, Agostino Tafuri, Michele Milella, Francesca Chiarini, Camilla Evangelisti, Lucio Cocco and Alberto M. Martelli

Abstract The Ras/Raf/MEK/ERK and PI3K/Akt/mTOR cascades are often activated by genetic alterations, either by mutations in upstream signaling molecules or by mutations in intrinsic pathway components. Upstream mutations in one signaling pathway or even in downstream components of the same pathway can alter the sensitivity of the

---

J. A. McCubrey (✉) · L. S. Steelman · W. H. Chappell · S. L. Abrams · R. A. Franklin
Department of Microbiology and Immunology, Brody School of Medicine at East Carolina University, Greenville, NC, USA
e-mail: mccubreyj@ecu.edu

G. Montalto
Department of Internal Medicine and Specialties University of Palermo, Palermo, Italy

M. Cervello
Consiglio Nazionale delle Ricerche, Istituto di Biomedicina e Immunologia Molecolare "Alberto Monroy", Palermo, Italy

F. Nicoletti · G. Malaponte · C. Massarino · M. Libra
Department of Biomedical Sciences, University of Catania, Catania, Italy

J. Bäsecke
Department of Medicine, University of Göttingen, Göttingen, Germany

A. Tafuri
Sapienza, University of Rome, Department of Cellular Biotechnology and Hematology, Rome, Italy

M. Milella
Regina Elena National Cancer Institute, Rome, Italy

F. Chiarini · C. Evangelisti · A. M. Martelli
Institute of Molecular Genetics, National Research Council-Rizzoli Orthopedic Institute, Bologna, Italy

L. Cocco · A. M. Martelli
Department of Biomedical and Neuromotor Sciences, Università di Bologna, Bologna, Italy

D. E. Johnson (ed.), *Cell Death Signaling in Cancer Biology and Treatment*,
Cell Death in Biology and Diseases, DOI: 10.1007/978-1-4614-5847-0_13,

X.-M. Yin and Z. Dong (Series eds.), *Cell Death in Biology and Diseases*

cells to certain small molecule inhibitors. These pathways have profound effects on proliferative, apoptotic, and differentiation pathways. Dysregulation of components of these pathways can contribute to: malignant transformation, resistance to other pathway inhibitors, and chemotherapeutic drug resistance. This chapter will first briefly describe these pathways and then evaluate potential uses of Raf, MEK, PI3K, Akt, and mTOR inhibitors that have been investigated in preclinical and clinical investigations.

13.1 Introduction

Since the discovery of the *RAS*, *RAF*, *MEK1*, *PIK3CA*, and *AKT* oncogenes and neurofibromin 1 (*NF*1), *PTEN*, *TSC1*, and *TSC2* tumor suppressor genes, the Ras/Raf/MEK/ERK, and Ras/PI3K/PTEN/Akt/mTOR signaling cascades have been extensively investigated with the ultimate goal of determining how these genes become activated/inactivated and whether it is possible to suppress their activity in human cancer and other diseases [1]. Furthermore, these pathways are also implicated in the resistance and sometimes sensitivity to therapy [2]. There have been breakthroughs in the discovery of complex interacting pathway components, and their genetic and epigenetic regulation. Furthermore, elucidation of the mechanisms by which mutations of components of the pathways can lead to aberrant signaling, uncontrolled proliferation, and in some cases confer sensitivity to targeted therapy has greatly advanced the field. This chapter will review some of the current inhibitors, their targets, and how they are being used to treat cancer and overcome therapeutic resistance.

Usually signaling commences upon ligation of a growth factor/cytokine/interleukin/mitogen (ligand) to its cognate receptor at the cell surface. This event can result in the activation of many downstream signaling cascades including the Ras/Raf/MEK/ERK and Ras/PI3K/PTEN/Akt/mTOR pathways. These pathways can further transmit their signals to different subcellular components, namely to the nucleus to control gene expression, to the translational apparatus to enhance the translation of "weak" mRNAs, to the apoptotic machinery to regulate apoptosis, or to other events involved in the regulation of cellular proliferation (e.g., interactions with the p53 pathway to regulate cell cycle progression). Regulation of the Ras/Raf/MEK/ERK and Ras/PI3K/PTEN/Akt/mTOR pathways is mediated by a series of kinases, phosphatases, GTP:GDP exchange, and scaffolding proteins. There are also many tumor suppressor proteins which interact with these cascades which frequently serve to fine tune or limit activity (e.g., NF1, PTEN, RKIP, PP2A, TSC1, and TSC2). Mutations occur in many of the genes in these pathways leading to uncontrolled regulation and aberrant signaling.

13.2 The Ras/Raf/MEK/ERK Pathway

An overview of the Ras/Raf/MEK/ERK pathway and the sites where small molecule inhibitors act is presented in Fig. 13.1. This figure serves to illustrate the flow of information through this pathway from a growth factor to a specific receptor

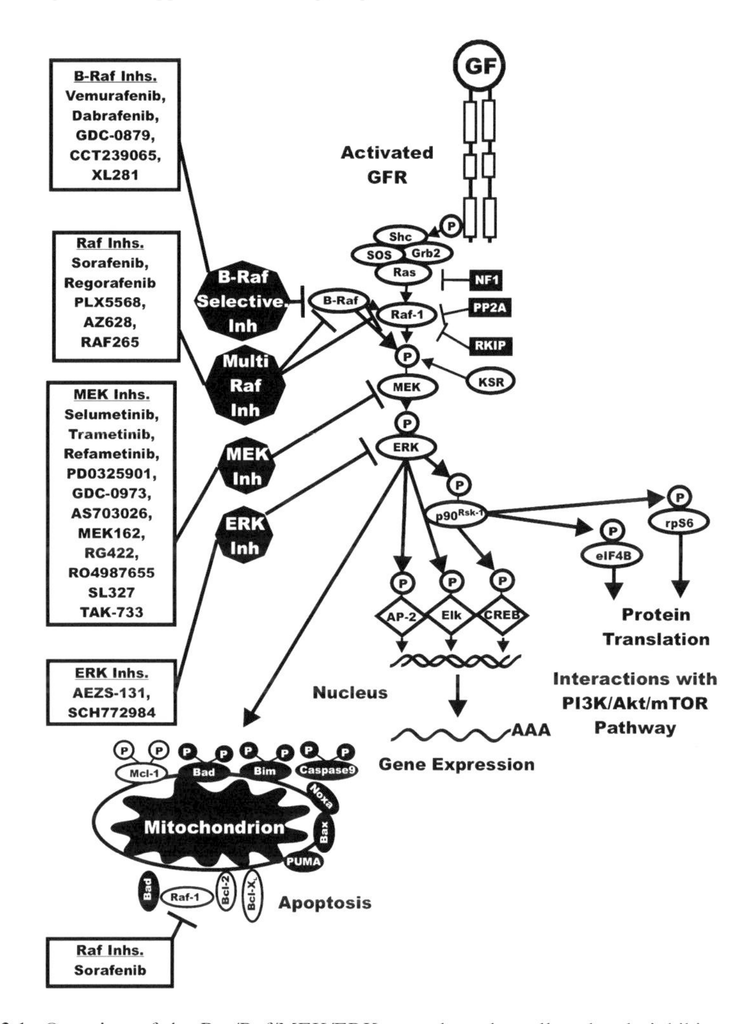

Fig. 13.1 Overview of the Ras/Raf/MEK/ERK cascade and small molecule inhibitors used for targeting this pathway. Activation of this pathway can occur by mutations in upstream Growth factor receptors (*GFR*) or by stimulation by the appropriate growth factors (*GF*). In addition, mutations can occur in intrinsic members of the pathway (*RAS*, *RAF*, *MEK1*, or the tumor suppressor Neurofibromin (*NF1*)). Sites where NF1, protein phosphatase 2A (*PP2A*), Raf kinase inhibitory protein (*RKIP*), kinase suppressor of Ras (*KSR*) interact with this pathway are on the right hand side of the Ras/Raf/MEK/ERK pathway. NF1, PP2A, and RKIP are depicted in black rectangles as they normally serve to dampen the activity of this pathway. Sites where various small molecule inhibitors function are in *black octagons* on the *left hand side* of the pathway. Representative inhibitors are listed in *boxes* next to the octagons

to phosphorylation of appropriate transcription factors as well as affect proteins involved in translation and apoptosis. Following the stimulation of a receptor with a growth factor/cytokine/mitogen, a Src homology 2 domain-containing protein (Shc) adaptor protein becomes associated with the C-terminus of the activated growth factor receptor (GFR), for example, epidermal growth factor receptor (EGFR), insulin-like growth factor-1 receptor (IGF-1R), vascular endothelial growth factor receptor (VEGFR) and many others [1, 2]. Shc recruits the growth factor receptor–bound protein 2 (Grb2) protein and the son of sevenless (SOS) homolog protein [a guanine nucleotide exchange factor (GEF)], resulting in the loading of the membrane-bound GDP:GTP exchange protein (GTPase). GEFs promote Ras activation by displacing GDP from Ras which leads to GTP binding. Ras activation is suppressed by the GTPase-activating proteins (GAPs) that stimulate the GTPase activity of Ras. There are two prominent GAP proteins, p120GAP and NF1. Ras can also be activated by growth factor receptor tyrosine kinases (GFRTK), such as insulin receptor (IR), via intermediates like insulin receptor substrate (IRS) proteins that bind Grb2 [3]. Ras:GTP then recruits the serine/threonine (S/T) kinase Raf to the membrane where it becomes activated [1, 2].

Both *RAS* and *RAF* are members of multi-gene families, and there are three *RAS* members (*KRAS*, *NRAS*, and *HRAS*) and three *RAF* members (*BRAF*, *RAF1* (a.k.a c-Raf), and *ARAF*) [1, 2]. Raf-1 and A-Raf are activated, in part, by a Src-family kinase, while B-Raf does not require the Src-family kinase for activation. Raf-1 can be regulated by dephosphorylation by the protein serine/threonine phosphatase 2A (PP2A). PP2A has been reported to positively and negatively regulate Raf-1 [4, 5].

Raf is responsible for S/T phosphorylation of mitogen-activated protein kinase kinase-1 (MEK1) (a dual specificity kinase (T/Y) [1, 2]. Other proteins such as kinase suppressor of Ras (KSR) have recently been shown to phosphorylate MEK1 [6]. KSR has scaffolding properties and interacts with Raf, MEK, and ERK which regulates ERK activation. KSR can form dimers with various Raf proteins which alter the effects of Raf inhibitors. KSR competes with Raf-1 for Raf inhibitor–induced binding to B-Raf which decreases the normal ERK activation observed after Raf-inhibitor treatment [7].

MEK1 phosphorylates extracellular signal-regulated kinases 1/2 (ERK1 and 2) at specific T and Y residues [1, 2]. MEK1 was originally not thought to be mutated frequently in human cancer. However, recent large-scale mutation screening studies and studies aimed at determining mechanisms of resistance to small molecule inhibitors have observed that MEK1 is mutated in certain human cancers and also is mutated in certain inhibitor-resistant cells [8].

Activated ERK1 and ERK2 serine S/T kinases phosphorylate and activate a variety of substrates, including p90 ribosomal six kinase-1 ($p90^{Rsk1}$) [2]. ERK also phosphorylates MAPK signal-integrating kinases (Mnk1/2) which can in turn phosphorylate eukaryotic translation initiation factor 4E (eIF4E), a key protein involved in the translation of difficult mRNAs [9].

$p90^{Rsk1}$ can activate the cAMP-response element-binding protein (CREB) transcription factor as well as proteins involved in regulation of protein translation (e.g., Mnk-1, p70 ribosomal S6 kinase (p70S6K), eukaryotic translation initiation factor 4B, (eIF4B), and ribosomal protein S6 (rpS6) [10].

The number of ERK1/2 substrate/targets is easily in the hundreds. These substrates/targets include different types of molecules including other kinases, transcription factors, or proteins involved in protein translation or apoptosis. Suppression of MEK and ERK can have profound effects on cell growth, inflammation, and aging. Activated ERK can also phosphorylate "upstream" Raf-1 and MEK1 which alter their activity. Depending upon the site phosphorylated on Raf-1, ERK phosphorylation can either enhance [11] or inhibit [12] Raf-1 activity. In contrast, some studies have indicated that when MEK1 are phosphorylated by ERK, their activity decreases [13]. Recent studies indicate that ERK does not negatively feedback-inhibit B-Raf [14].

These phosphorylation events induced by ERK serve to alter the stability and/or activities of the proteins. These examples of feedback loops become important in consideration of whether to just target MEK or to target both Raf and MEK in various cancers. It is important that the reader realize that certain phosphorylation events can either inhibit or repress the activity of the affected protein. This often depends on the particular residue on the protein phosphorylated which can confer a different configuration to the protein or target the protein to a different subcellular localization that may result in proteasomal degradation or association with certain scaffolding proteins.

13.3 The Ras/PI3K/PTEN/Akt/mTOR Pathway

An introductory overview of the Ras/PI3K/PTEN/Akt/mTOR pathway is presented in Fig. 13.2. Also outlined in this diagram are common sites of intervention with signal transduction inhibitors. Many of these inhibitors have been evaluated in various clinical trials, and some are currently being used to treat patients with specific cancers. Extensive reviews of many inhibitors targeting these pathways have been recently published [1, 15, 16].

Phosphatidylinositol 3-kinase (PI3K) is a heterodimeric protein with an 85 kDa regulatory subunit and a 110 kDa catalytic subunit (*PIK3CA*) [1, 2, 66–69]. *PIK3CA* is frequently mutated in certain cancers such as breast, ovarian, colorectal, endometrial, and lung [2, 17].

PI3K serves to phosphorylate a series of membrane phospholipids including phosphatidylinositol 4-phosphate (PtdIns(4)P) and phosphatidylinositol 4,5-bisphosphate (PtdIns(4,5)P_2), catalyzing the transfer of ATP-derived phosphate to the D-3 position of the inositol ring of membrane phosphoinositides, thereby forming the second messenger lipids phosphatidylinositol 3,4-bisphosphate (PtdIns(3,4)P_2) and phosphatidylinositol 3,4,5-trisPhosphate (PtdIns(3,4,5)P_3) [2, 15]. Most often, PI3K is activated via the binding of a ligand to its cognate receptor, whereby p85 associates with phosphorylated tyrosine residues on the receptor via a Src homology 2 (SH2) domain. After association with the receptor, the p110 catalytic subunit then transfers phosphate groups to the aforementioned membrane phospholipids [15]. It is these lipids, specifically PtdIns [3–5] P_3, that attract a series of kinases to the plasma membrane, thereby initiating the signaling cascade.

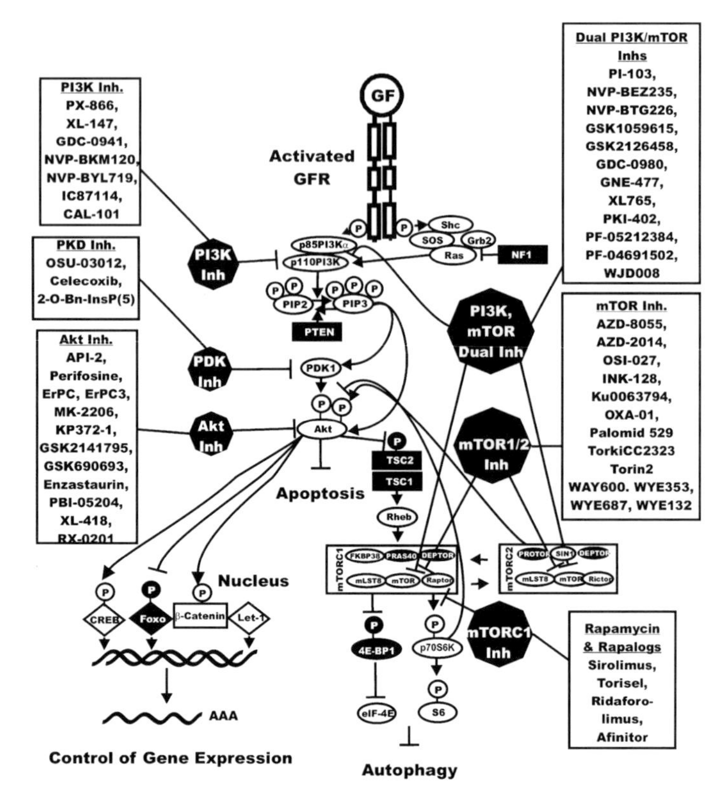

Fig. 13.2 Overview of the PI3 K/Akt/mTOR cascade and small molecule inhibitors used for targeting this pathway. Activation of this pathway can occur by mutations in upstream growth factor receptors (*GFR*) or by stimulation by the appropriate GF. In addition, mutations can occur in intrinsic members of the pathway (*RAS*, *PIK3CA*, *AKT*, or the tumor suppressors (*NF1*, *PTEN*, *TSC1*, *TSC2*). Sites where NF1, PTEN, TSC1, TSC2 are depicted in black rectangles as they normally serve to dampen the activity of this pathway. Sites where various small molecule inhibitors function are in black octagons. Representative inhibitors are listed in boxes next to the octagons

Downstream of PI3K is the primary effector molecule of the PI3K signaling cascade, Akt/protein kinase B (PKB) which is a 57 kDa S/T kinase that phosphorylates many targets on RxRxxS/T (R = Arginine) consensus motifs [18]. Akt was discovered originally as the cellular homolog of the transforming retrovirus AKT8 and as a kinase with properties similar to protein kinases A and C [19]. Akt contains an amino-terminal pleckstrin homology (PH) domain that serves to target the protein to the membrane for activation [15]. Within its central region, Akt has a large kinase domain and is flanked on the carboxyl-terminus by hydrophobic and proline-rich regions [15]. Akt-1 is activated via phosphorylation of two residues: T308 and S473. Akt-2 and Akt-3 are highly related molecules and have similar modes of activation. Akt-1 and Akt-2 are ubiquitously expressed, while Akt-3 exhibits a more restricted tissue distribution and is found abundantly in nervous tissue [20].

The phosphatidylinositol-dependent kinases (PDKs) are responsible for the activation of Akt. PDK1 is the kinase responsible for phosphorylation of Akt-1 at T308 [18]. Akt-1 is also phosphorylated at S473 by the mammalian target of *r*apamycin (mTOR) complex referred to as (Rapamycin-insensitive companion of mTOR/mLST8 complex) mTORC2 [15]. Therefore, phosphorylation of Akt is complicated as it is phosphorylated by a complex that lies downstream of activated Akt itself [15]. Thus, as with the Ras/Raf/MEK/ERK pathway, there are feedback loops that serve to regulate the activity of the Ras/PI3K/PTEN/Akt/mTOR pathway. Once activated, Akt leaves the cell membrane to phosphorylate intracellular substrates.

After activation, Akt is able to translocate to the nucleus [15] where it affects the activity of a number of transcriptional regulators. Some examples of molecules which regulate gene transcription that are phosphorylated by Akt include CREB [21], E2F [22], nuclear factor kappa from B cells (NF-κB) via inhibitor kappa B protein kinase (Iκ-K) [23], and the forkhead transcription factors [24]. These are all either direct or indirect substrates of Akt and each can regulate cellular proliferation, survival, and other important biologic processes. Besides transcription factors, Akt targets a number of other molecules to affect the survival state of the cell including the proapoptotic molecule Bcl-2-Associated Death promoter (BAD) [25] and *g*lycogen *s*ynthase kinase-3β (GSK-3β) [26].

Negative regulation of the PI3K pathway is primarily accomplished through the action of the phosphatase and TENsin homolog deleted on chromosome 10 (PTEN) tumor suppressor protein. PTEN encodes a lipid and protein phosphatase whose primary lipid substrate is PtdIns(3,4,5)P_3 [27]. The purported protein substrate(s) of PTEN are more varied, including focal adhesion kinase (FAK), the Shc exchange protein, the transcriptional regulators E-twenty six-2 (ETS-2) [28] and Sp1 and the platelet-derived growth factor receptor (PDGF-R) [29].

Next, we discuss some of the key targets of Akt that can also contribute to abnormal cellular growth by the regulation of protein translation. Akt-mediated regulation of mTOR activity is a complex multi-step phenomenon. Akt inhibits tuberous sclerosis 2 (TSC2 or tuberin) function through direct phosphorylation [30]. TSC2 is a GTPase-Activating protein (GAP) that functions in association with the putative tuberous sclerosis 1 (TSC1 or hamartin) to inactivate the small G protein Ras homolog enriched in brain (Rheb) [31]. TSC2 phosphorylation by Akt represses GAP activity of the TSC1/TSC2 complex, allowing Rheb to accumulate in a GTP-bound state. Rheb-GTP then activates, through a mechanism not yet fully elucidated, the protein kinase activity of mTOR which complexes with Raptor (Regulatory-associated protein of mTOR) adaptor protein, DEP domain-containing mTOR-interacting protein (DEPTOR) and mLST8, a member of the Lethal-with-Sec-Thirteen gene family, first identified in yeast, FK506-binding protein 38 (FKBP38) and proline-rich Akt Substrate 40 kDa protein (PRAS40). mTORC1 is sensitive to rapamycin and, importantly, inhibits Akt via a negative feedback loop which involves, at least in part, p70S6K [31]. This is due to the negative effects that p70S6K has on IRS1. DEPTOR may be a tumor suppressor gene as decreased expression of DEPTOR results in increased mTORC1 activity [32].

The mechanism by which Rheb-GTP activates mTORC1 has not been fully elucidated yet; however, it requires Rheb farnesylation and can be blocked by Farnesyl transferase (FT) inhibitors. It has been proposed that Rheb-GTP would relieve the inhibitory function of FKBP38 on mTOR, thus leading to mTORC1 activation [33].

As stated previously, TSC1 and TSC2 have important roles in the regulation of mTORC1. Two additional molecules important in this regulation are liver kinase B (LKB1 also known as STK11). LKB1 is an upstream activator of 5′AMP-activated protein kinase (AMPK) which activates TSC2 that negatively regulates mTORC1 [34]. LKB1 mediates the effects of the diabetes drug metformin [35]. Metformin has also been shown to be effective in suppressing the developments of certain cancers [36–38].

Akt also phosphorylates PRAS40, an inhibitor of mTORC1, and by doing so, it prevents the ability of PRAS40 to suppress mTORC1 signaling (recently reviewed in [15]). Thus, this could be yet another mechanism by which Akt activates mTORC1. Moreover, PRAS40 is a substrate of mTORC1 itself, and it has been demonstrated that mTORC1-mediated phosphorylation of PRAS40 prevents the inhibition of additional mTORC1 signaling [31].

Ras/Raf/MEK/ERK signaling positively impinges on mTORC1. Both $p90^{Rsk-1}$ and ERK 1/2 phosphorylate TSC2, thus suppressing its inhibitory function [31]. Moreover, mTORC1 inhibition resulted in ERK 1/2 activation, through $p70^{S6K}$/PI3K/Ras/Raf/MEK [39].

The relationship between Akt and mTOR is further complicated by the existence of the mTOR/Rictor complex (mTORC2), which, in some cell types, displays rapamycin-insensitive activity. mTORC2 is comprised of rapamycin-insensitive companion of mTOR (Rictor), mTOR, DEPTOR, mLST8, stress-activated protein kinase interacting protein 1 (SIN1), and protein observed with Rictor (Protor). mTORC2 phosphorylates Akt on S473 in vitro which facilitates T308 phosphorylation [15]. Thus, mTORC2 can function as the elusive PDK-2 which phosphorylates Akt-1 on S473 in response to growth factor stimulation [40]. Akt and mTOR are linked to each other via positive and negative regulatory circuits, which restrain their simultaneous hyperactivation through mechanisms involving $p70^{S6K}$ and PI3K [15, 31]. Assuming that equilibrium exists between these two complexes, when the mTORC1 complex is formed, it could antagonize the formation of the mTORC2 complex and reduce Akt activity. Thus, at least in principle, inhibition of the mTORC1 complex could result in Akt hyperactivation. This is one problem associated with therapeutic approaches using rapamycin or modified rapamycins (rapalogs) that block some actions of mTOR but not all.

mTOR is a 289 kDa S/T kinase. mTOR was the first identified member of the phosphatidylinositol 3-kinase-related kinase (PIKK) family [15]. mTOR has been referred to as the gatekeeper of autophagy [41]. mTOR regulates translation in response to nutrients and growth factors by phosphorylating components of the protein synthesis machinery, including $p70^{S6K}$ and eukaryotic initiation factor (eIF)-4E binding protein-1 (4EBP-1), the latter resulting in the release of the eukaryotic initiation factor-4E (eIF-4E) allowing eIF-4E to participate in the assembly of a translational initiation complex [2]. p70S6K phosphorylates the 40S rpS6, leading to translation of "weak" mRNAs. Integration of a variety of signals (mitogens, growth factors, hormones) by mTOR assures cell cycle entry only if nutrients and energy are sufficient for cell duplication [15].

Unphosphorylated 4E-BP1 interacts with the cap-binding protein eIF4E and prevents the formation of the 4F translational initiation complex (eIF4F) by competing for the binding of eukaryotic initiation factor 4G (eIF4G) to eIF4E. 4E-BP1 phosphorylation by mTORC1 results in the release of the eIF4E, which then associates with eIF4G to stimulate translation initiation [15].

eIF4E is a key component for translation of 5′ capped mRNAs, which include transcripts mainly encoding for proliferation and survival promoting proteins, such as c-Myc, cyclin D1, cyclin-dependent kinase-2 (CDK-2), signal activator and transducer of transcription-3 (STAT-3), ornithine decarboxylase, surviving, B-cell lymphoma 2 (Bcl-2), Bcl-xL, myeloid cell leukemia-1 (Mcl-1) [2].

13.4 Overview of Pathway Inhibitors

Sites of intervention with signal transduction inhibitors in the Ras/Raf/MEK/ERK are presented in Fig. 13.1. Some of the inhibitors are currently being used to treat patients with specific cancers, and others have been or are being evaluated in numerous clinical trials with many different types of cancer patients. Effective inhibitors, specific for many of the key components of the Ras/Raf/MEK/ERK and Ras/PI3K/PTEN/mTOR pathways, have been developed [1, 15]. In many cases, these inhibitors have been examined in clinical trials. Furthermore, inhibitors that target the mutant protein more than the wild-type (WT) protein of various genes (e.g., *BRAF* and *PIK3CA*) either have been or are being characterized.

13.4.1 Raf Inhibitors

Raf inhibitors have been developed, and some are being used for therapy while others are being evaluated in clinical trials. Raf inhibitors have, in general, exhibited greater response rates in clinical trials than MEK inhibitors which may be related to the broader therapeutic index of Raf inhibitors that suppress ERK activity in a mutant-allele-specific fashion as opposed to MEK inhibitors which suppress MEK activity in tumor and normal cells [42].

13.4.1.1 Sorafenib

Sorafenib (Bayer) was initially thought to specifically inhibit Raf but has been subsequently shown to have multiple targets (e.g., VEGF-R, Flt-3, PDGF-R) [43].

However, that does not preclude its usefulness in cancer therapy. Sorafenib is approved for the treatment for certain cancers (e.g., renal cell Carcinoma (RCC) and patients with unresectable HCC and was further evaluated in the Sorafenib Hepatocellular carcinoma Assessment Randomized Protocol (SHARP) trial, which demonstrated that the drug was effective in prolonging median survival and time to progression in patients with advanced HCC [44, 45]. Sorafenib is generally well tolerated in HCC patients with a manageable adverse events profile [44]. While sorafenib is not considered effective for the treatment for most melanomas with *BRAF V600E* mutations, it may be effective in the treatment for a minority of melanomas with *G469E* and *D594G* mutations which express constitutive ERK1/2 but low levels of MEK. These melanomas are sensitive to sorafenib, potentially because they signal through Raf-1. Raf-1 also exerts anti-apoptotic effects at the mitochondrion in association with Bcl-2 family members [46].

13.4.1.2 Vemurafenib

Vemurafenib (Zelboraf, PLX-4032, Plexxikon/Roche) is a B-Raf inhibitor that has and is being evaluated in many clinical trials [47–49]. Vemurafenib has been approved by the US Food and Drug Administration (FDA) for the treatment of patients with unresectable or metastatic melanoma carrying the *BRAF*(*V600E*) mutation. For vemurafenib to be clinically effective, it needs to suppress downstream ERK activation essentially completely [47].

13.4.1.3 Dabrafenib

Dabrafenib (GSK2118436) is an ATP-competitive inhibitor of mutant B-Raf, WT B-Raf, and WT Raf-1 developed by GlaxoSmithKlein (GSK) [50]. Dabrafenib is in clinical trials [51, 52].

13.4.1.4 CCT239065

CCT239065 is a mutant B-Raf inhibitor developed at the Institute of Cancer Research in London, UK. It inhibits mutant *BRAF V600E* signaling and proliferation more than those cells containing *WT BRAF* [53]. Its effects are more selective for cells containing mutant *BRAF* than *WT BRAF*. CCT239065 is well tolerated in mice and had good oral bioavailability. It suppressed tumors containing *BRAF*-mutant gene but not *WT BRAF* tumors in mice tumor xenograft studies.

13.4.1.5 GDC-0879

GDC-0879 is a *BRAF*-mutant allele-selective inhibitor developed by Genentech [54]. The efficacy GDC-0879 is related to the *BRAF V600E* mutational status in the cancer cells and inhibition of downstream MEK and ERK activity.

13.4.1.6 AZ628

AZ628 is a selective Raf inhibitor developed by Astra Zeneca. It has been shown that when *BRAF*-mutant melanoma cells, which are normally very sensitive to AZ628, are grown for prolonged periods of time, they become resistant to AZ628 by upregulating the expression of Raf-1 [55].

13.4.1.7 XL281

XL281 is an oral active wild-type and mutant RAF kinase-selective inhibitor developed by Exelixis and Bristol-Myers Squibb. It has been examined in clinical trials primarily with patients having *BRAF* mutations (colorectal cancers (CRC), melanoma, papillary thyroid cancers (PTC), and NSCLC) [56].

13.4.1.8 PLX5568

PLX5568 is a selective Raf kinase inhibitor developed by Plexxicon. It is being examined for the treatment for polycystic kidney disease (PKD). In the kidney, Raf-1 is localized to the tubular cells where it is linked with many physiologically important functions. PLX5568 suppressed cyst enlargement in a rat model of PKD but did not improve kidney function as fibrosis was not suppressed [57].

13.4.1.9 Raf-265

Raf-265 is an ATP-competitive pan-Raf inhibitor developed by Novartis. Treatment for bronchus carcinoid NCI-H727 and CM-insulinoma cells with Raf-265 enhanced sensitivity to TRAIL-induced apoptosis. These cells were normally resistant to PI3K/mTOR inhibitors when combined with TRAIL. Raf-265 was shown to decrease Bcl-2 levels which correlated with their sensitivity to TRAIL-mediated apoptosis. This approach may be effective in the therapy of neuroendocrine tumors [58].

13.4.1.10 Regorafenib

Regorafenib (BAY 73-4506) is an oral multi-kinase inhibitor of angiogenic, stromal, and oncogenic RTKs developed by Bayer. Regorafenib inhibits RTKs such as VEGF-R2, VEGF-R1/3, PDGF-Rβ, fibroblast growth factor receptor-1 as well as mutant Kit, RET, and B-Raf. The effects of regorafenib on tumor growth have been evaluated in human xenograft models in mice, and tumor shrinkages were observed in breast MDA-MB-231 and renal 786-O carcinoma models [59].

13.4.2 MEK Inhibitors

Most MEK inhibitors differ from most other kinase inhibitors as they do not compete with ATP binding (non-ATP competitive), which confers a high specificity [60–62]. Most MEK inhibitors are specific and do not inhibit many different protein kinases [62] although as will be discussed below, certain MEK inhibitors are more specific than others.

Molecular modeling studies indicate that many MEK bind to an allosteric binding site on MEK1/MEK2. The binding sites on MEK1/MEK2 are relatively unique to these kinases and may explain the high specificity of MEK inhibitors. This binding may lock MEK1/2 in an inactivate conformation that enables binding of ATP and substrate, but prevents the molecular interactions required for catalysis and access to the ERK activation loop [61].

A distinct advantage of inhibiting MEK is that it can be targeted without knowledge of the precise genetic mutation that results in its aberrant activation. This is not true with targeting Raf as certain Raf inhibitors will activate Raf and also certain B-Raf-specific inhibitors will not be effective in the presence of *RAS* mutations.

An advantage of targeting MEK is that the Ras/Raf/MEK/ERK pathway is a convergence point where a number of upstream signaling pathways can be blocked with the inhibition of MEK. For example, MEK inhibitors, such as Selumetinib (AZD6244), are also being investigated for the treatment for pancreatic cancers, breast cancers, and other cancers such as hematopoietic malignancies, including multiple myeloma [1, 63].

13.4.2.1 Selumetinib

Selumetinib inhibits MEK1 in vitro with an IC_{50} value of 14.1 ± 0.79 nM [64–67]. It is specific for MEK1 as it did not appear to inhibit any of the approximately 40 other kinases in the panel tested. Selumetinib is not competitive with ATP. Selumetinib inhibited downstream ERK1/ERK2 activation in in vitro cell line assays with stimulated and unstimulated cells and also inhibited activation in tumor transplant models. Selumetinib did not prevent the activation of the related ERK5 that occurs with some older MEK1 inhibitors, which are not being pursued in clinical trials. Inhibition of ERK1/2 suppresses their ability to phosphorylate and modulate the activity of Raf-1, B-Raf, and MEK1 but not MEK2 as MEK2 lacks the ERK1/ERK2 phosphorylation site. In essence, by inhibiting ERK1/2 the negative loop of Raf-1, B-Raf, and MEK phosphorylation is suppressed, and hence, there will be an accumulation of activated Raf-1, B-Raf, and MEK [67]. This biochemical feedback loop may provide a rationale for combining Raf and MEK inhibitors in certain therapeutic situations. Selumetinib has also been shown to suppress cetuximab-resistant CRCs which had *KRAS* mutations both in vitro and in vivo models [68].

13.4.2.2 PD-0325901

The PD-0325901 MEK inhibitor is an orally active, potent, specific, non-ATP-competitive inhibitor of MEK. PD-0325901 demonstrated improved pharmacological and pharmaceutical properties compared with PD-184352, including a greater potency for inhibition of MEK and higher bioavailability and increased metabolic stability. PD-0325901 has a K_i value of 1 nM against MEK1 and MEK2 in in vitro kinase assays. PD-0325901 inhibits the growth of cell lines that proliferate in response to elevated signaling of the Raf/MEK/ERK pathways [62]. PD-0325901 has undergone phase I clinical trials [62, 69–71]. Although the initial trial results were not encouraging, it was determined that some tumors which proliferate in response to the Raf/MEK/ERK pathways may be sensitive to PD0325901 [72]. Although the clinical trials with PD-0325901 were initially suspended, there are now some clinical trials with PD-0325901 in combination with other pathway inhibitors.

13.4.2.3 Refametinib

Refametinib (RDEA119) is a more recently described MEK inhibitor developed by Ardea Biosciences [73]. It is a highly selective MEK inhibitor that displays a >100-fold selectivity in kinase inhibition in a panel of 205 kinases. In contrast, in the same kinase specificity analysis, other recently developed MEK inhibitors (e.g., PD0325901) also inhibited the Src and RON kinases.

13.4.2.4 Trametinib

Trametinib (GSK1120212) is an allosteric MEK inhibitor developed by GSK. It has been shown to be effective when combined with dabrafenib in certain dabrafenib-resistant *BRAF V600* melanoma lines that also had mutations at *NRAS* or *MEK1* [52]. The combination of trametinib and the PI3K/mTOR dual inhibitor GSK2126458 also enhanced cell growth inhibition in these B-Raf inhibitor-resistant *BRAF*-mutant melanoma lines.

13.4.2.5 GDC-0973

GDC-0973 (XL518) is a potent and selective MEK inhibitor developed by Genentech [74]. The effects of combining GDC-0973 and the PI3K inhibitor GDC-0941 on the proliferation of *BRAF* and *KRAS* mutant cancer cells indicated the combination efficacy both in vitro and in vivo.

13.4.2.6 AS703026

AS703026 (MSC1936369B) is a MEK inhibitor developed by EMD Serono. AS703026 suppressed cetuximab-resistant CRCs which had *KRAS* mutations both in vitro and in vivo models [68]. AS703026 inhibited growth and survival of multiple myeloma (MM) cells and cytokine-induced differentiation more potently than selumetinib, and importantly, AS703026 was cytotoxic, where as most MEK inhibitors are cytostatic [75]. AS703026 sensitized MM cells to a variety of conventional (dexamethasone, melphalan) and novel (lenalidomide, perifosine, bortezomib, rapamycin) drugs used to treat MM.

13.4.2.7 RO4987655

RO4987655 (CH4987655) is an allosteric, orally available MEK inhibitor developed by Roche/Chiron. It has been tested in humans and determined to inhibit active ERK levels. At the levels of RO4987655 administered, it was determined to be safe in healthy volunteers [76].

13.4.2.8 TAK-733

TAK-733 is a potent and selective, allosteric MEK inhibitor developed by Takeda San Diego [77]. TAK-733 is being investigated in clinical trials.

13.4.2.9 MEK162

MEK162 (ARRY-162) is a MEK inhibitor developed by Novartis. It is in clinical trials.

13.4.2.10 SL327

SL337 is a MEK inhibitor that has been used in many neurological and drug addiction studies [78].

13.4.2.11 Other MEK Inhibitors

Other MEK inhibitors are being developed. RG422 is one such inhibitor.

13.4.3 Combining Raf and MEK Inhibitors

The possibility of treating certain patients with Raf and MEK inhibitors is a concept which is gaining more acceptance as it may be a therapeutic possibility to overcome resistance [42]. Raf inhibitors induce Raf activity in cells with *WT RAF* if Ras is active [79]. The addition of a MEK inhibitor would suppress the activation of MEK and ERK in the normal cells of the cancer patient. Thus, B-Raf would be suppressed by the B-Raf-selective inhibitor in the cancer patient, while the consequences of Raf activation in the normal cells would be suppressed by the MEK inhibitor. These concepts are being examined in clinical trials.

13.4.4 Combining MEK and Bcl-2 Inhibitors

The effects of combining MEK and Bcl-2/Bcl-XL inhibitors have been examined in preclinical studies with AML cell lines and patient samples [80]. The Bcl-2 inhibitor, ABT-737, was observed to induce ERK activation and Mcl-1 expression. However, when the ABT-737 inhibitor was combined with the MEK inhibitor PD0325901, a synergistic response was observed in terms of the induction of cell death both on AML cell lines and on primary tumor cells with the properties of leukemia stem cells. Furthermore, these studies were also extended into tumor transplant models with the MOLT-13 cell line, and synergy between ABT-737 and PD0325901 were also observed in vivo.

13.4.5 ERK Inhibitors

There are at least two ERK molecules regulated by the Raf/MEK/ERK cascade, ERK1 and ERK2. Little is known about the differential in vivo targets of ERK1 and ERK2. The development of specific ERK1 and ERK2 inhibitors is ongoing and may be useful in the treatment for certain diseases such as those leukemias where elevated ERK activation is associated with a poor prognosis (e.g., AML, ALL) [81]. ERK inhibitors have been described [82].

13.4.5.1 AEZS-131

AEZS-131 has been reported on the Internet to be a highly selective ERK 1/2 inhibitor developed by Aeterna Zentaris and has been examined on human breast cancer cells.

13.4.5.2 Pyrimidylpyrrole ERK Inhibitors

A novel series of pyrimidylpyrrole ERK inhibitors has been developed at Vertex Pharmaceuticals [82]. A lead compound, ERKi, has been evaluated for its ability to overcome MEK inhibitor resistance [83]. These studies performed in breast and CRCs demonstrated that dual inhibition of MEK and ERK by small molecule inhibitors was synergistic. Furthermore, inhibition of both MEK and ERK acted to suppress the emergence of resistance and overcome the acquired resistance to MEK inhibitors in these breast and CRC cell line models.

13.4.5.3 SCH772984

SCH772984 is reported to be an ERK inhibitor.

13.4.6 PI3K/Akt/mTOR Inhibitors

Numerous PI3K, Akt, mTOR, and dual PI3K/mTOR inhibitors have been developed and evaluated. The PI3K and mTOR inhibitors have been used in basic science studies for years and have provided much information about the role of the PI3K/Akt/mTOR pathway in many biologic and diseases processes. We will focus on the newer inhibitors of this pathway and how they are now being used in clinical trials.

13.4.6.1 PX-866

The modified wortmannin PX-866 has been evaluated as a PI3K inhibitor [84]. It is being evaluated in phase II clinical trials for patients with advanced metastatic prostate cancer by Oncothyreon.

13.4.6.2 GDC-0941

GDC-0941 is a PI3K inhibitor developed by Genentech. GDC-0941 inhibited the metastatic characteristics of thyroid carcinomas by targeting both PI3K and hypoxia-inducible factor 1α (HIF-1α) pathways [85]. GDC-0941 synergized with the MEK inhibitor UO126 in inhibiting the growth of NSCLC [86]. It is being evaluated in a clinical trial for advanced cancers or metastatic breast cancers which are resistant to aromatase inhibitor therapy.

13.4.6.3 IC87114

IC87114 is a selective p110δ PI3K inhibitor. It decreased cell proliferation and survival in AML cells and increased sensitivity to etoposide [87–90].

13.4.6.4 CAL-101

CAL-101(GS-1101) is a derivative of IC87114 [91–93]. CAL-101 is an oral p110δ PI3K inhibitor developed by Calistoga Pharmaceuticals and Gilead Sciences.

CAL-101 is currently undergoing clinical evaluation in patients with various hematopoietic malignancies including relapsed or refractory indolent B-cell NHL, mantle cell lymphoma or CLL. An additional clinical trial will examine the effects of combining CAL-101 with chemotherapeutic drugs and the αCD20 monoclonal Ab (MoAb).

13.4.6.5 XL-147 (SAR245408)

XL-147 (SAR245408) is a PI3K inhibitor developed by Exelixis/Sanofi-Aventis [94]. It is in clinical trials, either as a single agent or in combination with erlotinib, hormonal therapy, chemotherapy, or MoAb therapy for various cancers including lymphoma, breast, endometrial, glioblastoma, astrocytoma, or other solid cancers.

13.4.6.6 Novartis PI3K Inhibitors

NVP-BKM120 is an orally available pan-class I PI3-kinase inhibitor developed by Novartis [95]. It is in many clinical trials, either as a single agent or in combination with other drugs or signal transduction inhibitors [96]. NVP-BKM120 is clinical trial with patients having advanced cancers such as CRC, NSCLC, breast, prostate, endometrial, squamous cell carcinoma of the head and neck, GIST, RCC, melanoma, and advanced leukemias.

NVP-BYL719 (BYL719) is a PI3Kα-selective inhibitor developed by Novartis. It is in clinical trials for patients with advanced solid tumors, some containing mutations at *PIK3CA*. It is also being examined in a clinical trial in combination with the MEK-162 inhibitor for patients with advanced CRC, esophageal, pancreatic, NSCLC, or other advanced solid tumors containing *RAS* or *BRAF* mutations.

13.4.7 Dual PI3K/mTOR Inhibitors

The catalytic sites of PI3K and mTOR share a high degree of sequence homology. This feature has allowed the synthesis of ATP-competitive compounds that target the catalytic site of both PI3K and mTOR. Several dual PI3K/mTOR inhibitors have also been developed. In preclinical settings, dual PI3K/mTOR inhibitors displayed a much stronger cytotoxicity against leukemic cells than either PI3K inhibitors or allosteric mTOR inhibitors, such as rapamycin and its derivatives (rapalogs). In contrast to rapamycin/rapalogs, dual PI3K/mTOR inhibitors targeted both mTOR complex 1 and mTOR complex 2 and inhibited the rapamycin-resistant phosphorylation of eIF4B-1 and inhibited protein translation of many gene products associated with oncogenesis (enhanced proliferation) in leukemic cells. The dual inhibitors strongly reduced the proliferation rate and induced an important apoptotic response [16].

The kinase selectivity profile of the dual PI3K/mTOR modulators is consistent with the high sequence homology and identity in the ATP-catalytic cleft of these kinases. Dual PI3K/mTOR inhibitors have demonstrated significant, concentration-dependent cell proliferation inhibition and induction of apoptosis in a broad panel of tumor cell lines, including those harboring PI3K p110α (*PIK3CA*) activating mutations [97].

Moreover, the in vitro activity of these ATP-competitive PI3K/mTOR modulators has translated well in in vivo models of human cancer xenografted in mice. They were well tolerated and achieved disease stasis or even tumor regression when administered orally [98]. In spite of their high lipophilicity and limited water solubility, the pharmacological, biologic, and preclinical safety profiles of these dual PI3K/mTOR inhibitors supported their clinical development.

There may be some benefits to treating patients with an inhibitor that can target both PI3K and mTOR as opposed to treating patients with two inhibitors, that is, one targeting PI3K and another specifically mTOR. An obvious benefit could be lowered toxicities. Treatment with a single drug could have fewer side effects than treatment with two separate drugs. The effects of detrimental Akt activation by mTOR inhibition might be avoided upon treatment with a dual kinase inhibitor. Furthermore, the negative side effects of mTOR inhibition on the activation of the Raf/MEK/ERK pathway might be eliminated with the PI3K inhibitor activity in the dual inhibitor. There remains, however, considerable uncertainty about potential toxicity of compounds that inhibit both PI3K and mTOR enzymes whose activities are fundamental to a broad range of physiological processes. Although it should be pointed out that there are some clinical trials in progress to determine whether it is beneficial to treat cancer patients with a PI3K/mTOR dual inhibitor and an mTORC1 blocker such as NVP-BEZ235 and RAD001, preclinical studies have documented the benefits of combining RAD001 with NVP-BEZ235 [99].

13.4.7.1 PI-103

PI-103 was the first reported ATP-competitive kinase inhibitor of mTOR which also blocked the enzymatic activity of PI3K p110 isoforms. It was developed at UCSF in 2006. PI-103 exhibits good selectivity over the rest of the human kinome in terms of non-selective inhibition of other kinases [100, 101]. PI-103 is a pan-class I PI3K inhibitor with IC_{50} values in the 2 nm (p110α PI3K) to 15 nm range (p110γ PI3K) PI-103 inhibits both mTORC1 ($IC_{50} = 0.02$ μm) and mTORC2 ($IC_{50} = 0.083$ μm).

13.4.7.2 Novartis Dual PI3K/mTOR Inhibitors

NVP-BEZ235 is a dual PI3 K/mTOR inhibitor developed by Novartis. Importantly and in contrast to rapamycin, NVP-BEZ235 inhibited the rapamycin-resistant phosphorylation of 4E-BP1, causing a marked inhibition of protein translation in AML cells. This resulted in the reduced levels of the expression of c-Myc, cyclin D1, and Bcl-xL known to be regulated at the translation initiation level [102]. NVP-BEZ235 suppressed proliferation and induced an important apoptotic response in AML cells without affecting healthy $CD34^+$ cell survival. Importantly, it suppressed the clonogenic activity of leukemic, but not healthy, $CD34^+$ cells [103]. NVP-BEZ235 targeted the *s*ide *p*opulation (SP) of both T-ALL cell lines and patient lymphoblasts, which might correspond to Leukemia-Initiating Cells (LIC), and synergized with several chemotherapeutic agents (cyclophosphamide, cytarabine, dexamethasone) currently used for treating T-ALL patients [104]. Also, NVP-BEZ235 reduced chemoresistance to vincristine induced in Jurkat cells by co-culturing with MS-5 stromal cells, which mimic the bone marrow microenvironment [105]. In this study, NVP-BEZ235 was cytotoxic to T-ALL patient lymphoblasts displaying pathway activation, where the drug dephosphorylated 4E-BP1, in contrast to the results with obtained rapamycin. Taken together, these findings indicated that longitudinal inhibition at two nodes of the PI3K/Akt/mTOR network with NVP-BEZ235, either alone or in combination with chemotherapeutic drugs, may be an effective therapy for of those T-ALLs that have aberrant upregulation of this signaling pathway.

NVP-BEZ235 has been evaluated also in a mouse model consisting of BA/F3 cells overexpressing either WT *BCR-ABL* or its imatinib-resistant *BCR-ABL* mutants (*E255K* and *T315I*) [106]. NVP-BEZ235 inhibited proliferation of both cytokine-independent WT *BCR-ABL* and mutant *BCR-ABL* (*E255K* and *T315I*) overexpressing cells, whereas parental cytokine-dependent Ba/F3 cells were much less sensitive. The drug also induced apoptosis and inhibited both mTORC1 and mTORC2 signaling. Remarkably, the drug displayed cytotoxic activity in vivo against leukemic cells expressing the *E255K* and *T315I BCRABL* mutant forms. However, in this experimental model, NVP-BEZ235 induced an overactivation of MEK/ERK signaling, most likely due to the well-known compensatory feedback mechanism that involves p70S6K [39]. NVP-BEZ235 has been intensively investigated and is in clinical trials for patients with advanced cancers [107]. In some trials, NVP-BEZ235 is being evaluated in combination with either paclitaxel or trastuzumab (herceptin). NVP-BTG226 is a recently developed PI3K/mTOR inhibitor [15].

13.4.7.3 Pfizer Dual PI3K/mTOR Inhibitors

PKI-587, also known as PF-05212384, inhibited class I PI3Ks, PI3Kα mutants, and mTOR. PKI-587-suppressed proliferation of approximately 50 diverse human tumor cell lines at IC_{50} values less than 100 nmol/L. PKI-587-induced apoptosis in cell lines with elevated PI3K/Akt/mTOR signaling. PKI-587 inhibited the tumor growth in various models including breast (MDA-MB-361, BT474), colon (HCT116), lung (H1975), and glioma (U87MG). The efficacy of PKI-587 was enhanced when administered in combination with the MEK inhibitor, PD0325901, the topoisomerase I inhibitor, irinotecan, or the HER2 inhibitor, neratinib [108].

PF-04691502 is an ATP-competitive PI3K/Akt inhibitor which suppresses the activation of Akt. PF-04691502 suppressed the transformation of avian cells in response to either WT or mutant *PIK3CA*. PF-04691502 inhibited tumor growth in various xenograft models including U87 (*PTEN* null), SKOV3 (*PIK3CA* mutation) and gefitinib (EGFR inhibitor) and erlotinib-resistant NSCLC [109]. Both PKI-587 and PF-04691502 are in clinical trials to treat endometrial cancers.

PKI-402 is a selective, reversible, ATP-competitive, PI3K and mTOR inhibitor. It suppress PI3Ks, PI3Kα mutant, and mTOR equally. PKI-402 inhibited the growth of many human tumor cell lines including breast, glioma, pancreatic, and NSCLC [110].

13.4.7.4 XL765

XL765 (SAR25409) is a dual PI3K/mTOR inhibitor developed by Exelixis/Sanofi-Aventis. XL765 has been investigated in brain and pancreatic cancer models either as a single agent or in combination with temozolomide [111] or the autophagy inhibitor chloroquine [112]. XL765 downregulated the phosphorylation of Akt induced by PI3K/mTORC2 and reduced brain tumor growth [111]. Combining XL765 with chloroquine suppressed autophagy and induced apoptotic cell death in pancreatic tumor models [112]. Clinical trials are being performed with XL765 in combination with temozolomide to treat patients with glioblastoma or in combination with erlotinib to treat NSCLC patients.

13.4.7.5 Genentech Dual PI3K/mTOR Inhibitors

GNE-477 is a dual PI3K/mTOR inhibitor developed by Genentech [113]. GDC-0980 is similar to GNE-477 and has been shown to have high activity in cancer models driven by PI3K pathway activation [114]. GDC-0980 is in a clinical trial for patients with advanced cancers or metastatic breast cancers which are resistant to aromatase inhibitor therapy.

13.4.7.6 GSK Dual PI3K/mTOR Inhibitors

GSK2126458 is a dual PI3K/mTOR inhibitor developed by GSK [52]. It is in at least two clinical trials with advanced cancer patients. In one trial, it is being combined with the MEK inhibitor GSK1120212. GSK1059615 is a dual PI3K/mTOR inhibitor developed by GSK. It was in a clinical trial with patients with solid tumors, metastatic breast cancer, endometrial cancers, and lymphomas which was terminated.

13.4.7.7 WJD008

WJD008 (Chinese Academy of Sciences, Shanghai) is a dual PI3K/mTOR [115]. WJD008 inhibited the increased activity of the PI3K pathway normally induced by *PIK3CA H1047R* and suppressed proliferation and colony formation of transformed RK3E cells containing *PIK3CA H1047R*.

13.4.8 PDK Inhibitors

Some compounds have been reported to be PDK inhibitors, including the osteoarthritis drug celecoxib [116], the modified celecoxib, OSU-03012 [76, 117], and 2-O-BN-InsP(5) [118]. Celecoxib (Celebrex, Pfizer) obviously has other targets than PDK, such as cyclooxygenase-2 (Cox-2). Celecoxib is used to treat CRC patients to reduce the number of polyps in the colon. OSU-03012 is reported not to inhibit Cox-2 [117]. 2-O-BN-InsP(5) is based on the structure of inositol 1,3,4,5,6-pentakisphosphate, it may inhibit both PDK and mTOR [118].

13.4.9 Akt Inhibitors

Many attempts to develop Akt inhibitors have been performed over the years. In many of the earlier attempts, the various Akt inhibitors either lacked specificity or had deleterious side effects. Part of the deleterious side effects is probably related to the numerous critical functions that Akt plays in normal physiology. Namely, some Akt inhibitors will alter the downstream effects of insulin on Glut-4 translocation and glucose transport.

13.4.9.1 Triciribine

Triciribine (API-2) is an Akt inhibitor that has been used in many studies: at least 92 are listed on PubMed. Triciribine suppressed the phosphorylation of all three Akt isoforms in vitro and the growth of tumor cells overexpressing Akt in mouse xenograft models [119]. The mechanism(s) by which triciribine inhibits Akt activity are not clear. The drug has been evaluated in a phase I clinical trial in patients with advanced hematologic malignancies, including refractory/relapsed AML. In this trial, triciribine was administered on a weekly schedule. The drug was well tolerated, with preliminary evidence of pharmacodynamic activity as measured by decreased levels of activated Akt in primary blast cells [120]. Triciribine has also been examined in clinical trial with Akt^{+} metastatic cancers.

13.4.9.2 MK-2206

MK-2206 (Merck) is an allosteric Akt inhibitor which inhibits both T308 and S473 phosphorylation. It also inhibits the downstream effects of insulin on Glut-4 translocation and glucose transport [121]. MK-2206 decreased T-acute lymphocytic leukemia (T-ALL) cell viability by the cells in the G_0/G_1 phase of the cell cycle and inducing apoptosis. MK-2206 also induced autophagy in the T-ALL cells. MK-2206 induced a concentration-dependent dephosphorylation of Akt and its downstream targets, GSK-3α/β and FOXO3A. MK-2206 also was cytotoxic to primary T-ALL cells and induced apoptosis in a T-ALL patient cell subset ($CD34^+/CD4^-/CD7^-$) which is enriched in LICs. [122]. MK-2206 is in at least 43 clinical trials either as a single agent or in combination with other small molecule inhibitors or chemotherapeutic drugs with diverse types of cancer patients.

13.4.9.3 GSK Akt Inhibitors

GSK690693 is a pan-Akt inhibitor developed by GSK. GSK690693 is an ATP-competitive inhibitor effective at the low nanomolar range. Daily administration of GSK690693 resulted in significant anti-tumor activity in mice bearing various human tumor models including SKOV-3 ovarian, LNCaP prostate and BT474 and HCC-1954 breast carcinoma. The authors also noted that GSK690693 resulted in acute and transient increases in blood glucose level [123]. The effects of GSK690693 were also examined 112 cell lines representing different hematologic neoplasia. Over 50 % of the cell lines were sensitive to the Akt inhibitor with an EC_{50} of less than 1 μm. ALL, non-Hodgkin lymphomas, and Burkitt lymphomas exhibited 89, 73, and 67 % sensitivity to GSK690693, respectively. Importantly, GSK690693 did not inhibit the proliferation of normal human $CD4^+$ peripheral T lymphocytes as well as mouse thymocytes.

GSK2141795 is a GSK Akt inhibitor under development. It is reported by GSK to be an oral, pan-Akt inhibitor which shows activity in various cancer models, including blood cancer and solid tumor models. In addition, it is reported by GSK to delay tumor growth in solid tumor mouse xenograft models. It has been investigated further in clinical trials.

13.4.9.4 KP372-1

KP372-1 inhibits PDK1, Akt, and Fms-like tyrosine kinase 3 (Flt-3) signaling and induces mitochondrial dysfunction and apoptosis in AML cells but not normal hematopoietic progenitor cells [124]. It also suppressed colony formation of primary AML patient sample cells but not normal hematopoietic progenitor cells. It has also been investigated in other cancer types, including squamous cell carcinomas of the head and neck, thyroid cancers, and glioblastomas.

13.4.9.5 Enzasturin

Enzasturin (LY317615) is a protein kinase C-β (PKC-β) and Akt inhibitor developed by Lilly. It has been investigated in clinical trials either by itself or in combination with other agents in various types of cancer patients including brain [125] and NSC [126], CRC [127] as well as other cancer types. It is reported to be in approximately 48 clinical trials on the ClinicalTrials.gov website.

13.4.9.6 Perifosine

Perifosine (KRX-0401, Keryx/AOI Pharmaceuticals, Inc., and licensed to AEterna Zentaris) is an alkylphospholipid that can inhibit Akt [128]. The effects of perifosine have been examined on many different tumor types. Perifosine induces caspase-dependent apoptosis and downregulates P-glycoprotein expression in multi-drug-resistant T-ALL cells by a JNK-dependent mechanism [104]. Perifosine is or has been in at least 43 clinical trials to treat various cancer patients, with either blood cancers or solid tumors, either by itself, or in combination with other agents. It has advanced to phase III clinical trials for CRC and MM. In the USA, it has orphan drug status for the treatment for MM and neuroblastoma.

13.4.9.7 Erucylphosphocholine and Erucylphosphohomocholine

Erucylphosphocholine (ErPC) and Erucylphosphohomocholine (ErPC3) have been shown to inhibit Akt and induce apoptosis in malignant glioma cell lines which are normally resistant to the induction of apoptosis. They are structurally related to perifosine [129]. ErPC enhanced radiation-induced cell death and clonogenicity [130]. These effects on the induction of apoptosis were correlated with increased Bim levels and decreased Bad and Foxo-3 phosphorylation, potentially consequences of decreased Akt activity. ErPC3 is the first intravenously applicable alkylphosphocholine. ErPC3 was cytotoxic to AML cells through JNK2- and PP2-dependent mechanisms [131].

13.4.9.8 PBI-05204

PBI-05204 (oleandrin) is an Akt inhibitor. PBI-05024 is a botanical drug candidate derived from *Nerium oleander* and developed by Phoenix Biotechnology. It also has other targets including FGF-2, NF-κB, and p70S6K. PBI-05204 is in clinical trials for cancer patients with advanced solid tumors [132]. Interestingly, PBI-05204 also provides significant neuroprotection to tissues damaged by glucose and oxygen deprivation which occurs in ischemic stroke [133].

13.4.9.9 RX-0201

RX-0201 (Akt1AO, Rexahn Pharmaceuticals, Inc.) is an Akt-1 antisense oligonucleotide molecule. RX-0201 downregulated Akt-1 expression at nanomolar concentrations in multiple types of human cancer cells. RX-0201 also inhibited tumor growth in mice xenografted with U251 human glioblastoma and MIA human pancreatic cancer cells [134]. RX-021 is in a clinical trial in combination with gemcitabine for patients with metastatic pancreatic cancer [135].

13.4.9.10 XL-418

XL-418 is reported to be a dual Akt/p70S6K inhibitor by developed by Exelixis/GSK. It was in clinical trials for patients with advanced cancer; however, those trials were suspended.

13.4.10 mTORC1 Inhibitors

Rapamycin (Rapamune, Pfizer) was approved by the FDA in 1999 to prevent transplant rejection in organ transplant patients. Rapamycin/rapalogs act as allosteric mTORC1 inhibitors and do not directly affect the mTOR catalytic site [15]. They associate with the FK506-binding protein 12 (FKBP-12) and by so doing, they induce disassembly of mTORC1, resulting in the repression of its activity [136, 137]. The rapalogs have been examined in clinical trials of various cancers including brain, breast, HCC, leukemia, lymphoma, MM, NSCLC, pancreatic, prostate, and RCC [138, 139]. The rapalogs Torisel (Pfizer) and Afinitor (Novartis) were approved in 2007 and 2009 (respectively) to treat RCC patients [140]. In 2008, Torisel was approved to treat Mantle cell lymphoma patients. In 2010, Afinitor was approved to treat subependymal giant cell astrocytoma (SEGA) tumors in tuberous sclerosis (TS) patients. In 2011, Afinitor was approved to treat patients with pancreatic neuroendocrine tumors [141]. Ridaforolimus (also known as AP23573 and MK-8669; formerly known as deforolimus) is a rapalog developed by ARIAD and Merck. Ridaforolimus has been evaluated in clinical trials with patients having metastatic soft-tissue or bone sarcomas where it displays promising results in terms of the risk of progression or death [142]. Recently, the ability of rapamycin and rapalog to treat various viral infections including AIDS has been considered [143, 144]. Clearly, rapamycin has proven to be a very useful drug.

13.4.11 mTOR Inhibitors

Small molecules designed for inhibiting the catalytic site of mTOR have shown promising effects on the suppression of signaling downstream of mTOR. mTOR kinase inhibitor has been developed which directly inhibits mTORC1 and mTORC2. The mTOR kinase inhibitors have advantages over rapamycin and the rapalogs as mTOR inhibitors will inhibit both mTORC1 and mTORC2, while rapamycin and the rapalogs only inhibit mTORC1. Also, the mTOR kinases inhibitors do not induce the feedback pathways which result in Akt activation. In vitro studies with purified mTOR and PI3K proteins have demonstrated that the mTOR inhibitors selectively bind mTOR more than PI3K.

13.4.11.1 OSI-027

OSI-027 is a pan-TOR inhibitor developed by OSI Pharmaceuticals/Astellas Pharma Inc. OSI-027 has been shown to be effective in inducing apoptosis in different types of cancer, including breast and leukemias [145, 146]. OSI-027 has been shown to inhibit the growth of imatinib-resistant CML cells which contain the *BCR-ABL T315I* mutation that are resistant to all BCR-ABL inhibitors [147]. OSI-027 has been evaluated in clinical trials with patients with advanced solid tumors and lymphoma [148].

13.4.11.2 Intelllikine mTOR Inhibitors

PP-242 is a potent inhibitor of both mTORC1 and mTORC2. INK-128 is a derivative of PP-242 which has shown anti-tumoral effects on multiple cancer types including RCC, MM, NHL, and prostate [149, 150]. INK-128 is in phase I clinical trials for patients with relapsed or refractory multiple myeloma or Waldenstrom's macroglobulinemia or patients with solid malignancies.

13.4.11.3 AstraZenica mTOR Inhibitors

AZD8055 and AZD2014 are pan-mTOR inhibitors with potent anti-tumor activity [151]. They are being evaluated in clinical trials patients with gliomas who have not responded to standard glioma therapies as well as patients with other types of cancer.

13.4.11.4 Palomid 529

Palomid 529 (Paloma Pharmaceuticals) is a pan-mTOR inhibitor which has potent anti-tumor affects and reduces tumor angiogenesis and vascular permeability [152]. Palomid 529 is undergoing phase I clinical trials for patients with macular degeneration.

13.4.11.5 Pfizer mTOR Inhibitors

WAY600, WYE353, WYE687, and WYE132 were developed by Wyeth (Pfizer). These inhibitors were derived from WAY001 which was more specific for PI3Kα than either mTORC1 or mTORC2. These inhibitors were modified which resulted in WYE132 (WYE125132)/WYE132 has 5000-fold greater selectivity for mTOR over PI3K. It caused tumor regression in breast, glioma, lung, renal tumors [153].

13.4.11.6 Other mTOR Inhibitors

Many other TOR inhibitors have been described which include Ku0063794 (KuDOS Pharmaceuticals) [154] and OXA-01 (OSI Pharmaceuticals) [155]. Torin2 has been developed by optimizing from Torin1 [156]. TORKiCC223 is a pan-TOR inhibitor developed by Celgene. Other companies are developing mTOR inhibitors; clearly, this is a very competitive but important research and clinical area.

13.5 Increasing the Effectiveness of Targeting the Raf/MEK/ERK and PI3K/PTEN/Akt/mTOR Pathways by Simultaneous Treatment with Two Pathway Inhibitors

In the following section, we discuss the potential of combining inhibitors that target two pathways to more effectively limit cancer growth. Treatment for inducible murine lung cancers containing *KRAS* and *PIK3CA* mutations with PI3K/mTOR (NVP-BEZ235) and MEK (selumetinib) inhibitors led to an enhanced response [157]. Synergistic responses between sorafenib and mTOR inhibitors were observed in xenograft studies with a highly metastatic human HCC tumor [158]. Some recent studies in thyroid cancer have documented the benefit of combining Raf and PI3K/mTOR inhibitors [159].

Intermittent dosing of MEK and PI3K inhibitors has been observed to suppress the growth of tumor xenografts in mice [74]. This study demonstrated that continuous administration of MEK and PI3K inhibitors is not required to suppress xenograft growth. These important results were obtained by performing washout studies in vitro and alternate dosing schedules in mice with MEK and PI3K inhibitors with cancer cells having mutations at *BRAF* and *KRAS*.

The combined effects of inhibiting MEK with PD-0329501 and mTOR with rapamycin or its analog, the rapalog AP-23573 (ARIAD Pharmaceuticals/Merck) were examined in human NSCLC cell lines, as well as in animal models of human lung cancer. PD-0325901 and rapamycin demonstrated synergistic inhibition of proliferation and protein translation. Suppression of both MEK and mTOR inhibited ribosomal biogenesis and was associated with a block in the initiation phase of translation [160]. The pan-TOR inhibitor AZD-8055 has been examined as a single agent and in combination with the MEK inhibitor selumetinib in a NSCLC xenograft and increased cell death and tumor regression [151, 161]. These preclinical results support the suppression of both the MEK and mTOR pathways in lung cancer therapy and indicate that both pathways converge to regulate the initiation of protein translation. ERK phosphorylates Mnk1/2 and $p90^{Rsk}$, which regulate the activity of the eukaryotic translation initiation factor eIF4E. The phosphorylation of 4EBP1 is altered in cells containing *BRAF* mutations. It should also be pointed out that 4EBP1 is also regulated by Akt, mTOR, and p70S6K. This may result in the efficient translation of certain mRNAs in *BRAF*-mutant cells. This could explain how co-inhibition of MEK and PI3K/Akt/mTOR synergizes to inhibit protein translation and growth in certain lung cancer cells.

13.6 Clinical Trials Based on Inhibiting Both the Raf/MEK/ERK and PI3K/PTEN/Akt/mTOR Pathways

Combinations of Raf and PI3K/Akt/mTOR or MEK and PI3 K/Akt/mTOR inhibitors are in clinical trials. The results of a phase 1 clinical trial on patients with advanced solid tumors indicate that the combined dosing appears to be well tolerated, at least as well as single agent dosing. Some anti-tumor effects were observed, and dose-escalation trials were performed [162]. Clinical trial combining MEK and Akt inhibitors (GSK1120212 and GSK2141795, respectively) is in progress. A clinical trial for patients with advanced cancers combining the PI3K/mTOR inhibitors (PF-04691502 and PF-05212384) with the MEK inhibitor (PD-0325901) or irinotecan is in progress. The study will include patients with metastatic CRC patients who have received previous therapy for their disease and whose cancers have a mutant *KRAS* gene. The dual PI3 K/mTOR inhibitor NVP-BEZ235 is in a combination clinical trial with RAD001 (everolimus) in patients with advanced solid cancers. A phase 1 clinical trial is in progress combining the MEK1/2 inhibitor MEK162 and the PI3K/mTOR dual inhibitor NVP-BEZ235. This combination will be evaluated in various cancer patients, for example, NSCLC with mutations at *EGFR* who have progressed after treatment with EGFR inhibitors, triple-negative breast, CRC, melanoma, and pancreatic cancers. In addition, patients with other advanced solid tumors with *KRAS*, *NRAS*, and/or *BRAF* mutations will be included in the study. A trial is underway testing the effects of combining two experimental drugs, SD703026 (MSC1936369B) (a MEK inhibitor) and XL755 SAR245409 (a PI3K/mTOR inhibitor) for the treatment for locally advanced or metastatic solid tumors. Patients with breast, NSCLC, melanoma, and colorectal cancers will be treated with this inhibitor combination. A clinical trial is examining the effects of combining MK-2206 (an Akt inhibitor) with selumetinib (a MEK inhibitor) in cancer patients with advanced solid tumors. A combination clinical trial combining the MEK inhibitor selumetinib and the Akt inhibitor MK-2206 in patients with stage III or stage IV melanoma that previous failed after treatment with vemurafenib or dabrafenib is in progress.

13.7 Trials Based on Combining Raf/MEK, PI3K/Akt/mTOR Inhibitors with Chemotherapy or MoAbs

Treatment of mice xenografted with vemurafenib-resistant *BRAF*-mutant CRCs with various combinations of vermurafenib and chemotherapeutic drugs (capecitabine, irinotecan), MoAbs [bevacizumab (avastin, targets VEGFα), cetuximab (erbitux, targets EGFR)], the Akt inhibitor MK-2206, or the EGFR inhibitor erlotinib, increased survival [163]. Combination of the Akt inhibitor MK-2206 and either EGFR/HER2-targeted therapy [erlotinib or lapatinib (tykerb, a dual EGFR and HER2 inhibitor from GSK)] or chemotherpapeutic drugs doxorubicin, camptothechin, gemcitabine, 5-flurouracil, docetaxel or carboplatin resulted in synergistic responses in lung (NCI-H460) and ovarian (A2780) cancer cell lines. In some cases, the timing of drug addition was determined to be important as MK-2206 suppressed the Akt activation induced by carboplatin and gemcitabine [164]. The effects of combining the dual PI3K/mTOR inhibitor NVP-BEZ235 and various chemotherapeutic drugs as well as other targeted therapies are being examined (doxorubicin, melphalan, vincristine, bortezomib) [165, 166]. The anti-tumor effects of WYE132 (a mTOR inhibitor) could be enhanced upon combination with bevacizumab in lung and breast xenograft models [153]. A clinical trial with INK-128 in combination with paclitaxel, either in the absence or in the presence of trastuzumab, is in progress in patients with advanced solid malignancies.

Clinical trials are ongoing based on combining the dual PI3K/mTOR inhibitor NVP-BEZ235 with PI3K and MEK inhibitors (BKM120, MEK162) and chemotherapeutic drugs (paclitaxel, trastuzumab) to treat advanced solid cancers and metastatic breast cancers which are difficult to treat (see below). BKM120 is a pan-PI3 K inhibitor. It is being included in some clinical studies since NVP-BEZ235 does not inhibit PI3Kβ [167]. Furthermore, NVP-BEZ235 is not effective in suppressing the growth of tumors which have the *KRAS G12D* mutation [157]. Thus, to achieve effective suppression of cancer growth in some situations, it may be important to combine PI3K/mTOR inhibitors with pan-PI3K inhibitors.

Palomid 529, a pan-mTOR inhibitor, is in some circumstances is effective as a single agent. However, when Palomid 529 was combined with either cisplatin or docetaxel, it had a better effect on hormone-refractory prostate cancers [168]. It also improved the effects of radiotherapy on prostate cancer cells [169].

The effects of inhibiting Akt in combination with other pathways, inhibitors, and chemotherapy are being evaluated in numerous phase I clinical trials. These trials highlight the importance of targeting multiple molecules to suppress the growth of cancer which are resistant to most therapies. Combination clinical trials with the Akt inhibitor MK-2206 and the dual EGFR/HER2 inhibitor lapatinib are in progress with patients having advanced or metastatic solid tumors or breast cancer patients. The effects of combining MK-2206 and erlotinib, docetaxel, or carboplatin + paclitaxel are being examined in clinical trials in certain patients with advanced cancers. Clinical trials with NSCLC patients are underway to examine the effects of combining MK-2206 with gefitinib (iressa, EGFR inhibitor AstraZenica). Clinical trials with postmenopausal metastatic breast cancer patients are in progress to examine the effects of combining anastrozole, letrozole, exemestane (aromatase-inhibitors), or fulvestrant (an estrogen receptor antagonist). Clinical trials are also underway examining the effects of combining MK-2206 with bendamustin (nitrogen mustard alkylating agent) and Rituximab (chimeric monoclonal antibody targeting CD20 from IDEC Pharmaceuticals/Genentech) on CLL cancer patients who have relapsed or on cancer patients with small lymphocytic lymphoma. Clinical trials combining MK-2206 and various other drugs including dalotuzumab (a MoAb which targets IGF-1R from Merck) and MK-0752 (a Y-secretase inhibitor which inhibits the NOTCH pathway from Merck) are in progress. The effects of MK-8669 (ridaforolimus an mTORC1 inhibitor from Merck) and dalotuzumab are being examined in patients with advanced cancers. Clinical trials combining MK-2206 and paclitaxel in cancer patients with locally advanced, metastatic solid tumors, or metastatic breast cancers are in progress. The above mentioned clinical trials document the importance of targeting Akt and other signaling molecules as well as critical targets involved in cellular division. Furthermore, the clinical trials document how basic research on these pathways is being translated into clinical therapy for cancer and other types of patients.

13.8 Enhancing the Effectiveness of Raf/MEK and PI3K/mTOR Inhibitors with Radiotherapy

Radiotherapy is a common therapeutic approach for treatment for many diverse cancers. A side effect of radiotherapy in some cells is induction of the Ras/Raf/MEK/ERK cascade [2]. Various signal transduction inhibitors have been evaluated as radiosensitizers. The effects of pre-treatment for lung, pancreatic, and prostate cancer cells with selumetinib were evaluated in vitro using human cell lines and in vivo employing xenografts [170]. The MEK inhibitor treatment radiosensitized the various cancer cell lines in vitro and in vivo. The MEK inhibitor treatment was correlated with decreased Chk1 phosphorylation 1–2 h after radiation. The authors noticed the effects of the MEK inhibitor on the G_2 checkpoint activation after irradiation, as the MEK inhibitor suppressed G_2 checkpoint activation. Since ERK1/ERK2 activity is necessary for the carcinoma cells to arrest at the G_2 checkpoint, suppression of phosphorylated Chk1 was speculated to lead to the abrogated G_2 checkpoint, increased mitotic catastrophe, and impaired activation of cell cycle checkpoints. Mitotic catastrophe was increased in cells receiving both the MEK inhibitor and radiation when compared to the solo-treated cells. It was also postulated in this study that the MEK inhibitor suppressed the autocrine cascade in DU145 prostate cancer cells that normally resulted from EGF secretion and EGFR activation. Suppression of this autocrine cascade by the MEK inhibitor may have served as a radiosensitizer. The other two cancer cell lines examined in this study (A549 and MiaPaCa2) had *KRAS* mutations and both were radiosensitized by the MEK inhibitor. Although these studies document the ability of a MEK inhibitor to radiosensitize certain cells, clearly other cancer cell lines without activating mutations in the Ras/Raf/MEK/ERK pathway or autocrine growth stimulation should be examined for radiosensitization by the MEK inhibitor as the *KRAS* mutation may also activate the PI3K pathway which could lead to therapy resistance.

PI3K/Akt/mTOR inhibitors will sensitize the tumor vasculature to radiation both in vitro in cell lines and in vivo in xenografts [171, 172]. mTOR and radiation play critical roles in the regulation of autophagy [173, 174]. When mTOR is blocked by rapamycin, there is an increase in autophagy. This is important as apoptotic cell death is a minor component to cell death in many solid tumors. These studies document the potential beneficial use of combining mTOR inhibitors and radiation to improve the induction of autophagy in the treatment for solid tumors.

13.9 Conclusions

Inhibitors to the Ras/Raf/MEK/ERK and Ras/PI3K/PTEN/Akt/mTOR pathways. Initially, MEK inhibitors were demonstrated to have the most specificity. However, these inhibitors may have limited effectiveness in treating human cancers, unless the particular cancer proliferates directly in response to the Raf/MEK/ERK pathway. Moreover, MEK inhibitors are often cytostatic as opposed to cytotoxic; thus, their ability to function as effective anticancer agents in a monotherapeutic setting is limited, and they may be more effective when combined with other small molecule inhibitors, MoAbs, chemo- or radiotherapy. Raf inhibitors have also been developed, and some are being used to treat various cancer patients (e.g., sorafenib and vemurafenib). This particular Raf inhibitor also inhibits other receptors and kinases which may be required for the growth of the particular cancer. This promiscuous nature of sorafenib has contributed to the effectiveness of this particular Raf inhibitor for certain cancers. Raf inhibitors such as vemurafenib, dabrafenib, and GDC-0879 are promising for the treatment for melanoma, CRC, thyroid as well as other solid cancers, and leukemias/lymphomas/myelomas which have mutations at *BRAF V600E*. However, problems have been identified with certain Raf inhibitors as they will be ineffective if Ras is mutated/amplified, if an exon of *BRAF* is deleted, if *BRAF* is amplified, if there are mutations at *MEK1* and various other genetic mechanisms that result in resistance [175]. Combination therapies with either a traditional drug/physical treatment or another inhibitor that targets a specific molecule in either the same or a different signal transduction pathway are also key approaches for improving the effectiveness and usefulness of MEK and Raf inhibitors.

Modified rapamycins (rapalogs) are being used to treat various cancer patients (e.g., patients with RCC and some other cancers). While rapalogs are effective and their toxicity profiles are well known, one inherent property is that they are not very cytotoxic when it comes to killing tumor cells. This inherent property of rapamycins may also contribute to their low toxicity in humans.

Mutations at many of the upstream receptor genes or *RAS* can result in abnormal Raf/MEK/ERK and PI3K/PTEN/Akt/mTOR pathway activation. Hence, targeting these cascade components with small molecule inhibitors may inhibit cell growth. The usefulness of these inhibitors may depend on the mechanism of transformation of the particular cancer. If the tumor exhibits a dependency on the Ras/Raf/MEK/ERK pathway, then it may be sensitive to Raf and MEK inhibitors. In contrast, tumors that do not display enhanced expression of the Ras/Raf/MEK/ERK pathway may not be sensitive to either Raf or MEK inhibitors, but if the Ras/PI3K/Akt/mTOR pathway is activated, these cancers may be sensitive to specific inhibitors that target this pathway. Finally, it is likely that many of the inhibitors that we have discussed in this review will be more effective in inhibiting tumor growth in combination with MoAb, cytotoxic chemotherapeutic drugs, or radiation. This is documented by the large number of clinical trials combining signal transduction inhibitors with these various therapeutic agents.

Some scientists and clinicians have considered that the simultaneous targeting of Raf and MEK by individual inhibitors may be more effective in cancer therapy than just targeting Raf or MEK by themselves. This is based in part on the fact that there are intricate feedback loops from ERK which can inhibit Raf and MEK. For example, when MEK1 is targeted, ERK1,2 is inhibited, and the negative feedback loop on MEK is broken and activated MEK accumulates. However, if Raf is also inhibited, it may be possible to completely shut down the pathway. This is a rationale for treatment with both MEK and Raf inhibitors. Likewise, targeting both PI3K and mTOR may be more effective than targeting either PI3K or mTOR by themselves. If it is a single inhibitor which targets both molecules, such as the new PI3K and mTOR dual inhibitors, this becomes a realistic therapeutic option. Although it should be pointed out that some studies are examining the effects of combining rapamycins and PI3K/mTOR inhibitors. Finally, an emerging concept is the dual targeting of two different signal transduction pathways, Raf/MEK/ERK and PI3K/PTEN/Akt/mTOR, for example. This has been explored in some preclinical models as discussed in the text. The rationale for targeting of both pathways may be dependent on the presence of mutations in either/or both pathways or in upstream Ras in the particular cancer which can activate both pathways.

It is not always clear why a particular combination of a signal transduction inhibitor and chemotherapeutic drug works in one tumor type but not at all in a different tumor type. This has also been experienced with the development of individual chemotherapeutic drugs, some work in some cells but not others. This may result from many different complex interacting events. Some of these events could include percentage of cells in different phases of the cell cycle, persistence of cancer stem cells, presence of multiple mutated activated oncogenes, or repressed tumor suppressor genes, epigenetic modifications and many other factors. Finally, chemotherapeutic drug therapy and other types of therapy (radiotherapy, antibody therapy) may induce certain signaling pathways (e.g., the reactive oxygen species generated by chemotherapy and radiotherapy induce the Ras/Raf/MEK/ERK pathway). The induction of these signaling pathways may counteract some of the effects of the signal transduction inhibitors. These effects could indicate that the timing of each therapy will be important for effective treatments.

In summary, targeted therapy has advanced from basic research studies to the treatment of cancer patients in less than 25 years. We have learned a lot regarding how specific inhibitors exert their effects and how the Ras/Raf/MEK/ERK and PI3K/Akt/mTOR pathways function; we still need to discover more about resistance mechanisms and how we can overcome therapeutic resistance and improve the effectiveness of targeted therapy.

Acknowledgments MC and GM were supported in part by grants from the Italian "Ministero dell'Istruzione, dell'Università e della Ricerca (Ministry for Education, Universities and Research)—MIUR" PRIN 2008 and FIRB-MERIT n. RBNE08YYBM. MC was also supported in part by a grant to the CNR from the Italian Ministry of Economy and Finance for the Project FaReBio di Qualità. ML was supported in part by a grant from the Italian Ministry of Health, Ricerca Finalizzata Stemness 2008 entitled "Molecular Determinants of Stemness and Mesenchymal Phenotype in Breast Cancer". AMM was supported in part by grants from Fondazione del Monte di Bologna e Ravenna, MinSan 2008 "Molecular therapy in pediatric sarcomas and leukemias against IGF-IR system: new drugs, best drug–drug interactions, mechanisms of resistance and indicators of efficacy", MIUR PRIN 2008 (2008THTNLC), and MIUR FIRB 2010 (RBAP10447J-003) and 2011 (RBAP11ZJFA_001). MM was supported in part from the Italian Association for Cancer Research (AIRC), the Cariplo Foundation, and the Italian Ministry of Health. AT was supported in part by grants from the Italian "Ministero dell'Istruzione, dell'Università e della Ricerca (Ministry for Education, University and Research) —MIUR—PRIN 2008 and grant from "Sapienza", University of Rome 2009–11.

References

1. Chappell WH, Steelman LS, Long JM, Kempf RC, Abrams SL, Franklin RA et al (2011) Ras/Raf/MEK/ERK and PI3K/PTEN/Akt/mTOR inhibitors: rationale and importance to inhibiting these pathways in human health. Oncotarget 2(3):135–164
2. McCubrey JA, Steelman LS, Kempf CR, Chappell WH, Abrams SL, Stivala F et al (2011) Therapeutic resistance resulting from mutations in Raf/MEK/ERK and PI3 K/PTEN/Akt/mTOR signaling pathways. J Cell Physiol 226(11):2762–2781
3. Hayashi K, Shibata K, Morita T, Iwasaki K, Watanabe M, Sobue K (2004) Insulin receptor substrate-1/SHP-2 interaction, a phenotype-dependent switching machinery of insulin-like growth factor-I signaling in vascular smooth muscle cells. J Biol Chem 279(39):40807–40818
4. Mischak H, Seitz T, Janosch P, Eulitz M, Steen H, Schellerer M et al (1996) Negative regulation of Raf-1 by phosphorylation of serine 621. Mol Cell Biol 16(10):5409–5418
5. Abraham D, Podar K, Pacher M, Kubicek M, Welzel N, Hemmings BA et al (2000) Raf-1-associated protein phosphatase 2A as a positive regulator of kinase activation. J Biol Chem 275(29):22300–22304
6. Brennan DF, Dar AC, Hertz NT, Chao WC, Burlingame AL, Shokat KM et al (2011) A Raf-induced allosteric transition of KSR stimulates phosphorylation of MEK. Nature 472(7343):366–369
7. McKay MM, Ritt DA, Morrison DK (2011) RAF inhibitor-induced KSR1/B-RAF binding and its effects on ERK cascade signaling. Curr Biol 21(7):563–568

8. Pao W, Girard N (2011) New driver mutations in non-small-cell lung cancer. Lancet Oncol 12(2):175–180
9. Topisirovic I, Sonenberg N (2011) mRNA translation and energy metabolism in cancer: the role of the MAPK and mTORC1 pathways. Cold Spring Harb Symp Quant Biol 28:28
10. Xing J, Ginty DD, Greenberg ME (1996) Coupling of the RAS-MAPK pathway to gene activation by RSK2, a growth factor-regulated CREB kinase. Science 273(5277):959–963
11. Balan V, Leicht DT, Zhu J, Balan K, Kaplun A, Singh-Gupta V et al (2006) Identification of novel in vivo Raf-1 phosphorylation sites mediating positive feedback Raf-1 regulation by extracellular signal-regulated kinase. Mol Biol Cell 17(3):1141–1153
12. Dougherty MK, Muller J, Ritt DA, Zhou M, Zhou XZ, Copeland TD et al (2005) Regulation of Raf-1 by direct feedback phosphorylation. Mol Cell 17(2):215–224
13. Catalanotti F, Reyes G, Jesenberger V, Galabova-Kovacs G, de Matos Simoes R, Carugo O et al (2009) A Mek1-Mek2 heterodimer determines the strength and duration of the Erk signal. Nat Struct Mol Biol 16(3):294–303
14. Sturm OE, Orton R, Grindlay J, Birtwistle M, Vyshemirsky V, Gilbert D et al (2010) The mammalian MAPK/ERK pathway exhibits properties of a negative feedback amplifier. Sci Signal 3(153)
15. Martelli AM, Evangelisti C, Chappell W, Abrams SL, Basecke J, Stivala F et al (2011) Targeting the translational apparatus to improve leukemia therapy: roles of the PI3K/PTEN/Akt/mTOR pathway. Leukemia 25(7):1064–1079
16. Martelli AM, Chiarini F, Evangelisti C, Cappellini A, Buontempo F, Bressanin D et al (2012) Two hits are better than one: targeting both phosphatidylinositol 3-kinase and mammalian target of rapamycin as a therapeutic strategy for acute leukemia treatment. Oncotarget 3(4):371–394
17. Steelman LS, Chappell WH, Abrams SL, Kempf RC, Long J, Laidler P et al (2011) Roles of the Raf/MEK/ERK and PI3K/PTEN/Akt/mTOR pathways in controlling growth and sensitivity to therapy-implications for cancer and aging. Aging 3(3):192–222
18. Alessi DR, James SR, Downes CP, Holmes AB, Gaffney PR, Reese CB et al (1997) Characterization of a 3-phosphoinositide-dependent protein kinase which phosphorylates and activates protein kinase Balpha. Curr Biol 7(4):261–269
19. Coffer PJ, Woodgett JR (1992) Molecular cloning and characterisation of a novel putative protein-serine kinase related to the cAMP-dependent and protein kinase C families. Eur J Biochem 205(3):1217
20. Gonzalez E, McGraw TE (2009) The Akt kinases: isoform specificity in metabolism and cancer. Cell Cycle 8(16):2502–2508
21. Du K, Montminy M (1998) CREB is a regulatory target for the protein kinase Akt/PKB. J Biol Chem 273(49):32377–32379
22. Brennan P, Babbage JW, Burgering BM, Groner B, Reif K, Cantrell DA (1997) Phosphatidylinositol 3-kinase couples the interleukin-2 receptor to the cell cycle regulator E2F. Immunity 7(5):679–689
23. Kane LP, Shapiro VS, Stokoe D, Weiss A (1999) Induction of NF-kappaB by the Akt/PKB kinase. Curr Biol 9(11):601–604
24. Buitenhuis M, Coffer PJ (2009) The role of the PI3K-PKB signaling module in regulation of hematopoiesis. Cell Cycle 8(4):560–566
25. del Peso L, Gonzalez-Garcia M, Page C, Herrera R, Nunez G (1997) Interleukin-3-induced phosphorylation of BAD through the protein kinase Akt. Science 278(5338):687–689
26. Cross DA, Alessi DR, Cohen P, Andjelkovich M, Hemmings BA (1995) Inhibition of glycogen synthase kinase-3 by insulin mediated by protein kinase B. Nature 378(6559):785–789
27. Chalhoub N, Baker SJ (2009) PTEN and the PI3-kinase pathway in cancer. Annu Rev Pathol 4:127–150
28. Weng LP, Brown JL, Baker KM, Ostrowski MC, Eng C (2002) PTEN blocks insulin-mediated ETS-2 phosphorylation through MAP kinase, independently of the phosphoinositide 3-kinase pathway. Hum Mol Genet 11(15):1687–1696

29. Mahimainathan L, Choudhury GG (2004) Inactivation of platelet-derived growth factor receptor by the tumor suppressor PTEN provides a novel mechanism of action of the phosphatase. J Biol Chem 279(15):15258–15268
30. Krymskaya VP, Goncharova EA (2009) PI3K/mTORC1 activation in hamartoma syndromes: therapeutic prospects. Cell Cycle 8(3):403–413
31. Tamburini J, Green AS, Chapuis N, Bardet V, Lacombe C, Mayeux P et al (2009) Targeting translation in acute myeloid leukemia: a new paradigm for therapy? Cell Cycle 8(23):3893–3899
32. Peterson TR, Laplante M, Thoreen CC, Sancak Y, Kang SA, Kuehl WM et al (2009) DEPTOR is an mTOR inhibitor frequently overexpressed in multiple myeloma cells and required for their survival. Cell 137(5):873–886 (Epub 2009 May 14)
33. Sato T, Nakashima A, Guo L, Tamanoi F (2009) Specific activation of mTORC1 by Rheb G-protein in vitro involves enhanced recruitment of its substrate protein. J Biol Chem 284(19):12783–12791 (Epub 2009 Mar 19)
34. Shaw RJ, Bardeesy N, Manning BD, Lopez L, Kosmatka M, DePinho RA et al (2004) The LKB1 tumor suppressor negatively regulates mTOR signaling. Cancer Cell 6(1):91–99
35. Shaw RJ, Lamia KA, Vasquez D, Koo SH, Bardeesy N, Depinho RA et al (2005) The kinase LKB1 mediates glucose homeostasis in liver and therapeutic effects of metformin. Science 310(5754):1642–1646
36. Evans JM, Donnelly LA, Emslie-Smith AM, Alessi DR, Morris AD (2005) Metformin and reduced risk of cancer in diabetic patients. BMJ 330(7503):1304–1305
37. Shackelford DB, Shaw RJ (2009) The LKB1-AMPK pathway: metabolism and growth control in tumour suppression. Nat Rev Cancer 9(8):563–575
38. Martinelli A, Chiarini F, Evangelisti C, Ognibene A, Bressanin D, Billi AM, Manzoli L, Cappellini A, McCubrey JA (2012) Targeting the liver kinase B1/AMP-dependent kinase pathway as a therapeutic strategy for hematologic malignancies. Expert Opinion Therapeutic Targets (In Press)
39. Carracedo A, Ma L, Teruya-Feldstein J, Rojo F, Salmena L, Alimonti A et al (2008) Inhibition of mTORC1 leads to MAPK pathway activation through a PI3K-dependent feedback loop in human cancer. J Clin Invest 118(9):3065–3074
40. Hresko RC, Mueckler M (2005) mTOR.RICTOR is the Ser473 kinase for Akt/protein kinase B in 3T3-L1 adipocytes. J Biol Chem 280(49):40406–40416
41. Liang C (2010) Negative regulation of autophagy. Cell Death Differ 17(12):1807–1815
42. Poulikakos PI, Solit DB (2011) Resistance to MEK inhibitors: should we co-target upstream? Sci Signal 4(166)
43. Wilhelm SM, Carter C, Tang L, Wilkie D, McNabola A, Rong H et al (2004) BAY 43–9006 exhibits broad spectrum oral antitumor activity and targets the RAF/MEK/ERK pathway and receptor tyrosine kinases involved in tumor progression and angiogenesis. Cancer Res 64(19):7099–7109
44. Rimassa L, Santoro A (2009) Sorafenib therapy in advanced hepatocellular carcinoma: the SHARP trial. Expert Rev Anticancer Ther 9(6):739–745
45. Cervello M, McCubrey JA, Cusimano A, Lampiasi N, Azzolina A, Montalto G (2012) Targeted therapy for hepatocellular carcinoma: novel agents on the horizon. Oncotarget 3(3):236–260
46. Smalley KS, Xiao M, Villanueva J, Nguyen TK, Flaherty KT, Letrero R et al (2009) CRAF inhibition induces apoptosis in melanoma cells with non-V600E BRAF mutations. Oncogene 28(1):85–94
47. Bollag G, Hirth P, Tsai J, Zhang J, Ibrahim PN, Cho H et al (2010) Clinical efficacy of a RAF inhibitor needs broad target blockade in BRAF-mutant melanoma. Nature 467(7315):596–599
48. Flaherty KT, Puzanov I, Kim KB, Ribas A, McArthur GA, Sosman JA et al (2010) Inhibition of mutated, activated BRAF in metastatic melanoma. N Engl J Med 363(9):809–819

49. Chapman PB, Hauschild A, Robert C, Haanen JB, Ascierto P, Larkin J et al (2011) Improved survival with vemurafenib in melanoma with BRAF V600E mutation. N Engl J Med 364(26):2507–2516
50. Long G, Kefford RF, Carr PJA, Brown MP, Curtis M, Ma B, Lebowitz P, Kim KB, Kurzrock R, Flachook G(2010) Phase 1/2 study of GSK2118436, a selective inhibitor of V600 mutant (mut) BRAF kinase: evidence of activity in melanoma brain metastases (mets). Annals of Oncology. 21(LBA27 (supplement 8)):viii12
51. Kefford R, Arkenau H, Brown MP, Millward M, Infante JR, Long GV, et al (2010) Phase I/II study of GSK2118436, a selective inhibitor of oncogenic mutant BRAF kinase, in patients with metastatic melanoma and other solid tumors. ASCO Meeting Abstracts 2010 June 14, 28(15_suppl):8503
52. Greger JG, Eastman SD, Zhang V, Bleam MR, Hughes AM, Smitheman KN et al (2012) Combinations of BRAF, MEK, and PI3K/mTOR inhibitors overcome acquired resistance to the BRAF inhibitor GSK2118436 dabrafenib, mediated by NRAS or MEK mutations. Mol Cancer Ther 11(4):909–920
53. Whittaker S, Menard D, Kirk R, Ogilvie L, Hedley D, Zambon A et al (2010) A novel, selective and efficacious nanomolar pyridopyrazinone inhibitor of V600EBRAF. Cancer Res 70(20):8036–8044
54. Hoeflich KP, Herter S, Tien J, Wong L, Berry L, Chan J et al (2009) Antitumor efficacy of the novel RAF inhibitor GDC-0879 is predicted by BRAFV600E mutational status and sustained extracellular signal-regulated kinase/mitogen-activated protein kinase pathway suppression. Cancer Res 69(7):3042–3051
55. Montagut C, Sharma SV, Shioda T, McDermott U, Ulman M, Ulkus LE et al (2008) Elevated CRAF as a potential mechanism of acquired resistance to BRAF inhibition in melanoma. Cancer Res 68(12):4853–4861
56. Schwartz GK, Robertson S, Shen A, Wang E, Pace L, Dials H et al (2009) A phase I study of XL281, a selective oral RAF kinase inhibitor, in patients (Pts) with advanced solid tumors. ASCO Meeting Abstracts. 27(15S):3513 (2009 June 8)
57. Buchholz B, Klanke B, Schley G, Bollag G, Tsai J, Kroening S et al (2011) The Raf kinase inhibitor PLX5568 slows cyst proliferation in rat polycystic kidney disease but promotes renal and hepatic fibrosis. Nephrol Dial Transplant 26(11):3458–3465
58. Zitzmann K, de Toni E, von Ruden J, Brand S, Goke B, Laubender RP et al (2011) The novel Raf inhibitor Raf265 decreases Bcl-2 levels and confers TRAIL-sensitivity to neuroendocrine tumour cells. Endocr Relat Cancer 18(2):277–285
59. Wilhelm SM, Dumas J, Adnane L, Lynch M, Carter CA, Schutz G et al (2011) Regorafenib (BAY 73–4506): a new oral multikinase inhibitor of angiogenic, stromal and oncogenic receptor tyrosine kinases with potent preclinical antitumor activity. Int J Cancer 129(1):245–255
60. Davies SP, Reddy H, Caivano M, Cohen P (2000) Specificity and mechanism of action of some commonly used protein kinase inhibitors. Biochem J 351(Pt 1):95–105
61. Ohren JF, Chen H, Pavlovsky A, Whitehead C, Zhang E, Kuffa P et al (2004) Structures of human MAP kinase kinase 1 (MEK1) and MEK2 describe novel noncompetitive kinase inhibition. Nat Struct Mol Biol 11(12).1192–1197
62. Sebolt-Leopold JS (2008) Advances in the development of cancer therapeutics directed against the RAS-mitogen-activated protein kinase pathway. Clin Cancer Res 14(12):3651–3656
63. McCubrey JA, Steelman LS, Abrams SL, Chappell WH, Russo S, Ove R et al (2010) Emerging MEK inhibitors. Expert Opin Emerg Drugs 15(2):203–223
64. Davies BR, Logie A, McKay JS, Martin P, Steele S, Jenkins R, et al (2007) AZD6244 (ARRY-142886), a potent inhibitor of mitogen-activated protein kinase/extracellular signal-regulated kinase kinase 1/2 kinases: mechanism of action in vivo, pharmacokinetic/pharmacodynamic relationship, and potential for combination in preclinical models. Molecular Cancer Therapeutics. 6(8):2209–2219

65. Tai YT, Fulciniti M, Hideshima T, Song W, Leiba M, Li XF et al (2007) Targeting MEK induces myeloma-cell cytotoxicity and inhibits osteoclastogenesis. Blood 110(5):1656–1663
66. Yeh TC, Marsh V, Bernat BA, Ballard J, Colwell H, Evans RJ et al (2007) Biological characterization of ARRY-142886 (AZD6244), a potent, highly selective mitogen-activated protein kinase 1/2 inhibitor. Clin Cancer Res 13(5):1576–1583
67. Friday BB, Yu C, Dy GK, Smith PD, Wang L, Thibodeau SN et al (2008) BRAF V600E disrupts AZD6244-induced abrogation of negative feedback pathways between extracellular signal-regulated kinase and Raf proteins. Cancer Res 68(15):6145–6153
68. Yoon J, Koo KH, Choi KY (2011) MEK1/2 inhibitors AS703026 and AZD6244 may be potential therapies for KRAS mutated colorectal cancer that is resistant to EGFR monoclonal antibody therapy. Cancer Res 71(2):445–453
69. Mohammad RM, Goustin AS, Aboukameel A, Chen B, Banerjee S, Wang G et al (2007) Preclinical Studies of TW-37, a New Nonpeptidic Small-Molecule Inhibitor of Bcl-2, in Diffuse Large Cell Lymphoma Xenograft Model Reveal Drug Action on Both Bcl-2 and Mcl-1. Clin Cancer Res 13(7):2226–2235
70. Haura EB, Ricart AD, Larson TG, Stella PJ, Bazhenova L, Miller VA et al (2010) A phase II study of PD-0325901, an oral MEK inhibitor, in previously treated patients with advanced non-small cell lung cancer. Clin Cancer Res 16(8):2450–2457
71. LoRusso PM, Krishnamurthi SS, Rinehart JJ, Nabell LM, Malburg L, Chapman PB et al (2010) Phase I pharmacokinetic and pharmacodynamic study of the oral MAPK/ERK kinase inhibitor PD-0325901 in patients with advanced cancers. Clin Cancer Res 16(6):1924–1937
72. Solit DB, Garraway LA, Pratilas CA, Sawai A, Getz G, Basso A et al (2006) BRAF mutation predicts sensitivity to MEK inhibition. Nature 439(7074):358–362
73. Iverson C, Larson G, Lai C, Yeh LT, Dadson C, Weingarten P et al (2009) RDEA119/BAY 869766: a potent, selective, allosteric inhibitor of MEK1/2 for the treatment of cancer. Cancer Res 69(17):6839–6847
74. Hoeflich KP, Merchant M, Orr C, Chan J, Den Otter D, Berry L et al (2012) Intermittent administration of MEK inhibitor GDC-0973 plus PI3K inhibitor GDC-0941 triggers robust apoptosis and tumor growth inhibition. Cancer Res 72(1):210–219
75. Kim K, Kong SY, Fulciniti M, Li X, Song W, Nahar S et al (2010) Blockade of the MEK/ERK signalling cascade by AS703026, a novel selective MEK1/2 inhibitor, induces pleiotropic anti-myeloma activity in vitro and in vivo. Br J Haematol 149(4):537–549
76. Lee TX, Packer MD, Huang J, Akhmametyeva EM, Kulp SK, Chen CS et al (2009) Growth inhibitory and anti-tumour activities of OSU-03012, a novel PDK-1 inhibitor, on vestibular schwannoma and malignant schwannoma cells. Eur J Cancer 45(9):1709–1720
77. Dong Q, Dougan DR, Gong X, Halkowycz P, Jin B, Kanouni T et al (2011) Discovery of TAK-733, a potent and selective MEK allosteric site inhibitor for the treatment of cancer. Bioorg Med Chem Lett 21(5):1315–1319
78. Longoni R, Spina L, Vinci S, Acquas E (2011) The MEK inhibitor SL327 blocks acquisition but not expression of lithium-induced conditioned place aversion: a behavioral and immunohistochemical study. Psychopharmacology 216(1):63–73
79. Poulikakos PI, Zhang C, Bollag G, Shokat KM, Rosen N (2010) RAF inhibitors transactivate RAF dimers and ERK signalling in cells with wild-type BRAF. Nature 464(7287):427–430
80. Konopleva M, Milella M, Ruvolo P, Watts JC, Ricciardi MR, Korchin B et al (2012) MEK inhibition enhances ABT-737-induced leukemia cell apoptosis via prevention of ERK-activated MCL-1 induction and modulation of MCL-1/BIM complex. Leukemia 26(4):778–787
81. Ricciardi MR, Scerpa MC, Bergamo P, Ciuffreda L, Petrucci MT, Chiaretti S et al (2012) Therapeutic potential of MEK inhibition in acute myelogenous leukemia: rationale for "vertical" and "lateral" combination strategies. J Mol Med 8:8
82. Aronov AM, Tang Q, Martinez-Botella G, Bemis GW, Cao J, Chen G et al (2009) Structure-guided design of potent and selective pyrimidylpyrrole inhibitors of extracellular signal-regulated kinase (ERK) using conformational control. J Med Chem 52(20):6362–6368

83. Hatzivassiliou G, Liu B, O'Brien C, Spoerke JM, Hoeflich KP, Haverty PM et al (2012) ERK inhibition overcomes acquired resistance to MEK inhibitors. Mol Cancer Ther 11(5):1143–1154
84. Ihle NT, Williams R, Chow S, Chew W, Berggren MI, Paine-Murrieta G et al (2004) Molecular pharmacology and antitumor activity of PX-866, a novel inhibitor of phosphoinositide-3-kinase signaling. Mol Cancer Ther 3(7):763–772
85. Burrows N, Babur M, Resch J, Ridsdale S, Mejin M, Rowling EJ et al (2011) GDC-0941 inhibits metastatic characteristics of thyroid carcinomas by targeting both the phosphoinositide-3 kinase (PI3K) and hypoxia-inducible factor-1alpha (HIF-1alpha) pathways. J Clin Endocrinol Metab 96(12):12
86. Zou ZQ, Zhang LN, Wang F, Bellenger J, Shen YZ, Zhang XH (2012) The novel dual PI3K/mTOR inhibitor GDC-0941 synergizes with the MEK inhibitor U0126 in non-small cell lung cancer cells. Mol Med Report 5(2):503–508
87. Sujobert P, Bardet V, Cornillet-Lefebvre P, Hayflick JS, Prie N, Verdier F et al (2005) Essential role for the p110delta isoform in phosphoinositide 3-kinase activation and cell proliferation in acute myeloid leukemia. Blood 106(3):1063–1066
88. Billottet C, Grandage VL, Gale RE, Quattropani A, Rommel C, Vanhaesebroeck B et al (2006) A selective inhibitor of the p110delta isoform of PI 3-kinase inhibits AML cell proliferation and survival and increases the cytotoxic effects of VP16. Oncogene 25(50):6648–6659
89. Tamburini J, Chapuis N, Bardet V, Park S, Sujobert P, Willems L et al (2008) Mammalian target of rapamycin (mTOR) inhibition activates phosphatidylinositol 3-kinase/Akt by up-regulating insulin-like growth factor-1 receptor signaling in acute myeloid leukemia: rationale for therapeutic inhibition of both pathways. Blood 111(1):379–382
90. Workman P, van Montfort RL (2010) PI(3) kinases: revealing the delta lady. Nat Chem Biol 6(2):82–83
91. Berndt A, Miller S, Williams O, Le DD, Houseman BT, Pacold JI et al (2010) The p110delta structure: mechanisms for selectivity and potency of new PI(3)K inhibitors. Nat Chem Biol 6(3):244
92. Lannutti BJ, Meadows SA, Herman SE, Kashishian A, Steiner B, Johnson AJ et al (2011) CAL-101, a p110delta selective phosphatidylinositol-3-kinase inhibitor for the treatment of B-cell malignancies, inhibits PI3K signaling and cellular viability. Blood 117(2):591–594
93. Meadows SA, Vega F, Kashishian A, Johnson D, Diehl V, Miller LL et al (2012) PI3Kdelta inhibitor, GS-1101 (CAL-101), attenuates pathway signaling, induces apoptosis, and overcomes signals from the microenvironment in cellular models of Hodgkin lymphoma. Blood 119(8):1897–1900
94. Gale S, Croasdell G (2010) 28th Annual JPMorgan healthcare conference–forest laboratories and Icagen. IDrugs 13(3):145–148
95. Maira SM, Pecchi S, Huang A, Burger M, Knapp M, Sterker D et al (2012) Identification and characterization of NVP-BKM120, an orally available pan-class I PI3-kinase inhibitor. Mol Cancer Ther 11(2):317–328
96. Bendell JC, Rodon J, Burris HA, de Jonge M, Verweij J, Birle D et al (2012) Phase I, dose-escalation study of BKM120, an oral pan-Class I PI3K inhibitor, in patients with advanced solid tumors. J Clin Oncol 30(3):282–290
97. Brachmann S, Fritsch C, Maira SM, Garcia-Echeverria C (2009) PI3K and mTOR inhibitors: a new generation of targeted anticancer agents. Curr Opin Cell Biol 21(2):194–198
98. Molckovsky A, Siu LL (2008) First-in-class, first-in-human phase I results of targeted agents: highlights of the 2008 American society of clinical oncology meeting. J Hematol Oncol 1:20
99. Xu CX, Li Y, Yue P, Owonikoko TK, Ramalingam SS, Khuri FR et al (2011) The combination of RAD001 and NVP-BEZ235 exerts synergistic anticancer activity against non-small cell lung cancer in vitro and in vivo. PLoS One 6(6):14
100. Fan QW, Knight ZA, Goldenberg DD, Yu W, Mostov KE, Stokoe D et al (2006) A dual PI3 kinase/mTOR inhibitor reveals emergent efficacy in glioma. Cancer Cell 9(5):341–349

101. Fan QW, Cheng CK, Nicolaides TP, Hackett CS, Knight ZA, Shokat KM et al (2007) A dual phosphoinositide-3-kinase alpha/mTOR inhibitor cooperates with blockade of epidermal growth factor receptor in PTEN-mutant glioma. Cancer Res 67(17):7960–7965
102. Maira SM, Stauffer F, Brueggen J, Furet P, Schnell C, Fritsch C et al (2008) Identification and characterization of NVP-BEZ235, a new orally available dual phosphatidylinositol 3-kinase/mammalian target of rapamycin inhibitor with potent in vivo antitumor activity. Mol Cancer Ther 7(7):1851–1863
103. Chapuis N, Tamburini J, Green AS, Vignon C, Bardet V, Neyret A et al (2010) Dual inhibition of PI3K and mTORC1/2 signaling by NVP-BEZ235 as a new therapeutic strategy for acute myeloid leukemia. Clin Cancer Res 16(22):5424–5435
104. Chiarini F, Fala F, Tazzari PL, Ricci F, Astolfi A, Pession A et al (2009) Dual inhibition of class IA phosphatidylinositol 3-kinase and mammalian target of rapamycin as a new therapeutic option for T-cell acute lymphoblastic leukemia. Cancer Res 69(8):3520–3528
105. Chiarini F, Grimaldi C, Ricci F, Tazzari PL, Evangelisti C, Ognibene A et al (2010) Activity of the novel dual phosphatidylinositol 3-kinase/mammalian target of rapamycin inhibitor NVP-BEZ235 against T-cell acute lymphoblastic leukemia. Cancer Res 70(20):8097–8107
106. Schuster K, Zheng J, Arbini AA, Zhang CC, Scaglioni PP (2011) Selective targeting of the mTORC1/2 protein kinase complexes leads to antileukemic effects in vitro and in vivo. Blood Cancer J 1:e34
107. Shuttleworth SJ, Silva FA, Cecil AR, Tomassi CD, Hill TJ, Raynaud FI et al (2011) Progress in the preclinical discovery and clinical development of class I and dual class I/IV phosphoinositide 3-kinase (PI3K) inhibitors. Curr Med Chem 18(18):2686–2714
108. Mallon R, Feldberg LR, Lucas J, Chaudhary I, Dehnhardt C, Santos ED et al (2011) Antitumor efficacy of PKI-587, a highly potent dual PI3K/mTOR kinase inhibitor. Clin Cancer Res 17(10):3193–3203 (Epub 2011 Feb 15)
109. Yuan J, Mehta PP, Yin MJ, Sun S, Zou A, Chen J et al (2011) PF-04691502, a potent and selective oral inhibitor of PI3K and mTOR kinases with antitumor activity. Mol Cancer Ther 10(11):2189–2199
110. Mallon R, Hollander I, Feldberg L, Lucas J, Soloveva V, Venkatesan A et al (2010) Antitumor efficacy profile of PKI-402, a dual phosphatidylinositol 3-kinase/mammalian target of rapamycin inhibitor. Mol Cancer Ther 9(4):976–984
111. Prasad G, Sottero T, Yang X, Mueller S, James CD, Weiss WA et al (2011) Inhibition of PI3K/mTOR pathways in glioblastoma and implications for combination therapy with temozolomide. Neuro Oncol 13(4):384–392
112. Mirzoeva OK, Hann B, Hom YK, Debnath J, Aftab D, Shokat K et al (2011) Autophagy suppression promotes apoptotic cell death in response to inhibition of the PI3K-mTOR pathway in pancreatic adenocarcinoma. J Mol Med 89(9):877–889
113. Heffron TP, Berry M, Castanedo G, Chang C, Chuckowree I, Dotson J et al (2010) Identification of GNE-477, a potent and efficacious dual PI3K/mTOR inhibitor. Bioorg Med Chem Lett 20(8):2408–2411
114. Wallin JJ, Edgar KA, Guan J, Berry M, Prior WW, Lee L et al (2011) GDC-0980 is a novel class I PI3K/mTOR kinase inhibitor with robust activity in cancer models driven by the PI3K pathway. Mol Cancer Ther 10(12):2426–2436
115. Li T, Wang J, Wang X, Yang N, Chen SM, Tong LJ et al (2010) WJD008, a dual phosphatidylinositol 3-kinase (PI3K)/mammalian target of rapamycin inhibitor, prevents PI3K signaling and inhibits the proliferation of transformed cells with oncogenic PI3K mutant. J Pharmacol Exp Ther 334(3):830–838
116. Zhu J, Huang JW, Tseng PH, Yang YT, Fowble J, Shiau CW et al (2004) From the cyclooxygenase-2 inhibitor celecoxib to a novel class of 3-phosphoinositide-dependent protein kinase-1 inhibitors. Cancer Res 64(12):4309–4318
117. Ding H, Han C, Guo D, Wang D, Duan W, Chen CS et al (2008) Sensitivity to the non-COX inhibiting celecoxib derivative, OSU03012, is p21(WAF1/CIP1) dependent. Int J Cancer 123(12):2931–2938

118. Falasca M, Chiozzotto D, Godage HY, Mazzoletti M, Riley AM, Previdi S et al (2010) A novel inhibitor of the PI3K/Akt pathway based on the structure of inositol 1,3,4,5,6-pentakisphosphate. Br J Cancer 102(1):104–114
119. Yang L, Dan HC, Sun M, Liu Q, Sun XM, Feldman RI et al (2004) Akt/protein kinase B signaling inhibitor-2, a selective small molecule inhibitor of Akt signaling with antitumor activity in cancer cells overexpressing Akt. Cancer Res 64(13):4394–4399
120. Garrett CR, Coppola D, Wenham RM, Cubitt CL, Neuger AM, Frost TJ et al (2011) Phase I pharmacokinetic and pharmacodynamic study of triciribine phosphate monohydrate, a small-molecule inhibitor of AKT phosphorylation, in adult subjects with solid tumors containing activated AKT. Invest New Drugs 29(6):1381–1389
121. Tan S, Ng Y, James DE (2011) Next-generation Akt inhibitors provide greater specificity: effects on glucose metabolism in adipocytes. Biochem J 435(2):539–544
122. Simioni C, Neri LM, Tabellini G, Ricci F, Bressanin D, Chiarini F et al (2012) Cytotoxic activity of the novel Akt inhibitor, MK-2206, in T-cell acute lymphoblastic leukemia. Leukemia 22(10):136
123. Rhodes N, Heerding DA, Duckett DR, Eberwein DJ, Knick VB, Lansing TJ et al (2008) Characterization of an Akt kinase inhibitor with potent pharmacodynamic and antitumor activity. Cancer Res 68(7):2366–2374
124. Zeng Z, Samudio IJ, Zhang W, Estrov Z, Pelicano H, Harris D et al (2006) Simultaneous inhibition of PDK1/AKT and Fms-like tyrosine kinase 3 signaling by a small-molecule KP372-1 induces mitochondrial dysfunction and apoptosis in acute myelogenous leukemia. Cancer Res 66(7):3737–3746
125. Rampling R, Sanson M, Gorlia T, Lacombe D, Lai C, Gharib M et al (2012) A phase I study of LY317615 (enzastaurin) and temozolomide in patients with gliomas (EORTC trial 26054). Neuro Oncol 14(3):344–350
126. Vansteenkiste J, Ramlau R, von Pawel J, San Antonio B, Eschbach C, Szczesna A (2012) A phase II randomized study of cisplatin-pemetrexed plus either enzastaurin or placebo in chemonaive patients with advanced non-small cell lung cancer. Oncology 82(1):25–29
127. Wolff RA, Fuchs M, Di Bartolomeo M, Hossain AM, Stoffregen C, Nicol S et al (2011) A double-blind, randomized, placebo-controlled, phase 2 study of maintenance enzastaurin with 5-fluorouracil/leucovorin plus bevacizumab after first-line therapy for metastatic colorectal cancer. Cancer 27(10):26692
128. Kondapaka SB, Singh SS, Dasmahapatra GP, Sausville EA, Roy KK (2003) Perifosine, a novel alkylphospholipid, inhibits protein kinase B activation. Mol Cancer Ther 2(11):1093–1103
129. Pal SK, Reckamp K, Yu H, Figlin RA (2010) Akt inhibitors in clinical development for the treatment of cancer. Expert Opin Investig Drugs 19(11):1355–1366
130. Handrick R, Rubel A, Faltin H, Eibl H, Belka C, Jendrossek V (2006) Increased cytotoxicity of ionizing radiation in combination with membrane-targeted apoptosis modulators involves downregulation of protein kinase B/Akt-mediated survival-signaling. Radiother Oncol 80(2):199–206
131. Martelli AM, Papa V, Tazzari PL, Ricci F, Evangelisti C, Chiarini F et al (2010) Erucylphosphohomocholine, the first intravenously applicable alkylphosphocholine, is cytotoxic to acute myelogenous leukemia cells through JNK- and PP2A-dependent mechanisms. Leukemia 24(4):687–698
132. Bidyasar S, Kurzrock R, Falchook GS, Naing A, Wheler JJ, Durand J et al (2009) A first-in-human phase I trial of PBI-05204 (oleandrin), an inhibitor of Akt, FGF-2, NF-Kb, and p70S6K in advanced solid tumor patients. ASCO Meeting Abstracts. 27(15S):3537 (2009 June 8)
133. Dunn DE, He DN, Yang P, Johansen M, Newman RA, Lo DC (2011) In vitro and in vivo neuroprotective activity of the cardiac glycoside oleandrin from Nerium oleander in brain slice-based stroke models. J Neurochem 119(4):805–814
134. Yoon H, Kim DJ, Ahn EH, Gellert GC, Shay JW, Ahn CH et al (2009) Antitumor activity of a novel antisense oligonucleotide against Akt1. J Cell Biochem 108(4):832–838

135. Marshall J, Posey J, Hwang J, Malik S, Shen R, Kazempour K, et al (2007) A phase I trial of RX-0201 (AKT anti-sense) in patients with an advanced cancer. ASCO Meeting Abstracts. 25(18_suppl):3564 (2007 June 21)
136. Oshiro N, Yoshino K, Hidayat S, Tokunaga C, Hara K, Eguchi S et al (2004) Dissociation of raptor from mTOR is a mechanism of rapamycin-induced inhibition of mTOR function. Genes Cells 9(4):359–366
137. Bai X, Ma D, Liu A, Shen X, Wang QJ, Liu Y et al (2007) Rheb activates mTOR by antagonizing its endogenous inhibitor, FKBP38. Science 318(5852):977–980
138. Fouladi M, Laningham F, Wu J, O'Shaughnessy MA, Molina K, Broniscer A et al (2007) Phase I study of everolimus in pediatric patients with refractory solid tumors. J Clin Oncol 25(30):4806–4812
139. Owonikoko TK, Khuri FR, Ramalingam SS (2009) Preoperative therapy for early-stage NSCLC: opportunities and challenges. Oncology 23(10):892
140. Rini BI, Campbell SC, Escudier B (2009) Renal cell carcinoma. Lancet 373(9669):1119–1132
141. Benjamin D, Colombi M, Moroni C, Hall MN (2011) Rapamycin passes the torch: a new generation of mTOR inhibitors. Nat Rev Drug Discov 10(11):868–880
142. Chawla SP, Staddon AP, Baker LH, Schuetze SM, Tolcher AW, D'Amato GZ et al (2012) Phase II study of the mammalian target of rapamycin inhibitor ridaforolimus in patients with advanced bone and soft tissue sarcomas. J Clin Oncol 30(1):78–84
143. Donia M, McCubrey JA, Bendtzen K, Nicoletti F (2010) Potential use of rapamycin in HIV infection. Br J Clin Pharmacol 70(6):784–793
144. Nicoletti F, Fagone P, Meroni P, McCubrey J, Bendtzen K (2011) mTOR as a multifunctional therapeutic target in HIV infection. Drug Discov Today 16(15–16):715–721
145. Bhagwat SV, Gokhale PC, Crew AP, Cooke A, Yao Y, Mantis C et al (2011) Preclinical characterization of OSI-027, a potent and selective inhibitor of mTORC1 and mTORC2: distinct from rapamycin. Mol Cancer Ther 10(8):1394–1406
146. Grimaldi C, Chiarini F, Tabellini G, Ricci F, Tazzari PL, Battistelli M et al (2012) AMP-dependent kinase/mammalian target of rapamycin complex 1 signaling in T-cell acute lymphoblastic leukemia: therapeutic implications. Leukemia 26(1):91–100
147. Carayol N, Vakana E, Sassano A, Kaur S, Goussetis DJ, Glaser H et al (2010) Critical roles for mTORC2- and rapamycin-insensitive mTORC1-complexes in growth and survival of BCR-ABL-expressing leukemic cells. Proc Natl Acad Sci U S A 107(28):12469–12474
148. Tan DS, Dumez H, Olmos D, Sandhu SK, Hoeben A, Stephens AW et al (2010) First-in-human phase I study exploring three schedules of OSI-027, a novel small molecule TORC1/TORC2 inhibitor, in patients with advanced solid tumors and lymphoma. ASCO Meeting Abstracts. 28(15_suppl):3006 (2010 June 14)
149. Jessen K, Wang S, Kessler L, Guo X, Kucharski J, Staunton J et al Abstract B148: INK128 is a potent and selective TORC1/2 inhibitor with broad oral antitumor activity. Molecular Cancer Therapeutics. 8(Supplement 1):B148
150. Hsieh AC, Ruggero D (2010) Targeting eukaryotic translation initiation factor 4E (eIF4E) in cancer. Clin Cancer Res 16(20):4914–4920
151. Chresta CM, Davies BR, Hickson I, Harding T, Cosulich S, Critchlow SE et al (2010) AZD8055 is a potent, selective, and orally bioavailable ATP-competitive mammalian target of rapamycin kinase inhibitor with in vitro and in vivo antitumor activity. Cancer Res 70(1):288–298
152. Xue Q, Hopkins B, Perruzzi C, Udayakumar D, Sherris D, Benjamin LE (2008) Palomid 529, a novel small-molecule drug, is a TORC1/TORC2 inhibitor that reduces tumor growth, tumor angiogenesis, and vascular permeability. Cancer Res 68(22):9551–9557
153. Yu K, Shi C, Toral-Barza L, Lucas J, Shor B, Kim JE et al (2010) Beyond rapalog therapy: preclinical pharmacology and antitumor activity of WYE-125132, an ATP-competitive and specific inhibitor of mTORC1 and mTORC2. Cancer Res 70(2):621–631

154. Garcia-Martinez JM, Moran J, Clarke RG, Gray A, Cosulich SC, Chresta CM et al (2009) Ku-0063794 is a specific inhibitor of the mammalian target of rapamycin (mTOR). Biochem J 421(1):29–42
155. Falcon BL, Barr S, Gokhale PC, Chou J, Fogarty J, Depeille P et al (2011) Reduced VEGF production, angiogenesis, and vascular regrowth contribute to the antitumor properties of dual mTORC1/mTORC2 inhibitors. Cancer Res 71(5):1573–1583
156. Liu Q, Wang J, Kang SA, Thoreen CC, Hur W, Ahmed T et al (2011) Discovery of 9-(6-aminopyridin-3-yl)-1-(3-(trifluoromethyl)phenyl)benzo[h][1,6]naphthyridin-2(1H)-one (Torin2) as a potent, selective, and orally available mammalian target of rapamycin (mTOR) inhibitor for treatment of cancer. J Med Chem 54(5):1473–1480
157. Engelman JA, Chen L, Tan X, Crosby K, Guimaraes AR, Upadhyay R et al (2008) Effective use of PI3K and MEK inhibitors to treat mutant Kras G12D and PIK3CA H1047R murine lung cancers. Nat Med 14(12):1351–1356
158. Wang Z, Zhou J, Fan J, Qiu SJ, Yu Y, Huang XW et al (2008) Effect of rapamycin alone and in combination with sorafenib in an orthotopic model of human hepatocellular carcinoma. Clin Cancer Res 14(16):5124–5130
159. Jin N, Jiang T, Rosen DM, Nelkin BD, Ball DW (2011) Synergistic action of a RAF inhibitor and a dual PI3K/mTOR inhibitor in thyroid cancer. Clin Cancer Res 17(20):6482–6489
160. Legrier ME, Yang CP, Yan HG, Lopez-Barcons L, Keller SM, Perez-Soler R et al (2007) Targeting protein translation in human non small cell lung cancer via combined MEK and mammalian target of rapamycin suppression. Cancer Res 67(23):11300–11308
161. Marshall G, Howard Z, Dry J, Fenton S, Heathcote D, Gray N et al (2011) Benefits of mTOR kinase targeting in oncology: pre-clinical evidence with AZD8055. Biochem Soc Trans 39(2):456–459
162. Shapiro G, LoRusso P, Kwak EL, Cleary JM, Musib L, Jones C et al (2011) Clinical combination of the MEK inhibitor GDC-0973 and the PI3 K inhibitor GDC-0941: A first-in-human phase Ib study testing daily and intermittent dosing schedules in patients with advanced solid tumors. ASCO Meeting Abstracts. 29(15_suppl):3005 (2011 June 9)
163. Yang H, Higgins B, Kolinsky K, Packman K, Bradley WD, Lee RJ et al (2012) Antitumor activity of BRAF inhibitor vemurafenib in preclinical models of BRAF-mutant colorectal cancer. Cancer Res 72(3):779–789
164. Hirai H, Sootome H, Nakatsuru Y, Miyama K, Taguchi S, Tsujioka K et al (2010) MK-2206, an allosteric Akt inhibitor, enhances antitumor efficacy by standard chemotherapeutic agents or molecular targeted drugs in vitro and in vivo. Mol Cancer Ther 9(7):1956–1967
165. Baumann P, Mandl-Weber S, Oduncu F, Schmidmaier R (2009) The novel orally bioavailable inhibitor of phosphoinositol-3-kinase and mammalian target of rapamycin, NVP-BEZ235, inhibits growth and proliferation in multiple myeloma. Exp Cell Res 315(3):485–497
166. Manara MC, Nicoletti G, Zambelli D, Ventura S, Guerzoni C, Landuzzi L et al (2010) NVP-BEZ235 as a new therapeutic option for sarcomas. Clin Cancer Res 16(2):530–540
167. Zhang YJ, Duan Y, Zheng XF (2011) Targeting the mTOR kinase domain: the second generation of mTOR inhibitors. Drug Discov Today 16(7–8):325–331
168. Gravina GL, Marampon F, Petini F, Biordi L, Sherris D, Jannini EA et al (2011) The TORC1/TORC2 inhibitor, Palomid 529, reduces tumor growth and sensitizes to docetaxel and cisplatin in aggressive and hormone-refractory prostate cancer cells. Endocr Relat Cancer 18(4):385–400
169. Diaz R, Nguewa PA, Diaz-Gonzalez JA, Hamel E, Gonzalez-Moreno O, Catena R et al (2009) The novel Akt inhibitor Palomid 529 (P529) enhances the effect of radiotherapy in prostate cancer. Br J Cancer 100(6):932–940
170. Chung EJ, Brown AP, Asano H, Mandler M, Burgan WE, Carter D et al (2009) In vitro and in vivo radiosensitization with AZD6244 (ARRY-142886), an inhibitor of mitogen-activated protein kinase/extracellular signal-regulated kinase 1/2 kinase. Clin Cancer Res 15(9): 3050–3057 (Epub 2009 Apr 14)

171. Edwards E, Geng L, Tan J, Onishko H, Donnelly E, Hallahan DE (2002) Phosphatidylinositol 3-kinase/Akt signaling in the response of vascular endothelium to ionizing radiation. Cancer Res 62(16):4671–4677
172. Shinohara ET, Cao C, Niermann K, Mu Y, Zeng F, Hallahan DE et al (2005) Enhanced radiation damage of tumor vasculature by mTOR inhibitors. Oncogene 24(35):5414–5422
173. Paglin S, Lee NY, Nakar C, Fitzgerald M, Plotkin J, Deuel B et al (2005) Rapamycin-sensitive pathway regulates mitochondrial membrane potential, autophagy, and survival in irradiated MCF-7 cells. Cancer Res 65(23):11061–11070
174. Moretti L, Attia A, Kim KW, Lu B (2007) Crosstalk between Bak/Bax and mTOR signaling regulates radiation-induced autophagy. Autophagy 3(2):142–144
175. Kudchadkar R, Paraiso KH, Smalley KS (2012) Targeting mutant BRAF in melanoma: current status and future development of combination therapy strategies. Cancer J 18(2):124–131

Chapter 14
Activation of Immune-Mediated Tumor Cell Death by Chemotherapy

Melanie J. McCoy, Anna K. Nowak and Richard A. Lake

Abstract Like pathogens, tumor cells express a range of antigens that can be recognized by the immune system and are therefore susceptible to immune-mediated death. It is widely recognized that the immune system plays a significant role in preventing cancer development through the elimination of malignant and pre-malignant cells, and in controlling tumor growth, once this protective mechanism has failed. The capacity of the immune system to contribute to the success of anti-cancer therapy, however, is a more recent concept. Originally assumed to have a detrimental effect on anti-tumor immunity due to its indiscriminate targeting of proliferating cells, including lymphocytes, it has now emerged that chemotherapy can enhance anti-tumor immune responses through altering the level and context of antigen presentation to immune effectors and through altering the immunological milieu, creating a favorable environment for the generation of anti-tumor immunity. Activation of immune-mediated tumor cell death by chemotherapy opens the door to the possibility of novel treatment strategies combining standard chemotherapy with immunotherapy agents aimed at enhancing such responses. Preclinical studies and early phase trials of combination chemoimmunotherapy have produced promising results. However, it is becoming clear that synergy is dependent not only on the drugs selected, but also on the intrinsic properties of the host and the tumor. Development of such combination regimens will therefore require careful design and an individualized approach.

M. J. McCoy (✉) · A. K. Nowak · R. A. Lake
School of Medicine and Pharmacology, The University of Western Australia, Crawley, Western Australia
e-mail: melanie.mccoy@uwa.edu.au

M. J. McCoy
St John of God Health Care, Subiaco, Western Australia

A. K. Nowak · R. A. Lake
National Centre for Asbestos Related Diseases, Perth, Australia

A. K. Nowak
Department of Medical Oncology, Sir Charles Gairdner Hospital, Nedlands, Western Australia

D. E. Johnson (ed.), *Cell Death Signaling in Cancer Biology and Treatment*,
Cell Death in Biology and Diseases, DOI: 10.1007/978-1-4614-5847-0_14,

X.-M. Yin and Z. Dong (Series eds.), *Cell Death in Biology and Diseases*

14.1 Introduction

The primary aim of chemotherapy is to eradicate tumor cells through direct cytotoxicity. However, chemotherapy drugs also impact on the immune system, and this impact may not be entirely negative as has been widely thought. Instead, evidence is mounting for an immune-mediated component to chemotherapy-induced tumor cell death. In this chapter, we will discuss the role of the immune system in controlling tumor development, the clinical significance of generating anti-tumor immunity, the mechanisms and characteristics of chemotherapy-induced activation of immune-mediated tumor cell death, and the implications for the development of future anti-cancer treatment strategies.

14.2 Cancer and the Immune System

14.2.1 The Three Es of Tumor Development: Elimination, Equilibrium and Escape

The concept of cancer immunosurveillance, whereby the immune system recognizes and destroys malignant and premalignant cells to prevent the development of cancer, was first introduced by Paul Ehrlich in the early twentieth century and has remained a contentious issue since formally presented to the scientific community by Burnet over 50 years ago [1]. The theory brought into question the established and easily understood notion of tolerance to self and protection from non-self. While still not universally accepted, there is now overwhelming evidence in favor of a complex and ongoing relationship between the immune system and a developing tumor. This relationship can be divided into three distinct stages: elimination, equilibrium, and escape [2] (Fig. 14.1). During the elimination phase, the immune system remains in control, identifying and destroying all malignant cells, that is, there is effective cancer immunosurveillance. If, however, tumor cell variants evolve that are less immunogenic, or active suppression of immune effectors occurs, tumor development enters the equilibrium phase, where the immune system struggles to maintain the balance of power. The final phase, escape, occurs when the host anti-tumor immune response is no longer sufficient and the tumor expands uncontrollably.

14.2.2 Generating an Anti-Tumor Immune Response: Cross-Presentation of Tumor Antigens

Tumors bear a range of antigens that can be recognized by autologous T cells. Some are tumor or tumor type-specific, some are shared by several different cancers, others are also expressed on normal tissue but display increased or altered

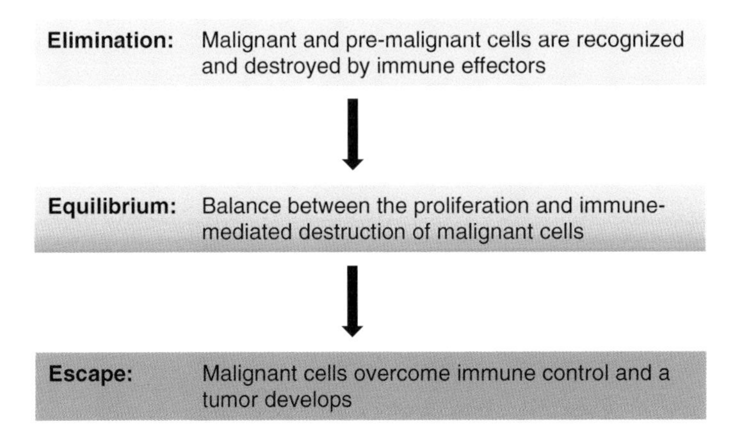

Fig. 14.1 Cancer immunosurveillance—the three Es of tumor development

expression on malignant cells [3]. Induction of tumor-specific $CD8^+$ cytotoxic T lymphocytes (CTL) through the *cross-presentation* of tumor antigens by dendritic cells (DC) is thought to be the primary mechanism by which anti-tumor immune responses are generated [4–6]. Cross-presentation refers to the process whereby exogenous antigen is processed and presented to $CD8^+$ T cells in association with major histocompatibility complex (MHC) class I, that is, exogenous antigen *crosses* into the MHC class I pathway of professional antigen-presenting cells (APC), which include DC, B cells, and macrophages. MHC class I molecules, constitutively expressed by all healthy cells except red blood cells, classically present endogenous peptides derived from proteins inside the cell to $CD8^+$ T cells. This is known as *direct presentation* and enables the detection and elimination of cells synthesizing foreign proteins, such as those infected by viruses [7]. Exogenous antigens, on the other hand, are usually processed and presented to $CD4^+$ T cells in association with MHC class II, expressed only by APC (*indirect presentation*) [8]. The phenomenon of cross-presentation was first observed by Bevan and colleagues in the 1970s using mouse strains differing in minor histocompatibility antigens [9]. Immunization of one mouse strain with lymphoid cells from the other resulted in the priming of $CD8^+$ T cells restricted by host MHC class I, which were able to lyse the transferred cells, thus demonstrating that CTL could be generated in the absence of direct antigen presentation. Precisely, how exogenous antigen enters the MHC class I presentation pathway is not completely understood. Currently, the most widely accepted mechanism is that following internalization into phagosomes, exogenous proteins escape into the cytosol where they enter the direct presentation pathway, undergoing proteosomal degradation into oligopeptides and transportation to the endoplasmic reticulum (ER) via the transporter associated with antigen processing (TAP) for MHC class I loading (reviewed in [10]). Depending on the context of tumor antigen cross-presentation, $CD8^+$ T cells are either stimulated to differentiate into CTL, which migrate to the tumor site and kill tumor cells displaying the relevant antigen via perforin/granzyme-induced apoptosis or ligation of death receptors (cross-priming), or are tolerized and subsequently deleted (cross-tolerization) [11].

14.2.3 CD8$^+$ T Cells: Key Players in Anti-Tumor Immunity

There are several lines of evidence that suggests CD8$^+$ T cells are critical to the development of anti-tumor immunity. Depletion of CD8$^+$ T cells prevents the rejection of chemically or UV-induced tumors, or transplanted tumors formed in immuno-compromised hosts, all of which are highly immunogenic in normal mice [12–14]. Mice deficient in the cytolytic molecule perforin are not only more susceptible to chemically induced tumor formation [15], but also develop spontaneous lymphoma in later life [13]. While perforin is required for both CTL and natural killer (NK) cell cytotoxicity, Smyth et al., demonstrated that CTL were essential for protection from lymphoma, since tumors that developed in perforin knockout mice were rejected when transplanted into wild-type mice depleted of NK cells, but not in those depleted of CD8$^+$ T cells. The efficacy of anti-tumor immunotherapy has also been shown to be dependent on the presence of CD8$^+$ T cells in a variety of murine models [16–19]. In humans, tumor infiltration by CD8$^+$ T cells has been associated with improved prognosis in many different cancers, as will be discussed in more detail later in this chapter.

The role of CD4$^+$ T cells in generating an anti-tumor immune response is not so well understood. Traditionally, activated CD4$^+$ T cells have been divided into two major subsets according to their cytokine secretion profile. T helper 1 (Th1) cells secrete cytokines usually associated with inflammation including interleukin-2 (IL-2), interferon gamma (IFNγ), tumor necrosis factor alpha (TNFα), and interleukin-12 (IL-12) and classically promote cell-mediated immunity against intracellular bacteria and viruses. T helper 2 (Th2) cells generally promote humoral immunity against extracellular pathogens and secrete IL-10, IL-4, and transforming growth factor beta (TGFβ) [20, 21] More recently, a third subset of T helper cells, characterized by secretion of the pro-inflammatory cytokines IL-17 and IL-22, has been defined. Termed Th17 cells, this subset protects against a variety of pathogens including intracellular and extracellular bacteria [22]. However, with CD4$^+$ T cells often showing non-classical cytokine expression profiles and increasing evidence that the tissue in which an immune response occurs, as well as the pathogen invading it, plays a major role in fine-tuning the class of response [23], the division of activated CD4$^+$ T cells into distinct helper subsets is perhaps less meaningful than once thought.

As their name suggests, CD4$^+$ helper T cells support CD8$^+$ T-cell-driven responses, not only through IL-2 production, which is required for T-cell expansion, but also through the *licensing* of APC via CD40–CD40L interactions [24–27]. Interaction of the co-stimulatory molecule CD40 (expressed on APC) with its ligand CD40L (expressed on CD4$^+$ T cells and mast cells) results in the production of IL-12 by APC and their upregulation of the co-stimulatory molecules CD80 and CD86, both of which are required for $CD8^+$ T-cell priming. Optimal $CD8^+$ T-cell cross-priming is dependent on the presence of $CD4^+$ T cells in several different animal models [27–29]. $CD4^+$ T cells can also kill tumor cells directly [30] and indirectly through the activation of eosinophils and macrophages, which can then lyse tumor cells via superoxide and nitric oxide production [31]. Yet $CD4^+$ T-cell depletion has frequently been found to have no effect on the ability of mice to reject tumors [14, 16, 18] and has even been shown to enhance the efficacy of anti-tumor immunotherapy [32]. This paradox may be explained by the presence of regulatory T cells (Treg), another subset of $CD4^+$ T cells that negatively regulate immune responses. Treg are vital for the maintenance of self-tolerance and prevention of immune-mediated pathology following infection [33, 34]. However, Treg also suppress anti-tumor immunity, with their depletion leading to immune-mediated tumor regression in many animal models (reviewed in [35]). The role of Th17 cells in anti-tumor immunity is not fully understood, but appears complex since their signature cytokine IL-17 can promote $CD8^+$ T-cell-driven anti-tumor immune responses but can also aid angiogenesis and metastasis, promoting tumor growth [36].

There is to date no evidence that B cells are required for the development of an anti-tumor immune response. In fact, B cells may inhibit the activation and expansion of tumor-specific CTL. Antigen presentation by B cells has been found to result in the tolerization and subsequent Fas-mediated deletion of $CD8^+$ T cells [37], while B-cell-deficient mice have been shown to be more resistant to tumor growth than their wild-type littermates [38]. Qin and colleagues hypothesized that the negative effect of B cells may be due to competition for tumor-derived antigen, which would otherwise be taken up by dendritic cells, thereby interfering with tumor-specific T-cell priming.

14.2.4 Interplay Between Innate and Adaptive Immune Effectors

T and B lymphocytes, the effectors of cell-mediated and humoral immunity, respectively, together form the adaptive arm of the immune system and provide a targeted response to specific antigens. When T and B cells encounter their cognate antigen and receive appropriate co-stimulatory signals, they become activated and undergo clonal expansion [39]. However, T- and B-cell clonal expansion and differentiation into effector/helper and plasma cells respectively, takes several days. Adaptive immune responses are therefore not immediate. In contrast, the innate arm of the immune system, comprised of macrophages, monocytes, NK cells, and mast cells, provides a rapid, non-specific, first-line defense against pathogens (reviewed in [40]). It is of note, however, that the identification of cell subsets with characteristics of both innate and adaptive immune cells, including NKT cells and

$\gamma\delta$ T cells, suggests that such a clear division between the two arms of the immune system may be overly simplistic [41].

Research into immunity against infection has shown that the generation of a potent adaptive immune response is usually dependent upon the initial stimulation of a strong innate immune response [42–44]. Consistent with this finding, it is now becoming clear that the generation of an effective anti-tumor immune response involves a complex interplay between the innate and adaptive arms of the immune system. In the early stages of neoplasia, it seems that NK cells play a vital role in the recognition of cancerous and precancerous cells. NK cells detect cell stress markers through receptors such as NK group 2 member D (NKG2D), and reduced MHC class I expression (lack of "self"), both of which are a common feature of malignant cells. NK cells, like CTL, can exert direct anti-tumor activity through secretion of perforin and granzymes, and via Fas and tumor necrosis factor-related apoptosis inducing ligand (TRAIL)-mediated cell death [45–48]. IFNγ secretion by NK cells stimulates DC maturation [49] and the induction of chemokines including IFN-inducible protein-10 (IP-10), monokine induced by IFNγ (MIG) and interferon-inducible T-cell α chemoattractant (I-TAC) [50, 51], which promotes lymphocyte recruitment. Mature, activated DC are then able to migrate to the draining lymph node, where they can cross-present antigen to tumor-specific $CD8^+$ T cells.

Type I IFN (IFNα/β) are known to play an important role in the initiation of an innate immune response [52]. Injection of activated plasmacytoid DC (pDC) has been shown to induce NK and $CD8^+$ T-cell-dependent rejection of transplanted tumors in a mouse melanoma model through secretion of type I IFN [53], illustrating the close collaboration between innate and adaptive immunity in generating an effective anti-tumor immune response. However, mice deficient in both an adaptive immune system and type I and type II IFN signaling ($RAG2^{-/-}$ $STAT^{-/-}$ double knockout mice) are more susceptible to spontaneous tumor development than either $RAG2^{-/-}$ or $STAT^{-/-}$ single knockouts [54]. This suggests that while both the innate and adaptive arms of the immune system are required for optimal immunosurveillance, each plays a partially independent role.

14.2.5 Avoiding Immune-Mediated Destruction

There are therefore three vital sequential steps in the generation of an anti-tumor immune response. Firstly, tumor cells must be recognized by innate and/or adaptive immune effectors. Secondly, recognition of tumor cells must lead to activation of an immune response rather than tolerance. Thirdly, immune effectors must effectively kill the tumor cells. Interruption of any one of these three steps can lead to immuno subversion and uncontrolled tumor growth, that is, transition from the elimination phase to escape. Tumors employ many mechanisms to subvert anti-tumor immunity (Table 14.1). One key means by which tumors avoid recognition by the immune system is through loss or reduced expression of tumor antigens.

Table 14.1 Mechanisms used by tumors to subvert anti-tumor immunity

Mechanism	Stage affected	Result
Loss/reduced expression of tumor antigens [55–59]	Tumor cell recognition	Prevents cross-presentation of antigen to $CD8^+$ T cells, indirect presentation to $CD4^+$ T cells and direct recognition by $CD8^+$ T cells.
Loss/altered expression of MHC class I [60, 61]	Tumor cell recognition	Impedes direct recognition by $CD8^+$ T cells.
Non-classical MHC expression (e.g. HLA-G) usually restricted to immune-privileged cells [173–175]	Tumor cell recognition/killing	Protects from NK and CTL-mediated death.
Reduced expression of NK cell ligands [176]	Tumor cell recognition	Inhibits NK-mediated killing.
Induction of NKG2D down-regulation [177]	Tumor cell recognition	Inhibits NK and CTL-mediated killing.
Secretion of IL-10 and TGF-β [62–64]	Immune effector activation	Inhibits DC recruitment, prevents cross-priming of $CD8^+$ T cells, and promotes Treg generation.
Expression of inhibitory molecules including PDL-1 [65, 66]	Immune effector activation/ tumor cell killing	Impaired T cell proliferation and function.
Expression of IDO [178]	Immune effector activation	Tryptophan-depletion, causing T cells to undergo cell cycle arrest.
Sterol synthesis [179]	Immune effector activation	Prevents CCR7 expression on maturing DC inhibiting their traffic to lymph node.
Mutations in death receptors including Fas and DR4/DR5 [68, 67]	Tumor cell killing	Protects from NK and CTL-mediated death.
Expression of Fas-L [69]	Tumor cell killing	Induces apoptotic cell death in NK cells and activated T cells.

CTL cytotoxic T lymphocyte, *DR* death receptor, *HLA* human leukocyte antigen, *IDO* indoleamine 2,3-dioxygenase, *PDL-1* programmed death ligand-1

This has been observed in human and mouse cell lines resistant to CTL lysis in vitro [55, 56], in outgrowing tumors following immunotherapy in mouse models [57, 58] and in recurrent melanoma in patients treated with antigen-specific immunotherapy [59]. Aberrant human leukocyte antigen (HLA) expression, which impedes direct recognition by $CD8^+$ T cells, has also been detected in a variety of human cancers (reviewed in [60]), and human tumor cell lines lacking HLA class I expression evade CTL-mediated cell death in vitro [61]. Both loss of tumor antigen and HLA expression likely represent passive selection rather than active adaptation, reflecting survival of the least immunogenic cells.

Secretion of soluble factors including IL-10 and TGFβ by tumors can skew the immune system toward tolerance rather than active immunity. IL-10 can inhibit DC recruitment [62], while TGFβ prevents the cross-priming of $CD8^+$ T cells [63] and promotes the generation of Treg [64]. Tumor cells can also express inhibitory molecules on their surface, which negatively regulate T-cell function. For example, a variety of human tumors express programed death ligand-1 (PDL-1), which inhibits T-cell proliferation and cytokine production through interaction with its receptor, programed death-1 (PD-1), expressed by activated T cells [65, 66].

Finally, tumor cell evolution can also select for cells less susceptible to CTL and NK cell-mediated cell death. Mutations in death receptors including Fas and TRAIL receptors, DR4 and DR5, have been observed in B-cell lymphoma [67] and in metastatic breast cancer [68]. Interestingly, tumor cells have also been found to express Fas ligand (Fas-L) [69], creating a role reversal situation in which tumor cells can induce apoptotic cell death in activated T cells.

14.2.6 Antitumor Immunity: Does it Matter?

While original evidence for the existence of anti-tumor immunity came from animal models, this has now been corroborated by data demonstrating that anti-tumor immune responses can be detected in cancer patients, even in the context of progressive disease, and that such responses have clinical significance.

Melanoma is considered one of the most immunogenic cancers with tumor-associated antigen (TAA)-specific $CD8^+$ T cells detectable in around 50 % of patients [70]. Vitiligo-like hypopigmentation, thought to result from the destruction of normal melanocytes by autoantibodies and autoreactive T cells, is also observed in melanoma patients and is associated with improved outcome [71]. While less frequently, circulating TAA-specific T cells have now been detected in most cancer types [72–77]. Cancer patients are also commonly found to have an increased proportion of peripheral Treg, consistent with the idea of an underlying anti-tumor immune response blocked by the concurrent expansion of Treg [78–81].

Perhaps, the most convincing evidence for the clinical relevance of anti-tumor immunity in patients is the strong prognostic significance of the balance of T-cell subsets within the local tumor environment. Tumor infiltration by $CD8^+$ T cells has been associated with an improved prognosis in many cancer types including ovarian [82], colorectal [83], esophageal [84], and lung [85]. Conversely, tumor-infiltrating Treg predict poorer survival in most solid cancers [86–89], again consistent with the suppressive role played by this T-cell subset. An exception, however, appears to be colorectal cancer, where a higher density of tumor-infiltrating (Ti) Treg has been identified as a good prognostic factor in several independent studies [90–93]. One explanation for this seemingly paradoxical finding is that expansion of Treg within the tumor occurs in response to the activation and expansion of tumor-specific $CD8^+$ T cells, thereby representing an indirect measure of an anti-tumor immune response. Alternatively, it has been suggested that Treg may limit the tumorigenic effects of Th17 cell-mediated inflammation [94]. In support of this, Th17 cells are more frequently identified in colorectal tumors than other cancer types, likely due to the abundance of pathogens present in the gut [95], and a lower density of Ti-Th17 cells has recently been associated with improved disease-free survival [96]. Ti-Treg have also been associated with improved outcome in nasopharyngeal and head and neck squamous cell carcinoma [97, 98], both of which also develop in a non-sterile environment. It may be therefore that it is the balance of T-cell subsets within a tumor, in relation to the stage of tumor development and the environment in which this occurs, that is important. In any case, this serves to highlight the complexity of the relationship between the immune system and a developing tumor and the potential value of harnessing anti-tumor immunity to improve the efficacy of cancer therapy.

14.3 Driving Anti-Tumor Immunity with Chemotherapy

14.3.1 Cytotoxic Chemotherapy: Mechanisms of Action

The major classes of cytotoxic drugs currently used for the treatment of cancer include anti-metabolites, anti-folates, alkylating agents, topoisomerase inhibitors, anthracyclines, vinca alkaloids, and taxanes (Table 14.2). While differing in their precise mode of action, most interfere with DNA synthesis resulting in apoptotic tumor cell death. The anti-metabolites gemcitabine and 5-fluorouracil (5-FU), for example, are synthetic nucleoside analogues, which upon intracellular conversion to their active forms, cause apoptosis by two different mechanisms. They are incorporated into DNA (and RNA in the case of 5-FU), disrupting DNA synthesis and/or RNA processing and function, and also inhibit enzymes essential for DNA synthesis and repair [99, 100]. Alkylating agents, including cyclophosphamide, and platinum-based drugs such as cisplatin and oxaliplatin, cross-link DNA, inhibiting DNA synthesis and RNA transcription, disrupting the cell cycle and activating several DNA damage-mediated signal transduction pathways, leading to apoptosis [101, 102]. Anthracycline antibiotics, including doxorubicin, idarubicin and epirubicin also intercalate DNA/RNA base pairs, but primarily cause apoptotic cell death through inhibition of topoisomerase II activity, which is vital for DNA structuring and repair [103]. Vinca alkaloids (e.g., vinorelbine and vinblastine) and taxanes (paclitaxel and docetaxel) differ in their mode of action in that they inhibit mitosis rather than DNA synthesis through stabilizing or destabilizing microtubules and, thereby preventing cell cycle progression [104]. However, like the other cytotoxics mentioned, this ultimately results in apoptosis. Most commonly used chemotherapy drugs are therefore not specifically toxic to tumor cells, but target all dividing cells, including lymphocytes. Consequently, chemotherapy has traditionally

Table 14.2 Major classes of cytotoxic drugs and mechanisms of action

Class	Examples	Mechanism(s) of DNA damage / apoptosis induction
Anti-metabolites	gemcitabine, 5-fluorouracil	Incorporation into DNA/RNA causing strand termination and inhibition of enzymes required for DNA synthesis/ repair including thymidylate synthase, ribonucleoside reductase and DNA polymerase [99, 100].
Anti-folates	pemetrexed, methotrexate	Inhibition of enzymes involved in folate metabolism and purine/pyrimidine synthesis [180, 181].
Alkylating agents	cyclophosphamide, ifosphamide, dacarbazine	Addition of an alkyl group to DNA resulting in inter and intra-strand cross-links [101].
Platinum-based drugs	cisplatin, carboplatin oxaliplatin	Binding of DNA bases resulting in inter and intra-strand cross-links [102].
Topoisomerase inhibitors	irinotecan, topotecan, etoposide	Inhibition of topoisomerase I/II preventing DNA structuring and repair [182].
Anthracyclines	doxorubicin, idarubicin epirubicin	Intercalation of base pairs and inhibition of topoisomerase II [103].
Vinca alkaloids	vinorelbine, vinblastine, vincristine	Binding of tubulin preventing microtubule formation and cell cycle progression [104].
Taxanes	docetaxel, paclitaxel	Stabilization of microtubules preventing their breakdown and causing cell cycle arrest [104].

been viewed as immunosuppressive, and combining chemotherapy with immunotherapy appeared counterintuitive. However, over the past decade, there has been a paradigm shift, with the potential of chemotherapy to induce or enhance anti-tumor immunity increasingly recognized.

It has now been demonstrated in a variety of animal models that optimal efficacy of many cytotoxic drugs is dependent upon an intact immune system. Oxaliplatin, for example, effectively controls tumor growth in immunocompetent wild-type mice bearing EL4 thymomas, but is ineffective in mice athymic nude mice lacking T cells, in $CD8^+$ T-cell-depleted wild-type mice, or in mice with a deficiency in the IFNγ signaling pathway [105]. Similarly, the efficacy of doxorubicin against CT26 colon cancer and methylcholanthrene-induced sarcoma is lost in the absence of IFNγ signaling [105]. Using a mouse model of

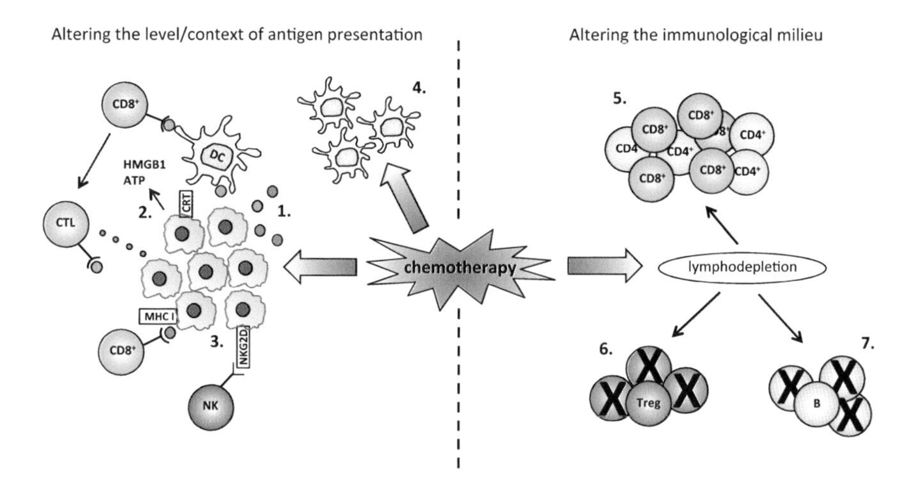

Fig. 14.2 Mechanisms by which chemotherapy can alter the level and/or context of antigen presentation and the immunological milieu to promote anti-tumor immunity: *1* Physical destruction of the tumor-releasing antigen into the cross-presentation pathway. *2* Induction of "eat me" and "danger" signals from dying tumor cells including CRT translocation and HMGB1/ATP release. *3* Upregulation of tumor or cell-surface receptors including MHC class I and NKG2D. *4* Promotion of DC expansion and activation. *5* Homeostatic T-cell proliferation. *6* Selective Treg depletion. *7* Selective B-cell depletion. *ATP* adenosine triphosphate, *CRT* calreticulin, *CTL* cytotoxic T lymphocyte, *DC* dendritic cell, *HMGB1* high-mobility group box 1, *MHC* major histocompatibility complex, *NK* natural killer, *Treg* regulatory T cell

mesothelioma, we have found gemcitabine and the anti-folate pemetrexed to be less effective in nude versus wild-type mice [106], and that the curative effect of the DNA alkylating agent cyclophosphamide we observe in wild-type mice, is $CD8^+$ T-cell-dependent [107]. Much evidence has arisen to support the existence of multiple mechanisms through which chemotherapy may induce or enhance anti-tumor immune responses. These can be divided into those that alter the level and/or context of antigen presentation through effects on tumor cells and APC, and those that alter the immunological milieu, through effects on lymphocyte populations, providing a favorable environment for the generation of anti-tumor immunity (Fig. 14.2).

14.3.2 Altering the Level and Context of Antigen Presentation

Perhaps, the most obvious way by which chemotherapy could increase the level of antigen presentation, is through physical destruction of the tumor, releasing antigens from the dead and dying cells for entry into the cross-presentation pathway. It is likely that the availability of some tumor antigens, normally present at levels too low to induce an immune response, could be increased above the required threshold in this process. In our murine model of mesothelioma, using hemagglutinin (HA) antigen-transduced tumor cells (AB1-HA) injected subcutaneously, gemcitabine significantly increases the level of antigen presentation to tumor-specific $CD8^+$ T cells for a given tumor size [108]. For example, proliferation of adoptively transferred HA-specific $CD8^+$ T cells in a mouse bearing a 40 mm^2 tumor treated with gemcitabine was almost double the level observed in animals with an untreated tumor of equivalent size. Vaccination with the HA antigen-bearing PR8 virus following gemcitabine treatment resulted in a marked decrease in tumor growth and increased survival compared with vaccination without gemcitabine. It is also plausible that so-called cryptic antigens, not usually visible to the immune system, could be released following chemotherapy-induced cell death, along with many of the ~90 neoepitopes generated by point mutations that are harbored by the average tumor [109]. In support of this, we found that animals cured of AB1-HA tumors following combination chemo-immunotherapy with gemcitabine and the CD40 activating antibody FGK45 were resistant to rechallenge with both AB1-HA and non-transfected AB1 tumor cells, demonstrating memory to tumor antigens other than HA [110].

With regard to altering the context of antigen presentation, chemotherapy can induce the expression or release of molecules associated with *dangerous*, and therefore immunogenic, cell death [111, 112]. Using an assay in which tumor cells are exposed to cytotoxic drugs in vitro, injected into immunocompetent naive mice, and assessed for their capacity to elicit a protective immune response against rechallenge with live tumor cells, Zitvogel, Kroemer, and colleagues have demonstrated that oxaliplatin and the anthracyclines doxorubicin, idarubicin, and mitoxantrone, induce immunogenic cell death in several tumor models [105, 113–115]. Characteristics of immunogenic versus non-immunogenic tumor cell death in their system include cell-surface exposure of the calcium ion-binding protein, calreticulin [115], secretion of the non-histone chromatin-binding nuclear constituent, high-mobility group box 1 (HMGB1) [113], and release of adenosine triphosphate (ATP) [105]. Calreticulin is usually sequestered within the endoplasmic reticulum. Its translocation to the cell surface occurs before cells become apoptotic and is required for uptake of dying tumor cells by DC [115]. Blockade of cell-surface calreticulin with specific antibodies, or knockdown of its expression with siRNA, inhibited phagocytosis of anthracycline-treated tumor cells, while addition of exogenous recombinant calreticulin, enabled uptake of treated tumor cells transfected with siRNA [115]. Whereas calreticulin exposure therefore provides an "eat me" signal for DC, binding of HMGB1 to toll-like receptor 4 (TLR4) on DC, and subsequent signaling through the MyD88 pathway is thought to represent a danger signal required to induce antigen processing and presentation. Mice deficient in TLR4 or the MyD88 adapter molecule were unable to mount a tumor-specific immune response following vaccination with oxaliplatin or doxorubicin-treated tumor cells and were not protected against re-challenge [113]. Although WT and $TLR4^{-/-}$ DC were equally efficient in engulfing dying tumor cells, $TLR4^{-/-}$ DC were defective in their capacity to process and present antigen in association with MHC class I [113]. Binding of ATP released from oxaliplatin-treated tumor cells to the P2rX7 receptor on DC results in activation of the NLRP3 inflammasome and subsequent secretion of IL-1β, which is required for effective $CD8^+$ T-cell priming [105]. It is extremely likely, however, that other factors distinguishing immunogenic from non-immunogenic cell death are yet to be discovered.

Some drugs, including gemcitabine, cisplatin, and paclitaxel, have also been shown to induce upregulation of tumor cell-surface receptors/ligands, resulting in increased antigen presentation and/or susceptibility to immune effector-mediated cell death. Liu et al., found that gemcitabine induced upregulation of HLA class I expression on human tumor cell lines [116], thereby increasing the opportunity for direct recognition by $CD8^+$ T cells, while Ramakrishnan et al, demonstrated that increased sensitivity of murine and human tumor cells to CTL lysis following cisplatin or paclitaxel exposure was mediated via the upregulation of mannose receptors, increasing their permeability to granzyme B [117]. This allowed *bystander killing* of antigen-deficient tumor cells by CTL activated through recognition of neighboring antigen-expressing tumor cells. Sensitization of tumor cells to TRAIL-mediated cell death can also contribute to the anti-tumor effects of cytotoxic drugs. In our murine mesothelioma model, the efficacy of cyclophosphamide is reduced in TRAIL-deficient mice, while agonistic anti-DR5 antibodies, which mimic TRAIL ligation, rescued the partial effect of the drug observed in athymic nude mice [118]. Increased susceptibility of tumor cells to NK cell-mediated death has also been observed following exposure to cytotoxic drugs, mediated both by upregulation of NKG2D ligands [119] and downregulation of c-type lectin receptor ligands [120], which leaves tumor cells vulnerable to NK cell-mediated killing in accordance with the "missing self hypothesis" [121].

Cytotoxic drugs may also enhance tumor antigen presentation through direct effects on DC. In several mouse models, cyclophosphamide has been shown to promote DC expansion from bone marrow progenitors [122–124], while exposure of DC to non-toxic concentrations of paclitaxel in vitro can lead to upregulation of co-stimulatory molecules and an increased capacity to present antigen [125, 126].

14.3.3 Altering the Immunological Milieu

Although seemingly paradoxical, chemotherapy-induced lymphodepletion is one of the major mechanisms through which the immunological milieu can be altered to favor the generation of anti-tumor immune responses. As the size and composition of the T-cell repertoire is under constant regulation in the periphery [127], lymphodepletion triggers a period of homeostatic T-cell proliferation, driven by IL-7 and IL-15 [128–130]. Being at least partially dependent upon self-antigen-MHC interactions, homeostatic T-cell proliferation results in a freshly reconstituted T-cell pool biased toward self-antigen reactivity [131, 132]. Since most tumor antigens are self-antigens, lymphopenia in a tumor-bearing host may result in increased anti-tumor immunity, as has been demonstrated in sub-lethally irradiated mice, whereby homeostatic expansion of transferred autologous or syngenic T cells inhibited the growth of established tumors [133]. This becomes even more important if the level of available antigen and the capacity for presentation to effectors is simultaneously increased.

Another indirect way by which chemotherapy has been shown to enhance anti-tumor immunity is through differential effects on lymphocyte subsets. Despite also functioning as a potent immunosuppressor, Maguire and Ettore discovered over 40 years ago that cyclophosphamide could, in fact, increase contact sensitivity reactions in guinea pigs [134]. More than a decade later, this finding was extended to humans, with antigen-specific delayed-type hypersensitivity reactions shown to be augmented in a cohort of patients receiving cyclophosphamide treatment for metastatic melanoma or colorectal cancer [135], the authors suggesting that this was due to inhibition of suppressor T-cell function, a distinct Treg population not having been characterized at the time. More recently, it has been demonstrated both in animal models and cancer patients that cyclophosphamide can be used to target Treg [136, 137]. Importantly, however, the effects of cyclophosphamide are dose-dependent, with lower doses (50–100 mg/day) causing selective Treg depletion and enhanced effector function, while higher doses (200 mg/day) result in indiscriminate loss of all lymphocyte subsets ([137] and our own unpublished data). This increased susceptibility of the Treg population may not be limited to cyclophosphamide, with gemcitabine now shown to cause transient selective Treg depletion in cancer patients receiving single agent treatment [138] and remaining Treg in paclitaxel-treated mice found to have reduced suppressive function [139]. Whether the effects of gemcitabine and paclitaxel on immune cell populations are dose-dependent is not yet clear. Since cytotoxic drugs target dividing cells, one explanation for this apparent selective Treg depletion is their increased proliferative state, relative to other lymphocyte subsets, which appears to be an intrinsic characteristic of this cell population [138, 140, 141]. In support of this theory, those Treg most actively proliferating are preferentially depleted by cyclophosphamide [107]. This may have important implications for anti-tumor immunity as cycling Treg also express higher levels of markers associated with maximal suppressive activity, including inducible co-stimulatory molecule (ICOS) and tumor necrosis factor receptor type 2 (TNFR2) [107, 142–144]. Depletion of a maximally suppressive Treg population by cytotoxic chemotherapy could favorably alter the context of antigen presentation, *removing the brakes* on anti-tumor immunity. In the same vein, B cells are more profoundly affected by some cytotoxics than T cells [145, 146]. With evidence to date indicating that B cells have a negative, rather than a positive impact on anti-tumor immunity, this again could potentially skew any ensuing immune response in the right direction.

14.3.4 Immune-Mediated Tumor Cell Death Contributes to Treatment Efficacy

Consistent with preclinical data demonstrating a requirement for an intact immune system for optimal treatment efficacy, it is becoming increasingly clear that anti-tumor effects of chemotherapy in cancer patients are not entirely due to direct cytotoxicity; activation of anti-tumor immunity may also play a role. In breast cancer, infiltration of $CD3^+$ T cells and disappearance of Treg in surgical specimens following adjuvant taxane/anthracycline-based chemotherapy have both been associated with complete pathological response (i.e., no residual tumor cells) [147, 148]. Similarly, the presence of Ti-T cells in samples taken from patients undergoing surgical debulking for ovarian cancer, was associated with complete response to adjuvant platinum-based chemotherapy [82], while in rectal cancer, density of both $CD8^+$ and $CD4^+$ T cells in pre-treatment biopsy samples was recently shown to predict complete response to 5-FU-based neoadjuvant chemoradiotherapy [149].

Factors indicative of immunogenic tumor cell death following chemotherapy in animal models have also been shown to influence clinical outcome in patients, with loss of function mutations in both TLR4 and P2xr7 genes associated with reduced disease-free survival following anthracycline-based chemotherapy in breast cancer patients [105, 113], and calreticulin expression on tumor cells found to correlate with the presence of tumor-infiltrating memory T cells and favorable prognosis in colorectal cancer [150].

The immune system may even contribute significantly to the efficacy of targeted therapies, a new class of anti-cancer drugs that interfere with specific molecules involved in tumor growth and progression. Examples of these include the human epidermal growth factor 2 (HER-2) inhibitor, trastuzumab, used in breast cancer; the vascular endothelial growth factor (VEGF) blocking antibody, bevacizumab used in lung, breast, colorectal, and renal cancer; the epidermal growth factor receptor (EGFR) tyrosine kinase inhibitor, erlotinib, used in lung and pancreatic cancer; and the multi tyrosine kinase inhibitor, imatinib, used to treat gastrointestinal stromal tumors (GIST) and certain types of leukemia. Approximately 80 % of human GISTs harbor an activating mutation in the c-kit gene that encodes the receptor tyrosine kinase KIT [151]. Imatinib is now the recommended first-line treatment for c-kit mutation positive GIST with around 50 % of patients achieving at least a partial response [152]. While the primary mechanism of action is thought to be a direct anti-proliferative effect on tumor cells through inhibition of KIT signaling, imatinib can prevent the in vivo growth of tumors refractive to the antiproliferative effects of imatinib in vitro in several animal models through enhancement of DC-mediated NK cell activation [153]. More recently, imatinib has been shown to induce $CD8^+$ T-cell activation and Treg apoptosis in a transgenic mouse model in which an activating c-kit gene mutation results in spontaneous development of GIST, through reducing tumor cell expression of the immunosuppressive enzyme indoleamine 2,3-dioxygenase (IDO) [154]. The authors also demonstrated that imatinib sensitivity in patients with GIST correlated with an increase in the intratumoral $CD8^+$ T cell to Treg ratio and reduced IDO protein expression [154].

14.4 Cancer Therapy: Looking Forward

14.4.1 Combination Chemoimmunotherapy

Immunotherapies, when used as single agents, have proved largely disappointing to date. However, given that chemotherapy, at least in certain situations, can enhance anti-tumor immunity, combining standard chemotherapy with immunotherapy treatment has the potential to increase the efficacy of anti-cancer therapy. Synergy between chemotherapy and immunotherapy has been demonstrated in a variety of animal models using several different strategies aimed at inducing or enhancing an anti-tumor immune response. In murine mesothelioma, combination chemoimmunotherapy with gemcitabine and the αCD40-activating antibody FGK45 can induce long-term cures and resistance to rechallenge in up to 80 % of animals, a dramatic improvement over either treatment alone, which only delays tumor growth [110]. Importantly, however, this superior effect is only observed when FGK45 therapy follows gemcitabine treatment; reversing the order of administration is actually less effective than chemotherapy alone. Combining gemcitabine or pemetrexed with Treg depletion using the αCD25 antibody PC61 also results in a synergistic treatment effect in this model [155].

Cisplatin, cyclophosphamide, and paclitaxel have also have also been shown to synergize with immunotherapy in preclinical studies [139, 156–160]. Using a model of human papilloma virus-associated malignancy, Tseng et al., demonstrated that cisplatin followed by a tumor antigen-encoding DNA vaccine led to reduced tumor growth and increased survival, associated with the development of tumor-specific CTL [158]. Again, this synergistic effect was only observed when chemotherapy preceded vaccination. Chemotherapy prior to vaccination was also reported by Wada et al., to be the most effective treatment schedule, inducing antigen-specific effector T-cell expansion and inhibiting tumor growth, using a granulocyte–macrophage colony-stimulating factor (GMCSF)-secreting tumor cell vaccine following cyclophosphamide treatment in an autochthonous prostate cancer model [159]. However, other studies have found immunotherapy prior to chemotherapy to be the best approach. For example, in a mouse lymphoma model, Correale et al, found peptide vaccination administered 13 days before combination chemotherapy gave the biggest synergistic effect [156]. Similarly, type I IFN gene therapy was most effective when given prior to chemotherapy with cisplatin and gemcitabine in murine mesothelioma and lung cancer models [157].

There have been few randomised clinical trials to date powered to evaluate the efficacy of chemoimmunotherapy over chemotherapy alone. A recent practice-changing study demonstrated that the anti-CTLA4 antibody ipilimumab in combination with dacarbazine improved survival by around 2 months in patients with metastatic melanoma, as compared to dacarbazine with placebo [161]. However, dacarbazine alone has extremely modest activity in melanoma, and ipilimumab has activity as a single agent [162]. While the combination appeared to show a "tail on the curve" for long-term survival, it is unclear whether this represents a synergistic or merely additive response. Impressive results have been reported in some early phase studies. For example, in a phase II trial of gemcitabine plus FOLFOX (5-FU, oxaliplatin, and levofolinic acid) chemotherapy combined with subcutaneous GMCSF and IL-2 injections for colorectal cancer, an objective response rate of almost 70 % and an average time to progression of 12.5 months were observed, significantly greater than previously reported for any other regimen [163]. In lung cancer, a randomised phase II trial of the TLR9 agonist PF-3512676 in combination with first-line taxane plus platinum chemotherapy demonstrated a higher objective response rate in patients who received combination therapy compared to chemotherapy alone [164]. While too small, or not designed to determine clinical benefit, several studies have demonstrated that combination chemoimmunotherapy is tolerable and results in enhanced anti-tumor immune responses [165–168].

Data from preclinical studies and early phase trials have therefore been encouraging. However, the success of chemoimmunotherapy will rely heavily on careful drug selection and treatment scheduling. Chemotherapy drugs vary significantly in their capacity to induce immunogenic tumor cell death [114] and to complicate things further; this may be model-dependent [106, 169]. There are also arguments to support both administration of immunotherapy prior to and following chemotherapy. Treg respond far more rapidly in the initial stages of an immune response than naive conventional T cells [170], and Treg activation and proliferation have been shown to precede and prevent the generation of tumor-specific effector T cells in animal models [171]. Memory T cells, however, respond to antigen with similar kinetics to Treg [170, 171] and are more resistant to cytotoxic chemotherapy than other T-cell subsets [172]. Immunotherapy given prior to chemotherapy therefore has the potential to generate tumor-specific memory T cells better placed to compete with Treg in the context of increased antigen availability following chemotherapy-induced tumor cell death. Alternatively, immunotherapy administered following chemotherapy, could capitalize on homeostatic T-cell proliferation following chemotherapy-induced lymphodepletion to skew newly generated T cells toward anti-tumor activity and/or enhance T-cell priming and differentiation. As discussed, evidence from animal models can be found to support either strategy.

14.4.2 Entering the Era of Personalized Medicine

While cytotoxic chemotherapy is still the mainstay of most cancer therapy, the emphasis is now shifting from a *one size fits all* approach to more individualized treatment strategies. Many factors contribute to chemotherapy efficacy and these pertain as much to the tumor and to the host as to intrinsic properties of the drug. Not least of these factors is the potential for development of an anti-tumor immune response, and the capacity for chemotherapy to induce or enhance this. Patients whose cancers have "druggable" targets may be treated with novel-targeted therapies, and our understanding of the interaction between targeted therapies and the immune response is even less mature. Nevertheless, many of these patients will also have chemotherapy before or after treatment with a targeted agent.

Chemoimmunotherapy using carefully selected drug combinations represents a promising treatment option. However, it is likely that this strategy will be most effective when used in an individualized setting. Adjunct immunotherapies may need to be tailored not only to the type of chemotherapy, but also to the immunogenicity of the patient's tumor (expression of tumor antigens) and any defects in the patient's immune system (e.g., TLR4/P2xr7 mutations). Measurement of immunological parameters prior to initiating therapy may prove a useful tool in predicting which patients may benefit from adjunct immunotherapy and in selecting the type of immunotherapy to administer, for example, tumor-specific CTL stimulating or Treg depleting. More work needs to be done to determine how the precise balance of T-cell subsets within the local tumor environment before and after chemotherapy affects response, and whether assessment of peripheral blood T-cell subsets could potentially represent a less invasive surrogate measure, which could be more easily translated into the clinic.

14.5 Conclusions

Although once a highly controversial topic, there is now overwhelming evidence to support a role for the immune system in controlling cancer development and influencing treatment response. It is also becoming increasingly recognized that chemotherapy, rather than purely suppressing anti-tumor immune responses due to lymphocyte toxicity, can actually enhance anti-tumor immunity through a variety of mechanisms. This relatively recent discovery has major implications for the development of cancer therapy. Harnessing the capacity of certain cytotoxic drugs to synergize with the immune system through combining chemotherapy with novel immunotherapy agents represents an exciting new treatment strategy. Rather than using immunotherapy as an alternative to chemotherapy, which has produced disappointing results to date, in this context, we may instead see a real benefit. However, it is imperative that combination chemoimmunotherapy regimens are carefully designed, not just with regard to the potential for synergy between the selected drugs, but also taking account the intrinsic properties of the tumor and the capacity of the patient to generate an anti-tumor immune response. It is therefore likely that the development of more effective treatment strategies for cancer patients will increasingly require a more individualized approach.

References

1. Burnet M (1957) Cancer: a biological approach. III. Viruses associated with neoplastic conditions. IV. Practical applications. Br Med J 1:841–847
2. Dunn GP, Bruce AT, Ikeda H et al (2002) Cancer immunoediting: from immunosurveillance to tumor escape. Nat Immunol 3:991–998
3. Neller MA, Lopez JA, Schmidt CW (2008) Antigens for cancer immunotherapy. Semin Immunol 20:286–295
4. Huang AY, Golumbek P, Ahmadzadeh M et al (1994) Role of bone marrow-derived cells in presenting MHC class I-restricted tumor antigens. Science 264:961–965
5. Marzo AL, Lake RA, Lo D et al (1999) Tumor antigens are constitutively presented in the draining lymph nodes. J Immunol 162:5838–5845
6. van Mierlo GJ, Boonman ZF, Dumortier HM et al (2004) Activation of dendritic cells that cross-present tumor-derived antigen licenses CD8 + CTL to cause tumor eradication. J Immunol 173:6753–6759
7. Rock KL, York IA, Goldberg AL (2004) Post-proteasomal antigen processing for major histocompatibility complex class I presentation. Nat Immunol 5:670–677
8. Watts C (2004) The exogenous pathway for antigen presentation on major histocompatibility complex class II and CD1 molecules. Nat Immunol 5:685–692
9. Bevan MJ (1976) Cross-priming for a secondary cytotoxic response to minor H antigens with H-2 congenic cells which do not cross-react in the cytotoxic assay. J Exp Med 143:1283–1288
10. Lin ML, Zhan Y, Villadangos JA et al (2008) The cell biology of cross-presentation and the role of dendritic cell subsets. Immunol Cell Biol 86:353–362
11. van der Most RG, Currie A, Robinson BWS et al (2006) Cranking the immunologic engine with chemotherapy: using context to drive tumor antigen cross-presentation towards useful antitumor immunity. Cancer Res 66:601–604
12. Boesen M, Svane IM, Engel AM et al (2000) CD8 + T cells are crucial for the ability of congenic normal mice to reject highly immunogenic sarcomas induced in nude mice with 3-methylcholanthrene. Clin Exp Immunol 121:210–215
13. Smyth MJ, Thia KY, Street SE et al (2000) Perforin-mediated cytotoxicity is critical for surveillance of spontaneous lymphoma. J Exp Med 192:755–760
14. Ward PL, Koeppen HK, Hurteau T et al (1990) Major histocompatibility complex class I and unique antigen expression by murine tumors that escaped from CD8 + T-cell-dependent surveillance. Cancer Res 50:3851–3858
15. van den Broek ME, Kagi D, Ossendorp F et al (1996) Decreased tumor surveillance in perforin-deficient mice. J Exp Med 184:1781–1790
16. Broomfield SA, van der Most RG, Prosser AC et al (2009) Locally administered TLR7 agonists drive systemic antitumor immune responses that are enhanced by anti-CD40 immunotherapy. J Immunol 182:5217–5224
17. Serba S, Schmidt J, Wentzensen N et al (2008) Transfection with CD40L induces tumour suppression by dendritic cell activation in an orthotopic mouse model of pancreatic adenocarcinoma. Gut 57:344–351
18. Slos P, De Meyer M, Leroy P et al (2001) Immunotherapy of established tumors in mice by intratumoral injection of an adenovirus vector harboring the human IL-2 cDNA: induction of CD8(+) T-cell immunity and NK activity. Cancer Gene Ther 8:321–332
19. Sutmuller RP, van Duivenvoorde LM, van Elsas A et al (2001) Synergism of cytotoxic T lymphocyte-associated antigen 4 blockade and depletion of CD25(+) regulatory T cells in antitumor therapy reveals alternative pathways for suppression of autoreactive cytotoxic T lymphocyte responses. J Exp Med 194:823–832
20. Constant SL, Bottomly K (1997) Induction of Th1 and Th2 CD4 + T cell responses: the alternative approaches. Annu Rev Immunol 15:297–322
21. Pulendran B, Palucka K, Banchereau J (2001) Sensing pathogens and tuning immune responses. Science 293:253–256

22. Martinez GJ, Nurieva RI, Yang XO et al (2008) Regulation and function of proinflammatory TH17 cells. Ann N Y Acad Sci 1143:188–211
23. Matzinger P, Kamala T (2011) Tissue-based class control: the other side of tolerance. Nat Rev Immunol 11:221–230
24. Bennett SR, Carbone FR, Karamalis F et al (1998) Help for cytotoxic-T-cell responses is mediated by CD40 signalling. Nature 393:478–480
25. Lanzavecchia A (1998) Immunology. Licence to kill. Nature 393:413–414
26. Ridge JP, Di Rosa F, Matzinger P (1998) A conditioned dendritic cell can be a temporal bridge between a CD4 + T-helper and a T-killer cell. Nature 393:474–478
27. Schoenberger SP, Toes RE, van der Voort EI et al (1998) T-cell help for cytotoxic T lymphocytes is mediated by CD40-CD40L interactions. Nature 393:480–483
28. Bennett SR, Carbone FR, Karamalis F et al (1997) Induction of a CD8 + cytotoxic T lymphocyte response by cross-priming requires cognate CD4 + T cell help. J Exp Med 186:65–70
29. Ossendorp F, Mengede E, Camps M et al (1998) Specific T helper cell requirement for optimal induction of cytotoxic T lymphocytes against major histocompatibility complex class II negative tumors. J Exp Med 187:693–702
30. Quezada SA, Simpson TR, Peggs KS et al (2010) Tumor-reactive CD4 + T cells develop cytotoxic activity and eradicate large established melanoma after transfer into lymphopenic hosts. J Exp Med 207:637–650
31. Hung K, Hayashi R, Lafond-Walker A et al (1998) The central role of CD4(+) T cells in the antitumor immune response. J Exp Med 188:2357–2368
32. Currie AJ, Prosser A, McDonnell A et al (2009) Dual control of antitumor CD8 T cells through the programmed death-1/programmed death-ligand 1 pathway and immunosuppressive CD4 T cells: regulation and counterregulation. J Immunol 183:7898–7908
33. Belkaid Y, Rouse BT (2005) Natural regulatory T cells in infectious disease. Nat Immunol 6:353–360
34. Sakaguchi S, Yamaguchi T, Nomura T et al (2008) Regulatory T cells and immune tolerance. Cell 133:775–787
35. Zou W (2006) Regulatory T cells, tumour immunity and immunotherapy. Nat Rev Immunol 6:295–307
36. Murugaiyan G, Saha B (2009) Protumor vs antitumor functions of IL-17. J Immunol 183:4169–4175
37. Bennett SR, Carbone FR, Toy T et al (1998) B cells directly tolerize CD8(+) T cells. J Exp Med 188:1977–1983
38. Qin Z, Richter G, Schuler T et al (1998) B cells inhibit induction of T cell-dependent tumor immunity. Nat Med 4:627–630
39. Mueller DL, Jenkins MK, Schwartz RH (1989) Clonal expansion versus functional clonal inactivation: a costimulatory signalling pathway determines the outcome of T cell antigen receptor occupancy. Annu Rev Immunol 7:445–480
40. Medzhitov R, Janeway C Jr (2000) Innate immunity. New Engl J Med 343:338–344
41. Borghesi L, Milcarek C (2007) Innate versus adaptive immunity: a paradigm past its prime? Cancer Res 67:3989–3993
42. Dalod M, Hamilton T, Salomon R et al (2003) Dendritic cell responses to early murine cytomegalovirus infection: subset functional specialization and differential regulation by interferon alpha/beta. J Exp Med 197:885–898
43. Fearon DT, Locksley RM (1996) The instructive role of innate immunity in the acquired immune response. Science 272:50–53
44. Megjugorac NJ, Young HA, Amrute SB et al (2004) Virally stimulated plasmacytoid dendritic cells produce chemokines and induce migration of T and NK cells. J Leukoc Biol 75:504–514
45. Bradley M, Zeytun A, Rafi-Janajreh A et al (1998) Role of spontaneous and interleukin-2-induced natural killer cell activity in the cytotoxicity and rejection of Fas + and Fas- tumor cells. Blood 92:4248–4255

46. Pardo J, Balkow S, Anel A et al (2002) Granzymes are essential for natural killer cell-mediated and perf-facilitated tumor control. Eur J Immunol 32:2881–2887
47. Takeda K, Hayakawa Y, Smyth MJ et al (2001) Involvement of tumor necrosis factor-related apoptosis-inducing ligand in surveillance of tumor metastasis by liver natural killer cells. Nat Med 7:94–100
48. van den Broek MF, Kagi D, Zinkernagel RM et al (1995) Perforin dependence of natural killer cell-mediated tumor control in vivo. Eur J Immunol 25:3514–3516
49. Mocikat R, Braumuller H, Gumy A et al (2003) Natural killer cells activated by MHC class I(low) targets prime dendritic cells to induce protective CD8 T cell responses. Immunity 19:561–569
50. Cole KE, Strick CA, Paradis TJ et al (1998) Interferon-inducible T cell alpha chemoattractant (I-TAC): a novel non-ELR CXC chemokine with potent activity on activated T cells through selective high affinity binding to CXCR3. J Exp Med 187:2009–2021
51. Farber JM (1997) Mig and IP-10: CXC chemokines that target lymphocytes. J Leukoc Biol 61:246–257
52. Barchet W, Cella M, Colonna M (2005) Plasmacytoid dendritic cells–virus experts of innate immunity. Semin Immunol 17:253–261
53. Liu C, Lou Y, Lizee G et al (2008) Plasmacytoid dendritic cells induce NK cell-dependent, tumor antigen-specific T cell cross-priming and tumor regression in mice. J Clin Invest 118:1165–1175
54. Shankaran V, Ikeda H, Bruce AT et al (2001) IFNgamma and lymphocytes prevent primary tumour development and shape tumour immunogenicity. Nature 410:1107–1111
55. Sanchez-Perez L, Kottke T, Diaz RM et al (2005) Potent selection of antigen loss variants of B16 melanoma following inflammatory killing of melanocytes in vivo. Cancer Res 65:2009–2017
56. Slingluff CL Jr, Colella TA, Thompson L et al (2000) Melanomas with concordant loss of multiple melanocytic differentiation proteins: immune escape that may be overcome by targeting unique or undefined antigens. Cancer Immunol Immunother 48:661–672
57. Singh R, Paterson Y (2007) Immunoediting sculpts tumor epitopes during immunotherapy. Cancer Res 67:1887–1892
58. Zhou G, Lu Z, McCadden JD et al (2004) Reciprocal changes in tumor antigenicity and antigen-specific T cell function during tumor progression. J Exp Med 200:1581–1592
59. Yee C, Thompson JA, Byrd D et al (2002) Adoptive T cell therapy using antigen-specific CD8 + T cell clones for the treatment of patients with metastatic melanoma: in vivo persistence, migration, and antitumor effect of transferred T cells. Proc Natl Acad Sci USA 99:16168–16173
60. So T, Takenoyama M, Mizukami M et al (2005) Haplotype loss of HLA class I antigen as an escape mechanism from immune attack in lung cancer. Cancer Res 65:5945–5952
61. Baba T, Hanagiri T, Ichiki Y et al (2007) Lack and restoration of sensitivity of lung cancer cells to cellular attack with special reference to expression of human leukocyte antigen class I and/or major histocompatibility complex class I chain related molecules A/B. Cancer Sci 98:1795–1802
62. Qin Z, Noffz G, Mohaupt M et al (1997) Interleukin-10 prevents dendritic cell accumulation and vaccination with granulocyte-macrophage colony-stimulating factor gene-modified tumor cells. J Immunol 159:770–776
63. Gorelik L, Flavell RA (2001) Immune-mediated eradication of tumors through the blockade of transforming growth factor-beta signaling in T cells. Nat Med 7:1118–1122
64. Chen W, Jin W, Hardegen N et al (2003) Conversion of peripheral CD4 + CD25- naive T cells to CD4 + CD25 + regulatory T cells by TGF-beta induction of transcription factor Foxp3. J Exp Med 198:1875–1886
65. Dong H, Strome SE, Salomao DR et al (2002) Tumor-associated B7–H1 promotes T-cell apoptosis: a potential mechanism of immune evasion. Nat Med 8:793–800
66. Okazaki T, Honjo T (2007) PD-1 and PD-1 ligands: from discovery to clinical application. Int Immunol 19:813–824

67. Takahashi T, Dejbakhsh-Jones S, Strober S (2006) Expression of CD161 (NKR-P1A) defines subsets of human CD4 and CD8 T cells with different functional activities. J Immunol 176:211–216
68. Shin MS, Kim HS, Lee SH et al (2001) Mutations of tumor necrosis factor-related apoptosis-inducing ligand receptor 1 (TRAIL-R1) and receptor 2 (TRAIL-R2) genes in metastatic breast cancers. Cancer Res 61:4942–4946
69. O'Connell J, O'Sullivan GC, Collins JK et al (1996) The Fas counterattack: Fas-mediated T cell killing by colon cancer cells expressing Fas ligand. J Exp Med 184:1075–1082
70. Letsch A, Keilholz U, Schadendorf D et al (2000) High frequencies of circulating melanoma-reactive CD8 + T cells in patients with advanced melanoma. Int J Cancer 87:659–664
71. Ram M, Shoenfeld Y (2007) Harnessing autoimmunity (vitiligo) to treat melanoma: a myth or reality? Ann N Y Acad Sci 1110:410–425
72. Andersen MH, Pedersen LO, Capeller B et al (2001) Spontaneous cytotoxic T-cell responses against survivin-derived MHC class I-restricted T-cell epitopes in situ as well as ex vivo in cancer patients. Cancer Res 61:5964–5968
73. Kokowski K, Harnack U, Dorn DC et al (2008) Quantification of the CD8 + T cell response against a mucin epitope in patients with breast cancer. Arch Immunol Ther Exp (Warsz) 56:141–145
74. Matsuzaki J, Qian F, Luescher I et al (2008) Recognition of naturally processed and ovarian cancer reactive CD8 + T cell epitopes within a promiscuous HLA class II T-helper region of NY-ESO-1. Cancer Immunol Immunother 57:1185–1195
75. Minev B, Hipp J, Firat H et al (2000) Cytotoxic T cell immunity against telomerase reverse transcriptase in humans. Proc Natl Acad Sci USA 97:4796–4801
76. Nagorsen D, Keilholz U, Rivoltini L et al (2000) Natural T-cell response against MHC class I epitopes of epithelial cell adhesion molecule, her-2/neu, and carcinoembryonic antigen in patients with colorectal cancer. Cancer Res 60:4850–4854
77. Nakamura Y, Noguchi Y, Satoh E et al (2009) Spontaneous remission of a non-small cell lung cancer possibly caused by anti-NY-ESO-1 immunity. Lung Cancer 65:119–122
78. Liyanage UK, Moore TT, Joo HG et al (2002) Prevalence of regulatory T cells is increased in peripheral blood and tumor microenvironment of patients with pancreas or breast adenocarcinoma. J Immunol 169:2756–2761
79. Miller AM, Lundberg K, Ozenci V et al (2006) CD4 + CD25high T cells are enriched in the tumor and peripheral blood of prostate cancer patients. J Immunol 177:7398–7405
80. Ormandy LA, Hillemann T, Wedemeyer H et al (2005) Increased populations of regulatory T cells in peripheral blood of patients with hepatocellular carcinoma. Cancer Res 65:2457–2464
81. Wolf AM, Wolf D, Steurer M et al (2003) Increase of regulatory T cells in the peripheral blood of cancer patients. Clin Cancer Res 9:606–612
82. Zhang L, Conejo-Garcia JR, Katsaros D et al (2003) Intratumoral T cells, recurrence, and survival in epithelial ovarian cancer. New Engl J Med 348:203–213
83. Galon J, Costes A, Sanchez-Cabo F et al (2006) Type, density, and location of immune cells within human colorectal tumors predict clinical outcome. Science 313:1960–1964
84. Cho Y, Miyamoto M, Kato K et al (2003) CD4 + and CD8 + T cells cooperate to improve prognosis of patients with esophageal squamous cell carcinoma. Cancer Res 63:1555–1559
85. Hiraoka K, Miyamoto M, Cho Y et al (2006) Concurrent infiltration by CD8 + T cells and CD4 + T cells is a favourable prognostic factor in non-small-cell lung carcinoma. Br J Cancer 94:275–280
86. Curiel TJ, Coukos G, Zou L et al (2004) Specific recruitment of regulatory T cells in ovarian carcinoma fosters immune privilege and predicts reduced survival. Nat Med 10:942–949
87. Perrone G, Ruffini PA, Catalano V et al (2008) Intratumoural FOXP3-positive regulatory T cells are associated with adverse prognosis in radically resected gastric cancer. Eur J Cancer 44:1875–1882
88. Petersen RP, Campa MJ, Sperlazza J et al (2006) Tumor infiltrating Foxp3 + regulatory T-cells are associated with recurrence in pathologic stage I NSCLC patients. Cancer 107:2866–2872

89. Zhou J, Ding T, Pan W et al (2009) Increased intratumoral regulatory T cells are related to intratumoral macrophages and poor prognosis in hepatocellular carcinoma patients. Int J Cancer 125:1640–1648
90. Correale P, Rotundo MS, Del Vecchio MT et al (2010) Regulatory (FoxP3 +) T-cell tumor infiltration is a favorable prognostic factor in advanced colon cancer patients undergoing chemo or chemoimmunotherapy. J Immunother 33:435–441
91. Frey DM, Droeser RA, Viehl CT et al (2010) High frequency of tumor-infiltrating FOXP3(+) regulatory T cells predicts improved survival in mismatch repair-proficient colorectal cancer patients. Int J Cancer 126:2635–2643
92. Lee WS, Park S, Lee WY et al (2010) Clinical impact of tumor-infiltrating lymphocytes for survival in stage II colon cancer. Cancer 116:5188–5199
93. Salama P, Phillips M, Grieu F et al (2009) Tumor-infiltrating FOXP3 + T regulatory cells show strong prognostic significance in colorectal cancer. J Clin Oncol 27:186–192
94. Ladoire S, Martin F, Ghiringhelli F (2011) Prognostic role of FOXP3 + regulatory T cells infiltrating human carcinomas: the paradox of colorectal cancer. Cancer Immunol Immunother 60:909–918
95. Su X, Ye J, Hsueh EC et al (2010) Tumor microenvironments direct the recruitment and expansion of human Th17 cells. J Immunol 184:1630–1641
96. Tosolini M, Kirilovsky A, Mlecnik B et al (2011) Clinical impact of different classes of infiltrating T cytotoxic and helper cells (Th1, th2, treg, th17) in patients with colorectal cancer. Cancer Res 71:1263–1271
97. Badoual C, Hans S, Rodriguez J et al (2006) Prognostic value of tumor-infiltrating CD4 + T-cell subpopulations in head and neck cancers. Clin Cancer Res 12:465–472
98. Zhang YL, Li J, Mo HY et al (2010) Different subsets of tumor infiltrating lymphocytes correlate with NPC progression in different ways. Mol Cancer 9:4
99. Longley DB, Harkin DP, Johnston PG (2003) 5-fluorouracil: mechanisms of action and clinical strategies. Nat Rev Cancer 3:330–338
100. Mini E, Nobili S, Caciagli B et al (2006) Cellular pharmacology of gemcitabine. Ann Oncol 17(Suppl 5):v7–v12
101. Hall AG, Tilby MJ (1992) Mechanisms of action of, and modes of resistance to, alkylating agents used in the treatment of haematological malignancies. Blood Rev 6:163–173
102. Siddik ZH (2003) Cisplatin: mode of cytotoxic action and molecular basis of resistance. Oncogene 22:7265–7279
103. Nielsen D, Maare C, Skovsgaard T (1996) Cellular resistance to anthracyclines. Gen Pharmacol 27:251–255
104. Perez EA (2009) Microtubule inhibitors: differentiating tubulin-inhibiting agents based on mechanisms of action, clinical activity, and resistance. Mol Cancer Ther 8:2086–2095
105. Ghiringhelli F, Apetoh L, Tesniere A et al (2009) Activation of the NLRP3 inflammasome in dendritic cells induces IL-1beta-dependent adaptive immunity against tumors. Nat Med 15:1170–1178
106. Nowak AK, Mahendran S, van der Most RG et al (2008) Cisplatin and pemetrexed synergises with immunotherapy to result in cures in established murine mesothelioma. In: Proceedings of the American association of cancer research annual meeting 487 (Abstract 2073, 2008)
107. van der Most RG, Currie AJ, Mahendran S et al (2009) Tumor eradication after cyclophosphamide depends on concurrent depletion of regulatory T cells: a role for cycling TNFR2-expressing effector-suppressor T cells in limiting effective chemotherapy. Cancer Immunol Immunother 58:1219–1228
108. Nowak AK, Lake RA, Marzo AL et al (2003) Induction of tumor cell apoptosis in vivo increases tumor antigen cross-presentation, cross-priming rather than cross-tolerizing host tumor-specific CD8 T cells. J Immunol 170:4905–4913
109. Sjoblom T, Jones S, Wood LD et al (2006) The consensus coding sequences of human breast and colorectal cancers. Science 314:268–274

110. Nowak AK, Robinson BW, Lake RA (2003) Synergy between chemotherapy and immunotherapy in the treatment of established murine solid tumors. Cancer Res 63:4490–4496
111. Matzinger P (2002) The danger model: a renewed sense of self. Science 296:301–305
112. Zitvogel L, Kepp O, Senovilla L et al (2010) Immunogenic tumor cell death for optimal anticancer therapy: the calreticulin exposure pathway. Clin Cancer Res 16:3100–3104
113. Apetoh L, Ghiringhelli F, Tesniere A et al (2007) Toll-like receptor 4-dependent contribution of the immune system to anticancer chemotherapy and radiotherapy. Nat Med 13:1050–1059
114. Casares N, Pequignot MO, Tesniere A et al (2005) Caspase-dependent immunogenicity of doxorubicin-induced tumor cell death. J Exp Med 202:1691–1701
115. Obeid M, Tesniere A, Ghiringhelli F et al (2007) Calreticulin exposure dictates the immunogenicity of cancer cell death. Nat Med 13:54–61
116. Liu WM, Fowler DW, Smith P et al (2010) Pre-treatment with chemotherapy can enhance the antigenicity and immunogenicity of tumours by promoting adaptive immune responses. Br J Cancer 102:115–123
117. Ramakrishnan R, Assudani D, Nagaraj S et al (2010) Chemotherapy enhances tumor cell susceptibility to CTL-mediated killing during cancer immunotherapy in mice. J Clin Invest 120:1111–1124
118. van der Most RG, Currie AJ, Cleaver AL et al (2009) Cyclophosphamide chemotherapy sensitizes tumor cells to TRAIL-dependent CD8 T cell-mediated immune attack resulting in suppression of tumor growth. PLoS One 4:e6982
119. Soriani A, Zingoni A, Cerboni C et al (2009) ATM-ATR-dependent up-regulation of DNAM-1 and NKG2D ligands on multiple myeloma cells by therapeutic agents results in enhanced NK-cell susceptibility and is associated with a senescent phenotype. Blood 113:3503–3511
120. Fine JH, Chen P, Mesci A et al (2010) Chemotherapy-induced genotoxic stress promotes sensitivity to natural killer cell cytotoxicity by enabling missing-self recognition. Cancer Res 70:7102–7113
121. Karre K (2008) Natural killer cell recognition of missing self. Nat Immunol 9:477–480
122. Radojcic V, Bezak KB, Skarica M et al (2010) Cyclophosphamide resets dendritic cell homeostasis and enhances antitumor immunity through effects that extend beyond regulatory T cell elimination. Cancer Immunol Immunother 59:137–148
123. Salem ML, Al-Khami AA, El-Naggar SA et al (2010) Cyclophosphamide induces dynamic alterations in the host microenvironments resulting in a Flt3 ligand-dependent expansion of dendritic cells. J Immunol 184:1737–1747
124. Salem ML, El-Naggar SA, Cole DJ (2010) Cyclophosphamide induces bone marrow to yield higher numbers of precursor dendritic cells in vitro capable of functional antigen presentation to T cells in vivo. Cell Immunol 261:134–143
125. Pfannenstiel LW, Lam SS, Emens LA et al (2010) Paclitaxel enhances early dendritic cell maturation and function through TLR4 signaling in mice. Cell Immunol 263:79–87
126. Shurin GV, Tourkova IL, Kaneno R et al (2009) Chemotherapeutic agents in noncytotoxic concentrations increase antigen presentation by dendritic cells via an IL-12-dependent mechanism. J Immunol 183:137–144
127. Goldrath AW, Bevan MJ (1999) Selecting and maintaining a diverse T-cell repertoire. Nature 402:255–262
128. Mackall CL, Hakim FT, Gress RE (1997) Restoration of T-cell homeostasis after T-cell depletion. Semin Immunol 9:339–346
129. Schluns KS, Kieper WC, Jameson SC et al (2000) Interleukin-7 mediates the homeostasis of naive and memory CD8 T cells in vivo. Nat Immunol 1:426–432
130. Tan JT, Ernst B, Kieper WC et al (2002) Interleukin (IL)-15 and IL-7 jointly regulate homeostatic proliferation of memory phenotype CD8 + cells but are not required for memory phenotype CD4 + cells. J Exp Med 195:1523–1532
131. Marleau AM, Sarvetnick N (2005) T cell homeostasis in tolerance and immunity. J Leukoc Biol 78:575–584

132. Theofilopoulos AN, Dummer W, Kono DH (2001) T cell homeostasis and systemic autoimmunity. J Clin Invest 108:335–340
133. Dummer W, Niethammer AG, Baccala R et al (2002) T cell homeostatic proliferation elicits effective antitumor autoimmunity. J Clin Invest 110:185–192
134. Maguire HC Jr, Ettore VL (1967) Enhancement of dinitrochlorobenzene (DNCB) contact sensitization by cyclophosphamide in the guinea pig. J Invest Dermatol 48:39–43
135. Berd D, Mastrangelo MJ, Engstrom PF et al (1982) Augmentation of the human immune response by cyclophosphamide. Cancer Res 42:4862–4866
136. Ghiringhelli F, Larmonier N, Schmitt E et al (2004) CD4 + CD25 + regulatory T cells suppress tumor immunity but are sensitive to cyclophosphamide which allows immunotherapy of established tumors to be curative. Eur J Immunol 34:336–344
137. Ghiringhelli F, Menard C, Puig PE et al (2007) Metronomic cyclophosphamide regimen selectively depletes CD4 + CD25 + regulatory T cells and restores T and NK effector functions in end stage cancer patients. Cancer Immunol Immunother 56:641–648
138. Rettig L, Seidenberg S, Parvanova I et al (2011) Gemcitabine depletes regulatory T-cells in human and mice and enhances triggering of vaccine-specific cytotoxic T-cells. Int J Cancer 129:832–838
139. Vicari AP, Luu R, Zhang N et al (2009) Paclitaxel reduces regulatory T cell numbers and inhibitory function and enhances the anti-tumor effects of the TLR9 agonist PF-3512676 in the mouse. Cancer Immunol Immunother 58:615–628
140. Fisson S, Darrasse-Jeze G, Litvinova E et al (2003) Continuous activation of autoreactive CD4 + CD25 + regulatory T cells in the steady state. J Exp Med 198:737–746
141. Vukmanovic-Stejic M, Zhang Y, Cook JE et al (2006) Human CD4 + CD25hi Foxp3 + regulatory T cells are derived by rapid turnover of memory populations in vivo. J Clin Invest 116:2423–2433
142. Chen X, Subleski JJ, Kopf H et al (2008) Cutting edge: expression of TNFR2 defines a maximally suppressive subset of mouse CD4 + CD25 + FoxP3 + T regulatory cells: applicability to tumor-infiltrating T regulatory cells. J Immunol 180:6467–6471
143. Chen Y, Shen S, Gorentla BK et al (2012) Murine regulatory T cells contain hyperproliferative and death-prone subsets with differential ICOS expression. J Immunol 188:1698–1707
144. Strauss L, Bergmann C, Szczepanski MJ et al (2008) Expression of ICOS on human melanoma-infiltrating CD4 + CD25highFoxp3 + T regulatory cells: implications and impact on tumor-mediated immune suppression. J Immunol 180:2967–2980
145. Nowak AK, Robinson BW, Lake RA (2002) Gemcitabine exerts a selective effect on the humoral immune response: implications for combination chemo-immunotherapy. Cancer Res 62:2353–2358
146. Wijayahadi N, Haron MR, Stanslas J et al (2007) Changes in cellular immunity during chemotherapy for primary breast cancer with anthracycline regimens. J Chemother 19:176–723
147. Demaria S, Volm MD, Shapiro RL et al (2001) Development of tumor-infiltrating lymphocytes in breast cancer after neoadjuvant paclitaxel chemotherapy. Clin Cancer Res 7:3025–3030
148. Ladoire S, Arnould L, Apetoh L et al (2008) Pathologic complete response to neoadjuvant chemotherapy of breast carcinoma is associated with the disappearance of tumor-infiltrating foxp3 + regulatory T cells. Clin Cancer Res 14:2413–2420
149. Yasuda K, Nirei T, Sunami E et al (2011) Density of CD4(+) and CD8(+) T lymphocytes in biopsy samples can be a predictor of pathological response to chemoradiotherapy (CRT) for rectal cancer. Radiat Oncol 6:49
150. Peng RQ, Chen YB, Ding Y et al (2010) Expression of calreticulin is associated with infiltration of T-cells in stage IIIB colon cancer. World J Gastroenterol: WJG 16:2428–2434
151. Hirota S, Isozaki K, Moriyama Y et al (1998) Gain-of-function mutations of c-kit in human gastrointestinal stromal tumors. Science 279:577–580
152. Blanke CD, Rankin C, Demetri GD et al (2008) Phase III randomized, intergroup trial assessing imatinib mesylate at two dose levels in patients with unresectable or metastatic

gastrointestinal stromal tumors expressing the kit receptor tyrosine kinase: S0033. J Clin Oncol 26:626–632
153. Borg C, Terme M, Taieb J et al (2004) Novel mode of action of c-kit tyrosine kinase inhibitors leading to NK cell-dependent antitumor effects. J Clin Invest 114:379–388
154. Balachandran VP, Cavnar MJ, Zeng S et al (2011) Imatinib potentiates antitumor T cell responses in gastrointestinal stromal tumor through the inhibition of Ido. Nat Med 17:1094–1100
155. Anraku M, Tagawa T, Wu L et al (2010) Synergistic antitumor effects of regulatory T cell blockade combined with pemetrexed in murine malignant mesothelioma. J Immunol 185:956–966
156. Correale P, Del Vecchio MT, La Placa M et al (2008) Chemotherapeutic drugs may be used to enhance the killing efficacy of human tumor antigen peptide-specific CTLs. J Immunother 31:132–147
157. Fridlender ZG, Sun J, Singhal S et al (2010) Chemotherapy delivered after viral immunogene therapy augments antitumor efficacy via multiple immune-mediated mechanisms. Mol Ther 18:1947–1959
158. Tseng CW, Hung CF, Alvarez RD et al (2008) Pretreatment with cisplatin enhances E7-specific CD8 + T-cell-mediated antitumor immunity induced by DNA vaccination. Clin Cancer Res 14:3185–3192
159. Wada S, Yoshimura K, Hipkiss EL et al (2009) Cyclophosphamide augments antitumor immunity: studies in an autochthonous prostate cancer model. Cancer Res 69:4309–4318
160. Zhong H, Han B, Tourkova IL et al (2007) Low-dose paclitaxel prior to intratumoral dendritic cell vaccine modulates intratumoral cytokine network and lung cancer growth. Clin Cancer Res 13:5455–5462
161. Robert C, Thomas L, Bondarenko I et al (2011) Ipilimumab plus dacarbazine for previously untreated metastatic melanoma. New Engl J Med 364:2517–2526
162. Wolchok JD, Neyns B, Linette G et al (2010) Ipilimumab monotherapy in patients with pretreated advanced melanoma: a randomised, double-blind, multicentre, phase 2, dose-ranging study. Lancet Oncol 11:155–164
163. Correale P, Cusi MG, Tsang KY et al (2005) Chemo-immunotherapy of metastatic colorectal carcinoma with gemcitabine plus FOLFOX 4 followed by subcutaneous granulocyte macrophage colony-stimulating factor and interleukin-2 induces strong immunologic and antitumor activity in metastatic colon cancer patients. J Clin Oncol 23:8950–8958
164. Manegold C, Gravenor D, Woytowitz D et al (2008) Randomized phase II trial of a toll-like receptor 9 agonist oligodeoxynucleotide, PF-3512676, in combination with first-line taxane plus platinum chemotherapy for advanced-stage non-small-cell lung cancer. J Clin Oncol 26:3979–3986
165. Emens LA, Asquith JM, Leatherman JM et al (2009) Timed sequential treatment with cyclophosphamide, doxorubicin, and an allogeneic granulocyte-macrophage colony-stimulating factor-secreting breast tumor vaccine: a chemotherapy dose-ranging factorial study of safety and immune activation. J Clin Oncol 27:5911–5918
166. Hegmans JP, Veltman JD, Lambers ME et al (2010) Consolidative dendritic cell-based immunotherapy elicits cytotoxicity against malignant mesothelioma. Am J Respir Crit Care Med 181:1383–1390
167. Ramlau R, Quoix E, Rolski J et al (2008) A phase II study of Tg4010 (Mva-Muc1-Il2) in association with chemotherapy in patients with stage III/IV non-small cell lung cancer. J Thorac Oncol 3:735–744
168. Walker DG, Laherty R, Tomlinson FH et al (2008) Results of a phase I dendritic cell vaccine trial for malignant astrocytoma: potential interaction with adjuvant chemotherapy. J Clin Neurosci 15:114–121
169. Ciampricotti M, Hau CS, Doornebal CW et al (2012) Chemotherapy response of spontaneous mammary tumors is independent of the adaptive immune system. Nat Med 18:344–346. Author reply 346

170. O'Gorman WE, Dooms H, Thorne SH et al (2009) The initial phase of an immune response functions to activate regulatory T cells. J Immunol 183:332–339
171. Darrasse-Jeze G, Bergot AS, Durgeau A et al (2009) Tumor emergence is sensed by self-specific CD44hi memory Tregs that create a dominant tolerogenic environment for tumors in mice. J Clin Invest 119:2648–2662
172. Turtle CJ, Swanson HM, Fujii N et al (2009) A distinct subset of self-renewing human memory CD8 + T cells survives cytotoxic chemotherapy. Immunity 31:834–844
173. Dunker K, Schlaf G, Bukur J et al (2008) Expression and regulation of non-classical HLA-G in renal cell carcinoma. Tissue Antigens 72:137–148
174. Lin A, Zhang X, Zhou WJ et al (2011) HLA-G expression is associated with a poor prognosis in patients with esophageal squamous cell carcinoma. Int J Cancer 129:1382–1390
175. Paul P, Cabestre FA, Le Gal FA et al (1999) Heterogeneity of HLA-G gene transcription and protein expression in malignant melanoma biopsies. Cancer Res 59:1954–1960
176. Chang CC, Ferrone S (2006) NK cell activating ligands on human malignant cells: molecular and functional defects and potential clinical relevance. Semin Cancer Biol 16:383–392
177. Clayton A, Mitchell JP, Court J et al (2008) Human tumor-derived exosomes down-modulate NKG2D expression. J Immunol 180:7249–7258
178. Uyttenhove C, Pilotte L, Theate I et al (2003) Evidence for a tumoral immune resistance mechanism based on tryptophan degradation by indoleamine 2,3-dioxygenase. Nat Med 9:1269–1274
179. Villablanca EJ, Raccosta L, Zhou D et al (2010) Tumor-mediated liver X receptor-alpha activation inhibits CC chemokine receptor-7 expression on dendritic cells and dampens antitumor responses. Nat Med 16:98–105
180. Adjei AA (2004) Pharmacology and mechanism of action of pemetrexed. Clin Lung Cancer 5(Suppl 2):S51–S55
181. Longo-Sorbello GS, Bertino JR (2001) Current understanding of methotrexate pharmacology and efficacy in acute leukemias. Use of newer antifolates in clinical trials. Haematologica 86:121–127
182. Hande KR (1998) Etoposide: four decades of development of a topoisomerase II inhibitor. Eur J Cancer 34:1514–1521

About the Editor

Dr. Daniel E. Johnson received his undergraduate training at North Park University and his doctoral degree in molecular biology from Princeton University. He was a postdoctoral fellow at the University of California, San Francisco. In 1993, he joined the faculty at the University of Pittsburgh and the University of Pittsburgh Cancer Institute, where he is currently a Professor in the Departments of Medicine and Pharmacology and Chemical Biology. Dr. Johnson has served as a standing member on study sections for the National Institutes of Health and American Cancer Society and is a long-standing Section Editor for the journal Leukemia. His research has focused on molecular mechanisms of apoptosis in leukemia and head and neck cancer, as well as the mechanisms of myeloid differentiation. He has placed particular emphasis on the translation of findings from his laboratory to the clinic and, together with physician scientist collaborators, has helped to develop ongoing trials in both acute myeloid leukemia and head and neck cancer.

D. E. Johnson (ed.), *Cell Death Signaling in Cancer Biology and Treatment*,
Cell Death in Biology and Diseases, DOI: 10.1007/978-1-4614-5847-0,

X.-M. Yin and Z. Dong (Series eds.), *Cell Death in Biology and Diseases*

Index

D. E. Johnson (ed.), *Cell Death Signaling in Cancer Biology and Treatment*, Cell Death in Biology and Diseases, DOI: 10.1007/978-1-4614-5847-0,

X.-M. Yin and Z. Dong (Series eds.), *Cell Death in Biology and Diseases*

P (*cont.*)

Printed in Great Britain
by Amazon